Race and Ethnicity in the United States:

Our Differences and Our Roots

Reid Luhman
Eastern Kentucky University

Harcourt College Publishers

Fort Worth   Philadelphia   San Diego   New York   Orlando   Austin   San Antonio
Toronto   Montreal   London   Sydney   Tokyo

Publisher   Earl McPeek
Acquisitions Editor   Bryan Leake
Developmental Editor   Lisa Hensley
Marketing Strategist   Laura Brennan
Project Manager   Barrett Lackey

ISBN: 0-15-503876-1
Library of Congress Catalog Number: 00-111243

Address for Domestic Orders
Harcourt College Publishers, 6277 Sea Harbor Drive, Orlando, FL 32887-6777
800-782-4479

Address for International Orders
International Customer Service
Harcourt College Publishers, 6277 Sea Harbor Drive, Orlando, FL 32887-6777
407-345-8000
(fax) 407-345-4060
(e-mail) hbintl@harcourtbrace.com

Address for Editorial Correspondence
Harcourt College Publishers, 301 Commerce Street, Suite 3700, Fort Worth, TX
76102

Web Site Address
http://www.harcourtcollege.com

Printed in the United States of America

0 1 2 3 4 5 6 7 8 9   039   9 8 7 6 5 4 3 2 1

Harcourt College Publishers

To the memory of Frederick and Helen Douglass (pictured seated)

The new millennium brought the United States a unique presidential election that riveted the nation for weeks before its final resolution. While the electoral vote was closer than the popular vote, the overall election could easily be called a tie. Both George W. Bush and Al Gore received 48 percent of the vote, but the tied outcome masks the divisions within the American electorate. Different racial and ethnic groups clearly had their preferences. White voters favored Bush 54 percent to 42 percent; by contrast, Gore received 54 percent of the Asian American vote, 67 percent of the Hispanic vote, and 90 percent of the African American vote. Bush received 63 percent of the Protestant American vote but only 19 percent of the Jewish American vote. In short, subgroups of Americans had no difficulty recognizing their interests and making up their minds. Such differences highlight the importance of understanding race and ethnicity in achieving an understanding of the United States. That nation of immigrants is still, in many ways, a nation of diversity.

Social scientists invariably approach race and ethnicity by focusing on cultural diversity and differences in power. Terms such as *minority group* suggest such an approach, conjuring up images of somewhat small, culturally distinctive groups that are excluded from the centers of power in a society. Studying such phenomena is clearly important within the agenda of the social sciences. But the question of how this study should be directed and organized, particularly in relation to the United States, remains.

Even though cultural diversity and power lie intertwined in race and ethnic studies, most textbooks in this area are organized only by cultural diversity, devoting a chapter or so to the historical and current circumstances each different racial or ethnic group. The danger lies in leaving the impression that each group in the United States lives in a vacuum, unaffected by the other groups with which it shares a society.

By contrast, the purpose of this book is twofold. First, it demonstrates the *interrelations* among different racial and ethnic groups. As these interrelations are examined, issues such as class differences and gender issues within and among groups are highlighted. A second and closely related goal is to provide students with the *historical and cultural context* of these events and issues.

Such an approach naturally requires a different structure from most other texts in this area. Therefore, the text is divided into three sections, covering *basic concepts*, *chronologically presented historical information*, and *contemporary issues*. In all three sections of the book, chapters are enhanced by *related readings* designed to expand upon chapter topics and to add a human face to the subject matter. many of these readings are autobiographical accounts that provide powerful and unique perspectives.

The first section of this text introduces the basic concepts and theories social scientists use to study the cultural and power differentials that characterize racial and ethnic groups. We will enter the world of prejudice and discrimination and explore how they occur within systems of social and ethnic stratification. We will also explore both psychological and sociological theories designed to shed light on why such systems of stratification occur and why particular forms of prejudice and discrimination follow (and create) the racial and ethnic dimensions of that stratification.

Section Two places us in touch with American society, chronologically tracing the racial and ethnic groups that entered into contact with one another as that society grew. Beginning with conflict between the first European explorers and the native peoples, this section focuses on the interrelations among different racial and ethnic groups. It will show why patterns of cooperation, competition, and sometimes conflict arose as groups varied from having common to opposing interests at different points in time. By its end, this section arrives at the most contemporary immigration to the United States of the early 21st century; by that point, many current attitudes and conflicts that erupt along racial and ethnic lines will be understood in a historical context.

Finally, the last section of this book provides a strikingly contemporary look at the racial and ethnic dimensions of current American institutions. The first two chapters of this section examine education and the economy respectively, showing both the relations between those institutions and also the way they function—sometimes positively and sometimes negatively—for specific minorities in the United States today. The following two chapters will lead us through the institutions of health/health care delivery and criminal justice, again taking a racial/ethnic focus. We will learn who suffers with the worst health, who winds up in prison, and, in both cases, why. By the end of this section, we will see how the racial and ethnic dimensions of different institutions interrelate. Groups not well served by the educational institution will be tracked through their problems with the economy, their failure to achieve health and health care parity with other groups, and their subsequent difficulties with the criminal justice system.

This book has been a long process from its inception; the final result reflects considerable efforts beyond those of the author. The research process was greatly aided by Pat New and her interlibrary loan staff at Eastern Kentucky University. Many of the more obscure pieces of research that gave the book its flavor would not have been available without their efforts. Many thanks are also due Steve Coe, Nancy Forderhase, Aaron Thompson, and Cynthia Benson for their insightful comments upon reading earlier drafts of this manuscript. I would also like to acknowledge a number of colleagues who reviewed all or part of this manuscript. this book has benefited from the comments and suggestions of the following individuals: Cathy Meeks, Mercer University; Harry Gold, Oakland University; Sylvia Culp, Western Michigan University; Homer Garcia, Baylor University; Clovis Semmes, Eastern Michigan University; Pyong Gap Min, Queens College of The City University of New York; Steven Ybarrola, Central College; Brenda Forster, Elmhurst College; Aaron Thompson, Eastern Kentucky University; Karen Baird-Olson, University of Central Florida; Cindy Benson, University of Central Florida; and Peter Adler, University of Denver.

Similarly, I must also thank my students who encountered earlier versions of this book in class and helped me improve it. At the publishing end, I extend my thanks to several individuals at Harcout College Publishers, including Bryan Leake, Acquisitions Editor; Lisa Hensley, Senior Developmental Editor; Caroline Robbins, Photo and Rights Editor; and Barrett Lackey, Production Manager.

Finally, I must thank my wife, Susan, who lived through the creation of this book, made many significant suggestions, and suffered through the difficult times. Her help and forbearance played significant roles in whatever quality lies between these covers.

This book is dedicated to Frederick and Helen Douglass. Beyond Frederick Douglass' significance as a 19th century community leader, his marriage to Helen reflects an added dimension to his individuality and foresight. Before her death in 1903, Helen Douglass considered the impact of her interracial marriage and commented to her relative, Lillian Butler: "In a hundred years from now, this will be quite common and people will think nothing of it." Such marriages today are indeed more common, but Helen's prediction has turned out to be far from accurate. The importance of racial and ethnic divisions in American life begs continued examination and reflection. I would like to extend the dedication of this book to the futures of my grandchildren, Tyler and Joysha.

CONTENTS

Chapter 3

Chapter 4

Chapter 5

Chapter 6

PART III

Theories and Dilemmas: Modern Racial and Ethnic Relations in the United States
267

I

Concepts and Theories

This book focuses on the relationships among racial and ethnic groups, particularly in the United States. Most generally, that focus will be social scientific. More specifically, it will be sociological. But it cannot be only sociological. We will need to explore the histories of the different groups to understand how they came to be. We will need to examine their political and economic positions to better comprehend the resources available to them. And we will also need a psychological perspective to learn how many thousands of individual feelings, aspirations, and decisions paint the picture we view. Sociology will be the glue that combines these parts of understanding into a whole.

As with all social sciences, sociology's foundation is composed of the concepts and theories that direct its attentions and explanations. Part I of this book will introduce both these conceptual building blocks and the theories they have built. Chapter 1 focuses on the most basic concepts necessary to this study—*ethnic groups, racial groups, minority groups, and dominant groups*. Together, these terms focus our attention on the interplay between culture and power in society. The opening chapter will also explore some of the political and cultural outcomes of ethnic group encounters in examining theories of assimilation and pluralism.

Chapter 2 turns to some classic sociological perspectives on social inequality—the study of social stratification. In order to understand racial or ethnic inequality and how it functions, we must first understand inequality in general. This chapter will introduce concepts such as *social class* and *social mobility, class societies*, and *caste societies*. Our goal here will be to understand how access to power is both granted and limited by one's position within a social stratification system. This is the study of discrimination. We will also want to understand how one's relative position within such a system affects one's interests, abilities, values, and world views. Finally, we will explore the position of ethnic and racial *groups* within a social stratification system. When a relationship exists between race or ethnicity and social class, our study turns from *social* stratification to *ethnic* stratification.

Chapter 3 introduces some of the theories social scientists have developed to understand the many facets of racial and ethnic group relationships. We will examine how and why stereotypes and prejudices develop, and how they damage groups that are affected. Of equal or greater importance, we will explore theories of discrimination. Some of these theories will attempt to explain why certain groups create discriminatory practices that affect other groups. Other theories will focus on the impact of such practices on groups affected and the political responses these practices often elicit from affected groups.

Taken together, these three chapters in Part I are designed to establish the scientific underpinnings for understanding both historical and contemporary relationships among racial and ethnic groups in the United States. The material in these first three chapters will reappear regularly in later chapters as we watch the arrival of a seemingly endless stream of new ethnic groups into American society; this material will continue to be important as we then attempt to understand how differences in ethnic background continue to be significant in separating otherwise equal American citizens from one another.

1

Race and Ethnicity: Some Basic Concepts

Our species has existed for approximately 90,000 years. In those years, people have wandered to every piece of real estate available, searching for new food sources and responding to climactic changes. But the geographical separation among those wanderers has not really been all that long in biological time. The fact that all people are members of the same species attests to that, since separation does produce biological differences. Although biological differences among people do exist, we will see that they are relatively minor and we only pay attention to some of them.

Geographical separation also has produced differences in the way people live. Technology, beliefs, and patterns of behavior have moved in as many different directions as the people themselves. The longer the separation, the greater those differences grow. Our main concern in this book is what happens when such different people become reunited.

How can we begin to make sense of these differences? How are these differences related to one another? I could make several predictions about you just from knowing what you are doing right now. This text is written in English, so you must know how to read English. You might have acquired this skill in a wide variety of circumstances—in non-English speaking countries or perhaps as an adult—but the odds are that you learned the language as an infant (when we learn languages quickly and easily) and were later taught to read. If you fall into the latter category, you were probably born in an area where English is a common language. And inhabitants of the areas where English is spoken also share many other habits, skills, and beliefs along with language. I might guess, for example, that you prefer not to eat insects in your diet—very few English speakers do.

There are things about you I do not know. I do not know, for example, whether you are male or female. Whatever your gender, however, I know it has been a critical factor in making you who you are today. Gender gives people different life experiences in every human society, whether they want those experiences or not. I also do not know whether or not you have wisdom teeth and, if you do not, whether you had them pulled or were born without them. But whichever the case, I can guess that the presence or absence of wisdom teeth has not had a major impact on your life (except, possibly, the day you had them removed). In American society, gender is an important social category, whereas possession of wisdom teeth is not.

Social categories occur whenever a collection of individuals shares some socially important characteristic in common. Which characteristics are important? The people with whom we share our lives dictate just which characteristics deserve our attention and which do not. American society does not attach any special significance to wisdom teeth, so we generally ignore them. We are not, however, invited to ignore our gender. Nor are we invited to ignore how we speak, how we worship (or if we do), the color of our skin, or how we earn our living. Other people will judge us on these characteristics and, as a result, we will probably come to judge ourselves. In short, social categories happen whenever differences become meaningful, and they become meaningful only when social agreement occurs. If that agreement changes, categories lose their importance and assume the significance of wisdom teeth; the differences may remain, but we choose to ignore them.

Assuming you already know something about racial groups and ethnic groups, you may well be wondering why you are reading about teeth. As with so many topics we try to understand, race and ethnic relations are often better grasped if we do not get too close to them. If you grew up in American society, you probably cannot imagine a society in which skin color would be totally ignored. But the importance people in American society give to skin color could just as easily be taken away. An important question facing us is how human differences are made important in our thinking about social relations. An even more important question is *why* we choose to make them important. As we will see, race and ethnicity together account for a great many of those characteristics that thrust different life experiences upon us.

This chapter provides a foundation as we focus on these questions in a journey toward an understanding of the concepts of *race* and *ethnicity*. When members of either group gain power in a society, or are denied access to it, they become *dominant groups* and *minority groups*, respectively. Finally, we explore some of the ongoing relationships that racial and ethnic groups develop as they interact over time. As we will see, a separatist minority wanting its own state might successfully assimilate into a dominant group over time; they also might become the victims of genocide if the dominant group should find their mass deaths convenient.

Ethnic and Racial Groups

Culture is the most basic concept used by social scientists. It refers to everything that people create, share with one another, and pass along to the next generation. This includes creations you can touch, such as clothes, buildings, or this book; these objects are part of **material culture**. Cultural creations also can be part of **nonmaterial culture**, which includes knowledge, language, beliefs, expectations, hopes, and dreams. We learn all of these things as we grow up in our societies, ultimately becoming much like the older members who teach us unless something or someone interferes with the process.

The more we learn about our culture, the more it becomes part of us. By the time you were a teenager, you not only *knew* about your culture, but you also *cared* about it. Cultural beliefs became your personal beliefs and cultural norms (or rules for behavior) became patterns you chose to follow. As you acquired all these beliefs, you also received the ultimate belief of every culture—the belief that your culture is the best of all possible cultures. This means that your food is the best food, your religion is the only true religion, your music is the most entertaining, and so on. When you reached this point, you had acquired **ethnocentricism**—the belief that your culture is superior to all others.

One curious aspect of culture is that we take so much of it for granted. If everyone you know wears clothes when they leave the house, you will not consciously notice the presence of clothing on people around you. Likewise, when you wake in the morning, you consciously decide *what* you should wear but do not consider whether you should leave the house clothed or nude. We notice different types of clothing because they vary; we do not notice clothing per se because everyone is wearing something. In short, the only parts of culture that receive our conscious attention are those that vary, confronting us with alternatives. This observation is extremely important as we follow the movements of people from many different cultures and note the outcomes when cultures collide. Meeting someone from a different culture allows us to contrast our way of life because now we have a means of comparison. More often than not, these confrontations bring out the ethnocentrism in which we have been steeped.

The Changing Boundaries of Ethnicity

The term *ethnic group* is both fundamental to our study and also elusive because social scientists define it in so many ways. These definitions, however, are derived from the many ways that real people think of themselves. We can incorporate that variation, however, with the following definition: An **ethnic group** is a group of people who

1. share a common culture and/or ancestry, or

2. are defined by themselves or others as sharing a common culture and/or ancestry.

In each case, boundaries exist among groups of people, and these boundaries are subject to change over time (see Barth, 1969). Although such boundaries are linked to culture and/or ancestry, the connection may be loose indeed. Ethnic groups can exist even when the "common ancestry" is totally imagined. It really does not matter. If people believe themselves (or are believed by others) to share such commonalities, that sharing becomes real in its consequences for all involved.

A personal example illustrates this point. When I was about nine years old, I was playing baseball with my best friend in the park one afternoon when a classmate rode by on his bicycle and shouted, "Hey, kike!" I had not heard the term before, but my friend explained that *kike* was an insulting term for a Jew, much like the term *nigger* was insulting to black people. I asked him why it had been shouted at us. My friend said it was because he was a Jew.

I was confused. I knew my friend's family quite well, and I knew his father's parents attended a place called a temple on Saturday because they were Jews, and I knew that Judaism was a religion. But I also knew my friend's parents attended neither temples nor churches, having no religious preference in particular. In addition, I knew no one was Jewish on his mother's side. It seemed to me that being Jewish should stop when one left the religion. To others, at least to non-Jews, my friend was Jewish no matter what he thought, said, or did. To many Jews, you have to have a Jewish mother to be considered Jewish, and my friend's mother was not Jewish in any sense of the word. As he tells me to this day, he is the opposite of whatever you are.

I received an early lesson in ethnic boundaries from that experience. Ethnic groups are similar to all human groups in that they have boundaries. There is a clear line, for example, between your relatives and nonrelatives. We keep track of such boundary lines because we want to know who is in our group and who is not. If we want to keep our definition of *ethnic group* in line with the real world, we have to pay attention to those boundaries, *regardless of who believes them to exist*. As with brick walls, ethnic group boundaries can be built to keep your people in or other people out. Ethnicity, in short, can be imposed upon us from the outside as long as those outside people have enough power to make their boundaries strong. Culture may have nothing to do with it.

That same power can be used to create one ethnic group from many. Consider the case of Native Americans. In North and South America in 1491, many different ethnic groups existed, many of them clearly fitting our first definition (i.e., sharing a common culture and/or ancestry). But the coming of Europeans changed all that because they had far more power than the native peoples and they tended not to distinguish among the different groups, lumping them all together as "Indians." Today, Native Americans identify themselves with their tribe as they would have centuries ago, but most have acquired a second, more general, self-identity as Native Americans. They have, in short, come to agree with the early Europeans that *all* the native peoples have something more in common with each other than any one group has with Europeans, even though they reached that conclusion by very different means.

This same process of merging ethnic groups can occur in a completely voluntary manner. A classic example in the United States is "white people" or, as referred to in this book, European Americans. In the nineteenth century, most Americans clearly drew lines to distinguish German Americans from English or Italian Americans. Today, their descendants barely notice such distinctions.

What about boundaries created from within? We can look to Jews once again for an illustration. A few thousand years ago, this group fit our first definition of an ethnic group; however, Jews have since scattered all around the world, acquiring different cultures. Some have even changed their religion or given it up completely. Yet many of these people prefer to identify themselves as Jews, basing their ethnic boundaries on their common ancestry and little else. Are they encouraged to do this through experiences such as the one my friend had on the playground? No doubt they are. But they might wish to maintain their identity anyway. And ethnic identity may arise from a complex selection process based on which ethnic backgrounds are perceived to be the most significant; the case of Afro-Amerasians such as Tiger Woods provides such complexity, as most Americans emphasize his African ancestry over his Asian ancestry (Williams and Thornton, 1998).

If any single defining characteristic can be derived from all these examples, it is this: Ethnic groups exist wherever and whenever people believe ethnic boundaries exist and act on those beliefs. Those boundaries may be based on cultural differences or differences in ancestral heritage, both either real or imagined. This definition allows us to include racial groups as ethnic groups.

Does Race Exist?

A racial group is generally thought to be a collection of people who share a common genetic heritage that distinguishes them biologically from other racial groups. Are racial groups so different? Although the concept of biological differences among races is questionable and its definition is rooted in the biological sciences, the social scientist has to deal with this concept. In this text, we use the term **racial group** to refer to ethnic groups *that have been labeled as* biologically different from other ethnic groups. Almost always, the group developing the label attaches biological superiority to itself. Groups so labeled receive different treatment. Keep in mind, however, that groups labeled as racially different always begin ethnically different. They also usually stay that way because racial group boundaries are always more solid than ethnic group boundaries. This is why we can fit racial groups within the definition of ethnic group offered earlier.

Race first appeared on the scientific scene with works such as *Systema Naturae* by Carolus Linnaeus published in 1758 and the 1775 *De Generis Humani Varietate Nativa* by Johann Friedrich Blumenbach (Gould, 1994). The latter suggested that humans belonged to five races: Caucasians (whites), Mongolians (Asians), Ethiopians (blacks), Americans (Native Americans), and the Malays (Polynesians, Melanesians, and Australian aborigines). Scientists who give credence to the concept of race use pretty much the same groupings today. The fact that we have seen little change in this area of science for over two centuries should perhaps make us a little suspicious; high quality scientific inquiry is typically marked by regular theoretical change.

We noted earlier that people have a history of wandering. Because most large groups of people in different corners of the globe did not search far for mates for so many centuries, different kinds of biological change occurred in different areas. If one travels on foot, the changes appear gradual. Travel by ship or plane, however, allows us to cover vast distances, magnifying the biological differences among geographical areas, and notions of race seem to make more sense. But even given such biological differences, we would still need to be able to compile long lists of genetic traits present in one race but absent from another for the concept to be of much scientific use (see Cartmill, 1998).

Jared Diamond (1994) offers some alternative views of race. He notes that racial characteristics are both genetically fairly trivial and not interconnected. For example, the grandchildren of a European American and African American couple could range from curly haired and light skinned to straight haired and darker skinned. This is not a roll with one set of dice, but rather many rolls with many sets in each generation. You can observe the same "mix and match" with any family's physical traits. But more important, argues Diamond, are the biological traits we ignore in categorizing races. For example, we could categorize on the basis of fingerprints, separating those with mostly loops from those with mostly whorls. This would place most Europeans and Africans in one race, Jews and some Indonesians in a second, and aboriginal Australians would occupy the third. Grouping by lactose tolerance, sickle-cell traits, or tooth shape would create three more (and totally different) sets of races. The traits we have picked out as significant for race may be more important for social reasons than biological.

In the coming chapters, we find racial categories used in all sorts of ways. But we generally find an interesting similarity in the social circumstances in which racial distinctions arise. Almost always with race classifications, a relatively powerless ethnic group (or groups) faces terrible abuse from a much more powerful ethnic group (or groups). Why race? Imagine biological differences between yourself and someone else that allow you to label the other person (and their descendants) as permanently inferior to you. If the differences were only cultural, they would be capable of becoming your equal, given the time to learn. If the abuse is bad enough—consider slavery, for example—your behavior is much easier to rationalize if your victim is thought of as biologically (thus permanently) different. Just as we have different rules for the treatment of human babies and kittens based on biology, *race* allows us the same flexibility with people. You cannot enslave your equal any more than you can drown a bag full of babies.

Minority and Dominant Groups

When Bill Clinton ran for President of the United States in 1992, his main campaign advisor was a seasoned professional named James Carville. A campaign advisor serves many purposes, but among the most important is to identify political issues that will attract the attention of voters and win their support. The best issue is the one that works. His candidate was running against George Bush, and a sitting president can be hard to beat. Carville began with a huge array of issues that appeared in various Clinton speeches. Finally, Carville woke up one morning and banged the palm of his hand against his forehead, having realized the obvious. The United States economy was in a recession and that was the only issue Clinton needed. Economic survival tops almost everyone's list of needs. Carville created a sign to hang over his desk that constantly reminded him, "It's the economy, stupid," lest he once again forget the obvious (Wayne, 1996).

We might want to borrow Carville's sign both for this chapter section and the rest of this text as a reminder to us all that economic factors are integral to understanding ethnic and racial differences. Although it would seem that we should focus only on human culture—language, religion, political systems, child rearing and so on—in our search for understanding, cultural variations comprise only part of the picture. Because cultural differences between groups of people have motivated some of the most horrendous atrocities imaginable, they are a focus of this text. Such atrocities create intense motivations for victims to relocate, far from their tormentors, only to encounter yet more cultural differences among the people with whom they finally settle.

But Carville's advice to himself should be kept in mind here. Undoubtedly the oldest human motivation for moving is economic. This is no less true for people who came to North and South America during the early days of European expansion than it was for early nomadic humans thousands of years ago. As we now turn to examine how different ethnic groups relate to one another when they meet, we need to keep in mind that their cultural interactions will have much to do with their varying economic positions. People who move because of poverty typically have few choices, either before or after they move. When they arrive at their destinations, they often find themselves powerless and unable to prevent yet further economic disadvantage.

Size versus Power

A **minority group** is an ethnic or racial group that occupies a subordinate position in a society, unable to gain equal access to wealth or influence. Their subordinate status is maintained through limitations placed on them by the structure of their society. Those limitations either are enforced at the level of national government or, at least, allowed to exist by that government. By contrast, a **dominant group** is an ethnic or racial group that occupies the dominant positions in a society, monopolizing positions of power and wealth. They are the creators of the limitations minority groups face, and they benefit from the lack of options available to minority group members.

The critical distinction between dominant and minority groups is power, not size. The term *minority* can be misleading because it suggests a group small in number rather than a powerless group. Minority groups often are small—lack of numbers can be one of the factors that limits power—but power alone is the key. For years, the Republic of South Africa was run by white South Africans with nineteenth century ties to England and Holland. Together, they represented a tiny minority of South Africa's population, but for over a century they controlled all the power and wealth of that country. It is easier to maintain dominance if the dominant group is large, but size is not the most critical factor.

Linking Power with Ethnicity

Imagine the following scene: A farmer is standing on a country road looking fondly at his most prized possession—several hundred acres of good farmland. A stranger comes walking down the road and strikes up a conversation.

"That's a mighty nice piece of land there," says the stranger.

"Yup," replies the farmer.

"Belongs to you?" asks the stranger.

"Yup," replies the farmer.

"How did you come by it?" asks the stranger.

"Well," replies the farmer, "I got it from my father, who worked it all his life and left it to me."

"Well, how did he get it?"

"From my grandfather."

"But how did he get it?"

"From his father before him."

"But how did he get it?"

"Well," replies the farmer, "he fought for it."

"That's fine," counters the stranger, "I'll fight you for it."

Why does the stranger not have a right to fight for the land? No piece of land on the face of the earth is occupied by its original owner or even the descendants of the original owner. All ownership is initially or subsequently fought for. At some point following a fight, a political state takes over to maintain the status quo. In our story, the farmer's great-grandfather took the land by force, but he subsequently received a legal deed to the property from the state. That deed means that the next person who wants to fight for the land will have to fight both the state and the farmer.

As this example suggests, the study of racial and ethnic groups in any society also will be the study of power and its centralization in a government. In most societies, the central government can be viewed as a stand-in for the dominant group, legislating its wishes and enforcing them. It is easy to think of dominant groups as those who make their way to the top, but the trick is to stay there. The backing of a government makes staying there much easier than the initial climb. The connection between dominant ethnic groups and governments is therefore critical if that ethnic group is to maintain its dominance. When we later examine some American history followed by contemporary American race and ethnic relations, we will pay close attention to the role of the central (or federal) government (Enloe, 1973).

Ethnic Group Encounters: From Genocide to Separatism

When ethnic groups migrate and encounter one another, their subsequent relationships vary widely. They may dislike each other immediately; if one of the groups is significantly less powerful than the other, it will undoubtedly suffer greatly. Economics also may play a major role in such outcomes. Weaker ethnic groups may be quite welcome if they are willing to work for lower wages than the current labor force. And immigrants may well arrive with their own agenda, determined to avoid all other ethnic groups in their adopted country in hopes of maintaining their ethnic distinctiveness. Although the reality of such ethnic group encounters covers a wide range of relationships, social scientists have created a basic typology of encounters that attempts to categorize those relationships into the most common occurrences.

Genocide and Expulsion

Genocide occurs when the dominant group deliberately acts to cause the deaths of many or all members of a minority group. In some cases, genocide is clear cut, as in Hitler's efforts to exterminate Jews from Europe—an intended mass murder that, for the most part, succeeded. In other cases, genocide is less obvious. Massive deaths among Native Americans after the arrival of Europeans occurred mainly through the spread of European diseases to which Native Americans had little immunity. This was not intentional. On the other hand, most Europeans were generally happy with the outcome and found numerous ways to cause Native American deaths over the years when nature did not succeed. All in all, was this genocide? Although the killing may not have been as evidently systematic as in the Holocaust, Native Americans were targeted for genocide.

Expulsion refers to the forcible removal of minority group members beyond the borders of a country. You might want to think about this as mass deportation. American history offers us several examples of this process. The first, dating back to the early 1800s, was the forcible removal of eastern Native American peoples by the federal government. They were relocated in Indian Territory in what is now the state of Oklahoma. Although land west of the Mississippi belonged to the United States, it was not then viewed as valuable, or the government would not have foolishly signed so many land treaties that would later come back to haunt it. Although the minorities in question did not exactly wind up outside the borders, the intent to remove ethnic groups from a society was clearly present.

Kosovo refugees endured crowded evacuation routes to escape genocide during the 1999 Serbian invasion.

Perhaps a better example of expulsion comes from the twentieth century in the United States. Since 1900, Mexican Americans had been welcomed in the United States when their labor was needed. During recessions or depressions, however, they would be out of work and a potential cost to social welfare systems. During the 1930s and again during the 1950s, Mexican Americans were rounded up and transported across the border to Mexico. Both times, some American citizens found themselves deposited outside the borders of their country.

Many historical events combine genocide and expulsion. For example, in the 1999 Serbian invasion of Kosovo, the initial actions of the Serbs appeared to be genocide. Many Kosovars were simply lined up and shot. But such events also greatly affected survivors, many of whom became immediate refugees as they made their way out of Kosovo and into neighboring Macedonia and Albania. Because Serbian troops interfered relatively little with the refugees, one can only conclude that their expulsion from Kosovo was an equally acceptable outcome. Indeed, genocide appears to have created the expulsion.

These two terms are discussed together because the motives behind their use are identical as are the outcomes for the dominant group. In either case, the dominant group cannot gain wealth from the presence of minority group members. As we will see, one of the main uses of minority groups for the dominant group is to supply inexpensive labor. If the minority cannot or will not work (or if no work currently needs doing), it is generally beneficial to the dominant group if they are removed. Genocide solves the problem neatly, but public opinion may make it a poor alternative. In addition, as was the case with Mexican labor, the minority may be wanted later. When considering long-standing cultural hatreds between Serbs and Kosovars, however, economic concerns may well be of little consequence to a strong ethnic group with a centuries-old vendetta motivating them.

A quick glance at American history shows us that the normal state of affairs involved neither genocide nor expulsion. Centuries of open door immigration coupled with slavery (forced immigration) suggests that minorities were very welcome indeed. When the need for labor decreased in the early twentieth century, the solution was to turn off the faucet of immigration rather than killing or expelling those already here. Nevertheless, both of these "solutions" have occurred around the world far more often than anyone would like, and continue to take place to the present day.

Assimilation vs. the "Melting Pot"

What happens over time to minority groups whose members are neither killed nor expelled? For many years, the experience of immigrants to the United States was described as their joining a melting pot. Popularized by playwright Israel Zangwill in his play of the same name, the **melting pot** refers to a multicultural society in which members of different groups intermarry and, in so doing, blend their respective cultures into a new culture that had never before existed. Just as a cooked dish receives its taste from the blend of its many ingredients, so too might a society take on a totally new character that would arise from the blend of its members. The melting pot is an appealing and romantic ideal, but it just does not happen.

Families who immigrate often become more acculturated with each new generation.

The melting pot rarely occurs because it is almost never in the interests of dominant group members. As we saw earlier, dominant groups use their power to impose their culture on the societies they control. They do this not only out of ethnocentrism but also for very practical reasons. Consider language. Every society contains more than one version of its primary language (termed dialects), but there is almost always just one prestige version. Mastery of the prestige version opens doors and leads to opportunities. How and when do we learn language? From our parents, as infants. It will be many generations before any immigrant group will produce many parents with this mastery; by contrast, infants in the dominant group learn this prestige version automatically. It is therefore in the dominant group's interests, even looking ahead to descendants, to maintain control over a society's culture, so that it continues to reflect their own. Giving into anything like the melting pot would be giving up these privileges. Since, by definition, a dominant group is in a position to keep the melting pot from occurring, the melting pot will not occur.

What will take place is some form of assimilation. **Assimilation** is the process by which minority group members come to acquire the dominant group's culture and receive acceptance to higher social status within that society. The first half of the definition—the acquisition of the dominant group's culture—is often termed **acculturation**. The second part of the definition calls our attention to the limitations that minority group members face and suggests that the reward for acculturation is either a partial or total removal of those limitations by the dominant group. If all limitations are lifted, the minority group in question ceases to be a minority and can be described as completely assimilated as, for the most part, are German Americans. Many millions immigrated to the United States, but you will not find much evidence of German culture. The people (or their descendants) are still in the United States, but the *group* is gone. All those immigrants ultimately learned to speak English and were rewarded with both social and economic acceptance.

In his study of assimilation, Milton Gordon (1964) renames acculturation—the first stage—**behavioral assimilation**. The second stage, which focuses on dominant group acceptance of the newly acculturated immigrant, he terms **structural assimilation**. He describes behavioral assimilation as being under the control of the minority group; its members have to (1) want to acquire the dominant group's culture and (2) be willing to put in the work necessary to do so. Much of this work can be accomplished by simply having the second generation come along—the first generation born in the host country. These children acquire language and other cultural elements more naturally and more effectively. But some minorities do not want this to happen. Some—the Amish, for example—care very deeply about not assimilating and have gone to great pains to keep their children away from the dominant culture. More often, however, the strength of the dominant culture practically kidnaps minority children. For groups whose members wish to maintain an immigrant culture, both physical and social isolation is necessary.

Gordon describes structural assimilation as being under the control of the dominant group. Because they control the centers of politics, education, and employment in any society, they can choose whether or not to open those doors to any given minority. Behavioral assimilation is prerequisite. A minority group member will not get far in education, for example, unless he or she has acquired at least the basics of the dominant group's language and knowledge. But behavioral assimilation guarantees nothing. If the dominant group dislikes a given minority for any reason whatsoever, all the behavioral assimilation in the world will not result in the integration of that group into the main structures of society.

Gordon (1978) further elaborates on the structural assimilation process by dividing it into two steps. The first step he terms structural assimilation at the level of secondary relationships. Secondary relationships refer to those social relationships we maintain that are more casual and/or businesslike than emotion filled. For example, the people with whom you work, attend PTA meetings, or meet in university classrooms are largely secondary relationships. You may enjoy their company, but it would not be the end of the world if you never saw them again. Gordon describes this first level of structural assimilation as easier for the dominant group to tolerate. Dominant group members may not think highly of a given minority, but they may well be able to work in the same factory with them. We should also note that this level of structural assimilation is probably the most important goal for a minority. Giving up one's culture for assimilation is a painful loss; at the very least, one would expect civil rights and occupational advancement in return.

The second level of structural assimilation is the level of primary relationships. Primary relationships are those social relationships characterized by strong emotional ties and the involvement of one's whole self. In general, these relationships are limited to family relationships and close friends. At this level of acceptance, minority group members will find themselves welcome in country clubs, at private parties, and ultimately as marriage partners. At this final and complete level of assimilation, the minority group may safely be said to have disappeared. We should also note that the members of many minority groups may have no desire for this acceptance. Having achieved political and economic rights at the preceding level, they may prefer to socialize within their own group so as to maintain some cultural distinctiveness, however quietly.

Before leaving Gordon's typology, a word of caution is in order. Real life never fits typologies neatly. In particular, Gordon's levels suggest that minority group members first finish the job of behavioral assimilation before structural assimilation begins. In the real world, some degree of structural assimilation is necessary before behavioral assimilation can get very far. You will not learn English well if you are confined on a reservation, for example. We should also keep in mind that some behavioral assimilation is expensive to acquire and/or requires social acceptance from the dominant group. Consider the following list of cultural skills and knowledge in the United States and consider the difficulties facing a poor immigrant group member in his or her efforts to acquire each item:

1. Knowledge of how to act at a formal dinner party.

2. Skill in playing the game of polo.

3. Computer skills.

4. Knowledge of how to write a formal research paper for a university course requirement.

When you begin thinking in these ways, you will understand why immigrant children who work hard in American schools excel much more rapidly in mathematics than in language arts: far less background knowledge is necessary.

Forms of Pluralism

As with assimilation, pluralism is a term developed to describe what people were already doing. In the coming chapters, you will see examples of it with early German immigrants who wanted to keep to themselves and retain their culture. You will also see it in the distinctive Chinatowns created by Chinese immigrants or "Little Havana" in Miami where many Cuban immigrants live. The idea of immigrant groups retaining both their group boundaries and their cultures while living inside a country dominated by a very different culture is what Horace Kallen had in mind when he wrote about pluralism in early twentieth century America.

Kallen's concern was the future of Jewish Americans. He wanted them to be in America but not of it. The important goal was for Jews to retain a clear sense of their own distinctiveness yet not be penalized for it. **Pluralism**, for Kallen (1915, 1924), meant that cultural minorities could retain their distinctive cultures, but also acquire the dominant group's culture and receive both respect and acceptance from the dominant group. This respect and acceptance often takes the form of special group recognition of the minority. For example, the government might be called upon to protect the group should they face persecution, or perhaps provide special services to that group if their cultural distinctiveness required it. Kallen's conclusion was that pluralism would produce both a more interesting and a more humane society.

The political battleground in pluralism arises over the special group recognition. Invariably, someone or some group will lose while another gains. A classic version of this battle involves the United States Constitution, which created a representative government. The argument revolved around how representation should occur. One solution was to have each state send a set number of representatives to Congress. This would provide equality for states but not for individuals; those people living in populous states would wind up having less clout. Another solution was to have one representative in Congress for every so many people back in a home district. This, of course, would be fair to individuals but would penalize states with small populations. The compromise, of course, was to do both. Think of the United States Senate as pluralism in action where groups (states) get special treatment. And what is so special about states? Nothing, except that we have decided they are. Kallen would have us decide that ethnic *groups* are also important and deserve this recognition.

In action, you can see the political battles that pluralism produces. The French in the Canadian province of Quebec, for example, saw their culture disappearing as English speakers grew in power and numbers in the province. The pluralistic solution from French-Canadians was to eliminate most English language schools in the province and require a knowledge of French for almost all employment. This special group recognition of one group simultaneously damages other groups. In this case, English speakers find that they are no longer qualified for most occupations in Quebec. To the extent that groups do not play the assimilation game but demand rights for their distinctive culture, the dominant group loses. Much of its power, you will recall, comes from its control of culture.

Yet another example of the battleground between pluralism and assimilation is provided by author Timothy Fong in the article that follows this chapter. Fong traces the racial and ethnic history of Monterey Park , California, showing interrelationships among whites, Latinos, African Americans, and Asians. An earlier period of general assimilation produced relative harmony; more recent immigration by pluralistically minded Chinese has produced some interesting conflicts and still more interesting group alliances.

Ethnic Nationalism and the Road to Separatism

The previously mentioned dispute between Quebec and the rest of Canada may eventually result in a political separation between the two. Much as the American South attempted to leave the Union in 1861, Quebec may well declare itself an independent state, albeit a civil war in Canada over this issue is extremely unlikely. Nevertheless, nationalism and separatism create major and serious changes in ethnic relations. They are included here because they are best thought of as extreme versions of pluralism (or where pluralism might lead once you head down that road).

Nationalism is the political goal of achieving a state to represent the interests of a particular ethnic group. **Separatism** refers to the actual political process through which such a state is constructed. They are probably the ultimate solutions to ethnic or racial strife. If you return to the definitions of dominant and minority groups, you will note that the entire problem stems from having two ethnic groups but only one government. If one of those ethnic groups controls that government, the rulings it makes will undoubtedly reflect its interests; the remaining group is therefore a minority. But if you create two governments, each with jurisdiction over two pieces of real estate (one for ethnic group A and the other for ethnic group B), any further disputes between them fall under the realm of international relations and are not in the scope of this book. Still, we have to study the processes by which such changes come about.

As a political movement, ethnic nationalism is always characterized by a glorification of the minority ethnic group's culture coupled with constant reference to the many injustices suffered at the hands of the dominant group. Both of these factors must be present for cultural nationalism to succeed. Members of the minority must acquire enough pride in and recognition of their distinctiveness to draw those ethnic boundaries a little more distinctly in their minds. They also have to be involved in the cause emotionally, because separatism is extremely disruptive and is not a move to be taken lightly.

Separatism can occur in only two ways: either people move or political boundaries move. Israel is probably the best modern example of the former. Jews from all over the world congregated at their ancestral home of Palestine and created a new state to be run by Jews. This was the only assurance, many felt, for Jews avoiding minority status in the future. The problem with this form of separatism is that unoccupied real estate is dwindling. When you move to form a new state, you will have to displace the people currently living there. It should come as no great surprise that Israel has been in a state of war or near war for the full half century of its existence.

The second method of achieving separatism is generally much more agreeable unless you have someone like Abraham Lincoln in the way. Moving political boundaries means that you take what had been one country and slice it up as one would a pie, setting aside this sector for that group and another sector for another group. Modern India and Pakistan were created this way out of the newly independent India following World War II. India at that time was composed of both modern India and Pakistan, but ethnic strife convinced all concerned that the formation of two separate countries was the only solution. This method of separatism can be bloodless, but it is a move never taken lightly. Very few governments look fondly on giving up large sectors of the land they control.

Summary

In terms of their species, people are all the same; if we view them as individuals, no two are alike. Between those extremes, certain characteristics of people are often selected to create subgroups. People can be separated according to gender, language, religion, skin color, or virtually any characteristic that varies. When these characteristics are treated as significant in any society, they are social categories.

Many such categories are culturally based. Cultural differences among groups of people elicit attention because the differences are usually numerous and varied, including language, religious beliefs, clothing styles, child-rearing practices, and food preferences. Although all of these are subject to change (and often do change from one generation to the next), they are impossible to ignore when they are present. When such differences create dislike or hostility between groups, those groups may be described as ethnocentric.

Cultural differences are typically described as ethnicity. By most definitions, *ethnicity* includes both cultural distinctiveness and an awareness by group members of that distinctiveness. That awareness is often coupled with an emphasis on ancestral distinctiveness. Boundaries between ethnic groups are in a state of constant change as both cultures and attitudes about group membership change over generations. Ethnic groups can both merge and separate over time. Whichever occurs, the roots of such changes are in the particular social contexts within which people live.

Racial groups are social categories based on presumed biological differences among people. Of the many biological differences that have developed over the years due to the physical separation of different peoples, only certain observable traits are usually selected to define racial differences. Ancestry also creates racial differences, even if the physical traits are not present. From a social standpoint, racial distinctions serve as efficient rationalizations or justifications for the mistreatment and subordination of others.

When racial and/or ethnic groups share a society (that is, one government), they often develop unequal power relationships. More powerful groups are dominant groups while the less powerful groups are minority (or subordinate) groups. Dominant groups typically use their power to force the minority to behave in ways beneficial to the dominant group. Depending upon circumstances, very different behaviors may be thrust upon (or opened to) the minority.

When minority groups are "in the way" from the dominant group's perspective, genocide or expulsion may occur. Genocide represents the systematic killing of minority group members; expulsion refers to the forced removal of minority group members from that society's boundaries, either returning them to the country of their original ancestry or anywhere else convenient. Both solutions remove the minority group from the society.

Some level of assimilation is a more common outcome for minorities because their continued presence in society is often useful to the dominant group. Assimilation refers to various degrees of the dominant group's acceptance of the minority, coupled with equally various degrees of inclusion into the society. Behavioral assimilation (or acculturation) refers to attempts by minority group members to acquire the dominant group's culture; structural assimilation measures the degree to which minority members are included in the mainstream society. This inclusion may range from complete inclusion (which generally makes future generations of the minority largely invisible with their culture jettisoned) to only partial inclusion.

Minorities may respond to their circumstances by seeking either pluralism or separatism. Pluralism represents minorities' attempts to maintain their cultural distinctiveness while simultaneously achieving dominant group acceptance of that distinctiveness. Separatism refers to the minority group's efforts to separate politically from the dominant group, either by partitioning the current state into two states or by leaving voluntarily to form a new country elsewhere.

Chapter 1 Reading
The First Suburban Chinatown: The
Remaking of Monterey Park, California

Timothy P. Fong

Many immigrants to the United States in the past assimilated into American culture. The choice was often made for them, however, by the necessity to make a living. Opportunity typically carried the price of acquiring the host culture. By contrast, some recent immigrants to the United States appear to be following a path of pluralism, finding economic opportunities in strong ethnic communities. Timothy Fong invites us to visit Monterey Park, California—a community near Los Angeles with a complex ethnic history. As you will see, new immigrants from China have created ethnic tension and generated complaints from whites, Latinos, African Americans and Asian Americans whose ancestors long ago assimilated.

A New and Dynamic Community

On an early morning walk to Barnes Memorial Park, one can see dozens of elderly Chinese performing their daily movement exercises under the guidance of an experienced leader. Other seniors stroll around the perimeter of the park; still others sit on benches watching the activity around them or reading a Chinese-language newspaper.

By now children are making their way to school, their backpacks bulging with books. They talk to each other in both English and Chinese, but mostly English. Many are going to Ynez Elementary, the oldest school in town.

When a nearby coin laundry opens its doors for business, all three television sets are turned on: one is tuned to a Spanish novella, another to a cable channel's Chinese newscast, and the third to Bryant Gumbel and the Today show.

Up the street from the park a home with a small stone carved Buddha and several stone pagodas in the well-tended front yard is an attractive sight. The large tree that provides afternoon shade for the house has a yellow ribbon tied around its trunk, a symbol of support for American troops fighting in the Persian Gulf. On the porch an American flag is tied to a crudely constructed flagpole. Next to it, taped to the front door, Chinese characters read "Happiness" and "Long Life" to greet visitors.

These sights and sounds are of interest not because they represent the routine of life in an ethnic neighborhood but because they signal the transformation of an entire city. Monterey Park, California, a rapidly growing, rapidly changing community of 60,000 residents, is located just eight miles east of downtown Los Angeles. An influx of immigrants primarily from Taiwan, Hong Kong, and the People's Republic of China has made Monterey Park the only city in the continental United States the majority of whose residents are of Asian background. According to the 1990 census,

Asians make up 56 percent of the city's population, followed by Hispanics with 31 percent, and whites with 12 percent.[1]

In the early 1980s Monterey Park was nationally recognized for its liberal attitude toward newcomers. In fact, on June 13, 1983, *Time* magazine featured a photograph of the city council as representative of a successful suburban melting pot. The caption read, "Middle-class Monterey Park's multiethnic city council: two Hispanics, a Filipino, a Chinese, and, in the rear, an Anglo."[2] Another national public relations coup came in 1985 when the National Municipal League and the newspaper *USA Today* named Monterey Park an "All-America City" for its programs to welcome immigrants to the community.[3] Nicknamed "City with a Heart," it took great pride in being a diverse and harmonious community. But despite these accolades, there were signs that the melting pot was about to boil over.

Tensions had begun to simmer with the arrival in the late 1970s of Chinese immigrants, many of whom were affluent and well educated. New ethnic-oriented businesses sprang up to accommodate them: nearly all the business signs on Atlantic Boulevard, the city's main commercial thoroughfare, conspicuously displayed Chinese characters with only token English translations. In 1985, the same year Monterey Park received its "All-America" award, some three thousand residents signed a petition attempting to get an "Official English" initiative on the municipal ballot; a local newspaper printed an article accusing the Chinese of being bad drivers; and cars displayed bumper stickers asking, "Will the Last American to Leave Monterey Park Please Bring the Flag?"[4]

In April 1986 the two Latinos and the Chinese American on the city council were defeated in their bids for reelection. Voted into office were three white candidates, one a proponent of controlled growth, the other two closely identified with the official-English movement in Monterey Park and the state. In June the new council passed Resolution 9004, which, among other things, called for English to be the official language of the United States of America.[5] Though the resolution was purely symbolic and carried no legal weight, it was immediately branded as a deliberate slap at the city's Chinese and Latino population. Undaunted, the council continued to take controversial actions that critics labeled "anti-Chinese," among them adopting a broad moratorium on new construction and firing the city planning commission that had approved many Chinese-financed developments. But it was rejection of the plans proposed by a Taiwanese group to build a senior housing project that prompted a rare display of public protest by the usually apolitical Chinese community. Four hundred people, mostly elderly Chinese, marched to City Hall carrying American flags and signs reading, "Stop Racism," "We Are Americans Too," and "End Monterey Park Apartheid."[6]

These high-profile controversies, lasting throughout the 1980s were not isolated or incidental cases of cultural conflict. Indeed, events in this community have received publicity in local, national, and even international media; recently, scholars too have become interested in Monterey Park, focusing primarily on ethnic politics and race relations.[7] Close study of the community is important for several reasons. To begin with, Monterey Park's Chinese residents reflect the changing pattern of Chinese immigration nationwide. Chinese newcomers to Monterey Park and elsewhere are not analogous to the historically persecuted and oppressed male laborers who came to this country in the mid-nineteenth century; they are men and women generally much better educated and more affluent than either their Chinese predecessors or their white counterparts.[8] Further, similar demographic and economic changes are occurring not just in Monterey Park but throughout southern California's San Gabriel Valley and Orange County, and in the northern California cities of San Francisco, Mountain View, and San Jose. Increasing Chinese influence is felt also in New York City's boroughs of Manhattan and Queens (particularly Flushing), in Houston, Texas, and Orlando, Florida. Outside the United States, recent examples of a rapid influx of Chinese people and capital are found in Sydney, Australia, and in Vancouver and Toronto, Canada.[9]

Next, because demographic change and economic development issues have created a complex controversy in Monterey Park, the intersection of ethnic, racial, and class conflict shows up quite clearly there. One prominent aspect of the social, economic, and political dynamics in Monterey Park is the popular call for controlled growth combined with a narrow nativist, anti-Chinese anti-immigrant tone in debates that crossed ethnic lines throughout the community. And again, these developments too are relevant nationwide, occurring as they did at a time of increasing concern over immigration: over statistics showing that almost 90 percent of all legal immigrants coming to the United States since 1981 have been from non-European countries,[10] and over the numbers of undocumented immigrants crossing the southern U.S. borders. Documented and undocumented immigrants are rapidly changing the face of many urban centers.

Finally, the conflicts in Monterey Park took place in a period of increased anti-Asian sentiment and violence. Debate occasioned by the large trade deficit between the United States and Japan, suspicion raised by large Asian investments throughout the nation, and envy generated by repeated headlines about Asian superachievers in education all fueled the fires of resentment throughout the 1980s. The 1982 killing of Vincent Chin in Detroit, a widely cited act of anti-Asian violence, prompted a U.S. Commission on Civil Rights investigation.[11] The commission concluded that the upswing in animosity toward Asians reflected a perception that all Asian Americans, immigrants, and refugees are "foreigners" and as such are responsible for the economic woes of this country.[12]

This study of Monterey Park examines the evolution of conflict in the city and locates the beginnings of its recovery from internal strife and unwanted negative media attention. I argue that what was generally seen by the media and outsiders as a "racial" conflict was in fact a class conflict. At the same time, I demonstrate the highly charged saliency of ethnicity and race in the political arena and show how they were used to obscure class interests and to further political interests.

Effects of Chinese Immigration

As the influx of Chinese to Monterey Park began, most community leaders and residents compared the newcomers with the American-born Japanese *nisei* who had moved to the community twenty years earlier and quickly assimilated. Together they welcomed the Chinese as yet another group of hardworking people who would naturally be more than happy to settle into the established wholesome life of the community. But because these Chinese were new immigrants, expectations for their immediate assimilation proved unrealistic, and several areas of friction developed—involving business and social organizations, schools, and even supermarkets.

Divided Organizations

When it became obvious that no one could stop the influx of Chinese immigrants to the community, Eli Isenberg wrote a conciliatory column in December 1977 titled, "A Call for Open Arms," which was later translated into Chinese and republished in the [Monterey Park] *Progress*:

> Twenty years ago, Monterey Park became a prestige community for Japanese. At first they settled in Monterey Hills. Today they live throughout and are active in the community. They were invited and accepted invitations to become involved. Today George Ige is our mayor, Keiji Higashi, a past president of chamber of commerce, is president-elect of Rotary. Fifty other Japanese men and women serve on advisory boards and in other leadership roles.
>
> Today we must offer the same hand of friendship to our new Chinese neighbors. They should be invited to join service clubs, serve on advisory boards, become involved in little theater and PTA. . . . To become and stay a good community, there must be a structured effort to assimilate all those who want to become a part of Monterey Park. The city itself should coordinate this effort through the community relations commission and call on all organizations in Monterey Park to play their part in offering a hand of friendship to our new neighbors.[13]

Isenberg may have written partly in response to the formation of an independent Monterey Park Chinese Chamber of Commerce in September 1977—much to the chagrin of the original chamber. A great deal of animosity and criticism were leveled at this separate group for their reluctance to cooperate with established merchants. Shortly after Isenberg's column appeared, a series of meetings between the two groups resulted in the admission of the Chinese organization to the regular city Chamber of Commerce and the formation of a new Chinese American committee. "Helping keep the doors open was Fred Hsieh," recalls Isenberg. "Fred played an important role in maintaining an integrated Monterey Park Chamber of Commerce."[14]

After the proposed "Chinatown theme" was rejected in 1978, however, some dissatisfied Chinese business people resurrected the idea of a separate Chinese business organization and grumbled about other aspects of their chamber membership. For one thing, few of the Chinese businessmen spoke much English and could understand little of what was being said during meetings. Chinese merchants also resented having to seek chamber approval for business decisions; they wanted more autonomy. Furthermore, unlike Frederic Hsieh, most of the Chinese saw little to be gained by interacting with established merchants who, they felt, were antagonistic. Though they remained in the chamber, the tension was not resolved, and flare-ups periodically occurred.

The Lions Club was even less successful at amalgamating with the newcomers. In the early 1980s an ad hoc group of Chinese asked Lions Club International to charter the Little Taipei Lions Club in Monterey Park. Given the historical prestige of the Lions Club in Monterey Park, its aging and dwindling membership was embarrassed by the formation of a separate club. Although they formally voted to sponsor the Chinese Lions organization in 1985, there was a great deal of reluctance. "The effort to recreate Little Taipei in Southern California," says Joseph Graves, was "unfortunate": "We would infinitely rather they had joined the existing, strong, long-time club with traditions." Graves spoke with pride of the original club's accomplishments, such as "screening all the children's eyes in Monterey Park.... [And] it looks like about 50 percent to 60 percent are Oriental."[15]

The projects of the Little Taipei Lions Club have been admirable, as well. Twice a year, during Chinese New Year's Day and on Thanksgiving, it sponsors a free lunch for senior citizens in Monterey Park's Langley Center, and it has raised considerable money for various non-profit organizations in the community—for example, making major donations to the city's public library to purchase Chinese-language books. But Graves objects that the Little Taipei Lions Club just gives out money rather than organizing work projects: "The Lions Club believed in the idea of going down and pouring cement to build a Memorial Bowl, or hammering nails to the roof of the pavilion at the park," he insists. "As older members, we look down our noses at any organization that doesn't get their hands dirty."[16]

In the mid-1980s the Monterey Park Kiwanis Club refused to sponsor a separate Chinese chapter, but one was formed anyway. To persistent rumors that a Chinese Rotary Club would soon be organized as well, long-time Rotary member Eli Isenberg responded in 1985: "Apartheid, whether in South Africa or in service clubs in Monterey Park, is a giant step back." In a tone quite different from that of his 1977 "Call for Open Arms," he continued: "Asians do not have a Constitutional right to form service clubs where they will be comfortable with members of their kind. All service clubs, from their international, should ban this happening. Provided, of course, that the Anglo clubs are willing to accept Asians as is the case in Monterey Park."[17]

Little Taipei Lions Club members interviewed during their Thanksgiving day luncheon in 1990, however, denied that they are separatist. While passing out plates of turkey and trimmings to senior citizens, many said they meant no disrespect toward the established Lions Club and had no intention of competing with it in service to the community. As a master of ceremonies in the background called out winning door prize numbers in both English and Chinese, one member asserted that there was plenty of room for both clubs. Another member found nothing surprising about preferring to be with people his own age who spoke his language: "What is wrong with a service club that happens to be sensitive and in touch with the Chinese community?" Angered by any perception that the Little Taipei Lions Club serves only the Chinese, he added: "Look around you. There are lots of different people here. We happily serve them [all].... But we do things for the Chinese in this city that no one else would."[18]

Bilingual Education

The impact of the newcomers on the local schools also generated a great deal of tension. Brightwood Elementary School is located in the heart of one of the most heavily concentrated Asian sections in Monterey Park (census tract 4820.02), and surrounded by well maintained middle-class homes built in the 1950s. In early 1978 a Chinese bilingual education plan initiated at Brightwood School opened what the PTA president called "a bucket of worms."[19]

On January 21, 1974, the United States Supreme Court had ruled in the landmark *Lau v. Nichols* case that the San Francisco Unified School District had failed to provide necessary assistance to nearly 2,000 Chinese American students who did not speak English. The district was ordered to implement "appropriate relief," subject to approval by the court. This precedent-setting case established bilingual education in public schools for students who speak limited or no English.[20]

In 1976 the school district of which Brightwood was a part was cited by the Department of Health, Education and Welfare's Office of Civil Rights for having an inadequate English-as-a-second language (ESL) program. The department ruled that affirmative steps should be taken to correct the language deficiency of many minority children, in order to give them equal educational opportunity. The district complied the following year with a Spanish bilingual program in elementary and secondary schools and planned to phase in a Chinese bilingual program in 1978.

The proposal divided the Brightwood School—which was 70 percent Asian at the time—along English- and non-English-speaking lines. The plan called for all students from kindergarten to third grade to be taught in Chinese *and* English. Opposition to the program was led by American-born parents of Japanese and Chinese ancestry who were fearful that implementation would impede their children's educational progress in the future. Some threatened to take their children out of Brightwood and place them in private schools, or move them out of the district entirely. Supporters of the plan, mostly immigrant parents, welcomed bilingual education because they believed it would help their children maintain their native language and provide them with emotional and psychological support and the acceptance they needed within a new environment. A small third group of more moderate parents supported bilingual education but wanted the district to consider a "transitional" program that would instruct children in their native language but at the same time teach them enough English to allow their eventual transfer to a regular classroom.

During meetings to discuss the plan, the debate became intense. "Let them talk English," cried out one angry mother. "Why don't they leave the whole damn school as it is?"[21] Eventually, even supporters of the program asked the school board to delay implementation until the district could provide parents with more information and options. The delay was granted, and the bilingual program at Brightwood School did not start until early the following year. The result of months of meetings by the Brightwood Bilingual Committee turned out to be a much weaker variation of the original plan. Only one second grade class offered Chinese bilingual instruction; other Chinese students were taught English by "traveling teachers" at the parents' request.

Asian Markets

The prominence of Chinese-owned and -operated businesses in town became an even greater source of resentment. Non-Asians in Monterey Park commonly complain that Chinese merchants quickly replaced many established businesses and catered almost exclusively to an Asian and Chinese-speaking clientele. The best examples are food stores and eateries. Chinese have taken over all but two of the town's major chain supermarkets. Bok choy is more common than lettuce in produce departments, and dim sum and tea more readily available than a hamburger and coffee in the restaurants.

The first Asian grocery in Monterey Park was opened in 1978 by Wu Jin Shen, a former stockbroker from Taiwan. Wu's Diho Market proved to be an immediate success because the owner hired workers who spoke both Cantonese and Mandarin, and sold such popular items as preserved eggs and Taiwan's leading brand of cigarettes. Wu built the Diho Market into a chain of stores with 400 employees and $30 million in sales.[22] Likewise, the Hong Kong Supermarket and the Ai Hoa, started in Monterey Park, were so successful that today they operate satellite stores throughout the San Gabriel Valley.

In Monterey Park there are now half a dozen large Asian supermarkets and about a dozen medium-sized stores. Their proprietors also lease out small spaces to immigrant entrepreneurs who offer videos, newspapers, baked goods, tea, ginseng, and herbs. Together, these enterprises attract Chinese and other Asian residents in large numbers to shop for the kinds of groceries unavailable or overpriced in "American" chain stores: fifty-pound sacks of rice, "exotic" fruits and vegetables, pig parts (arranged in piles of ears, snouts, feet, tails, and innards, as well as buckets of fresh pork blood), live fish, black-skinned pigeon, and imported canned products used in Chinese, Vietnamese, Indonesian, Thai, Philippine, and Japanese menus. In these markets, Chinese is the dominant language of commerce, and much of the merchandise is unfamiliar to non-Asian shoppers.

Growth and Resentment

For many residents, the redevelopment and replacement of businesses in the Garvey-Garfield district, along Atlantic Boulevard, and throughout other areas in the city seemed sudden and dramatic. In January 1979, under the headline "Monterey Park Is Due for Big Facelift," the *Monterey Park Progress* reported that a northern portion of Atlantic Boulevard was set to "be transformed so it's unrecognizable." Construction there was to include the completion of a shopping center, office, and theater complex developed by the Kowin Development Company; ground breaking for a new office building at the northeast corner of Atlantic and Newmark Avenue; and a hillside condominium project on the west side of Atlantic Boulevard. The article went on to state with great anticipation that "a large international concern" planned to "locate its international service center in Monterey Park," that substantial construction in anticipation of new tenants was to be done at McCaslin Industrial Park in the eastern section of town, and that several street and park improvement projects were in the works. In addition, a major city-sponsored Community Redevelopment Agency (CRA) project would erect a new civic center complex and make necessary improvements on a senior center, a school cafetorium, a community center, and the municipal library.[23]

Between the influx of new Chinese immigrants, the infusion of large amounts of capital, the rapid introduction of Chinese-owned and -operated businesses, and the disruptions caused by construction crews tearing up the city and starting new projects, rumblings of discontent among long-time established residents became quite audible.

"I Don't Feel at Home Anymore!"

At first the new Chinese-owned businesses seemed novel, innocuous, even humorous. "The gag was that if it wasn't a bank, it was going to be a real estate office, or another Chinese restaurant," says Lloyd de Llamas.[24] But as these and other Chinese businesses proliferated rapidly from 1978 on—taking over previously established merchants, displaying large Chinese-language signs, and seeming to cater only to a Chinese-speaking clientele—residents became increasingly hostile.

The famous Laura Scudder potato chip factory, converted into a Safeway store in the 1960s, became a bustling Chinese supermarket. Frederic Hsieh bought the Edwards Theater and began showing Chinese-language movies; when people complained he added such English-language films as *Gone with the Wind*, *Doctor Zhivago*, and *Ryan's Daughter* to the afternoon repertoire. Even the locally revered Garvey Hardware Store was sold to new Chinese owners who divided the premises into mini-shops, relegating the much-reduced hardware department to the back of the building. Kretz Motorcycle, Paris' Restaurant, and the Midtown Pharmacy were similarly redeveloped, engendering resentment among many residents, particularly older whites. For "old-timers" the loss of a familiar business could be akin to the loss of an old friend. "Just a few years before they sold Paris' Restaurant I walked in there for lunch alone," remembers Ed Rodman, "and . . . there wasn't a single person in there that I knew by name! That describes the changes in Monterey Park."[25]

Such losses were compounded when many long-time residents felt they were not welcomed by new businesses because they were not Chinese. Avanelle Fiebelkorn told the *Los Angeles Times*: "I go to the market and over 65 percent of the people there are Chinese. I feel like I'm in another country. I don't feel at home anymore." Emma Fry agreed: "I feel like a stranger in my own town. You can't talk to the newcomers because many of them don't speak English, and their experiences and viewpoints are so different. I don't feel like I belong anymore. I feel like I'm sort of intruding."[26]

Joseph Graves particularly remembers an incident that occurred in the late 1970s when he was a member of the Monterey Park Chamber of Commerce. A group of visiting dignitaries from Taiwan asked the chamber whether a statue of Confucius could be built in one of the parks to remind young Chinese to respect and honor his teachings. Graves had no objection but told them that "the people coming over here ought to be building Statues of Liberty all over town." Graves, who was born in Monterey

Park the year the city was incorporated, continues to live there and says he harbors no resentment toward the Chinese. "I ride my bike everywhere and I see all these Chinese people out there taking their walks. They are so warm and friendly. How can you end up with anger? And yet, [if] I look at something they're doing that forces me to change, then I can be temporarily angry. I reserve the right to be temporarily angry as long as I don't nurse grievances."[27]

Others, however, have nursed grievances, and white flight has been the most obvious reaction to the changes in the community. While the Asian population in Monterey Park has grown and the Latino population has remained relatively stable, the white population has plummeted. In 1960 the 32,306 white residents made up 85 percent of the population; by 1990 the number of whites had dropped to 16,245, or just 12 percent. When former Monterey Park resident Frank Rizzo moved out, he cited the large condominium complexes on either side of his house and the people in them as reasons he could no longer stay. Prior to the influx of Chinese, Rizzo said, his neighborhood had been a quiet and friendly block of mostly single-family homes with expansive yards. But his new neighbors lived in large extended families in cramped quarters, spoke little English, and seemed unwilling to give up their traditions and settle into an American way of life. Rizzo, who sold his home to a Chinese developer, was emphatic about leaving Monterey Park: "What I might do is hang a little American flag on my truck and drive through town on my way out and wave goodbye to all my old friends.... I'm moving far away from here."[28]

Latinos in Monterey Park too were concerned that they were losing the integrated community they thought they'd moved into. David Barron has lived in the city since 1964 and raised his family there. Previously, he attended nearby East Los Angeles Community College and California State University, Los Angeles. He still remembers when Monterey Park was referred to as the "Mexican Beverly Hills." Fluent in Spanish and proud of his heritage, Barron thought he had found the ideal integrated community. He is still involved in many of the city's social and civic activities and has no immediate plans to move, but he misses the diversity he initially found in the town. "I would like to see a balance maintained," he explains. "I cannot live in a mono-ethnic community. I wouldn't want to live in an all-Hispanic . . . or all-Chinese . . . or all-white community. I want to live in a mixed community."[29]

Similar sentiments were expressed by Fernando Zabala, a hair stylist who grew up in East Los Angeles and also found Monterey Park a stepping-stone out of the barrio. "It was very important that my children grow up in a racially diverse community," Zabala said. "When we moved to Monterey Park, we had a little bit of everybody: whites, blacks, Latinos, and some Chinese and Japanese. But we lost that mix. In my neighborhood alone, it went from twenty-five Latino families to three."[30] Unlike Barron, Zabala sold his house and moved out.

One woman, who asked not to be identified, said that she was one of the first Mexican Americans to move into a new hillside housing tract in Monterey Park in the late 1950s and that she had worked very hard to integrate into the community. Like many whites, she expressed anxiety about the rapid change in the commercial areas in town: "It wasn't like one business changing at a time, it was like two or three at a time. When they put in the Diho [supermarket], that right away changed the appearance of

Atlantic Boulevard." She recalled with particular sadness a Mexican restaurant she and her mother used to frequent. This small restaurant, greatly appreciated for its home-style cooking and family atmosphere, was forced to close when new owners bought the property. "The owner was very upset, and she put [up] a big sign.... 'I'm not leaving my friends because I want to, but the mall has been bought and my rent has been raised and I cannot afford it.' Things like that you would get upset about."[31]

Like the Latinos who had settled in Monterey Park, long-time Asian American residents had lived their entire lives believing in the "American Dream" that proclaimed just rewards for hard work and initiative. It was an affront to their sensibilities to see so many newcomers acquire the fruits of a material society seemingly without having to struggle. The newcomer Chinese were simply not playing by the rules of assimilation: they bought property, started businesses and banks, and built shopping malls as soon as they arrived—and many of them didn't even speak English! John Yee—whose great-great-grandfather had come to California during the gold rush, whose great-grandfather died building the transcontinental railroad, and whose grandfather and father owned a Chinese laundry business that served steel factory workers in Midland, Pennsylvania—is particularly articulate in this regard. "When I first came to L.A., I lived in Chinatown, went into the service, came out, worked in a lot of jobs, and step by step I moved to Monterey Park. It took how many years? Thirty, forty years? It seems like these immigrants . . . want to live in Monterey Park as soon as they get off the boat. Not the boat, now they come by airplane. Give them another forty years, they'll be in Beverly Hills. I won't ever get to that point.... Maybe I'm jealous like everybody else."[32]

The resentment of the older Latinos and Asian Americans who had experienced racial segregation and witnessed the civil rights struggles of the 1960s also stemmed from a feeling that Monterey Park's new Chinese immigrants were taking for granted the equality won by the struggles of others. Yee says: "I don't mind the people too much, don't get me wrong; I am of Chinese descent. But the thing is, you get these people with this attitude.... they think [everything] was like this all the time. It wasn't. I hear people say, 'China got strong and now the United States and the rest of the world has more respect for us.' Maybe so, but . . . if it wasn't for some of these guys [people of color born in the United States] who squawked about it, went into the service, these changes wouldn't happen. You got the blacks and Mexicans, they all helped change the government.... That attitude [among new Chinese immigrants] just burns me up."[33]

Particularly for Asian Americans born in the United States, the appearance of Chinese immigrants raised questions about their assumed assimilation and acceptance into American society. "When there were just Japanese people in Monterey Park, it was no problem because we were just like them [whites]," explains long-time resident Kei Higashi. "But now all of a sudden [with the arrival of the new immigrant Chinese] when we walk into a place and start talking perfect English, they [non-Asians] look at us like we're some foreign creature," he laughs. "That's what happened in Monterey Park."[34]

In the middle of all this are many of the Chinese immigrant professionals, who found themselves lumped together with the development- and business-oriented newcomers. Many express appreciation for the large Chinese population that makes them feel welcome, but at the same time, they say, had they wanted to live in a crowded, exclusively Chinese environment, they never would have left home. This is the case for Dr. Frances Wu, who moved to Monterey Park in 1971, after she was accepted in the doctoral program at the University of Southern California. Born and educated in China, Wu lived in Taiwan for four years following the Communist takeover; in 1953 she went to Canada to earn a master's degree from McGill University, then spent fifteen years in New York working in the Child Welfare Department.

When Wu came to southern California, she changed her social work specialty to gerontology, and shortly after earning her Ph.D. she started the Golden Age Village, a retirement center located in Monterey Park. Although the project is open to all elderly people who qualify, Wu told the *Monterey Park Progress*, "My motivation was to develop a social program for elderly Chinese and we selected Monterey Park because of its growing Chinese population," as well as its uncongested, small-town atmosphere.[35] The overall design of the Golden Age Village is obviously Asian, with its curved roofs and a courtyard that features a babbling brook surrounded by a decorative Oriental-style garden. The majority of residents are retired Chinese, many of whom speak little or no English, and the communal food garden grows bok choy and Chinese parsley among other vegetables. But the serene environment that Wu found in Monterey Park and recreated at the Golden Age Village is threatened by what she considers too much growth too fast. "I would rather keep this community a bedroom community," she says. "For retired people, we like a quiet environment.... People describe Monterey Park as 'Little Taipei,' but Taipei is horrible. I don't want Monterey Park to be like that."[36]

Notes

[1.]U.S. Bureau of the Census, "Monterey Park, City, California," 1990 Census of Population and Housing Summary Tape File 1, May 13, 1991.

[2]Kurt Anderson, "The New Ellis Island: Immigrants from All Over Change the Beat, Bop, and Character of Los Angeles," *Time*, June 13, 1983, p.21.

[3]Several newspapers have incorrectly cited this honor as the "All-American" award. According to the official entry form, the term is "All-America."

[4]Mike Ward, "Language Rift in 'All-American City,'" *Los Angeles Times*, November 13, 1985; Gordon Dillow, "Why Many Drivers Tremble on the Streets of Monterey Park," *Los Angeles Herald*, July 8, 1985; "English Spoken Here, OK?" *Time*, August 25, 1985.

[5]Monterey Park City Council Minutes, June 2, 1986.

[6]Mike Ward, "Racism Charged over Monterey Park Vote," *Los Angeles Times*, July 15, 1986; Ray Babcock, "'Sanctuary' Resolution Stays," *Monterey Park Progress*, July 16, 1986; Evelyn Hsu, "Influx of Asians Stirs Up L.A. Area's 'Little Taipei,'" *San Francisco Chronicle*, August 1, 1986.

[7]See Jose Calderon, "Latinos and Ethnic Conflict in Suburbia: The Case of Monterey Park," *Latino Studies Journal* 1 (May 1990): 23–32; John Horton, "The Politics of Ethnic Change: Grass-Roots Response to Economic and Demographic Restructuring in Monterey Park, California," *Urban Geography* 10 (1989): 578–592; Don Nakanishi, "The Next Swing Vote? Asian Pacific Americans and California Politics," in *Racial and Ethnic Politics in California*, ed. Bryan O. Jackson and Michael D. Preston (Berkeley: University of California, Institute of Governmental Studies, 1991), pp. 25–54; Mary Pardo, "Identity and Resistance: Latinas and Grass-Roots Activism in Two Los Angeles Communities" (Ph.D. diss., University of

California, Los Angeles, 1990); Leland Saito, "Politics in a New Demographic Era: Asian Americans in Monterey Park, California" (Ph.D. diss., University of California, Los Angeles, 1992); Charles Choy Wong, "Monterey Park: A Community in Transition" in *Frontiers of Asian American Studies*, ed. Gail M. Nomura, Russel Endo, Stephen H. Sumida, and Russell Leong (Pullman: Washington State University Press, 1989), pp. 113–126; Charles Choy Wong, "Ethnicity, Work, and Community: The Case of Chinese in Los Angeles" (Ph.D. diss., University of California, Los Angeles, 1979).

[8]U.S. Commission on Civil Rights, *The Economic Status of Americans of Asian Descent: An Exploratory Investigation*, Publication no.95 (Washington, D.C.: Clearinghouse, 1988), p. 109.

[9]See Marshall Kilduff, "A Move to Ease Racial Tensions in S.F. Neighborhood," *San Francisco Chronicle*, August 11, 1986; Tim Fong, "The Success Stereotype Haunts Asian-Americans," *Sacramento Bee*, July 4, 1987; David Reyes, "'Asiantown' Plan Taking Shape in Westminster," *Los Angeles Times*, March 22, 1987; "Chinese Enclaves Abound in New York," *Asian Week*, October 3, 1986; Kevin P. Helliker, "Chinatown Sprouts in and near Houston with Texas Flavor," *Wall Street Journal*, February 18, 1983; "$50 Million 'Orlando Chinatown' Features Hotel-Retail Complex and 30 Restaurants," *AmeriAsian News*, March-April 1987; Russell Spurr, "Why Asians Are Going Down Under," *San Francisco Chronicle*, December 7, 1988; Howard Witt, "British Columbia's Anti-Asian Feelings Suddenly Surface," *Chicago Tribune*, February 5, 1989.

[10]U.S. Immigration and Naturalization Service, 1989 *Statistical Yearbook of the Immigration and Naturalization Service* (Washington, D.C.: Government Printing Office, 1990), p. xiv.

[11]In June 1982 Vincent Chen, a Chinese American draftsman, was beaten to death by a Chrysler Motors supervisor and his stepson. One of the assailants was alleged to have yelled, "It's because of you motherfuckers we're out of work." The two men later confessed to the crime, were fined $3,780 each, and placed on three years' probation. Neither spent a day in jail. See Ronald Takaki, *Strangers from a Different Shore* (Boston, Little, Brown, 1989), p.481.

[12]U.S. Commission on Civil Rights, *Recent Activities against Citizens and Residents of Asian Descent*: Publication no. 88 (Washington, D.C.: Clearinghouse, 1986), p. 3.

[13]Eli Isenberg, "A Call for Open Arms," *Monterey Park Progress*, December 7, 1977.

[14]Interview with Eli Isenberg.

[15]Interview with Joseph Graves.

[16]Ibid.

[17]Eli Isenberg, "It Seems to Me," *Monterey Park Progress*, February 27, 1985.

[18]Fieldnotes from November 20, 1990.

[19]Art Wong, "Bilingual Plan Opens Up 'Bucket of Worms,'" *Monterey Park Progress*, June 7, 1978.

[20]L. Ling-chi Wang, "Lau v. Nichols: History of a Struggle for Equal and Quality Education," *Amerasia Journal* 2 (1974): 16–46.

[21]Wong, "Bilingual Plan."

[22]See Andrew Tanzer, "Little Taipei," *Forbes*, May 6, 1985, pp. 68–71; Mike Ward, "Cities Report Growth—and Some Losses—from Asian Business," *Los Angeles Times*, April 19, 1987; and Randye Hoder, "A Passion for Asian Foods," *Los Angeles Times*, June 5, 1991.

[23]Malcolm Schwartz, "Monterey Park Is Due for Big Facelift in 1979," *Monterey Park Progress*, January 3, 1979.

[24]Interview with Lloyd de Llamas by Tim Fong, for the Monterey Park Oral History Project, sponsored by the Monterey Park Historical Heritage Commission, March 29, April 13, and May 11, 1990.

[25]Interview with Ed Rodman by Tim Fong, for the Monterey Park Oral History Project, sponsored by the Monterey Park Historical Heritage Commission, October 17 and 24, 1990.

[26]Mark Arax, "Selling Out, Moving On," *Los Angeles Times*, April 12, 1987.

[27]Interview with Joseph Graves.

[28]Arax, "Selling Out, Moving On."

[29]Interview with David Barron by Tim Fong, for the Monterey Park Oral History Project, sponsored by the Monterey Park Historical Heritage Commission, October 9, 1990.

[30]Mark Arax, "Nation's 1st Suburban Chinatown," *Los Angeles Times*, April 6, 1987.
[31]Fieldnotes from August 16, 1990.
[32]Interview with John Yee by Tim Fong, for the Monterey Park Oral History Project, sponsored by the Monterey Park Historical Heritage Commission, May 31 and June 4, 1990.
[33]Ibid.
[34]Interview with Kei Higashi by Tim Fong, for the Monterey Park Oral History Project, sponsored by the Monterey Park Historical Heritage Commission, May 7 and 30, 1990.
[35]"Second Housing Project for Seniors on Horizon," *Monterey Park Progress*, Sept 13, 1978.
[36]Interview with Dr. Frances Wu by Tim Fong, for the Monterey Park Oral History Project, sponsored by the Monterey Park Historical Heritage Commission, June 22 and July 6, 1990.

References

Anderson, Kurt. "The New Ellis Island: Immigrants from All Over Change the Beat, Bop, and Character of Los Angeles." *Time*, June 13, 1983.

Calderon, Jose. "Latinos and Ethnic Conflict in Suburbia: The Case of Monterey Park." *Latino Studies Journal* 1 (May 1990): 23–32.

Horton, John. "The Politics of Ethnic Change: Grass-Roots Responses to Economic and Demographic Restructuring in Monterey Park, California." *Urban Geography* 10 (1989): 578–592.

Monterey Park, City of. City Council *Minutes*, June 2, 1986.

Nakanishi, Don. "The Next Swing Vote? Asian Pacific Americans and California Politics." In *Racial and Ethnic Politics in California*, ed. Bryan O. Jackson and Michael D. Preston, pp. 25–54. Berkeley: University of California, Institute of Governmental Studies, 1991.

Pardo, Mary. "Identity and Resistance: Latinas and Grass-Roots Activism in Two Los Angeles Communities ." Ph.D. diss., University of California, Los Angeles, 1990.

Saito, Leland. "Politics in a New Demographic Era: Asian Americans in Monterey Park, California." Ph.D. diss., University of California, Los Angeles, 1992.

Takaki, Ronald. *Strangers from a Different Shore: A History of Asian Americans*. Boston: Little, Brown, 1989.

Tanzer, Andrew. "Little Taipei." *Forbes*, May 6, 1985.

U.S. Bureau of the Census. *Census of Population and Housing Summary Tape File 1* (STF1): Monterey Park City, California, 1990.

U.S. Commission on Civil Rights. *The Economic Status of Americans of Asian Descent*. Publication no. 95. Washington, D.C.: Clearinghouse, 1988.

_____. *Recent Activities against Citizens and Residents of Asian Descent*. Publication no. 88. Washington, D.C.: Clearinghouse, 1986.

U.S. Immigration and Naturalization Service. *1989 Statistical Yearbook of the Immigration and Naturalization Service*. Washington, D.C.: Government Printing Office, 1990.

Wang, L. Ling-chi. "Lau v. Nichols: History of a Struggle for Equal and Quality Education." *Amerasia Journal* 2 (1974): 16–46.

Wong, Charles Choy. "Ethnicity, Work, and Community: The Case of Chinese in Los Angeles." Ph.D diss., University of California, Los Angeles, 1979.

_____. "Monterey Park: A Community in Transition:" In *Frontiers of Asian American Studies*, ed. Gail M. Nomura, Russell Endo, Stephen H. Sumida, and Russell Leong, pp. 113–126. Pullman: Washington State University Press, 1989.

Newspapers

AmerAsian News, March-April 1987.
Asian Week, October 3, 1986.
Chicago Tribune, February 5, 1989.

Los Angeles Herald, July 8, 1985.
Los Angeles Times, November 13, 1985-June 5, 1991.
Monterey Park Progress, December 7, 1977-July 16, 1986.
Sacramento Bee, July 4, 1987.
San Francisco Chronicle, August 11, 1986-December 7, 1988.
Wall Street Journal, February 18, 1983.

2

Social and Ethnic Stratification

The supposedly unsinkable British ocean liner *Titanic* sank rapidly after colliding with an iceberg in 1912. As with most luxury liners, services on the *Titanic* were divided into different "classes": passengers could purchase the amount of luxury they wanted in their accommodations. Consumers are frequently lectured that "you get what you pay for," and the fatality statistics on *Titanic* passengers are just one case in point. Three percent of the first-class female passengers died, compared with 14 percent of the second-class female passengers, and 54 percent of the third-class female passengers. Even though women and children were supposed to be first into the lifeboats, more third-class women perished than first-class men (His Majesty's Stationery Office, 1912). The *Titanic* tragedy illustrates that even survival can be purchased.

The separation of people into rich and poor classes through a system of social stratification permeates every aspect of life in modern societies such as the United States. It is not only one of the most distinctive features of such societies, but also one of the most important features in the everyday lives of their members. Social scientists have few thoughts that do not wander into the realm of economic differences sooner or later.

Social stratification refers to the arrangement of different activities of a society into a hierarchy whereby activities ranked high are rewarded and activities ranked low are rewarded poorly if at all. Individuals perform those activities and receive those rewards (or are denied rewards) as long as they occupy their positions. In a manner of speaking, social stratification is a vast game board (not unlike Monopoly) on which we are forced to play. Corporation executives are established as more important than janitors, just as Boardwalk is more important than Baltic Avenue; the game board is there, and you must do what you can according to the rules of the game. If you do not happen to land in the executive's position (or on Boardwalk), you will have to take orders from those who do.

The system of social stratification in the United States has an impact on Americans similar to the impact that the game board and rules of Monopoly have on Monopoly players, constantly offering reminders of relative social position and coordinating behavior. This chapter examines the "game board" of social stratification, paying particular attention to the effects of the game on the players. We look first at the basic features of the social stratification hierarchy, emphasizing the arrangements of the positions and the degree to which it is possible for individuals to move among them. Second, we take a close look at economic differences as they exist today in the United States. As was the case for the upper-class passengers aboard the *Titanic*, it is still true that being a member of the upper class adds years to your life. Finally, we examine the competition of social stratification; specifically, we see how that competition produces discrimination, exploitation, and ultimately ethnic stratification.

Basic Features of Social Stratification

The United States does not have the extreme economic divisions found in some countries, but the differences in living conditions between American rich and poor are nevertheless dramatic. However, many fine lines separate the positions in the hierarchy, making it difficult to tell just which social classes contain which positions (Strobel & Peterson, 1997). Rewards also come in many different packages: Money or property is clearly a reward, but so is the respect received from others. Which is more important? If an individual gives up his or her job as a traveling salesperson to become a college teacher, he or she will probably make less money but receive more respect. Is that individual now in a different social class or the same one? This section introduces the basic concepts social scientists use to answer these and other confusing questions about social stratification.

Class, Status, and Power

The rewards our positions bring us in a social stratification hierarchy are generally separated into economic rewards (**social class**), social rewards (**social status**, or prestige), and political rewards (**social power**, or the ability to influence the behavior of others). Different positions and activities in a society can be ranked in terms of how much of each kind of reward they produce for their occupants. By definition, positions at the very top of the hierarchy bring their occupants large amounts of money, prestige, and influence over others; positions at the very bottom, also by definition, bring their occupants poverty, ridicule, and the privilege of taking orders. The positions in the middle of the hierarchy bring their occupants varying amounts of each kind of reward. Traveling salespeople, for example, may rank a little higher economically but proportionately lower socially than college teachers.

Socioeconomic status (SES) is an average score assigned to an individual on the basis of (1) income level, (2) level of education, and (3) occupational prestige. Such a measure does not reflect the dimension of social power very accurately, but it does include a measure of wealth (income) and social status (occupational prestige). Educational level usually is an index of both, since education often leads to better-paying, more prestigious jobs and, in addition, increases social status by itself.

Income and education are pretty straightforward attributes to measure, but how do you measure occupational prestige? The easiest way is to ask people to rank lists of occupations from those they most respect to those they least respect. If you ask enough people to do this, you can produce an average ranking that reflects a social consensus on occupational prestige (Davis and Smith, 1984). Occupational income is not the sole determinant of occupational prestige. How else could college professors be ranked above dentists and plumbers (Merton, 1968)?

Social Mobility: Movement Between Classes

Social mobility is the movement of groups or individuals within the ranking systems of social stratification. Just as geographical mobility represents a change in physical space, social mobility represents a change in social "space" whereby someone's social status changes in the eyes of other social group members. Social mobility involves movement from an activity or position at one social class level to an activity or position at another social class level. Going from "rags to riches," or, less dramatically, from office clerk to head of the accounting department, illustrates this concept. Going from "riches to rags" would be downward social mobility.

Social mobility can be either intragenerational or intergenerational. **Intragenerational social mobility** is a change in social class that occurs within the lifetime of one individual (or within one generation). Starting one's occupational career as an office clerk and working up to the top of the company is intragenerational mobility. **Intergenerational social mobility** is a change in social class that occurs across generations. For example, a father might have worked a lifetime as a mail carrier, and a child might have gone straight through law school to a position as a successful corporate lawyer. Within the child's experience no mobility has occurred, yet a definite change has occurred from one generation to the next; the child's social class is very different from the father's social class. Although both kinds of social mobility are important to those affected, our interests are primarily with intergenerational mobility. Changes from one generation to the next tell us much about overall social change (or lack thereof) and the presence (or absence) of social barriers to advancement.

Class and Caste Societies: Social Mobility and Group Formation

The ideal form of a **class society** would be a society that permits wide-scale vertical social mobility; positions in the social stratification hierarchy would be open to all and achieved through an open competition process. An ideal **caste society** would be based entirely on ascribed status and would permit no vertical social mobility at all; social positions would be inherited as a right of birth. Neither of these two descriptions provides a very good fit for any real society, although it is possible to find fair approximations of caste societies in the real world. The term *caste*, which refers to a totally closed social level, comes to us from India, where the population used to be rigidly separated into specific groups of social roles; the Brahmin caste was at the top of the hierarchy and the "untouchables" were at the bottom, with no possibility of change. The American system of slavery came close to a caste system because children born to a slave mother were automatically slaves. A similar caste system operated in South Africa throughout most of the twentieth century. Our concern is how different patterns of social mobility affect group formation.

Patterns of Competition and Group Formation   True class societies are much harder to find in the real world. Modern industrial societies such as the United States have just about as much social mobility as can be found anywhere, but access to positions is clearly limited. Even though many positions (or jobs) are technically open to anyone, it still often helps to "know somebody." Less obvious but perhaps more important advantages come as an accident of birth. Parents automatically pass on to children certain skills, abilities, resources, and knowledge. If the parents are from the upper class, that "inheritance" will help their children obtain admission to good schools, the abilities and skills to get through those schools, the confidence to get ahead, the social skills for leaving a good impression on employers at interviews, and other advantages too numerous to mention. If the parents are members of the lower class, the child receives an "inheritance" of equal size but will find its content less useful for attacking dominant American institutions. Because those institutions are largely under the control of the upper classes, they reflect the knowledge, skills, and values of those upper classes. To form a true class society, you would have to remove children from their parents at birth, give them all an equal background at a state-run orphanage, and then fill jobs through a competition that would be truly open.

Social scientists use the concepts of *class* and *caste* to better understand the range of social mobility that occurs between the two extremes. Different societies can be placed at different points between the extremes and compared accordingly. The same society also can change location between these extremes over time, as with the emancipation of slaves in the United States. Class and caste societies provide some insight into the ways people form social groups.

Consider first a caste society. How would such a stratification system affect the individual? First of all, the individual would become extremely aware of his or her caste. Most individuals would probably accept the caste system as a basic and unchangeable fact of existence, since they would have no experience with alternatives. Members of the bottom castes might hate and envy the castes above, but they probably would not try to do much about it. The possibility of change might not even occur to them because they had never witnessed it. This description fits the American system of slavery fairly well, even though those slaves born in Africa had a clear picture of an alternative society. Slaves were extremely aware of the color line (as are most Americans still today), but they rarely attempted to do much about it. Slave revolts were relatively uncommon in the United States because most slaves were American born and apparently felt there was little hope for success. Such a response is characteristic of a caste system. When successful New World slave revolts did occur, they generally occurred in areas where (1) many slaves were African born, and (2) slaves greatly outnumbered the slave-owning population. The Caribbean island of Haiti was such a place.

An ideal class society would produce very different kinds of individual thoughts and behavior. The rapidity with which individuals moved into and out of social positions would prevent groups of individuals from becoming aware of a shared situation, and, unlike the caste system, there would not be a strong sense of group identity. In fact, members of a class society would be engaged in constant acts of individual competition. The openness of the stratification system would encourage individuals to do something about their current positions but would discourage them from acting as groups. Your best possibility of getting ahead would come from working alone. It is harder to find a real-world illustration of this case, but a classroom of students graded on the curve provides a rough approximation. A "grading curve" means that certain percentages of the class will receive certain grades, *regardless of the quality of the work they do.* Their work must only be equal in quality to the best work in the class for them to receive high grades; if that best work is of poor quality, then poor quality will receive top grades. This system is designed to encourage individual competition. What would you do if you found the answer sheet to a final examination in a class graded on the curve? Assuming you would be unscrupulous enough to use it at all, the most rational course would be to share the answers with no one; the lower the other grades, the better yours will look by comparison. You might want to share the answers with a few close friends (thereby competing as a group), but consider that the more people

Dr. Martin Luther King, Jr., and family, 1960.

you help, the worse your grade will be by comparison. If you share the answers with the entire class, it will help you not at all. A class system, like a grading curve, encourages individual competition but discourages people from competing in groups.

The social stratification system of the United States is somewhere between a caste system and a class system. These extreme ideal types help us understand the kinds of competition and group formation that come from such stratification systems. Because social mobility is possible to some extent in the United States, many Americans are competitive. But American society is not all that open. The children of the upper classes enter the competition with advantages that children of the lower classes do not have. This inequality tends to keep the same families at the top or at the bottom—a castelike situation that makes individuals aware of those who share the boat with them and leads to strong feelings of group identity. When the competition characteristic of mobility is combined with the group formation characteristic of a caste system, it produces the compromise characteristic of American society—group competition.

Competition and Group Identity: An Illustration   In 1955, Martin Luther King, Jr. was a 26-year-old Baptist minister living in Montgomery, Alabama. An African American college-educated person in a professional occupation, he lived economically in a middle-class lifestyle. He had, in short, many reasons to be satisfied and not want to rock many boats. But he also had many reasons to be dissatisfied. Neither his education nor his occupation permitted him to patronize many business establishments, nor was he allowed to ride in the front seats of Montgomery's public buses. He was constantly reminded of his lower status based on his racial group membership and the limitations this imposed on his social mobility.

Was King in a class situation or a caste situation? In some ways, he was in both. He could enjoy the potential of mobility offered by a class society while facing the frustration of limits imposed by a caste society. The former encouraged him to focus on himself and his family, working *as an individual* to better himself. The latter encouraged him to focus on his lower status racial group membership and to work *as a group* to better the overall position of all group members. He personally would gain either way, even though the nature of the gains would be very different. Most importantly, the gains of *group* advancement would be much more enduring in that the ceiling for social mobility would be permanently raised. King obviously came to that conclusion as he sacrificed his personal gains for group gains. But that conclusion may not have been all that hard to reach. Particularly in the American South in 1955, the limitations of caste were far more salient for African Americans than the opportunities of class.

Although King is known today as the original leader of the Civil Rights Movement, he did not bring about social change by himself. He was the leader of a **social movement**—the organized effort by a group of people to bring about social change, either in the members themselves or in members and society alike. King's movement—the Southern Christian Leadership Conference (SCLC)—brought together a wide range of social activists who determined clear goals and tactics for their movement. One of those goals was to elevate the status of all African Americans, benefiting African Americans such as King as well as those with less education and poorer-paying jobs. With the passage of federal civil rights legislation in the mid-1960s, that goal was partially achieved.

Although King did not live to enjoy his movement's successes for long, it is easy to see both his motivation and the motivations of other SCLC leaders. Had race not been an important reality in their lives—the castelike circumstances that limited them—they would have worked individually to better themselves. Had social mobility been absolutely out of the question—the situation of a true caste society—there would have been no point in their working together as a group. It is the combination of some classlike and some castelike circumstances that motivates individuals to form social movements and bring about social change.

Social Class in the United States

All modern industrial nations have elaborate systems of social stratification, and the United States is no exception. Everyday activities are firmly embedded in a class hierarchy. It is probably the single most important social factor in explaining the differences among Americans; people in different social classes differ from each other in skills, abilities, values, lifestyles, and almost every other way imaginable. This section explores some of those differences, the reasons for them, and how class differences can limit social mobility.

Inequality in the United States

Systems of social stratification introduce structured inequality into any society. By *structured inequality*, social scientists mean that the inequality is not only the result of individual differences in talent or ability but also is built into the roles people play and the rewards they receive. Any stratification system therefore produces inequality in that those at the top receive more rewards than those at the bottom. One question to be answered in regard to any particular stratification system is the degree of inequality. How much difference is there from top to bottom and just how unequal is the distribution of rewards (Stehr, 1999)?

Figure 2.1 shows how income was distributed among citizens of the United States in 1997. It arbitrarily divides the population into five equal segments, ranging from the top 20 percent of the population in income level to the poorest 20 percent. The top 20 percent received 47.2 percent of all the income earned, while the bottom 20 percent had to get by with 4.2 percent of the available income. If we were to look beyond income to the more important measure of wealth, we would encounter even greater inequality between the top and the bottom (Wolff, 1995).

Table 2.1 presents income distribution statistics for the years 1950–1997. Comparing Table 2.1 with Figure 2.1 illustrates that although the distribution of income in the United States has changed only slightly since 1950, the change results in greater income *in*equality. The top 20 percent grew from 42.7 percent of the income in 1950 to 47.2 percent in 1997; the bottom 20 percent changed less during this period, dropping in share in the 1990s after a slow but steady rise since 1960. Table 2.1 also provides a closer look at the very top of the stratification hierarchy—the top 5 percent in income earned. In 1997, the top 5 percent earned 20.7 percent of all income earned in the United States. American culture and technology changed quite a bit during that

FIGURE 2.1   PERCENT OF INCOME BY FIFTHS OF POPULATION IN THE UNITED STATES, 1997

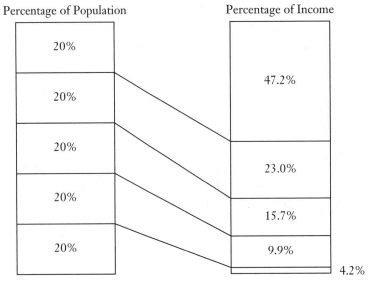

Percentage of Population

Percentage of Income

U.S. Bureau of the Census, 1999a.

TABLE 2.1   HOUSEHOLD INCOME AS A PERCENTAGE OF INCOME IN THE UNITED STATES BY FIFTHS OF THE POPULATION, 1950-1997.

Year	Lowest Fifth	Second Fifth	Middle Fifth	Fourth Fifth	Highest Fifth	Top 5 Percent
1997	4.2%	9.9%	15.7%	23.0%	47.2%	20.7%
1995	4.4	10.1	15.8	23.2	46.5	20.0
1991	4.5	10.7	16.6	24.1	44.2	17.1
1980	5.2	11.5	17.5	24.3	41.5	15.3
1970	5.4	12.2	17.6	23.8	40.9	15.6
1965	5.2	12.2	17.8	23.9	40.9	15.5
1960	4.8	12.2	17.8	24.0	41.3	15.9
1955	4.8	12.2	17.7	23.4	41.8	16.8
1950	4.5	11.9	17.4	23.6	42.7	17.3

U.S. Bureau of the Census, 1993c, 1996a, 1999a.

forty-year period (and average yearly income jumped from about $4,000 to over $27,000), but the ratio between the "haves" and the "have nots" increased in favor of the "haves." This change has occurred in spite of the fact that growing numbers of working wives have decreased income inequality, particularly for African American families (Cancian et al., 1993). Different people are certainly involved in the stratification system today than in 1950, but the system itself is no different; the same percentages of rewards are still dispersed to the same segments of the population.

Avenues and Barriers to Social Mobility

Most research on social mobility focuses on occupational change between generations. Stephen Thernstrom (1964) studied occupational changes between fathers and sons among nineteenth-century workers in New England. He found (perhaps not surprisingly) that the good old days in the land of opportunity were not all that good, because of lack of opportunity. Sons tended to follow pretty closely in their fathers' footsteps. Such findings raised questions about whether social mobility across generations had changed in the more contemporary United States.

A classic study of more modern American occupational mobility is *American Occupational Structure*, a 1967 study by Peter Blau and Otis Duncan. Working closely with census data, Blau and Duncan sought to determine just how much mobility occurred in the American occupational structure and, just as importantly, what caused it. They discovered a fair amount of occupational social mobility, but their data were largely limited to European Americans. Hauser and Featherman (1976) found a relation between a father and son's occupational prestige during the 1960s and 1970s: the most common occurrence was for white-collar fathers to have white-collar sons and for blue-collar fathers to have blue-collar sons.

The father's occupation appears to have a definite effect on the son's occupation, but Blau and Duncan wanted a more complete explanation. Using sophisticated statistical techniques designed to unearth such relationships, Blau and Duncan concluded that various factors determined the son's occupation, the most important of which was education. More education led to better jobs. But what determined the son's education? The more education and the higher the occupational level of the father, the more education the son would get (Blau and Duncan, 1967). Similar results have been found by more recent research. Jencks and associates (1979) found that an individual's income and occupational status are best explained by looking to the educational level of the individual's family and, in particular, at the skills and values for achievement acquired from the family. In short, social mobility is possible, but differences in family background have a strong impact on keeping the same kinds of jobs in the same kinds of families.

The avenues and barriers to social mobility have complex and intertwined roots that trail back into the overall workings of the social stratification system. Why, for example, should the father's level of education affect the son's level of education? A father who went to college might be more apt to encourage the same behavior from the son, but the full explanation is more complicated. Fathers who went to college (1) make more money and are better able to send their sons to college; (2) have acquired

As in all modern industrial nations, social stratification systems exist in the United States, resulting in important differences among social classes.

certain skills necessary for doing well in school that they pass on to their sons; and (3) probably have jobs that require a college degree and have friends with similar jobs; the sons grow up in this environment, become aware of those kinds of occupations, and probably want to work in one themselves (for which college will be necessary). Fathers who went to college may belong to alumni associations and have some pull in getting their sons accepted; this is particularly true with high-prestige universities. Such explanations could continue indefinitely. The coordination of life within social stratification tends to keep values, skills, abilities, personal contacts, general information, and individual motivations passed on as a cluster from one generation to the next. This coordination is loose enough to allow for some social mobility but tight enough to prevent very much. Overall, the barriers to social mobility are high enough to give the United States clear differences in social class that persist over generations.

Individual Experience of Social Class

We come finally to the individual's perspective on social stratification. How exactly does social stratification make people so different from each other? This very important question is unfortunately difficult to answer through any kind of systematic research. Social class affects every aspect of life, beginning at the moment of birth. Because of it, individuals will have daily experiences slightly different from those in other social classes. They will see different places and people, acquire different skills and abilities, be treated differently by others, learn different norms, and acquire different values (Inozemtsev, 1999; Mohr and DiMaggio, 1995). By adulthood, this accumulation of experience shapes people differently, just as the experience of growing up in London would make someone different from those who call New York home.

Although it is difficult to trace the exact connection between individual differences and social class differences, we can see many links. For example, people of different social classes are generally uncomfortable around each other in social situations. Of course, there are social settings in which people of different classes commonly meet, but most of them are formal. Interactions between an employer and employee, for example, are necessary to get on with the business at hand, but do the two individuals ever really come to know each other well? Most such settings have fairly strict rules as to how each person should talk and the kinds of topics that may be discussed. If the employer were to invite the employee to a party (or vice versa), both would probably be uncomfortable. The different classes have different kinds of parties, for one thing, but their interests and styles of expressing those interests also differ. Common ground is often inadequate for such interactions to occur easily.

Most people do not consciously see their experience as shaped by social class. Individuals encounter people and situations, some of which they like and some of which they do not. People rarely say, "I'm comfortable with this person in this situation because both are part of my social class." Instead they say, "This is a good person and a lot of fun." Part of the reason that we find certain people "good" and "fun," however, is that they fit in with the way we see the world and the people in it. Our worldview is acquired as part of social class (Bryson, 1996; Gimlin, 1996). Left to their own devices, most individuals organize their lives so that they can spend as much time as possible with the people and situations they prefer. It is therefore no surprise that people cluster together within social classes.

The Experience of Social Stratification: The Perspective of the Players

We have seen how social stratification can create social mobility, encourage (or hinder) mobility for members of different social classes, and create differences among people that extend well beyond different economic situations. It is time to return to our Monopoly game, conveniently patterned after social stratification, to see how social stratification produces competition.

In this section, we examine how competition encourages the construction of barriers to social mobility for those at the bottom of the hierarchy. We will come to view those barriers as discriminatory.

Competition and Discrimination: Security and Exploitation

Imagine yourself in the following parable. You have to run a foot race six months from now against one opponent. The winner and loser of this race will have their lives permanently altered. The winner will be given wealth and prestige that can be passed on to the next generation. The loser will be placed in hard labor with low wages for the remainder of his or her life, this too to be passed on. You have only six months to prepare. What might you do? One option would be to train hard every day for the next six months, practicing your starting skills and improving your speed to achieve personal excellence. On the day of the race, you would have to hope that your personal best was good enough. Another option would be to spend the next six months relaxing. Then, on the day of the race (and just before the starter's gun goes off), go over and break your opponent's leg. You may now walk to the finish line.

Consider the huge advantages of this second option. First, it requires no effort, outside of the actual leg breaking. You need not work hard to achieve excellence. Second, and even more important, option two is guaranteed. You cannot lose. When the stakes are extremely high, most rational people would choose a course with less worry and chance, even it seems like "cheating" in this parable. Do you have any moral compunctions about breaking someone else's leg? Is that unfair? If history teaches us anything, it is that people will find ways to rationalize virtually any behavior that works to their betterment. As we will see in the coming chapters, killing Native Americans and enslaving Africans caused some restless nights for a few people but many found ways to live with both. In general, history teaches us that the higher the stakes, the easier rationalizations are to accept.

This parable illustrates two typical responses to competition: Competition can produce excellence and is often introduced into various social situations with precisely that end in mind; however, competition also encourages cheating. The two previous observations about the advantages of cheating are fairly obvious. We can call it cheating in this parable, but when large numbers of people take advantage of this strategy, it requires a new name—discrimination.

Discrimination: Damage, Differentiation, and Competitive Security Discrimination refers to actions designed to hinder the competitive abilities of members of another group. The key words in this definition are *action* and *hinder*. Discrimination always involves taking some concrete steps to damage others. It should never be confused with **prejudice**—maintaining negative attitudes about members of another group. Not only are bad thoughts far less damaging, but attitudes and actions do not always go together. As for *hinder*, this term focuses attention on the *actions* that cause the damage. If those actions do not reduce the ability of the affected group to compete, then the actions are not discriminatory, regardless of intent.

This definition of *discrimination* is similar to those found in most dictionaries, but it is never the first definition offered. The first definition always involves distinguishing, differentiating, or choosing. And whenever you choose or differentiate some from others, you do so on the basis of some criterion or criteria. For example, when purchasing tomatoes, you would not select the first tomato you saw. It might be unripe or overripe. Instead, you pick up several and compare them according to your criterion (proper ripeness). You choose (or differentiate) those of proper ripeness while leaving the others in the bin. By the same token, a woman's basketball team would differentiate among those allowed to try out based on the criterion of gender. As with tomatoes, people can be classified according to whatever criteria the classifier thinks are important.

With racial and ethnic groups, we need to continually focus on *both* definitions of discrimination. It is essential never to lose sight of the damage, which is the point of social discrimination, and this will help us understand why it exists when and where it does. But we also need to understand the choosing process. The manner of choosing has been pretty forthright in most of American history. If you did not want to hire Catholic Irish immigrants, for example, you would put out a sign to that effect. This practice has been illegal since the passage of the Civil Rights Act of 1964. As a result, the "choosing" has gone underground and become much more complicated. Discrimination is still with us but may only be obvious through its resultant damage.

In the parable of the foot race, the primary motive for employing discrimination was to gain security through eliminating competitors. The competition introduced by the foot race motivated efforts to eliminate that competition. We might expect similar occurrences whenever we find two or more racial or ethnic groups seeking the same employment in a shrinking job market. Providing increased security through the elimination of competition is indeed a strong motivation for discrimination, but perhaps the parable does not go far enough in helping us understand the real world. Damaging the competitive abilities of another can help you—make you more secure—but consider the now diminished options for the person damaged. If, for example, many occupations have been closed to members of a certain group, those individuals will have to work at whatever occupations remain. The more their options are limited, the more dependent they become on whatever remains. This dependence leaves them open to labor exploitation.

Discrimination: The Route to Labor Exploitation   Labor exploitation is difficult to define (for all but Karl Marx) because it is a relative concept. We might best define **labor exploitation** as the creation of appreciably more value through labor than is received in compensation by a worker with limited options. But how does one apply this perhaps overly general definition? How limited do options have to be? Slaves would clearly fit this definition, as would indentured servants. Contemporary illegal immigrants also might fit because their illegal status makes it difficult for them to complain about working conditions or low wages. On the other hand, their immigration was voluntary. Did they have the option not to immigrate? Of course, but the alternative was probably poverty at home. Perhaps, like indentured servants in the colonies, a poor economy at home presents little choice. "Limited options" are not easily measured.

We have similar problems with the balance of compensation. It is in the interest of every employer to receive as much work for as little money as possible. Labor is a business cost and businesses profit when costs are lower. At what point is labor paid so little that it is exploited? The best answer to this question comes from combining it with the previous question on options. Exploitation occurs when labor appears to be relatively poorly paid *and* workers have few options. Our interest in exploitation is greatest when those employing the labor also are involved in limiting those options. When

Differences in employment opportunities are one way that social stratification manifests itself.

discrimination is imposed *consciously* for the purpose of labor exploitation, its motivation cannot be doubted. This situation is descriptive of the first several centuries of development in the New World when employers faced extreme labor shortages. Their solution was to artificially limit the options of indentured European labor and African slaves so they would have to work. The clearest example, by far, is slavery.

Labor exploitation is most easily seen when we limit our view to the very poor and the very rich. The former have very few options while the latter are in the strongest position both to impose discrimination and to benefit from the exploitation that then becomes possible. Under slavery, large plantation owners bought the most slaves, controlled the political system in the South that maintained slavery, and became quite wealthy as a result. But what of middle- and working-class European Americans during the antebellum South? Did they benefit from the exploitation of African slaves. Similarly, do middle-class European Americans in the late twentieth century benefit from the exploitation from those in the lower classes? The answer is, "Yes, usually."

When a business owner is in a position to enjoy low labor costs, he or she has two choices. The preferable one is simply to reap higher profits. This route will be taken unless business competition exists in whatever product or service is involved. When business competition exists, lower labor costs allow prices to be lowered for the consumer while still permitting the owner to make a tolerable profit. If those consumers are in the middle class, they are clearly benefiting from the exploitation of those below them in the social stratification system through lower prices.

An illustration may clarify this point. One of the more expensive sections in a modern grocery store is the produce section. Most fruits and vegetables require hand farmwork in addition to machines, which introduces labor costs. How much are farmworkers paid and where do farmworkers come from? The answers to both questions are related. Farmworkers receive relatively little compensation because they tend to be poorer minorities and/or immigrants, both legal and illegal. They can be paid little because they have few other options. If farmworkers received annual incomes of $35,000, how much would a salad cost? Anyone who has ever purchased a head of lettuce benefits from low farmworker wages. Perhaps the lettuce grower benefits more, but as a general rule, everyone in the social stratification system benefits somewhat from the exploitation of workers below them. Even middle-class workers are exploited (limited options are relative), which benefits those at the top.

Let us stay with those middle-class workers for a moment. What would happen to them if current farmworkers began to enjoy more and better options in the labor market? Those options, if they were indeed better, would be middle-class options. Farmworkers not only benefit the middle class through exploitation, but if that exploitation were to end, they could be competitive threats as well, puncturing the security of current middle-class workers. Discrimination thus acts as a double-edged sword. It provides economic security through decreased competition *and* it makes life in the middle and higher classes more pleasant through lower prices due to lower labor costs at the bottom of the system. As the opening parable suggests, discrimination is a highly rational act given a competitive economic situation. It is not a social behavior that will disappear on its own. It is far too beneficial.

Prejudice: A Brief Look   Prejudice literally means to pre-judge (presumably reaching conclusions without the facts). This definition hits the mark, but it is incomplete. For example, a Christian American might believe that Jewish Americans are cheap. This is not yet prejudice; this is a **stereotype**—an overgeneralization about a group of people that may be partly or entirely inaccurate. It becomes prejudice when the Christian American declares "cheapness" to be a negative character trait and the justification of why he or she dislikes Jewish Americans. What happens when this prejudiced individual meets a very generous Jewish American? In many cases, that newly met individual is simply considered an exception to the stereotype; neither the stereotype nor the overall negative feelings have changed. That generous Jewish American was "pre-judged" and then later found to have been judged inaccurately as an individual. Our interest with prejudice lies more with the larger picture. If violated stereotypes are written off as "exceptions," very little social change has occurred, and the prejudice lives on.

Do discrimination and prejudice always go together? For that matter, is prejudice always accompanied by stereotypes? The threesome so often appears as a unit that it would be tempting to answer those questions positively. When they do occur together, it is usually because a person who already discriminates is seeking justifications for actions that stereotypes and prejudice so conveniently provide. For this purpose, stereotypes are by far the more useful. U.S. slaveowners could conveniently justify their labor system by arguing that, as a race, Africans lacked the intelligence that would qualify them for a higher station in life. It would be easy to argue that such a stereotype was a "negative attitude" about a group of people, but slaveowners would be the first to claim that they loved their slaves. Note also that it is not required that discrimination be justified to oneself or others. A slaveowner could just as easily point out that slavery was an efficient economic arrangement and therefore a good thing.

It is more difficult to argue that prejudice or stereotypes lead to discrimination. The world is filled with people who hold negative attitudes toward groups of others but have never acted so as to damage those they dislike. It is important to remember that prejudice is an attitude and is therefore an emotional outlook. It may or may not be rational. Discrimination, on the other hand, is an act that requires forethought and some effort. That forethought may seem as irrational as prejudice often is—lynching comes to mind as an emotional act of discrimination—but discrimination more typically is a rational act. It usually provides very clear benefits. One could even argue that as emotional an act as lynching has a rational side to it. Beyond causing the death of its victim, it also provokes fear among potential victims and affects their willingness to compete.

Prejudice can be taught and learned. This characteristic means that prejudice also can be unlearned. Individuals raised to dislike members of a certain group may rethink their views, depending on their adult experiences. Even generations of prejudice can disappear overnight in the right circumstances. Conversely, we might also expect to see prejudice develop almost overnight when two newly met groups seek to damage one another. The antiblack prejudice among newly arrived Catholic Irish immigrants in the mid-nineteenth century grew because the two groups competed for jobs and housing in northern cities. Although prejudice can be learned and unlearned, discrimination will almost certainly always be with us. It simply works too well in competitive circumstances.

Types of Discrimination

Summarizing many of the previous points, discrimination must (1) select one group of people, separating them from others, (2) damage their competitive abilities, and (3) result in reduced economic options for the group affected. As we turn now to more contemporary forms of discrimination, our emphasis is somewhat different. First, we need to focus much more on the selection process. Discrimination is currently illegal in the United States, so those who wish to discriminate must do it covertly. Second, we also need to become more aware of discrimination *without conscious intent*. Individuals in positions of power may make decisions that discriminate in their consequences even if that was not the intent of the decision. Of these two concerns, however, the way in which individuals are selected for discriminatory damage is the key issue, separating overt discrimination from institutional discrimination.

Overt Discrimination   Overt discrimination is characterized by the selection of those affected simply because of their group membership. If, for example, you wished to damage the competitive abilities of women, you would select all women but no men for the damage. All women might be barred from certain educational experiences or job opportunities. They might be denied housing or loans. Men, on the other hand, would be offered these opportunities. Overt discrimination, in short, refers to the most common and traditional forms of discrimination. We use the term *overt* because the discrimination is usually obvious, either through denying the affected group some right or opportunity, or through the results of the damaging action. If one hundred percent of a given group's members are prevented from achieving particular goals, it is generally because group membership (and nothing else) is holding them back.

When discrimination is legal, there is no reason to hide overt discrimination. Of all the ways particular groups may be damaged, this method is the most effective—it affects all. When overt discrimination is illegal, some of it may disappear, but much of it simply goes underground. Examples we encounter in coming chapters include the overt discrimination of realtors who tell prospective minority home buyers that houses in European American neighborhoods are already sold when in fact they are not. We would include corporation executives who never seem to find women or minorities qualified for top corporate positions. In neither case is the discrimination admitted, but in both cases it is based on group membership.

Overt discrimination has a long tradition in American history and it persists to this day. Forcing behaviors underground does not stop them. Although it may seem strange to preserve a term such as *overt* for a behavior so often hidden, keep in mind that this form of discrimination is definitely alive and well. However well hidden, it is still overt discrimination if group membership is the only criterion for the selection of who receives the damage.

Institutional Discrimination **Institutional discrimination** selects those affected based on skills, abilities, or attributes that individuals in particular groups may or may not possess. What would happen to the student body of a university if admission standards required all students to be taller than 5' 10"? Obviously, the university would have nothing but tall students, but what kinds of *groups* of individuals would be adversely affected? Relatively few women would meet that height requirement. In addition, Asians and Hispanics tend to be shorter than European Americans and African Americans. They too would be disproportionately affected, as would short men, regardless of their race or ethnicity. When all was said and done, however, men and/or certain racial/ethnic groups would be the percentage winners.

While a height requirement for university admission seems absurd, many occupations require certain heights. Some police departments prefer taller individuals while the Navy submarine service prefers shorter individuals. Later we discuss whether or not given requirements can be justified. For now, note how one single requirement can have greater repercussions on one group than another. A height requirement is gender blind and race/ethnic blind but it nevertheless produces different results for different groups. What about dwarfs? A height requirement would be *overt* discrimination against them, since their group is essentially height defined. For women and certain racial or ethnic groups, height is not a critical distinction—it is just something that usually comes with the package.

Height requirements are rare as discriminatory criteria but other skills, abilities, and attributes are used every day to decide who gets ahead and who does not. A much more common attribute serving as a gateway or hurdle is formal education. Most employers attach minimum educational credentials when they hire. Does this requirement benefit by gender? Not much, although men do tend to achieve more of the advanced credentials as well as the most highly regarded and rewarded academic credentials on the job market. Does this requirement select for race or ethnicity? As we see in Chapter 8, differences in educational achievement are significant among racial and ethnic groups in the United States. Any employment requiring higher educational credentials benefits Asian Americans and European Americans more than others.

What about skills not always connected to educational credentials? An individual may speak grammatically nonstandard English or speak with a nonstandard pronunciation. The former might well prevent the individual from getting educational credentials in the first place, but many people can follow standard grammatical rules in their writing more easily than in their speech. And nonstandard pronunciation should have nothing to do with educational credentials. Will these skills (or lack of skills) have an impact during a job interview? Pronunciation differences alone can elicit negative stereotypes regarding critical qualities such as intelligence (Luhman, 1990). An individual speaking standard English with an Appalachian accent, for example, may regularly face institutional discrimination. This is not overt discrimination because not everyone from Appalachia speaks with an Appalachian accent; they are just more likely to.

The illegality of overt discrimination has made understanding institutional discrimination increasingly important in the study of race and ethnic relations. First, some individuals who previously discriminated in an overt fashion now consciously achieve similar results by carefully selecting skills, abilities and attributes rarely found in groups they wish to damage. Racial and ethnic groups vary in so many ways that such results are not difficult to achieve. They may differ culturally, in social class, or both. An upper-class European American and a working class Mexican American have very different skills and abilities derived from their different backgrounds. Any one of those differences can come to serve as a gatekeeper.

Second, the illegality of overt discrimination has brought institutional discrimination into the spotlight. Institutional discrimination has always existed but has been overshadowed by the more obvious and more effective overt discrimination. In the 1950s, African Americans with college degrees might have been forced to work as janitors because of overt discrimination. Still, relatively few African Americans in the 1950s had college degrees. The removal of one hurdle simply brought a new hurdle to light.

Third, institutional discrimination is often unconscious on the part of the discriminator. Using the earlier example of dialect difference, an employer may not be conscious of having a bias toward a particular regional dialect but, for some reason, leaves the job interview unimpressed with the applicant. That same employer also may have job requirements in place that are not critical to job performance but are qualities preferred by the employer in a general sense. Since institutional discrimination can be unconscious, the intent to discriminate cannot easily be proven.

Fourth, procedures for hiring and firing tend to become more formalized in older industrialized societies. Instead of a craftsman taking in a nephew as an apprentice, large corporations set up complex hiring and firing procedures, often with increasingly rigid specifications regarding skills and credentials. Every specified skill or credential opens a potential door to institutional discrimination. The lack of overt bias in such procedures can ironically increase the potential for institutional bias.

As a final note, it is not the intent here to suggest that all requirements for education or employment are inherently unfair. Obviously, most requirements are necessary. Rocket scientists should indeed be up on rocket science. We do, however, in Chapter 9, critically examine requirements to determine just how essential they may be. More important, institutional discrimination raises questions concerning the accessibility of skills and credentials. If members of particular racial or ethnic groups tend to disproportionately lack those characteristics, it is important that we explore why they do.

Ethnic and Racial Stratification

Ethnic and racial stratification are essentially the same thing, varying only when we want to emphasize the circumstances of an ethnic or a racial minority. In this section, differences between racial and ethnic status are unimportant. For the sake of simplicity, only the more general term *ethnic stratification* is used. **Ethnic stratification** refers to the clustering of particular ethnic groups at particular levels of the social stratification system; any distribution of a group *other than* equal representation in all social classes could be considered ethnic stratification. Some illustrative figures might make this easier to understand.

The Structure of Ethnic Stratification

For purposes of illustration, imagine a microsociety of only 16 people. This is a multiethnic society in that those 16 people come from four separate ethnic groups—A, B, C, and D. And to keep things symmetrical, let us have our 16 societal members equally distributed among those four groups, giving us four members from each of the four groups.

Because ethnic stratification requires a social stratification system, let us create a simple system with four social classes and four positions within each of those four classes. This gives us four upper class positions, four upper middle-class positions, and so on. At this point, we have 16 individuals from four different ethnic groups and a social stratification with 16 open positions. Our only task is to assign those individuals to those positions.

Figure 2.2 illustrates one of the many ways we might distribute those 16 individuals in a social stratification system. Note that each ethnic group is equally distributed throughout the system. Any given individual from any group is as likely to be found in the lower class as the upper class. In the real world, this would mean that knowing someone's ethnic background would provide no clues as to their social position; you could not make an educated guess based on the knowledge that members of that group are more likely to be poor or rich. In short, Figure 2.2 illustrates a social stratification system totally lacking any ethnic stratification. If one were to find something similar to this in the real world, it would almost certainly be a pluralistic society such as was described in Chapter 1.

Figure 2.3 is the exact opposite of Figure 2.2. The social stratification system has not changed whatsoever but the distribution of people within it is very different. Members of Group A now monopolize all upper class positions. To be an A is to be upper class and to be upper class is to be an A. The same is true for the other three

---

FIGURE 2.2 A MULTIETHNIC SOCIETY WITHOUT ETHNIC STRATIFICATION

Upper class	A	B	C	D
Upper-middle class	A	B	C	D
Lower-middle class	A	B	C	D
Lower class	A	B	C	D

FIGURE 2.3	A MULTIETHNIC SOCIETY WITH COMPLETE ETHNIC STRATIFICATION

Upper class	A	A	A	A
Upper-middle class	B	B	B	B
Lower-middle class	C	C	C	C
Lower class	D	D	D	D

groups in their respective social classes. This figure is ethnic stratification at its most extreme—the clustering of ethnic groups is total. In the real world, such a society would be a caste system as described earlier in this chapter. Only a caste system, with its lack of social mobility, could approximate this system. Individuals living in these circumstances would probably not distinguish social class from ethnicity. Since the two overlap so completely, there would be no point in describing someone as a "lower class D." The closest the United States has been to such a circumstance is slavery in which 90 percent of African Americans were slaves.

Figure 2.4 represents the middle ground between the first two figures, more likely to be found in the real world. Note again that the social stratification *system* has not changed—only the distribution of the 16 people within it. Figure 2.4 also clearly illustrates a case of ethnic stratification although not as extreme as in Figure 2.3. Groups A and D are the most strongly ethnically stratified, while Groups B and C are clustered more loosely. Still, if you had to live in this society, you would probably rather be in Group B than C. That might not be a sound economic choice, depending upon which B or C you wound up being, but the odds would be with you. Whenever odds like this exist, you are looking at ethnic stratification.

Ethnic stratification allows us to examine class and ethnic differences simultaneously. With Figure 2.4, for example, we note that 75 percent of Group D's members are in the lowest social class; the only individual not so situated has only ascended into the next higher class. If members of Group D should have difficulty with educational institutions, for example, the degree of their ethnic stratification would be invaluable information. Their problems might stem from their cultural distinctiveness or from their disproportionate poverty status. There also might be an interaction effect (Wilson and Sparks, 1996; Spreitzer and Snyder 1991). Poverty can enhance cultural differences through social isolation. In the case of education, it also can produce

FIGURE 2.4	A MULTIETHNIC SOCIETY WITH MODERATE ETHNIC STRATIFICATION

Upper class	A	A	A	B
Upper-middle class	A	B	B	C
Lower-middle class	B	C	C	D
Lower class	C	D	D	D

negative attitudes among students toward school—an issue we explore more fully in Chapter 8.

Ethnic Stratification and Life Chances: Class, Ethnicity, and Institutional Discrimination

At the same time that ethnic stratification allows us to examine class and ethnicity simultaneously, it also makes their effects hard to separate. Consider the subject of health, which we revisit in Chapter 10. Poorer people tend to eat foods that lack proper nutrition and promote various diseases and premature death. Why do they eat what they eat? Fruits and vegetables are expensive. Less nutritious foods make the grocery dollar go further. But food preferences are also part of culture. We like to eat what we were brought up eating. Many minorities in the United States prefer diets high in animal fats, regardless of cost. Did the cultural preferences come from ethnic traditions, low social class status, or both? When minority individuals become upwardly mobile, do they change their eating habits? Often they do, but that may not be solely the result of a larger bank account. Those mobile individuals now live in a new social class and, as we saw earlier, social classes promote lifestyle differences every bit as much as ethnic groups.

Social scientists are hesitant to speak of social classes as the creators of culture. Nevertheless, we cannot ignore the fact that virtually every human behavior in a social stratification system varies somewhat from one social class to another. Being socially mobile places an individual in somewhat the same circumstances as the immigrant, facing a new culture and needing to learn quickly how to survive. And like the immigrant, the newly mobile individual may never achieve either comfort or acceptance from those raised in that social class; in both cases, the second generation is the first generation to feel "at home."

The study of social class differences becomes particularly important when distinguishing between overt and institutional discrimination. Overt discrimination selects ethnic groups for its damage, regardless of their social class position. But institutional discrimination selects on the basis of skills, abilities, or attributes. As we have seen, both ethnic and class differences lead people into different activities. People naturally tend to become skilled at whatever they do frequently. Because institutions in any society tend to reward some sets of skills but not others, both ethnic *and* class differences can create a vulnerability to institutional discrimination.

An overriding theme throughout this book is the ability of dominant groups to dominate societies with all aspects of their cultures. We might also observe that upper class individuals, regardless of ethnicity, are also in a position to build social structures in their own images. The result of both is that skills and abilities found in the lower classes and/or among minorities will not be rewarded in social institutions. Lower class people of any ethnic background face institutional discrimination just as people with only ethnic differences do. Individuals who are both lower class and ethnically different tend to have the fewest of those rewarded skills and abilities. Throughout the remainder of this book, it is critical that we keep the social allocation of skills and abilities foremost in our minds.

Summary

Social stratification systems organize the activities or occupational roles of any society into a hierarchy in which different activities receive different rewards. These rewards may consist of economic rewards (social class), social rewards or prestige (social status), or political rewards (social power). While these rewards are tightly connected to the activities themselves, individuals in those activities benefit from the rewards as long as they remain there. When individuals change activities with the result of a noticeable change in rewards, it is called social mobility.

Societies with social stratification systems vary in the amount of social mobility allowed. Ideal class societies allow a considerable amount, whereas ideal caste societies allow almost none. The former produce high levels of competition but generally only competition by individuals; there is no advantage to competing *as groups*. The latter produce very little competition of any kind because there are no advantages short of staging a revolution. These societies also vary in the way individuals view themselves. Caste societies make group membership and group boundaries very clear because mobility is generally prohibited based on group membership; slavery is such an example. At the other extreme, class societies do not reward or punish on the basis of group membership and therefore make such distinctions less important.

A society with a stratification system between these extremes (and one found in the real world) would have both some mobility and some discrimination based on group membership. In such a system, we would expect to find both competition and strong group boundaries. The result of these two outcomes is group competition for mobility and rewards.

Social stratification within the United States has remained relatively stable in income inequality over the last four decades. The changes that have occurred have been in the direction of greater inequality in recent years. In general, the United States contains greater inequality than other Western industrial nations (as is illustrated in the article at the end of this chapter).

Social scientists have long studied mobility in the United States, hoping to learn which social factors predict the greatest likelihood of mobility. Parent's occupational status and, in particular, parent's educational attainment both lead to high mobility for children. To the extent that mobility does not occur, social classes tend to add social isolation to their already existing geographical isolation, creating significant lifestyle differences from one social class to the next.

The competition induced by social stratification can spur individuals toward excellence; it can also spur them to damage their opponents to ensure their own success. This damage is discrimination. It can be directed either at whole ethnic or racial groups (overt discrimination) or it can selectively apply damage to individuals with or without specific skills, abilities, or attributes. When those selected skills, abilities, or attributes vary significantly from one ethnic or racial group to another, the discrimination is described as institutional.

Because discrimination prevents the upward mobility of members of specific groups, it clearly removes many economic options open to others. Discrimination thus not only removes competitive threats but also produces a group of people vulnerable to labor exploitation. Having no other meaningful economic options, they must work at what remains.

Although discrimination produces clear competitive damage to those affected, prejudice and stereotypes are attitudes and do most of their damage at the psychological level. Prejudice is the possession of negative attitudes directed toward a specific group of people; stereotypes are the intellectual generalizations that support or justify the prejudice. Prejudice and stereotypes can accompany discrimination, but are not necessary conditions.

Ethnic or racial stratification results from ongoing discrimination directed against specific ethnic or racial groups. It exists when those groups are found clustered at specific social class levels, either high or low. Typically, dominant groups monopolize upper-class positions while minorities are clustered in the lower social classes. When groups become so clustered, their preexisting ethnic differences can become even greater through the social and geographical isolation that exists between social classes. The more such differences occur, the more susceptible a group is to various forms of institutional discrimination.

Chapter 2 Reading
Imagine A Country

Holly Sklar

Author Holly Sklar focuses attention on social stratification in the following article. In particular, she shines a critical light on the degree of social inequality in the United States in relation to other countries around the world. While social stratification creates inequality by definition, that inequality can vary in the social and economic distance between those on the top and those on the bottom. Of the modern industrial countries, the United States has some of the richest and some of the poorest people to be found. Since ethnic stratification places many minority individuals towards the bottom of the stratification hierarchy, the level of inequality in any society becomes very important to the study of race and ethnic relations. The comparisons Sklar offers will appear in every chapter that remains in this book.

Imagine a country where one out of four children is born into poverty, and wealth is being redistributed upward. Since the 1970s, the top 1 percent of families have doubled their share of the nation's wealth—while the percentage of children living in extreme poverty has also doubled.

Highlighting growing wage inequality, the nation's leading business newspaper acknowledges, "The rich really are getting richer, and the poor really are getting poorer."

Imagine a country where the top 1 percent of families have about the same amount of wealth as the bottom 95 percent. Where the poor and middle class are told to tighten their belts to balance a national budget bloated with bailouts and subsidies for the well-off.

It's not Mexico.

Imagine a country which demands that people work for a living while denying many a living wage.

Imagine a country where wages have fallen for average workers, adjusting for inflation, despite significant growth in the economy. Real per capita GDP (gross domestic product) rose 33 percent from 1973 to 1994, yet real weekly wages fell 19 percent for nonsupervisory workers, the vast majority of the workforce.

It's not Chile.

Imagine a country where the stock market provides "payoffs for layoffs."

Imagine a country where workers are downsized while corporate profits and executive pay are upsized. The profits of the 500 leading corporations rose a record 23 percent in 1996 and CEO compensation (including salary, bonus, and long-term compensation such as stock options) shot up 54 percent, while workers' wages and benefits barely kept pace with inflation. The average CEO of a major corporation was paid as much as 42 factory workers in 1980, 122 factory workers in 1989, and 209 factory workers in 1996.

A leading business magazine says, "People who worked hard to make their companies competitive are angry at the way the profits are distributed. They think it is unfair, and they are right."

It's not England.

Imagine a country where living standards are falling for younger generations despite the fact that many households have two wage earners, have fewer children, and are better educated than their parents. Since 1973, the share of workers without a high school degree has been cut in half. The share of workers with at least a four-year college degree has doubled.

The entry-level hourly wages of male high school graduates fell 27.3 percent between 1979 and 1995, and the entry-level wages of women high school graduates fell 18.9 percent.

A college degree is increasingly necessary, but not necessarily sufficient to earn a decent income. Between 1989 and 1995, the entry-level wages of male college graduates fell 9.5 percent, and the entry-level wages of women college graduates fell 7.7 percent.

Imagine a country where the percentage of young full-time workers (ages 18-24) earning low wages doubled from 23 percent in 1979 to 47 percent in 1992. Where families with household heads ages 25 to 34 had 1994 incomes that were $4,611 less than their 1979 counterparts.

It's not Russia.

Imagine a country where leading economists consider it "full employment" when the official unemployment rate reaches 6 percent (over 7 million people). You're not counted officially as unemployed just because—you're unemployed. To be counted in the official unemployment rate you must have searched for work in the past four weeks. The government doesn't count people as "unemployed" if they are so discouraged from long and fruitless job searches they have given up looking. It doesn't count as "unemployed" those who couldn't look for work in the past month because they had no child care, for example. If you need a full-time job, but you're working part-time—whether 1 hour or 34 hours—because that's all you can find, you're counted as employed.

A leading business magazine observes, "Increasingly the labor market is filled with surplus workers who are not being counted as unemployed."

Imagine a country where there is a shortage of jobs, not a shortage of work. Millions of people need work and urgent work needs people—from creating affordable housing, to repairing bridges and building mass transit, to cleaning up pollution and converting to renewable energy, to staffing after-school programs and community centers.

Imagine a country where for more and more people a job is not a ticket out of poverty, but into the ranks of the working poor. Between 1979 and 1992, the proportion of full-time workers paid low wages jumped from 12 percent to 18 percent—nearly one in every five full-time workers.

Imagine a country where one out of four officially poor children live in families in which one or more parents work full time, year round. The official poverty line is set well below the actual cost of minimally adequate housing, health care, food, and other necessities.

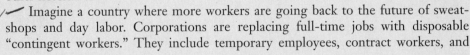Imagine a country where more workers are going back to the future of sweatshops and day labor. Corporations are replacing full-time jobs with disposable "contingent workers." They include temporary employees, contract workers, and

"leased" employees—some of them fired and then "rented" back at a large discount by the same company—and involuntary part-time workers, who want permanent full-time work.

It's not Spain.

How do workers increasingly forced to migrate from job to job, at low and variable wage rates, without health insurance or paid vacation, much less a pension, care for themselves and their families, own a home, pay for college, save for retirement, plan a future, build strong communities?

Imagine a country where after mass layoffs and union-busting, less than 15 percent of workers are unionized. One out of three workers were union members in 1955.

Imagine a country where the concerns of working people are dismissed as "special interests" and the profit-making interests of globe-trotting corporations substitute for the "national interest."

Imagine a country whose government negotiates "free trade" agreements that help corporations trade freely on cheap labor at home and abroad.

One ad financed by the country's agency for international development showed a Salvadoran woman in front of a sewing machine. It told corporations, "You can hire her for 33 cents an hour. Rosa is more than just colorful. She and her co-workers are known for their industriousness, reliability and quick learning. They make El Salvador one of the best buys." The country that financed the ad intervened militarily to make sure El Salvador would stay a "best buy" for corporations.

It's not Canada.

Imagine a country where more than half of all women with children under age 6, and three-fourths of women with children ages 6-17, are in the paid workforce, but affordable child care and after-school programs are scarce. (Families with incomes below the poverty line spend nearly one-fifth of their incomes on child care.) Apparently, kids are expected to have three parents: Two parents with jobs to pay the bills, and another parent to be home in mid-afternoon when school lets out—as well as all summer.

Imagine a country where women working year round, full time earn 74 cents for every dollar men earn. Women don't pay 74 cents on a man's dollar for their college degrees or 74 percent as much to feed or house their children.

Imagine a country where instead of rooting out discrimination, many policy makers are busily blaming women for their disproportionate poverty. Back in 1977, a labor department study found that if working women were paid what similarly qualified men earn, the number of poor families would decrease by half. A 1991 government study found that even "if all poor single mothers obtained [full-time] jobs at their potential wage rates," given their educational and employment background and prevailing wages, "the percentage not earning enough to escape from poverty would be 35 percent."

Two out of three workers who earn the miserly minimum wage are women. Full-time work at minimum wage pays below the official poverty line for a family of two.

Imagine a country where discrimination against women is pervasive from the bottom to the top of the payscale, and it's not because women are on the "mommy track."

In the words of a leading business magazine, "at the same level of management, the typical woman's pay is lower than her male colleague's—even when she has the exact same qualifications, works just as many years, relocates just as often, provides the main financial support for her family, takes no time off for personal reasons, and wins the same number of promotions to comparable jobs."

It's not Japan.

Imagine a country where the awful labeling of children as "illegitimate" has again been legitimized. Besides meaning born out of wedlock, illegitimate also means illegal, contrary to rules and logic, misbegotten, not genuine, wrong—to be a bastard. The word illegitimate has consequences. It helps make people more disposable. Single mothers and their children have become prime scapegoats for illegitimate economics.

Imagine a country where violence against women is so epidemic it is their leading cause of injury. So-called "domestic violence" accounts for more visits to hospital emergency departments than car crashes, muggings, and rapes combined. About a third of all murdered women are killed by husbands, boyfriends and ex-partners (less than a tenth are killed by strangers). Researchers say that "men commonly kill their female partners in response to the woman's attempt to leave an abusive relationship."

The country has no equal rights amendment.

It's not Algeria.

Imagine a country where homicide is the second-largest killer of young people, ages 15-24; "accidents," many of them drunk driving fatalities, are first. Increasingly lethal weapons designed for hunting people are produced for profit by major manufacturers and proudly defended by a politically powerful national rifle association. Half the homes in the country contain firearms, and guns in the home greatly increase the risk of murder and suicide for family members and close acquaintances.

Informational material from a national shooting sports foundation asks, "How old is old enough?" to have a gun, and advises parents:

Age is not the major yardstick. Some youngsters are ready to start at 10, others at 14. The only real measures are those of maturity and individual responsibility. Does your youngster follow directions well? Would you leave him alone in the house for two or three hours? Is he conscientious and reliable? Would you send him to the grocery store with a list and a $20 bill? If the answer to these questions or similar ones are "yes" then the answer can also be "yes" when your child asks for his first gun.

Imagine a country where children are taught violence is the way to resolve conflict through popular wars and media "entertainment." "In the media world, brutality is portrayed as ordinary and amusing" and often merged with sex, observes a prominent public health educator. The screen "good guys" not only use violence as a first resort, but total war is the only response to the dehumanized "bad guys" who often speak with foreign accents. War cartoons and violent "superhero" shows are created expressly to sell toys to children. Video and computer games showcase increasingly graphic and participatory "virtual" violence. The strong consensus of private and government research is that on-screen violence contributes to off-screen violence.

It's not Australia.

Caste society class society.

Imagine a country whose school system is rigged in favor of the already-privileged, with lower caste children tracked by race and income into the most deficient and demoralizing schools and classrooms. Public school budgets are heavily determined by private property taxes, allowing higher income districts to spend much more than poor ones. In one large state in 1991-92, spending per pupil ranged from $2,337 in the poorest district to $56,791 in the wealthiest.

In rich districts kids take well-stocked libraries, laboratories, and state-of-the-art computers for granted. In poor schools they are rationing out-of-date textbooks and toilet paper. Rich schools often look like country clubs—with manicured sports fields and swimming pools. Poor schools often look more like jails—with concrete grounds and grated windows. College prep courses, art, music, physical education, field trips, and foreign languages are often considered necessities for the affluent, luxuries for the poor.

Discrimination. Wealthier citizens argue that lack of money isn't the problem in poorer schools—family values are—until proposals are made to make school spending more equitable. Then money matters greatly for those who already have more.

It's not India.

Imagine a country where Black unemployment and infant mortality is more than twice that of whites, and Black life expectancy is seven years less. The government subsidized decades of segregated suburbanization for whites while the inner cities left to people of color were treated as outsider cities—separate, unequal, and disposable. Recent studies have documented continuing discrimination in employment, banking, and housing.

Imagine a country whose constitution once defined Black slaves as worth three-fifths of whites. Today, median Black per capita income is three-fifths of whites.

It's not South Africa.

Imagine a country which pretends that anyone who needs a job can find one, while its federal reserve board enforces slow growth economic policies that keep millions of people unemployed, underemployed, and underpaid.

Imagine a country with full prisons instead of full employment. The prison population has more than doubled since 1980. The nation is Number One in the world when it comes to locking up its own people. The bureau of justice statistics reports that in 1985, 1 in every 320 of the nation's residents were incarcerated. By the end of 1995, the figure had increased to 1 in every 167.

Imagine a country where prison labor is a growth industry and so-called "corrections" spending is the fastest growing part of state budgets. Apparently, the government would rather spend $25,000 a year to keep someone in prison than on cost-effective programs of education, community development, addiction treatment, and employment to keep them out. In the words of a national center on institutions and alternatives, this nation has "replaced the social safety net with a dragnet."

Imagine a country that has been criticized by human rights organizations for expanding rather than abolishing use of the death penalty—despite documented racial bias and numerous cases of innocents being put to death.

It's not China.

Imagine a country that imprisons Black men at a rate nearly five times more than apartheid South Africa. One out of three Black men in their twenties are either in jail, on probation or on parole. Meanwhile, one out of three Black men and women ages 16-19 are officially unemployed, as are nearly one out of five ages 20-24. Remember, to be counted in the official unemployment rate you must be actively looking for a job and not finding one. "Surplus" workers are increasingly being criminalized.

A 1990 Justice Department report observed, "The fact that the legal order not only countenanced but sustained slavery, segregation, and discrimination for most of our Nation's history—and the fact that the police were bound to uphold that order—set a pattern for police behavior and attitudes toward minority communities that has persisted until the present day." A 1992 newspaper article is titled, "GUILTY...of being black: Black men say success doesn't save them from being suspected, harassed and detained."

Imagine a country waging a racially biased "War on Drugs." More than three out of four drug users are white, but Blacks and Latinos are much more likely to be arrested and convicted for drug offenses and receive much harsher sentences. Almost 90 percent of those sentenced to state prison for drug possession in 1992 were Black and Latino.

A study in a prominent medical journal found that drug and alcohol rates were slightly higher for pregnant white women than pregnant Black women, but Black women were about ten times more likely to be reported to authorities by private doctors and public health clinics—under a mandatory reporting law. Poor women were also more likely to be reported.

It is said that truth is the first casualty in war, and the "War on Drugs" is no exception. Contrary to stereotype, "The typical cocaine user is white, male, a high school graduate employed full time and living in a small metropolitan area or suburb," says the nation's former drug czar. A leading newspaper reports that law officers and judges say, "Although it is clear that whites sell most of the nation's cocaine and account for 80% of its consumers, it is blacks and other minorities who continue to fill up [the] courtrooms and jails, largely because, in a political climate that demands that something be done, they are the easiest people to arrest." They are the easiest to scapegoat.

Imagine a country which intervenes in other nations in the name of the "War on Drugs," while it is the number one exporter of addictive, life-shortening tobacco. It is also number four in the world in alcohol consumption—the drug most associated in reality with violence and death—and number one in drunk-driving fatalities per capita. Those arrested for drunk driving are overwhelmingly white and male and typically treated much more leniently than illicit drug offenders.

It's not France.

Imagine a country where the cycle of unequal opportunity is intensifying. Its beneficiaries often slander those most systematically undervalued, underpaid, underemployed, underfinanced, underinsured, underrated, and otherwise underserved and undermined—as undeserving, "underclass," impoverished in moral and social values and lacking the proper "work ethic." The oft-heard stereotype of deadbeat poor people masks the growing reality of dead-end jobs and disposable workers.

Imagine a country abolishing aid to families with dependent children while maintaining aid for dependent corporations.

Imagine a country slashing assistance to its poorest people, disabled children, and elderly refugees to close a budget deficit produced by excessive military spending and tax cuts for corporations and the rich. Wealthy people—whose tax rates are among the lowest in the world—not only benefited from deficit spending and tax breaks, they earn interest on the debt as government bond holders.

Imagine a country with a greed surplus and justice deficit. According to a former secretary of labor, "were the tax code as progressive as it was even as late as 1977," the top 10 percent of income earners "would have paid approximately $93 billion more in taxes" than they paid in 1989. How much is $93 billion? About the same amount as the combined 1989 government budget for all these programs for low-income persons: aid to families with dependent children, supplemental security income, general assistance, food and nutrition benefits, housing, jobs and employment training, and education aid from preschool to college loans.

Imagine a country where state and local governments are rushing to expand lotteries, video poker and other government-promoted gambling to raise revenues, disproportionately from the poor, which they should be raising from a fair tax system.

Imagine a country whose military budget continues consuming resources at nearly average Cold War levels although the Soviet Union no longer exists. In the post-Cold War world, the "Peace Dividend" means the congress gives the military more than it asks for. This nation also leads the world in arms exports.

Imagine a country that ranks first in the world in wealth and military power, and 26th in child mortality (under five). If the government were a parent it would be guilty of child abuse. Thousands of children die preventable deaths.

Imagine a country where health care is managed for healthy profit. In many countries health care is a right, but in this one 42 million people have no health insurance and another 29 million are underinsured, according to the nation's college of physicians. Lack of health insurance is associated with a 25 percent higher risk of death.

Imagine a country where descendants of its first inhabitants live on reservations strip-mined of natural resources. Life expectancy averages in the forties—not the seventies. Infant mortality is seven times higher than the national average and a higher proportion of people live in poverty than any other ethnic group. An Indian leader is the country's best known political prisoner.

Imagine a country where 500 years of plunder and lies are masked in expressions like "Indian giver." Where the military still dubs enemy territory, "Indian country."

Imagine a country which has less than 5 percent of the world's population, but uses 25 percent of the world's oil resources. Only 3 percent of the public's trips are made by public transportation. It has felled more trees since 1978 than any other country. It is the number one contributor to acid rain and global warming.

It's not Brazil.

Imagine a country where half the eligible voters don't vote. The nation's House of Representatives is not representative of the nation. It is overwhelmingly male and disproportionately white. The senate is representative of millionaires.

Imagine a country where white men who are "falling down" the economic ladder are being encouraged to believe they are falling because women and people of color are climbing over them to the top or dragging them down from the bottom. That way, they will blame women and people of color rather than the system. They will buy the myth of "reverse discrimination." Never mind that white males hold 95 percent of senior management positions (vice president and above).

Imagine a country where on top of discrimination comes insult. It's common for people of color to get none of the credit when they succeed—portrayed as undeserving beneficiaries of affirmative action and "reverse discrimination"—and all of the blame when they fail. A study of the views of 15-to-24-year-olds found that 49 percent of whites believe that it is more likely that "qualified whites lose out on scholarships, jobs, and promotions because minorities get special preferences" than "qualified minorities are denied scholarships, jobs, and promotions because of racial prejudice." Only 34 percent believed that minorities are more likely to lose out.

Imagine a country where scapegoating thrives on misinformation. The majority of whites in a national 1995 survey said that average Blacks held equal or better jobs than average whites. Survey respondents also wrongly estimated the white share of the population to be under 50 percent—rather than 74 percent.

Imagine a country where a former presidential press secretary boasted to reporters: "You can say anything you want in a debate, and 80 million people hear it. If reporters then document that a candidate spoke untruthfully, so what? Maybe 200 people read it, or 2,000 or 20,000."

Imagine a country where a far-right television commentator-turned-presidential candidate—whose heroes include U.S. Senator Joe McCarthy, Spanish dictator Franco, and Chilean dictator Pinochet—told the national convention of one of the two major parties: "There is a religious war going on in this country. It is a cultural war." Delegates waved signs saying "Gay Rights Never"—the 90s version of segregation forever. Referring to recent rioting in a major city, following the acquittal of police officers who had severely beaten a Black man, the once and future candidate said: "I met the troopers of the 18th Cavalry, who had come to save the city...And as those boys took back the streets of [that city], block by block, my friends, we must take back our cities and take back our culture and take back our country."

It's not the former Yugoslavia.

Imagine a country where scapegoating fuels fear and fear fuels scapegoating. The list of scapegoats grows rapidly with homeless people, women and children receiving welfare, people of color, gays and lesbians, Jews, undocumented immigrants, longtime legal immigrants, people with disabilities. More and more children are declared illegitimate. More and more people are treated as disposable.

It's not Germany.

It's the disUnited States.

Decades ago Martin Luther King Jr. warned, in *Where Do We Go From Here: Chaos or Community?* (Harper & Row, 1967), "History is cluttered with the wreckage of nations and individuals who pursued [the] self-defeating path of hate." King declared:

A true revolution of values will soon cause us to question the fairness and justice of many of our past and present policies. We are called to play the good samaritan on life's roadside; but...one day the whole Jericho road must be transformed so that men and women will not be beaten and robbed as they make their journey through life....

A true revolution of values will soon look uneasily on the glaring contrast of poverty and wealth....There is nothing but a lack of social vision to prevent us from paying an adequate wage to every American citizen whether he be a hospital worker, laundry worker, maid or day laborer. There is nothing except shortsightedness to prevent us from guaranteeing an annual minimum—and *livable*—income for every American family. There is nothing, except a tragic death wish, to prevent us from reordering our priorities, so that the pursuit of peace will take precedence over the pursuit of war.

Selected Sources

Donald L. Barlett and James B. Steele, *America: Who Really Pays the Taxes?* (New York: Touchstone, 1994).

Business Week, annual reports on executive pay, "The Real Truth About the Economy," November 7, 1994.

Center on Budget and Policy Priorities: Isaac Shapiro and Robert Greenstein, *Trends in the Distribution of After-Tax Income: An Analysis of Congressional Budget Office Data* (Washington, DC, August 14, 1997).

Ira J. Chasnoff, et al., "The Prevalence of Illicit-Drug or Alcohol Use During Pregnancy and Discrepancies in Mandatory Reporting in Pinellas County, Florida," *New England Journal of Medicine*, April 26, 1990.

Children's Defense Fund: Arloc Sherman, *Rescuing the American Dream: Halting the Economic Freefall of Today's Young Families with Children* (1997); Children's Defense Fund and Northeastern University's Center for Labor Market Studies, *Vanishing Dreams: The Economic Plight of America's Young Families* (Washington, DC: Children's Defense Fund, 1992).

Steven B. Duke and Albert C. Gross, *America's Longest War: Rethinking Our Tragic Crusade Against Drugs* (New York: Jeremy P. Tarcher/Putnam, 1993).

Economic Policy Institute: Lawrence Mishel, Jared Bernstein and John Schmitt, *The State of Working America 1996-97* (Washington, DC: Economic Policy Institute, 1996).

Anne B. Fisher, "When Will Women Get To The Top?" *Fortune*, September 21, 1992.

Susan Glick and Josh Sugarmann, "Why Johnny can shoot," *Mother Jones*, January/February 1995, excerpted from Violence Policy Center (Washington, DC), *Use the Schools: How Federal Tax Dollars Are Spent to Market Guns to Kids*.

Ron Harris, "Blacks Feel Brunt of Drug War," *Los Angeles Times*, April 22, 1990.

Arthur L. Kellermann, et al., "Gun Ownership as a Risk Factor For Homicide in the Home" and Jerome P. Kassirer, "Guns in the Household," editorial, *New England Journal of Medicine*, October 7, 1993.

Arthur L. Kellermann and James A. Mercy, "Men, Women, and Murder: Gender-Specific Differences in Rates of Fatal Violence and Victimization," *The Journal of Trauma* 33:1, July 1992.

Jonathan Kozol, *Savage Inequalities: Children in America's Schools* (New York: Crown Publishers, 1991).

Peter Medoff and Holly Sklar, *Streets of Hope: The Fall and Rise of an Urban Neighborhood* (Boston: South End Press, 1994).

Richard Morin, "Across the Racial Divide," *Washington Post Weekly*, October 16-22, 1995.

National Center on Institutions and Alternatives, *Hobbling a Generation: Baltimore, Maryland* (September 1992).

National Labor Committee (New York), *Free Trade's Hidden Secrets: Why We Are Losing Our Shirts* (November 1993) and *Paying to Lose Our Jobs* (September 1992).

People for the American Way, *Democracy's Next Generation II: A Study of American Youth on Race* (Washington, DC: 1992).

Michael M. Phillips, "The Outlook: Inequality May Grow for Lifetime Earnings," *Wall Street Journal*, December 23, 1996.

Deborah Prothrow-Stith, *Deadly Consequences: How Violence is Destroying Our Teenage Population and a Plan to Begin Solving the Problem* (New York: Harper Perennial, 1991/1993).

Robert B. Reich, *The Work of Nations* (New York: Alfred A. Knopf, 1991).

Albert J. Reiss Jr. and Jeffrey A. Roth, eds., National Research Council, *Understanding and Preventing Violence* (Washington, DC: National Academy Press, 1993).

Jeffrey A. Roth, "Psychoactive Substances and Violence," U.S. Department of Justice, National Institute of Justice, *Research in Brief*, February 1994.

John E. Schwarz and Thomas J. Volgy, *The Forgotten Americans: Thirty Million Working Poor in the Land of Opportunity* (New York: W.W. Norton, 1992).

The Sentencing Project: Marc Mauer and Tracy Huling, *Young Blacks Americans and the Criminal Justice System: Five Years Later* (Washington, DC: Sentencing Project, 1995); Cathy Shine and Marc Mauer, *Does the Punishment Fit the Crime? Drug Users and Drunk Drivers: Questions of Race and Class* (1993).

Andrew L. Shapiro, *We're Number One: Where America Stands—and Falls—in the New World Order* (New York: Vintage Books, 1992).

Holly Sklar, *Chaos or Community? Seeking Solutions, Not Scapegoats for Bad Economics* (Boston: South End Press, 1995).

Carol Stocker and Barbara Carton, "GUILTY...of being black," *Boston Globe*, May 7, 1992.

John J. Sweeney, *America Needs a Raise* (Boston: Houghton Mifflin, 1996).

Lester Thurow, *The Future of Capitalism* (New York: William Morrow and Co., 1996).

Margery Austin Turner, et al., *Opportunities Denied, Opportunities Diminished: Racial Discrimination in Hiring* (Washington, DC: Urban Institute Press, 1991).

United Nations Children's Fund, *The State of the World's Children: 1996* (New York: UNICEF/Oxford University Press, 1996).

U.S. Bureau of the Census, *Statistical Abstract of the United States: 1996; Money Income in the United States: 1996; Poverty in the United States: 1996;* "The Earnings Ladder: Who's at the Bottom? Who's at the Top?" *Statistical Brief*, March 1994.

U.S. Department of Health and Human Services, National Institute on Drug Abuse (NIDA), *National Household Survey on Drug Abuse*, various reports.

U.S. Department of Justice, Bureau of Justice Statistics, various reports.

U.S. Department of Labor, Bureau of Labor Statistics, *Employment and Earnings* (monthly).

U.S. General Accounting Office, *Mother-Only Families: Low Earnings Will Keep Many Children in Poverty* (April 1991); *Workers At Risk: Increased Numbers in Contingent Employment Lack Insurance, Other Benefits* (March 1991).

Hubert Williams and Patrick V. Murphy, "The Evolving Strategy of Police: A Minority View," *Perspectives on Policing*, U.S. Department of Justice (January 1990).

Edward N. Wolff, "Time for a Wealth Tax?" *Boston Review*, February-March 1996; *Top Heavy: A Study of the Increasing Inequality of Wealth in America* (New York: Twentieth Century Fund, 1995).

3

Theories of Discrimination and Prejudice

All scientists approach their subject matter with the goal of explaining behavior. Just as a botanist wants to know how plants grow under varying conditions or a biologist explores how the human body responds to an invading bacteria, the social scientist seeks to understand human behavior.

All scientific explanations are presented in the form of a theory. A scientific **theory** specifies relations among concepts, predicting the degree to which they affect each other, and why. The building blocks of theories—**concepts**—are intellectual constructs that specify (or direct our attention to) certain similarities among different observations. The terms *theory* and *concept* require such general definitions because we use them to discuss a vast range of topics, for example, comets, wheat, bacteria, or human behavior. *Discrimination* is an example of a concept. It focuses our attention on all human behaviors that share the similarity of hindering competitive abilities among social group members. (The concept simultaneously instructs us to ignore extraneous details such as who is involved, when the act occurred, and so on.) So when we observe human behavior, concepts instruct us to group a wide variety of behaviors (the "different observations") under one heading.

Once concepts are in place, theories specify how concepts affect each other. We might combine two concepts in a simple theory: economic recessions tend to increase discriminatory behavior; the more the first increases, the more the second will increase. To do this properly, we need good measurements of our concepts. The field of economics provides us with many ways to measure an economic recession, but how would we measure a concept like discrimination? Lynching is an act of discrimination. We can even count the number of occurrences, and lynching certainly represents a greater degree of discrimination than a landlord's refusal to rent an apartment near a place of employment. But how much greater of a discriminatory act would a lynching be? Such problems plague all forms of science. We should not expect the study of race and ethnic relations to be any easier.

This chapter explores the various theories that have developed over the years in the study of race and ethnic relations. Whenever racial or ethnic issues figure prominently in human behavior, that behavior enters our subject matter. Because so much behavior involves significant racial and ethnic characteristics, it should come as no surprise that the variety of theories we encounter is large indeed.

Theory and Race/Ethnic Studies: What Do You Want to Know?

Scientific theories do not operate in vacuums. Science developed as a form of knowledge in the Western world because it helps people control the world around them. We generally want to be free from disease and for our corn to grow high. Such human interests usually direct the sciences; however, not every scientific theory has a clearly practical concern driving it. Nevertheless, research questions generally reflect whatever societal members perceive to be "problems." For the social scientific study of race and ethnic relations, this becomes more complex because different groups perceive different behaviors to be problems; indeed, many racial and ethnic groups consider one another to be the problem. Obviously, slaves and slaveowners would hand different research agendas to the social scientist. Theories vary depending on what the researcher wants to know. Let us begin by compiling two lists of questions, one seeking to explain attitudes, the other behavior.

Questions relating to attitudes:

- Under what social conditions or among what kinds of people is prejudice likely to occur?

- Under what social conditions or among what kinds of people is existing prejudice likely to disappear?

- How and why do stereotypes develop?

- Why do stereotypes disappear?

- How can acts of discrimination be justified or rationalized by individuals to the point no guilt is felt?

- What social circumstances encourage individuals to think of themselves as ethnic group members?

- What social circumstances increase an individual's sense of solidarity with his or her ethnic group?

Questions relating to behaviors:

- What social conditions lead to increasing acts of discrimination among groups?

- Under what circumstances does labor exploitation become more necessary for employers?

- What roles do economic security and exploitation play in the creation and maintenance of discriminatory acts?

- What social conditions lead to the categorization of ethnic groups in racial terms?

- What social conditions increase the probability of group violence directed at minorities?

- Under what circumstances are minority racial or ethnic groups likely to organize in political movements to improve their situation?

- Under what circumstances are such political movements likely to employ moderate or extreme tactics?

These questions form only the tip of the iceberg. Researchers have moved well beyond these few issues. The reason for the wide scope is that, like gender, race and ethnicity enter into almost all social arenas. If your neighbor, coworker, friend, or daughter-in-law is racially different from you, that fact will affect virtually all interactions between the two of you, one way or another. And should either of the two of you forget, there will be many around to remind you. Such impacts on social interaction may seem to be a very small part of the picture, but multiplied by all occurrences with people everywhere, they can play a major role in increasing ethnic consciousness, solidarity, and even political organizations.

As we turn to examine some of the major theories relating to race and ethnicity, keep in mind that the theories look different. In part, they differ because they come from different theoretical "schools" within the social sciences. These theoretical schools make fundamentally different assumptions about the social world. The theories also look different, however, because they try to explain many different things. A theory about the development of prejudicial attitudes necessarily looks quite different from an attempt to explain the development of minority political organizations.

Theories of Prejudice

How do you explain a negative attitude? Prejudice is directed specifically at racial and ethnic group members, but in many ways, it is much like all social values. We learn to prefer one type of music or food over another just as we learn to apply evaluations to people—we pick up values through our associations with others. We often think of prejudice as learned, passed on from parent to child, much like other values (Allport, 1954). However, the uniqueness of prejudice has led to some distinctive theories. Because prejudice is more aggressive than most values (hating people is not quite like hating broccoli), some theorists look to personality features of prejudiced individuals; we examine these as *psychologically based theories*. Other theorists try to connect

Prejudicial attitudes can be displayed openly, as in this photo of an American Nazi in Chicago, or can be much more subtle.

prejudice with societal changes (see Gaines and Reed, 1995); we examine these as *society based theories*.

 Psychologically Based Theories

One of the earliest of the psychologically focused theories was the work of Adorno and others (1950) on the authoritarian personality. Adorno and his colleagues hypothesized that a combination of various character traits—the authoritarian personality—predisposes some individuals to prejudicial thinking toward other groups. The assumption is that one or two of these traits alone would not predict prejudiced thinking, but the combination and interaction of these traits would. They concluded that such a combination of traits results from a particularly harsh and restricted childhood socialization.

What traits are necessary to create an authoritarian personality? Some of the most important include a preoccupation with power and "toughness," rigid adherence to social norms, an uncritical intellectual attitude toward dominant social values, a punitive attitude toward those who break social norms, and a general sense of hostility toward other people. To such an individual, prejudice would seem a logical response to any kind of group difference.

The work on the authoritarian personality received some support from an earlier study by E.L. Hartley (1946) on prejudiced attitudes. Hartley questioned individuals about their attitudes toward many racial and ethnic groups in the United States. Added to that list were some nonexistent groups such as Pirenians and Danerians. Hartley found that individuals prejudiced against real groups also indicated prejudiced attitudes toward groups that did not exist. More recent research continues to link authoritarianism with prejudice, particularly in predicting high levels of European

American stereotyping of African Americans as criminals, coupled with high respect for police officers (Oliver, 1996).

Jack Levin (1975) attempts to simplify such research by comparing the different ways individuals assess the value of their own achievements. He concludes that some individuals tend to be self-evaluators; that is, they gauge their own achievements against some internal standard of excellence. Such people are only satisfied when they meet their own standards and are little influenced by the achievement of others. Conversely, Levin argues, some people are relative evaluators who gauge their own achievements against the achievements of others in their immediate surroundings. Levin collected individuals of both types in an experimental setting and subjected all subjects to situations in which their abilities were questioned. As predicted, the relative evaluators were more likely to mend their damaged self-esteem through developing prejudiced attitudes toward others.

An earlier but somewhat similar approach linked self-esteem to prejudice through examining individual responses to frustration. In what has come to be known as the frustration-aggression hypothesis, John Dollard and his colleagues (1939) argue that aggression in some form is a natural response to the experience of frustration. Normally, an individual's preference is to be aggressive toward the source of that frustration, but that is not always possible. High regard for your income may keep you from hitting your boss, but the more general issue is, whom do you blame for an economic recession? In a similar light, Fein and Spencer (1997) find that measured levels of prejudice are directly related to the self-image of the prejudiced person; efforts to positively raise the self-esteem of subjects resulted in decreased prejudice. Interestingly, prejudice is more common among European Americans than minorities and among men more than women (Mills et al., 1995); at least theoretically, European American men should face fewer attacks on their self-images than members of other groups.

When the source of a frustration cannot be located, aggression can be displaced. A bad day at work can result in coming home to beat up one's children or kick the family dog. What about recessions? Beck and Tolnay (1990) relate historical increases in American southern lynchings to poor economic times for southern whites. The notion of displaced aggression has generally entered American popular culture, appearing regularly as a "common sense" explanation for aggressive behavior. But however accurate a motivator, there is nothing in the frustration-aggression hypothesis that predicts which way aggression will be displaced. Obviously, it tends to be directed at more powerless figures—hence the connection with prejudice against powerless minorities—but why would hatred of another ethnic group be preferable to kicking Spot? These and other concerns about the psychological theories of prejudice led many researchers to examine the social contexts within which prejudice occurs.

Society Based Theories

Looking beyond individual personalities, many social scientists have attempted to locate prejudiced attitudes in particular social contexts, the most common being social stratification. Prejudice is only occasionally linked to exploitation. When it is, its function is generally to justify or rationalize an unfree labor situation. But prejudice is much more emotional than discrimination. Most researchers focus their efforts on competition between ethnic or racial groups, viewing prejudice largely as responding to threatened security. If you or your group is downwardly mobile, attacking other upwardly mobile groups might seem in order. A more common situation is for a stationary ethnic group to fear the mobility of a group rising from below. Such assumptions and hypotheses have led social scientists into various research areas.

Two confusing and interrelated issues in studying social stratification and prejudice stem from the combined nature of lower class status and heightened competition with other groups. Minorities, by definition, are clustered toward the bottom of the stratification hierarchy. If they threaten anyone, it will be others also located in those general positions. Working class European Americans, for example, might be threatened by mobility among minorities. But are working class people more likely to be prejudiced regardless of the competition they face? Could there be elements of the working class lifestyle that lead in that direction?

Considerable research has linked educational levels with prejudiced attitudes. In general, less educated persons tend to have more prejudiced attitudes, in the United States and elsewhere (Allport, 1954; Case et al., 1989; Dyer et al., 1989). What might the connection be? Perhaps education produces critical thinking and a wider perspective on the world, increasing tolerance of group differences. Perhaps education teaches people to keep their prejudices to themselves. Or perhaps education leads to higher social class status, placing educated people out of direct competition with aspiring minorities. Could we clear this up by comparing those with high education and social class with those equally well educated but economically unsuccessful? People in the latter group might be the most frustrated people of all, feeling cheated that their educational achievements did not produce the promised results. In short, the role of education in prejudice is difficult to isolate.

We also know that people with lower incomes, regardless of educational level, tend to have higher levels of prejudice (Dyer et al., 1989). Such findings clearly raise questions about the competition-prejudice connection (Case et al., 1989). Many, if not most, researchers view lower class prejudice as a direct result of the increased competition between ethnic and racial groups at that level (Levin, 1975). Taylor (1998) reports that the immediate presence (hence potential competition) of African Americans increases prejudice among European Americans, but that nearby Latino and Asian populations do not produce the same reaction. Although most modern social scientists search for evidence in contemporary competitive situations, historical data probably have more to offer here.

Except for the very recent past, few Americans have been shy about expressing their prejudices. Historical data offer few worries about nonexpressed attitudes. History also provides the time spans necessary to view competition. We can see just which groups were in which places at which times competing for scarce resources. African Americans as slaves, for example, were not a competitive threat to European Americans because slavery prevented their competition. By contrast, Reconstruction brought a major threat as newly freed slaves changed into potential competitors. European American attitudes changed drastically, along with the stereotypes that supported them. African Americans were transformed from happy-go-lucky, singing, dancing, stupid, lazy people into razor-wielding, conspiring, thieving people who spent considerable time planning their next rape. When we see prejudice change so quickly and so drastically, combined with similarly drastic social change, the connection seems impossible to ignore.

Theories of Discrimination: Macro and Micro Approaches

The material in this section is organized to help the reader differentiate among the numerous theories of discrimination. We first focus on macro (or wide scale) theories. Many of these more general approaches form the basis of the micro (or smaller scale) theories discussed later. The macro approach emphasizes theories related to the original formation of ethnic or racial stratification. The micro approach contains theories that tend to take such ethnic stratification for granted, focusing instead on how dominant and minority group members deal with their respective circumstances. The questions that arise vary greatly from theory to theory.

The Formation of Ethnic or Racial Stratification

The presence of many ethnic groups around the world does not guarantee that any two or more will become involved in a system of ethnic stratification. First, ethnic stratification requires a multiethnic political entity (such as a modern state) before ethnic groups can compete within one stratification system. If every political entity contained only one ethnic group, ethnic stratification could not exist. Although this observation may seem obvious, keep in mind that such was largely the case many centuries ago when people seldom migrated and large scale invasions (with territory annexation) were not everyday occurrences. For ethnic stratification to be possible, either people have to move across boundaries (where they share a state with some other ethnic group) or the boundaries themselves have to move through invasion and/or treaty, engulfing diverse groups into a whole.

Second, a multiethnic political entity does not guarantee the formation of ethnic stratification. As we saw in Chapter 1, forms of assimilation and pluralism are possible outcomes. The former eliminates minority cultures while the latter maintains them; either way, however, the domination of ethnic stratification is not present. Sometimes ethnic groups initially occupy a lower position within ethnic stratification and then move into assimilation or pluralism over time; other times ethnic groups avoid ethnic stratification altogether. Before ethnic stratification can occur, it must be in the interests of a stronger, dominant ethnic group to create it. What creates those interests? Many theories attempt an explanation.

World-System Theory: An Overview   World-system theory is not specifically designed to explain racial and ethnic matters. It does, however, serve as a springboard for many other theories. Based generally on Marxist theory, world-system theory is both global and historical. It paints a theoretical picture of world history from the late fifteenth century to the present, showing how different parts of the world have become increasingly bound together both politically and economically.

Regarded as the father of world-system theory, Immanual Wallerstein (1974) bases his theory on European social changes that occurred between the late 1400s and the early 1600s—a time of pivotal change. This period was the beginning of world exploration and colonization by European empires, each attempting to outdo the others in land and wealth acquisition. It was a period of capitalism and economic growth such as the world had never seen. However, Europe's economic domination also required political domination.

Wallerstein conceptualizes economic and political expansion as the core and the periphery. The **core** is the central area of economic and political strength. This part of the world system bases its wealth on manufacturing and is dependent on less industrialized regions of the world for inexpensive labor and raw materials. Those less industrialized regions are labeled the **periphery** in relation to the core. Wallerstein observes that economic and political competition within the core requires core members to exploit peripheral resources if they are to succeed. Without a source of inexpensive labor and raw materials, core manufacturers cannot remain competitive. It is therefore in the interests of the core that peripheries remain peripheral, under the domination of outside interests.

In these terms, the American colonies were once peripheral to European nations. Later, regions within the United States became peripheral to other regions (e.g., the South supplied raw materials to the manufacturing North during the 1800s) (see Rokkan and Urwin, 1983). By the twentieth century, the United States was clearly a core region, beginning to look outside its boundaries for resources.

Having noted that world-system theory bears many similarities to Marxism, we might add that many of these same observations can be found under the headings of modernization theory, dependency theory, and internal colonialism theory. All three share world-system theory's emphasis on political and economic domination for the purpose of controlling labor and raw materials. This clearly affects geographically concentrated racial or ethnic minorities: if a given dominated peripheral region is home to one ethnic group, it automatically becomes a minority group. But can we apply any of this to circumstances in which many racial and ethnic groups share a region but only a few of such groups face discrimination? In such a case, the region itself would not be peripheral, but certain groups within it would appear to have peripheral status. As we will soon see, that leap is made by internal colonialism theory and dual labor market theory.

The Impact of Industrialization on Discrimination   One does not need to visit world-system theory to conclude that industrialization changes more than just technology:

- Technological change permits the mass production of manufactured goods. This creates a high demand for raw materials, laborers, and markets for those goods. Manufacturers must expand around the world to obtain enough of all three.

- The growth of factories forces the growth of cities, usually near water for convenient transportation. Manufacturing is more efficient when factories are located near each other and when laborers also are located nearby.

- This social change produces large scale labor migration as workers find less work in nonurban centers and increasing opportunities in urban centers. People who would otherwise never meet each other make contact.

- Large nation states often form from this process. Large business coupled with the international trade it requires is more efficient when based in larger and more powerful states. Industrialization has a tendency to move political boundaries as larger states form from many smaller ones.

Although these observations only scratch the surface of the social change that accompanies industrialization, they should provide a starting point for understanding how industrialization motivates the formation of ethnic stratification.

Wagley and Harris (1958) note many of the previous points in their effort to show how nation states lead to ethnic exploitation and conflict. They focus in particular on the large number of boundary shifts and growing migration of people. Their theoretical interests are as much cultural as economic, however. They note the idea of the *nation* state, which suggests not only a political entity but one with a distinctive cultural unity—a nation. Because almost all large states are multiethnic in the modern world, the only way to achieve such a result is through wide scale assimilation of minority ethnic groups into the dominant culture. Sometimes, they point out, pressures to assimilate result in smooth cultural transitions. At other times, however, those pressures result in ongoing ethnic conflict within the state. Such conflict strives toward pluralism and can just as easily move toward separatism.

Noel (1968) focuses his attention more on the economic competition that is increased through growing industrialization. Drawing data from American slavery, he notes the two end results of discrimination within a competitive setting. Discrimination can indeed occur as a tactic between two groups in competition as one attempts to gain an advantage, but this was not the primary case during slavery. European American farmers were competing with other European American farmers. Rather than attempting to lessen the direct competition by damaging each other, they sought to improve their advantage through discriminating against another group—Africans. The exploitation of African labor through slavery reduced labor costs and was a practical response to the competitive threat posed by other European American farmers.

Van den Berghe (1967) took this path a little further. He had a particular interest in the role of racism and racial categories that developed in the course of ethnic stratification. He notes that racism is most common when physical differences between groups coincide with vast differences in status and culture. Groups with such extreme differences are, of course, more likely to meet in the modern era with the increased migration of people. Once established, racism invites and supports massive exploitation such as slavery. Over time, it also adds rigidity to such systems, hindering the future upward mobility of groups so labeled.

Both Schermerhorn (1970) and Shibutani and Kwan (1965) incorporated industrialization into their theories in similar ways. Schermerhorn noted that symbiotic (as opposed to exploitative) relations between ethnic groups are much more likely to occur in nonindustrial societies. Friendly trading relations, for example, often occur between nomadic traders and sedentary agricultural peoples. Shibutani and Kwan itemize the many changes Europeans imported into their colonies, including money economies, taxes, private property, and regularized wage labor systems. The result of all these changes was an increasingly rigid set of economic relationships based on race or ethnicity.

The Role of First Contact Theorists also have paid considerable attention to types of migration and early contacts among ethnic groups. Schermerhorn (1970) categorizes migratory experiences according to the duress under which people migrate. He argues that greater freedom in migration leads to greater immigrant control over their future circumstances. Slavery (or the movement of forced labor) falls at one extreme as the least attractive form of migration for those being moved and tends to produce the most rigid systems of labor exploitation. The huge power inequality that allows the forced movement in the first place leaves open many doors of discrimination down the road.

At the other extreme are voluntary immigrants. As we have seen in American history, many voluntary immigrants believed they had little choice in their decision, often fleeing threatening economic or political situations. Still, voluntary immigrants tend to arrive at their destinations with far more freedom to develop and carry out survival strategies. Discrimination may be common, but the rigidity of ethnic stratification is usually much less apparent. Social mobility for future generations is much more likely.

Lieberson (1961) views the same process from the perspective of relative power between the "host" ethnic group and the migrant" ethnic group. Most European immigration to the United States, for example, involved weaker migrants joining stronger hosts. In such circumstances, the host is less threatened and is often less oppressive—a circumstance more likely to result in assimilation for the migrant group. By contrast, a strong migrant group joining a weak host group tends to produce high levels of conflict, more rigid forms of ethnic stratification for the host group, and often a numerical decline among the hosts. An example of this is the original European immigrants to the New World in their confrontations with the Native American population. For Lieberson, getting off on the right or the wrong foot often sets the future of ethnic relations.

You may have noted some redundancy among the theories we have discussed. These similarities stem from the overwhelming importance of power and economic change in creating multiethnic societies with ethnic stratification. However, different theorists provide unique slants on this issue, and they augment each other well in filling out different aspects of global change; at the same time, they all need to touch many of the same bases.

Responses to Ethnic Stratification: Micro Perspectives on Ethnic Relations

The theories we have examined so far are not only global in scope, but also they tend to focus on why ethnic stratification occurs in the first place. The theories included in this section differ in that they emphasize how people behave *after they find themselves* in an ethnic stratification system. Just as social stratification in general produces competitive social behavior, individuals who find themselves within ethnic stratification must take additional social factors into account before they act. Since discrimination is in place, they must assess the degree to which they benefit from it or are injured by it. You have to measure the slant of the playing field before you can design an intelligent play.

As noted earlier, expect various questions to be posed from among these theories. Some pursue dominant group activities, attempting to track how their members are affected by discrimination. This is obviously important because dominant group membership does not guarantee personal success. Most of these theories focus more on minority group activities as they attempt to either level the playing field somewhat or design shrewder plays. Although cultural concerns weave in and out of these theories (particularly with theories on enclave economies), the primary concern of all is ethnic economic competition.

Enclave Economies, Middleman Minorities, Segmented Labor Systems, and Ethnic Economies   The theories examined here are most commonly called (1) enclave economy theory (Portes and Manning, 1986), (2) middleman minority theory (Bonacich, 1973), (3) segmented labor systems (Hechter 1978; Hechter and Levi, 1979), and (4) ethnic economy theory (Light et al., 1994). They have three common features. First, they all attempt to explain the relative economic success of some minority groups through the high levels of entrepreneurship among their members. Second, these minority group businesses are clearly ethnic, from the owner down to the

The "Little Havana" section of Miami is an example of an enthnic enclave.

lowliest employee. In many cases, the businesses are run by an extended family. Third, minority group members involved in such activities are generally less dependent on the dominant group and therefore are more insulated from discrimination by the dominant group. The following list illustrates the wide variety of actual cases that meet these criteria:

- The development of Chinatowns in the United States during the last half of the nineteenth century and throughout the twentieth century

- The specialization in the laundry business (serving largely European Americans), first by Chinese immigrants and later by Japanese immigrants in California

- The domination of clothing manufacturing by both German Jews and Eastern European Jews in the late nineteenth and early twentieth centuries

- The creation of small, labor-intensive truck farms by Japanese immigrants in the 1920s and 1930s on the Pacific Coast

- The Amish farms of Pennsylvania

- The creation of "Little Havana" in Miami by Cuban immigrants from the late 1950s to the present

- The creation of small stores—many grocery and liquor stores—by Korean immigrants since the mid-1960s, serving largely African American customers

- The establishment of restaurants, hotels, and other medium-sized businesses by Asian Indians since the mid-1960s

Are these examples all the same? According to the three dimensions previously noted, they are similar, but let us look at some of the important differences.

First, notice the range of economic ventures from the tight-knit ethnic community structures of Chinatowns to the Korean example in which individual families live and work far from ethnic communities. The former circumstance tends to place a cultural cocoon around the minority business, which may well minimize efforts to assimilate. The latter example forces the minority business owner to interact outside his or her ethnic group to achieve success.

A second and related point is that the presence of a strong ethnic community increases the potential of having co-ethnic customers. Not only does this minimize potential conflict between the owners and customers, but also it can offer the possibility of a protected clientele. Ethnic communities almost always contain businesses that cater to specific ethnic needs such as ethnic grocery stores and restaurants. Because those goods and services are not available outside the ethnic community, the entrepreneur faces less competition.

Third, those minority business ventures that operate outside ethnic communities can vary greatly in the degree to which they compete with or complement dominant group business interests. Japanese truck farmers did not compete with California agriculture in the 1920s, and Koreans face little competition in operating most of their stores today. The Jewish control of the clothing industry filled a need for ready-to-wear clothing that was not currently being met elsewhere. It is possible to avoid competition with the dominant group outside the confines of a closed ethnic community.

Fourth, strong ethnic communities tend to be economically diverse, paralleling the production of goods and services found outside the community. It is possible, in short, to live one's entire life inside the community without ever venturing out. By contrast, a minority specialization (or "niche") in one area of the economy requires a much greater economic integration with the dominant society. Jews in the clothing business, for example, bought cloth from dominant group–controlled factories and sold clothes to dominant group customers. They also had to buy everything they needed outside their ethnic group. This is a very different strategy from a Chinatown or a Little Havana; it also produces very different results down the road.

Fifth, the strength of culture in ethnic communities contains a potential downside for residents. The support and economic gain that stems from the enclave economy may leave working class members of those communities with few options. Because they can depend on a minority culture, they may well not gain enough experience with the dominant culture to live and work elsewhere. When their options are limited to employment within the ethnic enclave, they become vulnerable to labor exploitation from co-ethnic employers (see Portes and Jensen, 1989).

Researchers in this theoretical area have focused primarily on the conditions under which such economic strategies occur and the degree to which such strategies benefit minority group members involved. The most important necessary preconditions found in most research are prior minority experience in entrepreneurship and the availability of capital (Light and Rosenstein, 1995; Light et al., 1994; Portes and Manning, 1986). As for benefits, research on Chinese, Japanese, Cubans, Koreans, and Iranians in the United States all report short-term and long-term economic advantages for those involved; individuals working outside strong ethnic enclaves generally faced more short-term problems but experienced positive long-term results (Bonacich and Modell, 1980; Jiobu, 1988; Light et al., 1994; Min, 1990; Pérez-Stable and Uriarte, 1995; Portes and Manning, 1986).

The Dual Labor Market: Discrimination and Exploitation   The dual (or split) labor market refers to ethnic or racial specialization in different sectors of the economy. Just as many types of employment are dominated by either men or women, other types are dominated by one ethnic group or another. So far, this should sound quite familiar. How is this perspective different from enclave economy and middleman minority theories? The theories of the previous section apply to minorities that voluntarily (more or less) adopt that strategy and benefit from it. The dual labor market perspective applies to minorities essentially forced into specific economic sectors because of discrimination and who subsequently do not benefit. Indeed, this distinction is almost always present in the ways these theories are applied in actual research.

What concrete examples would fit under the dual labor market? We look at employment segregation in much more detail in Chapter 9. For the present, consider the people you might expect to meet working as farmworkers, janitors or housekeepers, skycaps, and personal servants. Depending on where you are in the United States, their race or ethnicity will vary, but if you are in a region with a sizable proportion of nonwhite Americans, you will no doubt see many of them in these occupations. In general, low-level service occupations in the United States are increasingly becoming the employment sector for minorities.

Internal Colonialism. As we have seen, similar theories often appear under different names. At the same time Edna Bonacich (1972) was introducing the split labor market, Robert Blauner (1972) proposed the internal colonialism theory for understanding discrimination. Borrowing from a long tradition of global theory (neatly summarized by world-system theory), Blauner argues that the dependent status of former European colonies is remarkably similar to the contemporary treatment of minorities in the United States. Just as members of those previous colonies lived with employment discrimination, rampant prejudice, and social isolation, contemporary African Americans, Hispanics, and Native Americans receive much the same fate. Although many of these peoples are not geographically concentrated within the United States, they are nevertheless socially concentrated. Blauner suggests we not lose sight of this larger picture.

This perspective was quite popular in the 1970s. Rising minority expectations throughout the 1960s led to high levels of ethnic consciousness, high awareness of discrimination, and interest among many minorities in pluralism and ethnic nationalism. These changes occurred within the United States but were more pronounced around the world. Hechter (1975) and Nairn (1977) turned this perspective to the study of England's "colonies" as ethnic politics arose in Wales, Scotland, and Northern Ireland. Hechter introduced the term "cultural division of labor" to describe the clear ethnic divisions in the economy. Nairn prefers the term "uneven development."

The point of these theories is to call attention to the combined factors of discrimination and social isolation of ethnic minorities. Minorities that work separately and live separately build strong bonds with one another. Those bonds heighten their overall ethnic awareness, their awareness of the discrimination they face, and their anger directed at the dominant group. One direction for this anger is the formation of ethnic nationalist political movements.

While Hechter and others focus on the formation of ethnic nationalism, Bonacich has different questions about the dual labor market. Given the fact that the dual labor market is produced by discrimination, there is no question that minorities so affected are placed at a disadvantage. But what about dominant group members? Do they benefit, and if so, how much? More importantly, do middle-class and working-class dominant group members both benefit from that discrimination? When we looked at the discrimination in ethnic stratification in Chapter 2, it was argued that all middle-class individuals benefit both in economic security and a lower cost of living from labor exploitation. It is time to examine the basis for that argument.

The Dual Labor Market and Capitalism: Who Benefits? At first glance, a dual labor market produced through discrimination would seem irrational for a capitalist economy. An employer should seek the lowest possible labor costs to increase profits. If discrimination prevents a sizable section of the labor force from particular employment, the resulting reduced labor force would be a scarce commodity and could command higher wages. Capitalism with a truly free market would have no discrimination (Lundberg, 1994). Thus, the upper classes within the dominant group should seek to prevent discrimination. Since they obviously do not do that, we need to take another look at how capitalist economies operate.

We have seen that discrimination is economically beneficial to the upper classes in a capitalist economy. Discrimination decreases *some* labor pools but simultaneously increases others. If employers do have to pay some workers more, they can simultaneously benefit from other workers who have limited employment options and are thus vulnerable to exploitation. This was extremely important during the early days of the United States when labor was scarce everywhere; only discrimination would have produced a labor force at all. But the same principles apply in the contemporary United States. A lot of low wage work still needs doing and only workers with few choices will do it.

In addition, discrimination provides employers with a "second string" of workers should their current employees become obstinate or scarce. The historical use of racial minorities as strikebreakers in the United States is a classic case in point in the first instance (see Whatley, 1993). As for scarcity, that same pool of minority labor can be temporarily upgraded during times of severe labor shortages. An example would be the hiring of southern African American workers for northern manufacturing work during World War I and World War II.

A more difficult question concerns whether dominant group members in the middle and working classes benefit from discrimination. Szymanski (1976) argues that discrimination against minorities works to the disadvantage of *all* working class people. Such discrimination, he argues, prevents labor union formation that would strengthen the working class as a whole; in addition, he finds that the *absence* of discrimination increases European American income and lessens income inequality among European Americans.

Tomaskovic-Devey and Roscigno (1996) offer a more complicated picture with the help of historical data from the American South. While all researchers agree that upper class (or elite) members of the dominant group benefit from discrimination, they suggest that the nature of this elite determines whether working class dominant group members benefit. Specifically, the incomes of working class European Americans dropped in their analysis only in areas dominated by a few, large industries. The more diverse the economy, the more European American working class individuals benefited from discrimination.

Most research in this area focuses on relative levels of income. As was just stated, discrimination lowers dominant group working class income only under certain circumstances. If we expand our considerations of dominant group benefits to include such factors as a lower cost of living and the psychological benefits associated with favored ethnic status pride, a less confusing picture emerges. Most members of the dominant group prefer the status quo of discrimination because it is usually in their interests to do so.

Ethnic Competition Theories We come finally to a collection of theories with different names but many common assumptions and characteristics. These theories— loosely termed *ethnic competition theories* here—differ strongly from dual labor market theories in that exploitation largely drops out of the picture as an important cause of behavior. Indeed, the ethnic competition theories more often find that the *least* exploited members of a minority group may be the most politically active and maintain the strongest ethnic self-consciousness. On the other hand, ethnic competition theories do have much in common with enclave economy and middleman minority theories.

Ethnic competition theories tend to focus on specific research questions. They seek to explain (1) why ethnic consciousness changes or becomes stronger and (2) why a stronger ethnic consciousness may result in ethnic political organizations that seek to change the status quo. The influence of Fredrik Barth (1969; see Chapter 1) and his emphasis on why ethnic boundaries change is evident. A basic assumption guiding these theories is that social stratification causes competition but ethnic stratification often makes ethnically based *group* competition the most rational route to upward mobility. Most of the social conditions that make such a choice more rational are inherent in the many changes brought by modernization and industrialization.

These theories appear under various names, changing according to what forms of political organization most interest the theorist. A political organization could be a pressure group or an organized political constituency within a democratic society. Political organizations also can have the goal of political separatism and may employ violent strategies to achieve that end. Whatever form the organization takes, however, ethnic competition theories tend to view these organizations as springing from many of the same social circumstances. The following theories share most of these traits: resource competition theory (Olzak and Nagel, 1986), subnationalism theory (Ragin, 1980; 1986), elite competition theory (Brass, 1985), resource mobilization theory (McCarthy and Zald, 1977), and some instances of world-system theory (Hannan, 1979).

The assumptions common to these theories connecting modernization to ethnic consciousness and political organization can be found among the theoretical propositions listed in Box 3.1.

As should be obvious from the information in the box, ethnic competition theories view both ethnic consciousness and political organization in two ways. First, that consciousness and organization both increase or decrease in direct response to other social change. Neither can grow stronger in the wrong social environment. Second, they both can become conscious and rational strategies for minority group members—especially minority elites—in the right social environment. In general, upper class minority elites in any society tend to be the most assimilated—the price generally paid for such mobility. But these same people may discover that they can increase their economic and political clout even more by working to organize poorer members of their ethnic group. Those poorer members may not be politically astute and they are not likely to have much time or energy to invest in politics. Their involvement can, however, increase the size and importance of a political movement.

Ethnic competition theories allow researchers to investigate areas of race and ethnic relations sometimes ignored by social scientists. For example, race and ethnic relations are often viewed as a one-on-one conflict between a dominant group and a minority group. Ethnic competition theories allow us to examine many ethnic

BOX 3.1: ETHNIC COMPETITION THEORIES

A. Overall Societal Economic Development

1. Economic development (modern industrialization) in a society increases contact among different ethnic groups and often decreases the cultural differences among them. In spite of the decreasing cultural differences, the increased contact and competition promote ethnic conflict (Barth, 1969; Hannan, 1979; Olzak, 1992).

2. Economic development creates larger economic organizations as corporations grow in size and national governments become more involved in the economic sector. Growth in the economic sector promotes a parallel growth in the political sector with the formation of larger political organizations. This applies to all such organizations, whether ethnically based or not (Hannan, 1979; Nagel and Olzak, 1982).

B. The Impact of State Policies

1. Centralized decision making in larger and more powerful states can be threatening to regional differences within such states. This increases regional identity and political activity. If ethnic groups are regionally concentrated, that political activity is ethnically based (Nagel, 1986; Nagel and Olzak, 1982).

2. The official recognition of ethnic differences in state policies (e.g., affirmative action in the United States) increases ethnic awareness and competition, promoting ethnic political organization (Nagel, 1986; Nagel and Olzak, 1982).

3. State policies regarding official languages increase awareness of language differences within the population and awareness of ethnic differences. Those speaking unofficial languages will be motivated to respond politically (Brass, 1985; Nagel, 1986).

C. Economic Development in Peripheral Regions

1. If peripheral regions experience economic development and are no longer limited to the production of raw materials, barriers to social mobility are removed and regional ethnic groups are likely to respond politically as they realize potential opportunities (Olzak and Nagel, 1986).

2. Economic development in peripheral regions produces new resources that tend to become most available to regional ethnic elites through ethnic political organization (Brass, 1985; Nagel and Olzak, 1982).

D. Urbanization

1. Urbanization increases contact among different ethnic groups, which makes ethnic differences among people more obvious. This results in increased ethnic consciousness (Bonacich, 1972; Olzak and Nagel, 1986).

2. Urbanization increases the concentration of ethnic group members, which improves communication and makes political organization easier (Brass, 1985; McCarthy and Zald, 1977; Nagel and Olzak, 1982; Olzak, 1992).

minorities simultaneously. We can investigate conflict within one ethnic minority group, between two or more ethnic minorities, or among all ethnic minorities and the dominant group.

Ethnic competition theories address the difficult theoretical question of why upper class or elite members of a minority group are motivated to both risk what they have attained *and* identify with an ethnic group from which they seem to be far removed. From the perspective of ethnic competition, we can see that they gain personally from such political organization.

Summary

All social sciences approach their subject matter with scientific theories and concepts. Theories are attempts to explain or predict the behavior of the subject matter. Specifically, they go about this process through creating concepts and specifying the relationships among them. Concepts are intellectual constructs that identify (or direct our attention to) certain similarities among different observations. The concept *ethnic group*, for example, calls our attention to shared cultural traditions and/or ancestry among a group of people while simultaneously ignoring all the individual differences among those group members. Because a wide variety of human behavior occurs in relation to ethnic and racial distinctions, the theories social scientists employ in this area differ widely.

Theories of prejudice have roots both in psychology and sociology. The idea of an authoritarian personality, for example, stems from the former in predicting that people preoccupied with toughness, authority, and obedience tend to develop prejudiced attitudes. Other psychological theories link prejudice to the degree to which individuals value or rate their own behavior against others near them rather than against a fixed standard of excellence.

Most sociologically based theories focus on the social position of those with prejudiced attitudes. Poorly educated and low-income individuals tend to hold more prejudiced attitudes than those with higher education and incomes. Some theories link this outcome to a lack of critical thinking abilities among those poorly educated while others connect it to the greater competitive economic threats faced by such people.

Theories of discrimination and the formation of ethnic stratification are more complex and tend to dominate the general study of race and ethnic relations. World-system theory—a Marxist based perspective—links discrimination and ethnic stratification to the expansion of capitalist Europe and its search for cheaper raw materials and new markets. Related theories are modernization, dependency, and internal colonialism theories. All emphasize the advantages for expanding corporations to have large subordinate regions and/or minority ethnic groups to provide necessary inexpensive labor. Some focus more specifically on competition between and among ethnic groups and the role of first contact between and among ethnic groups.

Once ethnic stratification systems have been established, other theories help explain how such systems create specific minority and dominant group behaviors. These theories focus on the ways minorities attempt to circumvent their subordinate status through effective economic strategies. Enclave economy theory notes the competitive advantages obtained through tight-knit urban ethnic communities in which ethnic business development creates employment opportunities for co-ethnics. Middleman minority theory explains minority success through high levels of entrepreneurship in small businesses that mediate between two or more other ethnic groups but compete with neither.

One area of theoretical discussion lies in the potential advantages or disadvantages for the dominant group of ongoing discrimination against a particular group (the dual labor market). Although the previously noted theories emphasize the advantages of such discrimination through the creation of cheap labor, other theorists argue that poorer members of the dominant group also suffer in the job market from discrimination against minorities. Although research in this area has produced differing results, working class dominant group income appears to drop when coupled with minority discrimination only under specific circumstances.

Ethnic competition theories seek to explain (1) why ethnic consciousness changes or becomes stronger within ethnic stratification systems, and (2) why a stronger ethnic consciousness may result in ethnic political organizations that seek to change the status quo. They conclude that both consciousness and corresponding political organizations will grow with increased societal economic development, increased urbanization, and the recognition of ethnic differences in state policies.

II

The Cast of Characters: Entrances and Exits

The history of United States immigration is distinctive in many ways from other historical subject matter. First, we will focus on the topic of human migration. Who are all these people, and why did they come to the United States? Some were already here, and others were forced to come. Still others faced starvation if they did not move. We will have to account for all of them. Second, North and South America contained much unpopulated land from the 16th century through the 18th century. When your potential labor force has an option to farm their own land instead of working for you, you will have to be creative to find workers. And finally, we will need to pay some attention to why we are looking at history in the first place. The primary purpose of this book is to understand race and ethnic relations in the United States *today*. How can looking toward the past help us do that?

Migration: Who Moves and Why?

Human migration is usually conceptualized in two basic categories—push and pull factors (Ravenstein, 1889; Hoerder, 1994). **Push** refers to social factors that encourage people to leave their current homes and move elsewhere. In general, the most important "push" factors are economic. If the way you have traditionally survived no longer seems to work very well, you will definitely need to experiment with alternatives, and one such alternative is moving. Another important push factor is politics. There may something about your personal behavior (or your racial or ethnic group membership) that is bringing the wrath of your current government down upon you. Perhaps you have been involved in some revolutionary activities. Perhaps your government just doesn't like your ethnic group and chooses to tax you more than other citizens (a combined economic/political push factor). Perhaps your group is being systematically put to death. Being in political disfavor is less common than struggling for survival but, when it occurs, it can be one of the strongest of all push factors.

Pull factors share the economic and political facets with push factors, but the force here is attraction rather than repulsion; push factors include any and all reasons why relocating to a *particular* new location might be beneficial. A classic pull situation would be the California gold rush of 1849. It attracted people from all over the world who hoped to become rich quickly and easily. Other pull situations would include less striking economic opportunities, such as cheaper land or more plentiful jobs with higher pay. Political pull situations might include societies known for religious or political tolerance. The United States has clearly been an attractive destination for all of the above reasons. Since 1500, sixty million Europeans left Europe for a variety of push factors; the great majority of those people chose the United States as their destination (Daniels, 1990).

Abundant Land and Not Enough Labor: The New World Dilemma

A review of some basic economics might be in order here. Imagine that you manufacture shoes. You can only wear so many yourself, so you naturally hope to sell the rest to your barefooted neighbors. How much can you charge? The answer to that question depends on two things. First, just how barefooted are your neighbors? If they are truly shoeless, they will beat a path to your door. That circumstance allows you to raise your prices. Economists call that *demand*. But there is a second factor at work here. How many other people in your community are also making shoes? If there are several, your shoeless neighbors will be able to beat paths to many doors and you will find fewer at yours. That circumstance tends to push prices down. Economists call it *supply*. As you can see, the two operate at cross purposes; they have opposite effects on how much you can charge.

If we watch your shoe sales over time, we will see another aspect of the supply–demand relationship. Let us assume that you are the only shoe manufacturer around, and your neighbors truly need shoes. This situation of high demand with low supply means you can charge more, and you probably will, unless you are making shoes for your health. But is it not likely that some other people might notice your growing riches? Won't some of them take up shoe manufacturing themselves to cash in on this bounty? Of course they will. And the more people who begin making shoes, the more supply goes up. If people still need the same number of shoes but can now buy them at many different places, the sellers will have to charge less to attract business, and prices will go down. If they go down far enough, some people might stop making shoes altogether and prices might edge up some. Ultimately, some kind of balance might occur.

The above description of economic forces goes back many centuries. You could read about it in Adam Smith's (1776) classic, *The Wealth of Nations*. Economies of course, are affected by many other factors in addition to simple supply and demand. Also, the twentieth century has witnessed a wide variety of governmental experiments designed to smooth out the edges of the market. After all, we don't want either shoe-less people or starving manufacturers. But the basic rules of supply and demand still apply. The question here is, how do they apply to questions of land and labor?

To begin with, think about labor as if it were shoes. You can sell your labor on the market just as you can sell a pair of shoes. And just as the price of shoes varies according to supply and demand, so does the price of labor. If everyone is looking for workers and few people are looking for work, those few that are looking can charge more. If more prospective workers appear, the wage rate will drop. In the early days of the European colonization of North and South America, those who wanted to hire workers—in agriculture or mines, for example—faced a major problem. First, there was a vast shortage of labor to hire. Native Americans were either not interested or were dying of diseases imported from Europe. That left European immigrants as potential workers. There were not all that many of them in the early days, and there was also a second problem: There was so much free or cheap land available that working for someone else seemed pointless when you could work for yourself. Potential employers therefore faced a frustrating situation. They could make a lot of money on this new land if they only had the labor, but the pool of workers was just not there (Salinger, 1987; Engerman, 1986).

A dilemma such as that described above brings us to one of the many ways that people can intervene with the laws of supply and demand. Those laws apply to labor when people are free to choose how and where they work. What if they are not free to choose? Of course you know where I am going with this, as virtually everyone knows something about slavery. But have you ever thought about it in terms of the supply of, and demand for, labor? If there is enough money to be made and free labor is just not available, it may make economic sense to capture people many thousands of miles away and import them under guns and whips to do your bidding. But you will be reading about far more than slavery in the coming chapters. There are many degrees between totally free labor at one extreme and slaves with no rights whatsoever at the other extreme. Many Europeans, for example, were brought over to the New World to supply labor under contract to various masters. It was not as bad as slavery, nor was it permanent, but it was a solution when free labor was not available. Much of American history involving immigration requires us to focus on this relationship between employer and employee. Workers in different situations enjoy different degrees of freedom of choice, and employers can often limit that freedom when necessary. And no, this is not ancient history. It continues to the present day.

4

Early Colonial Development and the Early Republic: 1500–1845

Although populated in the same way as most other land masses—human wanderings—North and South America are unique in the human history of this planet. They were populated more recently than any other continent, and the first wanderers almost certainly walked from Asia to Alaska when they were still connected by land. Since then, both continents have been cut off by water from the rest of the world (see Guerra, 1993). For thousands of years, they were isolated from new peoples and animal species, and the bacteria and viruses they carry with them. They were introduced to no new lifestyles or technologies except for the homegrown variety. All of this made 1492 significant, not just for the European adventurers who came up short in their search for India, but also for the native peoples they encountered in what is now the Caribbean. The Europeans had to alter their trading plans, replacing the spices they hoped to acquire with gold and other riches of the "New World" they had "discovered." But the native peoples would be finding their own adaptation less lucrative and much more painful.

Native "Americans" in the "New World"

Columbus called the native peoples he met *Indians*, either in honor of his original destination or more likely because he was lost. The islands in the Caribbean later came to be called the *West Indies*. Although the term "Indians" is still generally used, we achieve a clearer sense of American ethnicity with the term *Native Americans*. Because our study later introduces us to European Americans, African Americans, and Asian Americans, a categorization by continent of origin is logical. Referring to the Native

Americans' land as the *New World* is appropriate because although it was new to European explorers, it also became new and almost unrecognizable to its native inhabitants.

First Encounters: Native Americans and Europeans

In 1492, hundreds of Native American ethnic groups occupied the lands of the future United States. Upon the Europeans' arrival, these groups were engaged in a wide variety of survival techniques under an equally wide array of societal structures. The eastern forests were home to hunters and farmers, the northwest offered fine fishing, and the southwest included wide scale agriculture and herding (Quintana, 1990; Wright, 1981). Almost all groups were involved with trading as evidenced by the existence of pidgin-like languages, such as Native American sign languages, used to circumvent language boundaries. Some ethnic groups were organized with strong hierarchical authorities, while others were democratic in structure. In many cases, Native American women were given far more responsibility and respect than women in any European society of the day (although this was perhaps not difficult to accomplish) (Wright, 1981; McPike, 1991).

With so little European labor available in the sixteenth century, many Europeans turned to Native Americans as a labor source. The Spanish were already experimenting with the forced labor of Native Americans in 1503 with the *encomienda* system in the West Indies (Reich, 1994). Much like serfs in the Middle Ages, Native American workers were essentially tied to the estate on which they worked and to the overlord to whom it belonged. The Spanish later spread this system throughout South America. They also introduced African slaves at about the same time.

In English North America, efforts were made to convince Native Americans to live and work in small communities sometimes called "praying towns" (where work and religious conversion would hopefully be linked). When slavery appeared in the British colonies toward the late 1600s, efforts were made to enslave Native Americans. The colony of South Carolina had the most Native American slaves (1,400 slaves as compared with 4,000 European inhabitants in 1708), but disease and escape made native peoples unsuitable for this purpose (Daniels, 1991). In addition, many of the eastern tribes were still too strong and powerful in the Carolina colony to be enslaved. But even though slavery within Carolina was kept somewhat low, Charleston merchants maintained a thriving Native American slave trade, shipping those slaves outside the colonies. They operated almost identically to the Atlantic slave trade by trading with coastal Native American peoples who would then conduct inland slave raiding expeditions for them (Grinde, 1977).

Native Americans were stereotyped in many ways. Some of these stereotypes came from the insecurity felt by the first Europeans who were outnumbered by native peoples. Read between the lines in this early comment from an English colonist:

> The Aboriginal Americans have no Honesty, no Honour, that is, they are of no Faith, but meer Brutes in that Respect. They generally have great Fortitude of Mind; without any Appearance of Fear or Concern, they suffer any

Torture and Death. In Revenge they are barbarous and implacable; they never forget nor forgive Injuries; if one Man kills another, the nearest in Kindred to the murdered, watches an Opportunity to kill the Murderer; and the Death of one Man may occasion the Deaths of many; therefore when a Man is guilty of Murder, he generally leaves the Tribe, and goes into a voluntary Kind of Banishment. They are a sullen close People. The Indian Wars ought to be called Massacres, or inhumane barbarous Out-rages, rather than necessary Acts of Hostility (quoted in Douglass, 1972:155).

Disease and Warfare

Estimating the numbers of Native Americans in 1492 lures historians and geographers into a gambler's paradise. Obviously, there was no census and very few of the ethnic groups had a written language. Records of European explorers are also of little help because the diseases they brought with them killed unknown numbers of Native Americans, and by the time those explorers got around to counting, untold numbers were already dead. Low estimates for both continents run about 8.5 million with approximately 1 million living in what is now the continental United States (Dobyns, 1966; Thornton, 1987). More accepted (moderate) estimates of the population of the continental United States raise the figure closer to 5 million, with some as high as 8 million (Lewis, 1993; Thornton, 1987). High estimates include those by Dobyns (1966) who suggests that both continents may have contained over 80 million residents in 1492. In his more focused research, Dobyns (1983) estimates numbers between 750,000 and 900,000 for Florida alone. Jennings (1992) proffers a population of 250,000 in greater Tenochtitlán, the Aztec capital city.

Where do these figures come from? We do have some reports from the very earliest of European explorers, some of whom noted tremendous population changes upon return trips to the same area (Dobyns, 1983; Sterba, 1996; Thornton, 1987). These stories are augmented by matching the technology of a particular ethnic group to the potential productivity of its environment (Dobyns, 1983). Archeologists can provide needed data by unearthing Native American towns or cities and examining skeletons for evidence of particular diseases; however, the massive death rates suggested by Dobyns have not been supported (see Blakely and Detweiler-Blakely, 1989). Whatever the numbers, arguing over them is a little like trying to distinguish between a medium-sized and a large holocaust.

Most Native American deaths were caused by diseases introduced by Europeans. The most lethal new diseases (in approximate order of lethality for Native Americans) included smallpox, measles, influenza, bubonic plague, diphtheria, typhus, cholera, scarlet fever, and malaria (Dobyns, 1983). These diseases were devastating to Native Americans because they had no built-in resistance; European Americans, who had lived with these diseases for years, had some biological defenses. In a European community, a smallpox epidemic could cause over 10 percent mortality, whereas in a Native American community, mortality rates were often over 90 percent (Thornton, 1987). European communities always had members who had once contracted the disease and survived; they were around to tend the sick and helpless. But in Native

American communities, everyone was sick at the same time. Even those who might have survived did not for lack of care (McNeill, 1976). Europeans noticed these epidemics, of course. Documented reports show that smallpox-contaminated blankets were donated to Native Americans to help spread the disease (McNeill, 1976; Thornton, 1987).

From 1500 to 1700 massive numbers of Native Americans died from disease introduced by traveling and settled Europeans, especially those who settled. Native Americans living on the eastern coast of North America were some of those hardest and earliest hit by this curse. Estimates are that the population of Massachusetts Native Americans dropped from 24,000 to 750 by 1638 (Wood, 1991). Not looking a gift horse in the mouth, Puritan leader John Winthrop commented, "The natives are neare all dead of the small Poxe, so as the Lord hathe cleared out title to what we possess" (quoted in Wood, 1991:96). Dead Native Americans, after all, do not inconveniently occupy real estate.

This time period also witnessed many small scale wars between the English and eastern Native Americans. Because the English were outnumbered, skirmishes were few in the early days, but increased immigration coupled with disease made warfare a more reasonable undertaking by the late 1600s. Still, the English were concentrated on the coast. Most eastern warfare would wait until after the Revolutionary War. In the meantime, French, English, and Spanish colonists often looked to Native Americans as allies as these imperialist countries fought each other (Worcester and Schilz, 1984).

European Immigration and Domination

The New World attracted the same European countries already involved in colonizing other parts of the world. North and South America joined Africa and Asia as real estate ripe for the picking. The Americas' uniqueness, as already noted, was its lack of an indigenous population suitable for a labor force. (It is important to keep in mind that the primary interest in colonization is not settling but making money.) The main players in the New World stakes were England, Spain, and France, with Portugal running close behind. Holland showed some interest, but its New World colonies never came close to matching its efforts around the globe. If we focus just on North America, we leave behind the Portuguese colonization (primarily Brazil) along with most of the Dutch settlements. All of these European countries came to have possessions among the islands of the Caribbean. On the mainland, the English began on the east coast, the Spanish started their explorations more to the west and south, and the French located in the north (although they fanned out considerably from there, particularly into the inland areas of what is now the United States).

From the British Isles: Puritans and Aristocrats

English interest in the New World came relatively late compared with other European countries, but since we know the end of the story, this is as good a beginning as any. England began on the mainland by laying claims to blocks of oceanfront property, at times stepping on toes other than Native American. These blocks (or grants) of land were typically granted to "companies" that then became responsible for settling the land with hopes of turning a profit. Thus, we find the Virginia Company of London making the first inroads at Jamestown (1607), followed shortly by settlements in what is now New England by the Virginia Company of Plymouth (soon to be the Massachusetts Bay Company). Although all of these settlers were English, they came for different reasons and had decidedly different lifestyles. In Puritan New England, settlers held strict religious beliefs and operated small farms or businesses. The Puritans were actually a numerical minority in New England, even in the early days, but they made up for their small numbers with enthusiasm. The area did not contain resources that immediately lent themselves to business enterprise. Timber was plentiful, for example, but wood is a little heavy to export (Bailey, 1990). This region later found ways to become wealthy, but not during the 1600s.

By contrast, the Chesapeake Bay area (generally Virginia and Maryland today) was populated by English nobles and businessmen, all of whom had hopes of making the new colony pay dividends to investors (themselves being first in line). Not surprisingly, the lack of a colonial labor force was first felt in the southern colonies; we shall shortly encounter their solutions. All the early colonies were slow to grow, however, facing high death rates among the colonists and harder conditions than were anticipated (Figure 4.1).

In the last half of the 1600s, the Carolina colony was founded; it was composed of pieces from today's North Carolina, South Carolina, and Georgia. The seaport of Charles Town (today's Charleston) was founded in 1670, giving the colony its hub for years to come. Finally, Pennsylvania got its start in 1681. Unlike any other colony, Pennsylvania soon collected a wide variety of nonmainstream religious groups who learned the economic possibilities of the New World.

Throughout the seventeenth century in all the English colonies, life was hard and creature comforts were few. Travelers sometimes commented that the rich and poor were not all that far apart because even the wealthier lived in roughly constructed houses and wore homespun clothes. In general, colonial health was poor, transportation was slow, and work was hard (Galenson, 1991). By 1649, for example, only 15,000 people lived in the Virginia colony. Yet the beginnings of wealth and power were already present before the century ended. By the time of the Revolutionary War, the agricultural elite of Virginia and other southern colonies were well housed and clad. From their numbers came leaders such as George Washington, Thomas Jefferson, James Monroe, and James Madison.

A Broader Base of European Immigration

The growth of the Pennsylvania colony brings us the first interesting change in European newcomers. In 1681, a colony charter was granted to William Penn, an English Quaker. Quakers began in England when George Fox began preaching in 1649, ultimately forming a new sect within Protestant Christianity. By 1665, he had 80,000 followers (Reich, 1994). In spite of its growth (or maybe because of it), Quakerism was

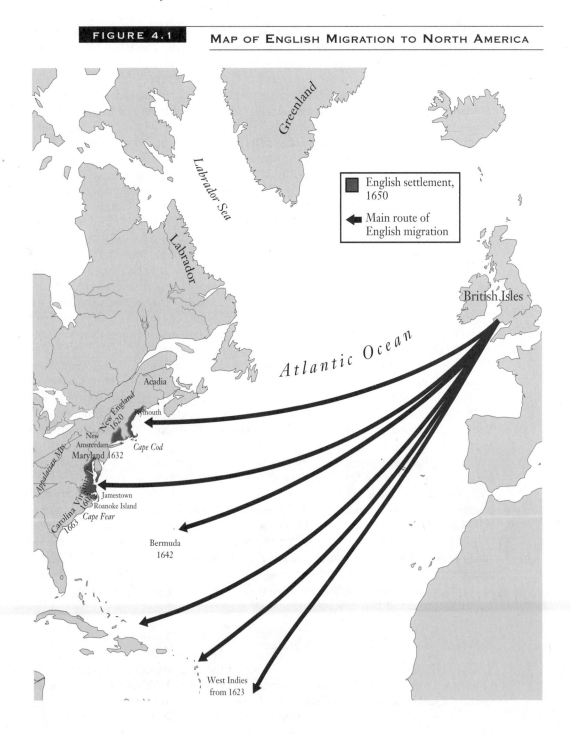

FIGURE 4.1   MAP OF ENGLISH MIGRATION TO NORTH AMERICA

perceived as a radical and strange religious cult in England. Penn's hope was that his new colony would provide a persecution-free environment for Quakers. Beyond that, he hoped that Pennsylvania might offer the same promise to other religious minorities.

As Quakerism was developing as a new offshoot of Protestantism in England, similar religious movements were growing in the German states. Penn encouraged immigration from these groups as well, populating the new colony with Amish and Mennonite settlers as well as his English Quakers. As Quakerism spread beyond England, more ethnic diversity entered. Dutch Quakers and Irish Quakers, for example, immigrated to the Pennsylvania colony in the late 1600s and early 1700s (Wabeke, 1983; Salinger, 1987). Descendants from all of these groups can be found in the state of Pennsylvania today. For example, Pennsylvanians called the "Pennsylvania Dutch," are not of Dutch but of German ancestry, gaining their name from a variation of the German word for "German"—*Deutsch*.

After Pennsylvania began to thrive, the colony of Georgia was formed in 1732. Although it would soon follow other southern colonies into cash crop slavery, it was founded with completely different (and very high) ideals. The hope in England was that Georgia would provide a colony for the English poor to become free farmers and better themselves; a less lofty ideal was the plan that Georgia would keep the Spanish in Florida from moving north (Gray and Wood, 1976).

The Scots-Irish (or the Protestant Irish) began coming in increased numbers during the eighteenth century (initially mostly to Pennsylvania). The Scots-Irish are an

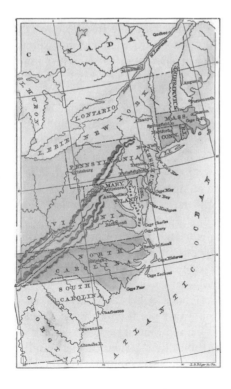

Map of the colonies.

ethnic group with sixteenth century roots in southern Scotland. Most entered at ports along the Delaware River and moved into both Virginia and Pennsylvania; later they would spread across the Alleghenies and Appalachians into North Carolina, South Carolina, Ohio, Kentucky, and Tennessee (Crozier, 1984). They tended to form extremely close-knit communities, often nearby but separate from German immigrant communities (Hofstra, 1990). When they later moved throughout the eastern colonies and states, they tended to move in extended kin groups (Reid, 1988). They remained relatively poor for quite some time, but some, like Patrick Henry, achieved notoriety.

While the Germans and Scots-Irish represented the largest of the non-English European colonists, many other ethnic groups were represented in much smaller numbers during the late colonial period. The descendants of the first Sephardic Jews in New Netherlands (New York) along with new Jewish arrivals brought the colonial Jewish population up to approximately 1,500 by the Revolutionary War (Reich, 1994). They were located primarily in the larger cities and went largely unnoticed by the almost exclusively Protestant Christian population at that time. A small Huguenot population—Protestant French who lived in predominantly Catholic (and increasingly hostile) France—also found refuge in the colonies (Bosher, 1995). Closer to the Revolutionary War came immigrants from Wales and Scotland (Daniels, 1991). To put all of this in perspective, between 1760 and 1776, 137,000 European immigrants arrived in the colonies (along with 84,000 slaves). Of these, 55,000 were Scots-Irish, 40,000 were Scots, 30,000 were English, and 12,000 were German (Reich, 1994).

Ireland had spent many centuries under English control, and the early 1800s brought frequent famines to Ireland, which spurred many Irish to move (McCaffrey, 1976). Between 1825 and 1830, 100,000 Catholic Irish left Ireland. Some moved to England, but most headed for North America (Jackson, 1984). Their numbers were small compared to what was to come, but Protestant America was already preparing for Catholic immigrants.

Many states in the new American republic passed anti-Catholic codes shortly after the Revolutionary War (McCaffrey, 1992a). Such laws were indeed interesting considering only 25,000 Catholics were in the United States in 1790; by 1812, there would be only 100,000 Catholics (Daniels, 1990). On August 11, 1834, the Ursuline Convent just outside of Boston was burned to the ground; in 1844, the homes of Irish immigrants in Philadelphia were burned along with Catholic churches. So when the Catholic Irish *did* arrive in significant numbers, it is not surprising that they faced hostility and intolerance.

Meanwhile, the first Jews from Germany and western Europe arrived in the early 1800s and were greeted with general toleration. As with the Sephardic Jews before them, their numbers were small and they were interested in rapid acculturation to their new country. Their motives for leaving Germany were both political and economic—they simply did not have the same civil rights as Christian Germans. At times, this placed them in political difficulties, but it always put them at an economic disadvantage. By 1840, 15,000 Jews were in the United States, many of them recent immigrants from Germany (Cohen, 1984). Ironically, the newcomers found themselves looked down upon as much by the Sephardic Jewish community in the United States as by the Christian community (Cohen, 1984). But, as with the Catholic Irish, their immigration was just beginning as was their impact upon the United States.

Debt Servitude and Slavery

Both the Virginia and Carolina colonies contained hopeful English businessmen facing a labor shortage. Virginia was colonized first and so first faced this dilemma. The Carolinas were not founded until the 1660s, and the area was still fairly undeveloped toward the end of that century, so the Carolina colony was able to learn from the earlier Virginia experiments. For its part, Virginia could look to labor shortage solutions in the West Indies and South America. Labor in the West Indies in the mid-1600s was either indentured labor from the British Isles (see European Indentured Servants) or slave labor from Africa. The English in Virginia found African slave labor a little difficult to get used to. Their reluctance is hard to pin down. They may have viewed slavery as a "last resort" in solving the labor shortage, or they may have been bothered by the extreme foreignness of African cultures. But probably the most important factor in Virginia was the crop they wanted to grow.

In the 1600s, the crop of choice in Virginia was tobacco (Engerman, 1986). Tobacco, along with sugar from the Caribbean, had become extremely fashionable in Europe and demand was high (Knapp, 1988; Austen and Smith, 1990). Virginia's climate was not suited to sugar but was perfect for tobacco. However, growing any kind of cash crop for export is a very different undertaking from subsistence farming. With the latter, you raise many different crops and/or animals with a focus on providing for a wide variety of family needs. This can be accomplished with relatively few people on a small amount of land. A cash crop, however, means that you plan to grow large quantities of it to ship elsewhere. Your profits must be large enough to purchase your subsistence elsewhere and pay for the costs of production. You will need a larger amount of land (although tobacco does not require huge acreage) and, most importantly, abundant labor that is cheap enough to allow you a profit after their payday. We have already seen that abundant labor was unavailable anywhere in the colonies on the free labor market. This leads to a very simple conclusion: anyone in the colonies who wanted to make profits with a cash crop would have to use some kind of unfree labor (Salinger, 1987). There was simply no other choice. The only question was what kind of unfree labor to use. This is where tobacco cultivation enters the equation. Unlike sugar (and later cotton), tobacco does not require huge teams of workers to be produced profitably. The profit can be obtained on a smaller scale.

European Indentured Servants

Indentured servitude is perhaps best thought of as voluntary slavery with an end point. By *voluntary*, I mean that the laborer enters into the labor contract (or indenture) of his or her own free will. Although a starving laborer may not seem to have much choice in the matter, the arrangement is still technically voluntary. Two exceptions to this are (1) the kidnapping of individuals followed by selling them into an indentured status and (2) legal actions by courts, such as ordering a guilty but poor person into indentured status to pay a legal fine. At times, fines were created for the sole purpose of creating labor.

Defining indentured servitude as voluntary *slavery* may be a slight exaggeration, but it does come very close to the mark. An indentured servant could be sold from master to master, would be returned by law if he or she ran away, could be prohibited from marrying or having children, and could be mistreated to some extent. There were generally laws prohibiting grosser forms of mistreatment in the colonies, but as we shall see, there were also some laws protecting slaves. In addition, not all indentured servants, particularly in the West Indies, received all that much protection, as illustrated by the following item from a newspaper in 1640: "two Barbados servants lodged complaints against their master, Capt. Thomas Stanhope, for maltreatment, after examination, the local magistrate found them to be malicious and had them publicly flogged" (quoted in Beckles, 1985:42). Distinctions can get a little fuzzy in practice.

The end point of indentured service is the most important distinction between this form of labor and slavery. The indentured servant entered into the labor contract with a clear term of service specified. He or she might belong to a particular master for three or four years, but then was allowed complete freedom. During the period of indenture, their labor belonged to the master twenty-four hours a day, but when the term was over, it was over.

This engraving, titled "Dinner in the Steerage," shows the conditions below deck on an emigrant vessel at mealtime.

From the master's standpoint, this arrangement was a clear way to acquire labor, with the fee guaranteed over time and the laborer forced to do whatever work was assigned. In the colonies, the most typical arrangement was for the master to pay for the contract up front in one lump sum. This often worked out to be very close to the costs of transportation for the laborer from Europe to the New World. Laborers in England and later Germany who were not finding work at home could mortgage their future for a ticket to what they hoped would be a better life following their indenture. Thus, people who otherwise could not get to the New World had the opportunity, but only if they were willing to accept the labor status offered to them.

During the seventeenth century, the average indentured servant from England was a young, single male. Men composed over 75 percent of the total indentured immigration during the century and that percentage would only rise after 1700 (Gemery, 1986; Salinger, 1983; Souden, 1978). Although most of them were not rural dwellers in England, almost all wound up in unskilled agricultural work in the New World (Souden, 1978). The demand in Virginia was so high that by 1671 there were 6,000 indentured servants in a total population of 40,000 people (Reich, 1994).

The average length of an indentured contract was three to four years with over 60 percent of the contracts falling into this time frame. Length of service could drop to one or two years and could rise to as many as ten years, depending on several factors. In general, shorter terms were given to younger, male, literate and/or skilled individuals who chose the least popular destinations (Galenson, 1977, 1981b; Salinger, 1987). The destination became an important factor in that, over time, word got back to England that the West Indies combined the worst working and social conditions coupled with the fewest opportunities for newly freed servants (Galenson, 1981d).

Eighteenth Century Indentured Servants

In the late 1600s and early 1700s, northern colonies ventured into the indentured labor market. The demand in New England was, of course, minimal; the few indentured servants in use there were in urban centers, some working in early industrial employment (Salinger, 1987). Pennsylvania and New York had more need of labor, however, and picked up the trade where the Chesapeake colonies left off (McKee, 1965). Of the two, Pennsylvania was far and away the larger player. The servant trade in the middle colonies continued throughout the 1700s and into the next century, ending in the 1820s (Alderman, 1975; Grubb, 1992, 1994b; Heavner, 1978). At its peak, the servant trade was impressive: over the years, 50 to 65 percent of all immigrants to the colonies from Great Britain were indentured servants (Salinger, 1987). Of all German immigrants to the United States between 1785 and 1804, 44.8 percent were bound servants (Grubb, 1985a).

Reasons for the end of the indentured servant trade are many. On the colonial side, demand dropped because of an increase in the free labor supply—the result of population growth among free immigrants and freed servants and a rising birth rate. The number of free immigrants rose partly because the cost of passage dropped and partly because freed indentured servants earned money to pay passage costs for other family members (Grubb, 1994b). On the European side, life was changing also. The early chaos caused by industrialization was starting to settle down, and an expanding economy produced more jobs. In addition, the poor and indebted English who had often been imprisoned found indentured servitude a better alternative. However, the end of debt imprisonment in England removed a large "push" factor in indentured migration (Heavner, 1978). The upshot of all this was that the price of indentured servants, which had remained relatively stable throughout the 1700s, began to rise steadily after the Revolutionary War (Grubb, 1992). Servants quickly lost their economic appeal. As just one example of this change, the labor force in colonial Philadelphia in 1767 consisted of 16.9 percent unfree labor (servants and slaves combined); by 1800, only 1 percent of that city's labor force was unfree (Salinger, 1981).

Unfree labor from Europe also came as convict labor. Convict labor was fundamentally different from the other two in that individuals were bound into labor and transported against their will (Smith, 1947). Although the British had used this technique in the West Indies earlier, they did not begin shipping convicts to the colonies until the passage of the Transportation Act of 1717. This act allowed for the export to the colonies of criminals convicted of noncapital crimes. Once they arrived in the colonies, they apparently worked in jobs very similar to voluntary indentured labor, being sold to the same masters in much the same way (Morgan, 1985). Between 1718 and 1776 (when it ended), estimates of the British convict trade to the colonies ranged between 30,000 and 50,000 individuals (Ekirch, 1984, Morgan, 1985). Most of these convict laborers wound up in Virginia at a time when indentured labor was dying out; many other colonies refused to accept them (Salinger, 1987).

Africans: From Servants to Slaves

The first encounter between Africans and mainland colonists occurred in 1619 at Jamestown in the Virginia colony. A Dutch ship stopped by and traded twenty Africans for needed supplies. Because the only form of unfree labor that currently existed in any of the English mainland colonies was indentured servitude, that status was applied to the twenty newcomers. It would take Virginia close to half a century to come to terms with slavery. As the title of this section suggests, the seventeenth century story of Africans on the mainland is the legal journey from servant status to slave status, a transformation that quickly took root in all colonies with African populations. But the seventeenth century is actually quite late in the overall story of African slavery. We need to go back farther and back to Africa to understand slavery in the colonies and later in the United States.

Africa and Slavery: The Value of a Human Export Slavery has almost certainly been around as long as humans have. Slaves appeared in the Old Testament, during the Roman empire and throughout the Middle Ages. They were found in Christian societies along with Islamic societies, and they came from any ethnic group that happened to be weak and handy. Often, they were the losers of the last war.

Hundreds of years before any contact with Europeans, Africa maintained a thriving slave trade, both internally and with Islamic countries in the Middle East (Ewald, 1992; Lewis, 1990; Mahadi, 1992; Ricks, 1989). Europeans would add the Atlantic crossing to the other slave trade routes. To put some perspective on the magnitude of slave trade, from 1500 to 1900, probably eleven to twelve million Africans left Africa for the New World as slaves; during that same period, six million are estimated to have been shipped to the Middle East and eight million were enslaved but remained within Africa (Manning, 1990b). If that slave trade period is increased from 650 to 1900, the Middle East trade probably equaled the Atlantic trade in numbers of slaves shipped (Piersen, 1996). Although these numbers are difficult to comprehend, they demonstrate that humans were export commodities from Africa for centuries.

The first Europeans to appear on the African trading scene were the Portuguese in the 1480s. They did not come for slaves, although they were available for sale. Africa also contained gold, ivory, pepper, and other attractive items. However, when the New World market opened, the Portuguese took advantage of Africa's slave market. Dutch traders arrived on the heels of the Portuguese, challenging them for the trade from the 1590s on (Boogaart, 1992). In 1635, the Dutch West India Company began purchasing slaves for the Brazil market, not to mention their own North American colony of New Netherlands (Emmer, 1991). Denmark and Sweden also showed some minor interest in the slave trade, but the big players were to be the English and French who became dominant traders in the mid-1600s (Law, 1991c; Emmer, 1991).

Although Atlantic slave trading would not peak until the late 1700s, the 1600s laid the foundation. For Europeans, the most important factor in trading slaves was to build alliances and trading relationships with coastal groups in western Africa who would provide slaves in exchange for European goods. Slaves destined for Islamic trade came from eastern Africa and were transported either through the desert and across the Red Sea or along the African coastline in the Indian Ocean (Jwaideh and Cox, 1989). But Europeans wanted ports as close to the New World as possible. From even the most western African port, the transportation of slaves across the Atlantic Ocean (which came to be called the "middle passage") ranged from five to eight or twelve weeks (Daniels, 1991). Long maritime voyages in the seventeenth century were killers; whether passengers were slave or free, poor nutrition and epidemics could easily produce 15 to 20 percent mortality, which was fairly common in the slave trade (Cohn, 1989; Curtin, 1969; Manning, 1990b). Thus, the procurers had to be coastal people in western Africa, and the slaves needed to come from the same general area. Ultimately the Atlantic slave trade came to affect Africa's entire west coast, ranging from near Senegal in the north to Angola in the south. Of all the eleven to twelve million Africans that were shipped to the New World, estimates are that only about 300,000 came from East Africa (Fage, 1989).

Historically, the coastal peoples of western Africa had been neither powerful nor wealthy. Much of Africa's trade had occurred on the eastern coast and inland. So when Europeans arrived on the west coast wanting to trade, they were generally welcomed (Falola, 1994; McGowan, 1990; Piersen, 1996). The slave trade with Europeans allowed western African peoples to gain power, wealth, and prestige through the acquisition of European products (Curtin, 1975). Most popular among these were cloth, guns, gunpowder, food, alcohol, iron, and assorted other items; cloth, guns, tobacco, and alcohol accounted for 50 percent of the trade (Eltis, 1991; Klein, 1990; Piersen, 1996). These incoming goods clearly altered the power relationships among West African peoples and changed the internal political and social relationships within the trading groups. Traditional leaders often gave way to new leaders whose power was based on their weapons and the size of the military raiding parties they maintained (Fage, 1989). "Trading firms" also were established as slave raiding military leaders formed alliances (Colchester, 1993). Initially, slaves were very inexpensive by European standards; after 1680, however, prices began to steadily rise and the trading advantage shifted somewhat in favor of the African traders (Eltis, 1991).

Slavery within Africa had long been a by-product of wars because captives were enslaved. This continued with the coming of the Europeans. The African trading partners most certainly did not sell their friends and families into slavery, but rather victimized nearby ethnic groups with whom they did not get along in the first place. Wars continued to produce slaves, but it is not known how many extra "wars" occurred in response to European slave demand. It seems clear that many such wars could more accurately be described as slave raids. Slaves also could be obtained through smaller scale kidnapping and by judicial proceedings through which enslavement was the sentence (Manning, 1983). Slave raiding ventured as far as 250 miles inland (Fage, 1989). From the captives' viewpoints, this new form of slave traffic was particularly horrific. Many reports indicate that western African victims of the slave raids expected to be taken across the ocean to be eaten (McGowan, 1990).

Although the Atlantic slave trade was growing in the late 1600s, probably ten times as many slaves were shipped to the Islamic world (Webb, 1993; Austen, 1988, 1992). The Atlantic trade grew in the next century as the New World's demand for slaves increased.

Early Slavery in the Colonies The earlier reference to the twenty African slaves who arrived in Jamestown in 1619, suggested that they were originally considered indentured servants, just like the European indentured servants in the colony. However, this depiction is not entirely accurate. Although no slaves were in British North America in 1619 and no laws dealt with the legality of slavery in the colonies (the first would not come for forty more years), the status of the African arrivals is questionable. Boles (1983) notes that although early (1624 to 1625) Virginia censuses are a bit unclear, only white servants were listed with their arrival dates (which were later used to calculate their term of service). The absence of such information for African members of the colony raises questions of how long their terms might have been. Yet Boles also notes that numbers of Africans arriving in the colony did have clear terms of service and ultimately achieved freedom. Virginia law was clearly not a finely oiled machine in the mid-1600s (see also Roper and Brockington, 1984).

Africans in Virginia achieved freedom in numerous ways. Some entered in that status, particularly if they had been baptized in the Christian faith (usually in the West Indies) and could speak English (Boles, 1983; Reich, 1994). Conversion also could occur after arrival and was a possible legal route to freedom in Virginia as late as 1641 (Finkelman, 1993). If Africans had a free father (in Maryland) or mother (in Virginia), they might be declared as free. And yet others might purchase their own freedom or have it bestowed upon them for meritorious service (Boles, 1983). Once free, Africans and African Americans had all the rights of other colonial freemen. They could, for example, buy indentured servants themselves. One prominent free black in Virginia— Anthony Johnson—owned a 250-acre farm and owned black servants. When one of his black servants ran away and took refuge with a white neighbor, claiming free status, Johnson took the neighbor to court. He won his suit, receiving his servant back plus damages from the neighbor (Berlin, 1974). Free blacks also were able to own white servants in Virginia up until 1670 (Boles, 1983).

As Johnson's story suggests, being free in colonial Virginia was more important than being white. It was certainly perceived this way by the people who lived then. Other historical data support this. Most colonial socializing took place within status rank. Servants, both black and white, worked and played together while the few wealthy free blacks were welcomed socially by wealthy free whites (Myrdal, 1944). Socializing often leads to sexual relations; we know this also occurred by the large number of laws that were passed later in the century dealing with the offspring of interracial pairings (Billings, 1991; Finkelman, 1993). In addition, white and black servants often ran away together; however, if caught, the black participants faced different penalties. As one such case, a black servant—John Punch— ran away with two white servants. When caught, the white servants were given an extra year's service to their master plus three additional years of service to the colony; John Punch was permanently enslaved (Finkelman, 1993). This occurred in 1640 when legal slavery had yet to be mentioned in the Virginia legal code. Also, a Virginia uprising in 1663 was planned by slaves and white indentured servants. The plot was uncovered before the rebellion could take place, but the event suggests how far interracial contact had developed by that time (Reich, 1994).

If all of this is confusing, keep in mind that the number of Africans grew slowly in the Virginia colony. By 1649, only 300 blacks were listed in a total Virginia population of 15,000. Africans were still more expensive to purchase than white servants at that time, and more to the point, the supply of white servants from England was more than large enough for the needs of tobacco. But as we saw earlier, circumstances changed. The supply of English servants began to decrease along with the price of tobacco. At the same time, the price of Africans slowly dropped throughout the seventeenth century. The English became increasingly involved in the slave trade, making those Africans even easier to obtain. In 1674, the British Crown gave the Royal African Company a monopoly to supply slaves to the New World. Even though Virginia did not raise a large scale cash crop like sugar, owning slaves came to make more and more economic sense to the planters. By the 1690s, slaves outnumbered indentured servants in Virginia (Galenson, 1977).

As the Virginia planters became accustomed to the idea of owning slaves, it became clear to them that the previously described loose legal ends would need some attention: doors to freedom must be shut if permanent slavery is the goal. Consider the problem of religious conversion and baptism illustrated by the case of Elizabeth Key. Key was the illegitimate daughter of an African slave/servant woman who successfully sued for her freedom in 1656 on the grounds that (1) she had been baptized and (2) she had once been sold for a fixed term of years, which made her a now free indentured servant (Galenson, 1991; Billings, 1991).

The very first Virginia statutes on slavery occurred in 1660 and 1661, making Virginia the first American colony to legalize permanent chattel slavery (Finkelman, 1993). Other colonies would soon follow. The problem of baptism was cleared up in a 1667 Virginia law that proclaimed:

"It is enacted . . . that the conferring of baptisme doth not alter the condition of the person as to his bondage or freedome; that diverse masters, ffreed from this doubt, may more carefully endeavour the propagation of christianity by permitting children, though slaves, or those of greater growth if capable to be admitted to that sacrament" (quoted in Galenson, 1991:271).

The Virginia Assembly did not want to discourage the spread of Christianity to slaves, but they also wanted to make slaves a safer investment. The motivation here is perhaps clearer in the title of a very similar Maryland law passed in 1671: "An Act for the Encourageing the Importacon [importing] of Negroes and Slaves into this Province." A similar law was passed in New York in 1706. Religions supported such laws by ensuring baptism would not be used as a means to claim freedom. In 1709, the Anglican Church's Society for the Propagation of the Gospel in Foreign Parts added an oath for slaves to be included during baptism declaring they did not expect freedom (Wood, 1991).

Perhaps more important than the role of Christianity was the problem of slaves or servants of mixed parentage (**mulattos**) who might claim freedom on the basis of a free parent. These problems were not unique to Virginia. Other colonies with significant African populations also were turning to slavery at this time. The Carolina colony, for example, was not settled until the 1660s, but Africans there were treated as slaves from the beginning (Boles, 1983).

Children of slave and free parents created a more complex legal problem, both for currently bound labor and their offspring in the future. Most mulatto children in the colony had an indentured or free European father and a black mother (Berlin, 1974). Considering that 75 percent of the indentured servants were young, single males, this should come as no surprise. In 1662, Virginia passed a law entitled, "Negro womens children to serve according to the condition of the mother" (Finkelman, 1993). This long-winded but descriptive title makes it clear that the children of slave women would be slaves regardless of their father's status. In 1691, Virginia added a penalty for white women who bore mulatto children, fining them fifteen pounds or five years indentured labor coupled with servitude for the child until he or she was 30-years-old (Finkelman, 1993). As slavery spread through the colonies and later the states, the status of slave women's children would always be slavery.

Interestingly, Virginia did provide slaves with minimal protections while limiting their other options. For example, it was illegal for a master to murder a slave. If the murder occurred accidentally during the correction (beating) of a slave, however, the master was not culpable. In addition, slaves were allowed to kill whites in self-defense if they could provide convincing evidence that their own life was in danger (Genovese, 1974). That evidence would be hard to provide, however, since it was generally against the law for slaves to testify against whites in court. Slaves would not be afforded the protection of self-defense until later. In 1705 Virginia, for example, even a free black could not physically defend himself or herself if attacked by a white—the penalty was thirty lashes (Finkelman, 1993). Indeed, the status of free blacks steadily declined, right along with that of slaves, from this period on as increasing limitations were placed on them.

While slavery spread throughout the southern colonies in the seventeenth century to provide agricultural labor, it also spread throughout the middle and northern colonies. New York inherited a population of 10 percent slaves when it was taken from the Dutch. All other northern colonies had fewer slaves, which can be attributed to the northern economy. There just was not appropriate work for which slaves made economic sense. We have already seen that the family-oriented small farms and high population growth of the northern colonies did not create the labor shortage experienced by the southern colonies. Instead of agricultural work, seventeenth century northern slaves were more likely to be found in the wealthier households working as servants. They might also be found working in lumberyards, shipyards, or on docks (Reich, 1994). In New York City in particular, having slave servants was a mark of high respectability. By 1715, 12,000 Africans and African Americans lived in the northern colonies, most as slaves (Piersen, 1996). Although northerners did not have as great an immediate interest in slave ownership, they were very interested in slave transportation and brokering. As early as the 1640s, Boston merchants and shippers were involved in the slave trade (Piersen, 1996). In the eighteenth century when the slave trade really began to grow and these shippers also learned to transport the products of slave labor, we see that slavery became as important an economic institution in the North as the South.

The Growth and Profitability of Slavery

In 1700 the Atlantic slave trade was still very much in its infancy. Most European powers possessing colonies around the world were involved already, but major changes were in store. In the period from 1700 to 1845, the following events occurred and drastically changed the Atlantic slave trade (Figure 4.2):

1. The number of slaves sent to the New World increased enormously. Before 1700, slightly over 1.3 million Africans are estimated to have arrived in the New World; after that date, at least another 8 million entered New World slavery (Fogel and Engerman, 1974; Eltis and Jennings, 1988). These figures do not include the numbers of Africans who died during capture, on their trip to the African coast, while in captivity on the African coast, and during the crossing. At least several million more met their deaths before arriving.

2. The English and French came to dominate the slave trade while other nations—most notably the Portuguese and Dutch—became less important.

3. The slave trade became more organized and businesslike with long-term trading relationships developing between European traders and African raiders. In addition, the traders became specialists in handling slaves.

4. African slaves came to dominate the more lowly economic positions in the New World while Europeans moved into other areas. Racism became more dominant and black and white social relationships almost completely disappeared as they had existed among Virginia servants and slaves.

FIGURE 4.2 MAP OF SHIPPING ROUTES OF THE AFRICAN SLAVE TRADE, 1451–1870

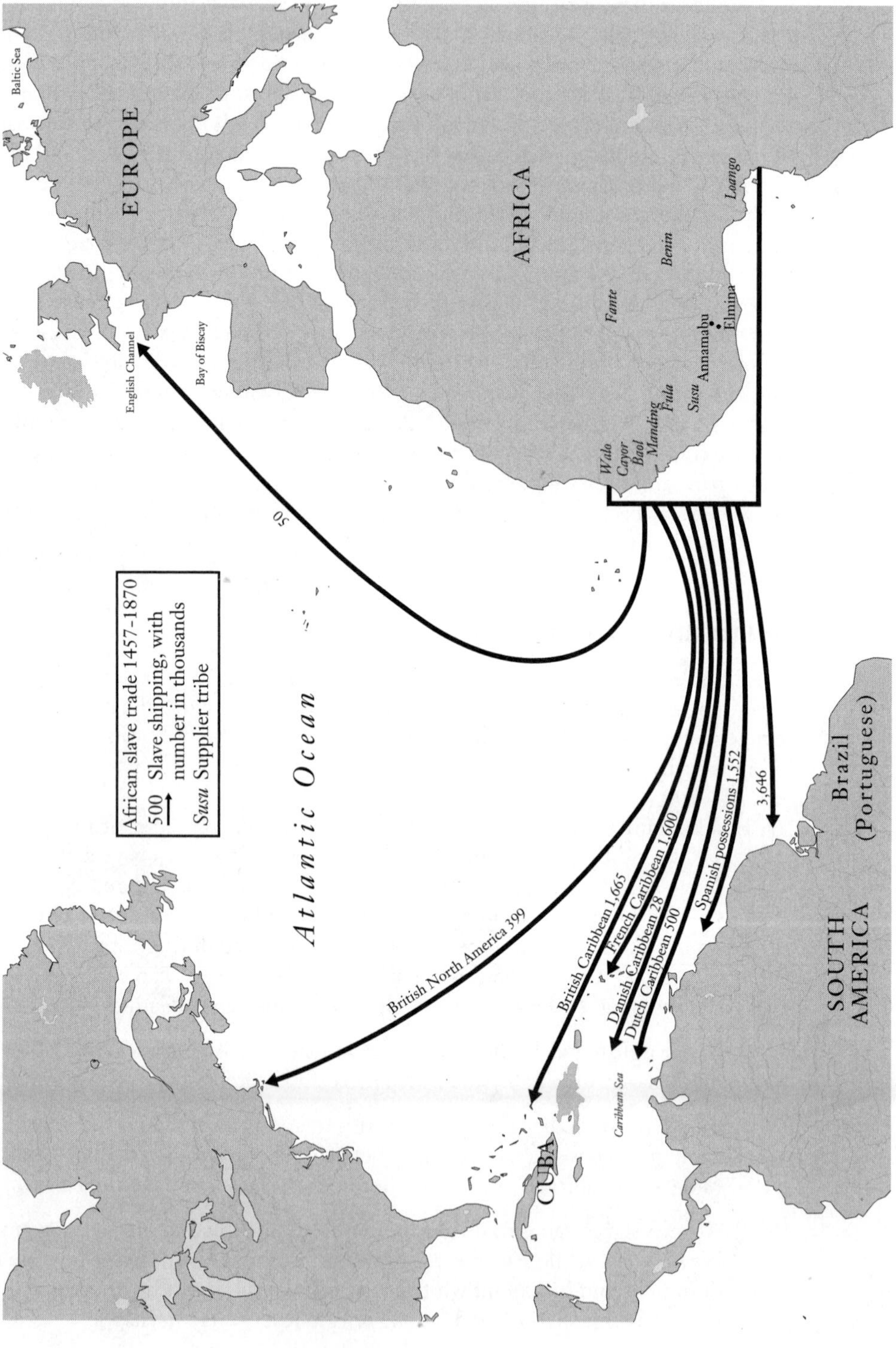

5. The American South found its "slave crop"—cotton—which made large scale slavery more profitable than it ever had been with tobacco crops in the Chesapeake. The South came to have the same labor needs that the West Indies faced for years with sugar. In short, slavery became much more important in the United States than it ever had been or otherwise would have been.

The New World Slave Trade

Were the New World colonists troubled by slavery? Some obviously were, and still more were likely bothered by the foreign cultures of Africans just as they had been troubled by the differences of Native Americans. Nevertheless, most colonists managed to get beyond their moral discomfort and the cultural differences to reap the potential economic benefits. The most striking example occurred with the Quaker population of Pennsylvania. From their beginnings in England, Quakers outdid all other groups in their opposition to slavery. They accepted no compromise and worked hard in England to abolish it. But in the 1700s and 1800s, slaves were not destined for England, they were destined for the New World. How did Quakers in Pennsylvania act in the early 1700s? Some maintained their opposition, but many more lined up at the slave auctions to purchase slaves (Salinger, 1987; Piersen, 1996). If Quakers could live with slavery, it should come as no surprise that every other group found ways to rationalize it as well.

Nineteenth Century African Slavery and the Slave Trade   As the Atlantic slave trade began to increase during the 1700s, both the internal African slave trade and slave trade to the Middle East continued (McDougall, 1992; Webb, 1993). They not only continued but also increased when the Atlantic trade ended in the 1800s, lasting into the twentieth century (Clarence-Smith, 1988; Harries, 1981; Manning, 1990a; Sheriff, 1989). The addition of the Atlantic trade served to increase the importance of slave exports from the continent. With Africa focused on selling slaves to meet their needs, they were not being economically creative in other areas.

The increased slave trade between Africa and Europe created wealth on both continents, but the European wealth was more permanent (Darity, 1990, 1992). The slave trade created wealth through the trade itself and through the products of slave labor. By contrast, African wealth was immediate because slaves were traded for all the European manufactured goods available at the time. The whole world wanted what Europe had to sell and Africa was no exception. But the results in Africa carried some long-term negative effects. First, Africans became dependent on European goods, which increased their dependency on slaves as a commodity (Eltis and Jennings, 1988). Because they focused on the slave trade in such a single-minded way, Africans neglected to develop any other part of their economy; they simply did not have the time or energy, nor was it in the interests of Europeans to help them do so (Darity, 1992; Inikori, 1988–1989). With just one commodity on the world market, Africa was only able to maintain its economic position through selling more slaves (Eltis, 1989b). When the Atlantic slave trade ended during the early to mid-1800s, it was a European decision; African slavetraders had no interest in seeing that portion of their trade end and attempted to pick up the slack elsewhere (Eltis, 1986; Renault, 1988). The increased trade is reflected in the stable price of slaves during this period. Had the demand for slaves fallen with the end of the Atlantic trade, so too would the prices of slaves (Lovejoy and Richardson, 1995).

The English and French dominated the Atlantic slave trade during the 1700s and into the 1800s; Dutch involvement continued until the late 1700s (Emmer, 1991). The English ended their role in the slave trade in 1807 and ended slavery in their possessions in 1834. The French would abolish their colonial slavery in 1848, but their African indentured labor replacements were barely distinguishable from slavery (Schuler, 1986). In the meantime, both countries would ship a lot of slaves. The English alone were responsible for half of the Africans shipped in the Atlantic slave trade during the 1700s, bringing 3,120,000 to the New World. From 1710 through 1793, the French were responsible for an additional 1,017,000. New England shippers brought somewhere between 167,000 and 294,000, depending on which estimates you use (Richardson, 1989a). (If your eyes are glazing over from all these numbers, pause a minute to recall that we are talking about numbers *of people*.) All of this trade occurred through relatively few but very large firms; the capital investment necessary to ship slaves was so great that smaller firms were unable to compete (Inikori, 1981). The peak years of the slave trade occurred in the 1780s when 921,000 Africans were shipped to the New World, representing 90 percent of total African exports (Eltis and Jennings, 1988).

After their capture by slaveraiders, Africans were held on the coasts of West Africa waiting to be shipped. The ships used were specialized for their purpose. (Richardson, 1987; Daniels, 1991). This specialization explains in part why large amounts of capital were needed to be successful in this business. Shelves built below deck were deep enough to hold an individual from head to foot. Shelf depth varied throughout the ship because some shelves were designed for men, others for women, and still others for children. The shelves were built in tiers so that many layers of slaves could be packed in, one layer over another. Initially, the shelves were constructed four feet apart, but later, for economic reasons, the space was reduced to two feet (Boles, 1983). A slave would thus be lying between two other slaves with a ceiling (the next shelf) about twelve inches above his or her nose.

Given these conditions and the length of the voyage, it is surprising that death rates were not higher than they were. Boles (1983) lists the mortality rates at 15 to 20 percent, noting that European indentured servants probably faced similar rates in their voyages of the same time. Richardson (1987) estimates a 15 percent death rate, but notes that rates varied considerably from voyage to voyage. Ship captains, he argues, often had to pack their ships more densely than they would have preferred because of the numbers of slaves awaiting shipping on the coast when they arrived. He also calculated that a profit could be made with up to 45 percent deaths during the voyage.

New World Demand and the Slave Trade   Slavery was a response to the need for gang labor on a cash crop. The first such crop in the New World was sugar, which came to dominate the West Indies and parts of South America by the mid-1600s. The continuation of sugar cultivation throughout the 1700s plus the addition of cotton in the American South produced the massive slave trade previously described. Slaves had proved to be effective in Old World sugar cultivation, and the practice was transplanted to the New World (Phillips, 1991). Between 60 to 70 percent of all slaves shipped to the New World were destined for labor in sugar cultivation. Brazil (Portuguese) was a major sugar producer by 1600 and 40 percent of all slaves in that colony worked in sugar production. In the Spanish colonies, between 30 and 50 percent of slaves were used to produce sugar as its cultivation spread throughout Mexico, Peru, and Cuba. In midcentury, British- and French-held Caribbean islands followed suit (Beckles, 1985; Beckles and Downes, 1987; Berleant-Schiller, 1989; Fogel and Engerman, 1974; Galenson, 1982). To gain additional perspective, the average-sized sugar plantation in British Jamaica contained 285 slaves in the late 1700s and the Barbados population was 70 percent slave; at the same time, the average tobacco growing slaveowner in the Chesapeake owned 13 slaves (Fogel and Engerman, 1974; Galenson, 1982).

The growth of sugar cultivation created much of the initial demand for slaves, but continuing demand was created by disease. The same climate that is ideal for sugar also harbors malaria, yellow fever, and other tropical diseases. Even if slaves were well treated, they could not be expected to live long. Mortality rates in the sugar growing colonies were extremely high (Clarence-Smith, 1989; Galenson, 1982). Even though more than 60 percent of the slaves brought to the New World were men, six slaves died in the Caribbean for every slave child born (Fogel and Engerman, 1974; Piersen, 1996).

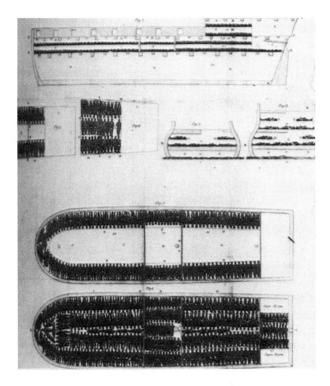

Diagram of a slave ship, 1808.

Between 1669 and 1823, 300,000 slaves were imported into Dutch Guiana in South America. In 1823, 50,000 were still alive (Voeks, 1993). Although this was tragic for the slaves, it was not a problem for the slaveowner, at least economically. In the Caribbean, the cost of a slave could be returned within the slave's first year's labor in sugar production (Piersen, 1996). The solution for the slaveowner was simply to buy more slaves.

By contrast with the sugar regions of the Caribbean and South America, the United States came to have a much smaller demand because of the lack of disease. While accounting for only 6 percent of the total slave imports to the New World, the United States contained 36 percent of all New World slaves by 1825 (Fogel and Engerman, 1974). Of all the millions of Africans shipped to the New World, only about 500,000 arrived in what is now the United States. Unlike other parts of the New World, slaves in the United States survived and reproduced, giving that country the only sizable native-born slave population in the region. This is important to remember because it means fewer native Africans were arriving in the United States with African culture fresh in their minds. By contrast, the Caribbean slave population retained a much more African culture because of the constant arrival of new slaves.

Slavery in the United States did not reach its ultimate level of profitability until the 1800s. Nevertheless, the growth of slavery in the Caribbean provided great opportunities for New England firms. Both Boston and Newport, Rhode Island became major centers for slave trade in British North America. Although they first became involved in the slave trade in the 1600s, their boom years occurred toward the end of the 1700s and into the 1800s. Sugar can be used to make molasses and molasses can be used to make rum. New England slavetrading firms traded rum for slaves in Africa and then traded those slaves for molasses and rum in the Caribbean. Later these firms would export raw cotton from the United States, again benefiting from slavery. By 1856, raw cotton comprised 56 percent of the exports from the United States. Without its connection with slavery and the slave trade, New England would probably not have developed industrially as early or as powerfully as it did (Bailey, 1990; Piersen, 1996; Richardson, 1991).

Although demand for slaves within the United States was increasing, Congress followed England in abolishing the slave trade in 1808. Even though the slave population of the United States was reproducing naturally, demand still forced slave prices up and made slavery even more a necessity in the American South. In spite of the English and American abolition, other countries continued the slave trade (as did some English and American traders on the sly). There is evidence that slaves were still being shipped into the Sea Islands off the Georgia and Carolina coasts as late as 1858 (Jackson et al., 1974). Cuba probably acquired 500,000 slaves after 1821 (Clarence-Smith, 1984, 1989).

Laws, Customs, and Artful Dodging: Life and Work Under Slavery Southern colonies began to pass the legislation that would formally create slavery in the late seventeenth century. "White" servants became clearly distinguished from "black" slaves. Keeping the status of freedom straight was far easier than keeping track of race. One's status of free or slave was legally defined in the seventeenth century by the status of one's mother. But how does one define race?

Lawmakers had particular difficulties with the definition of race. How much African ancestry made you "black"? Because the South had three distinct statuses—

black slave, free black, and white—race became an issue. The problem was that far too many "white" people had African ancestry, but they thought of themselves as white and were treated by others as white. This was particularly true during the seventeenth century when the colonies contained a significant number of free blacks and fraternization between white and black servants was considerable. After taking long hard looks at their own family trees, early Southern lawmakers settled on tracing ancestry only two generations to one's great grandparents. If any one of the those eight people was clearly recognized as having African ancestry, you were black; otherwise, you were white. Virginia put this rule on the books in 1723, and it was copied throughout the South, first in North Carolina (which added one more generation back) and throughout the southern states after the Revolutionary War (Berlin, 1974; Finkelman, 1993). By 1785, Virginia narrowed the definition of "black" to anyone with one or more black grandparents. Those states with larger numbers of whites with shaky family trees still avoided even this safer legal definition (Berlin, 1974). In one interesting example of such laws in practice, a court case in 1845 Virginia included the testimony of two free men who were at most one-eighth African in ancestry. A higher court later overturned the guilty verdict because the judge had not stressed the presence of the "Negro blood" to the jury in his instructions (Finkelman, 1993).

The time of the Revolutionary War was a curious period in colonial slavery. Even though the slave trade was enormous both before and after the war, slavery was clearly proving not to be cost effective in the northern colonies. Only New York City contained significant numbers of slaves, with 40 percent of the white households owning one or more slaves in 1790. Overall at that time, New York City was 25 percent black (Piersen, 1996). Although many people owned slaves, few owned very many. These slaves were employed either as household servants or sometimes as helpers with smaller businesses and artisans. The strong supply of European indentured servants at this time made more economic sense to middle colony employers at midcentury. As for the northern colonies, we have already seen far more money was to be made in the slave trade than in actually owning slaves. In the southern colonies, demand for slaves was still strong, although dropping tobacco prices were steadily making slaveowning a difficult long-term economic commitment. The end of the 1700s was really the beginning of the end of slavery in what we will soon refer to as the Upper South—those areas north of the Carolinas. Delaware and Maryland, for example, will soon be slave states in name only.

Life Under Slavery. Between 1790 and 1860, slaves comprised an average of 33.4 percent of the Southern population. Some states, such as South Carolina and Mississippi, were 50 to 60 percent slave. While 25 percent of the Southern white population owned slaves, half of that 25 percent owned fewer than five slaves. Twelve percent of all slaveholders owned between twenty and one hundred slaves; only 1 percent of all slaveholders owned more than 100 slaves (Boles, 1983). In short, most Southern whites owned no slaves at all, and of those who did, most owned very few. Most slaves lived on smaller farms with 20 or fewer slaves; only 25 percent of all slaves lived on plantations with 50 or more slaves (Genovese, 1974). As you might imagine, slavery was very different depending on the number of slaves present. Consider what life might have been like on a small southern farm with between one and five slaves. Undoubtedly, the owner worked next to his slaves in the field, doing the same work. One can only guess at the social relationships that might have developed under these circumstances.

The plantation system was, by necessity, more organized and less personal. Much like the systems that operated on the sugar plantations of the West Indies, the plantation was focused on a large cash crop that required gang labor. Slaves lived physically apart from owners and, on the larger plantations, might seldom even see their owner. Orders given in the field would either come from a white overseer (usually a working class white) or a black driver (who was always a slave). Black drivers were often slaves of importance within the slave community. As with the modern factory foreman, this was a difficult position, halfway between the owner and the workers. A driver who was respected off the job received more obedience on the job (Boles, 1983; Genovese, 1974).

Plantations typically also employed slaves as house servants, taking care of various tasks including cooking, cleaning, serving, baby-sitting, and even entertaining. Much has been written about the differences between house slaves and field slaves. The former were housed better, living in or very near the main house (to be handy for their duties), dressed better (they were seen by outsiders), ate better (often leftovers from the kitchen), and often carried higher status both with their owner and in the slave community at large. Genovese (1974) estimates that only 20 percent of house slaves actually carried such an elite status. He notes that field slaves might actually have had more leisure time when the cash crop was between plantings. He also notes, however, that house slaves were less likely to be sold, suggesting something of a more permanent and personal relationship.

Fogel and Engerman (1974) argue that the special status of house slaves is largely a myth in that house chores were generally done by slaves too young or too old to be of much use in the fields. They view this position as a temporary status for a great many slaves. Boles (1983) suggests that the very largest plantations probably did have a permanent status differential. As is true for much of the research about slavery, these observers are probably all right. A division of labor would certainly make sense, depending on the size of the operation, but it could also vary by the concerns and desires of individual owners. A dominant white vision of slavery was that owners had a paternalistic relationship with slaves, combining correction with love and loyalty. As we have seen, the economy of slavery was a far more important motivation, but some owners clearly felt personally involved.

Although popular images of slave work are limited to house chores and cotton, slaves actually did just about everything short of keeping the books. Slaveowners often found it in their interests to train slaves as artisans so that, for example, one might always have a blacksmith handy. Urban slaves were likely to be working as house servants but they also were more likely to be involved in a wide variety of urban work, including industrial work and skilled trades (Newton, 1977). It was a common practice in the South to lease slaves during slack times to anyone who might have work and was willing to pay. Near the Civil War, typically five to ten percent of slaves were working under lease at any given time. This practice also served to widen the range of slave occupations and job skills (Fogel and Engerman, 1974; Genovese, 1974).

Slavery was an understandably volatile situation in which millions of people were forced to work against their will. A wide variety of laws were generated to rigidly control slaves and prevent the ultimate fear of Southern whites—a slave revolt.

According to state laws throughout the South, slavery was a permanent status of service to whoever owned the slave. The individual slave had no rights beyond (1) protection against being flagrantly murdered by an owner or, (2) murder in self-defense if the slave could prove his or her own life was in immediate danger (Genovese, 1974). Slaves could not legally vote, testify against a white person, marry, care for their own children, own any kind of property, learn to read and write, have any control over work they were assigned, or travel anywhere without their owner's permission. As slavery moved into the 1800s, it also became illegal for owners to free their own slaves (manumission). Every once in a while, these restrictions worked against slaveowners rather than slaves. In an 1840 minor slave rebellion in Louisiana, for example, the slaves involved were led by an antislavery white man. When the insurrection failed, the slaves were willing to testify against their leader but by law they could not. Unfortunately for the prosecution, no one *but* slaves knew of his involvement, and he could not be prosecuted (Genovese, 1974).

In spite of the above irony, most laws regarding slaves accomplished what they were supposed to accomplish. Slaves who wished more control over their lives needed creativity to obtain that control. In some cases, owners saw fit to aid and abet. For example, it was in everyone's interest for slaves to have their own garden plots and time available for hunting. The more food slaves created for themselves, the less owners would have to provide. From the slave's point of view, this practice allowed them a larger and more varied diet. Interestingly, owners seem to have paid little attention to these gardens. Slaves were notorious for doing poor quality farmwork unless closely supervised; they also had a tendency to break tools or inadvertently leave them out in the rain. This, after all, fit right in with white stereotypes about racial intellectual inferiority. With their own gardens, however, both weeds and tools received meticulous attention (Boles, 1983).

The time allowed for slave "free time" included Sundays and holidays. Slaves also received a long vacation over Christmas complete with owner-provided food and drink. Over time, owners had found that paying some attention to slave morale created less damage to tools and higher levels of productivity. Just as with modern businesses and organizations in which people at lower levels find ways to manipulate those higher up, slaves encouraged and discouraged various owner behaviors through the way they worked. Although slaveholders held some big cards in this game, rebellious acts can be better controlled if the actor can be convinced to cooperate. For example, slaves were whipped as a "correction" of bad behavior and to provide a message to other slaves. But how often should one whip slaves? Too often might prove counterproductive; slaves were known to go beyond damaged tools to acts of arson and poisoning (Genovese, 1974). Adequate evidence supports that rules were sometimes bent in the favor of slaves, illustrating that slaves could exercise some control. Overly strict overseers received the least productive work from their slaves. On the other hand, if a slaveowner who the rented out a slave blacksmith to a neighbor allowed the slave to keep a little of the rental fee, the slave generally provided higher quality work and was more easily rented. Although this violated the law about a slave owning private property, from a slaveowner's perspective, the best course of action was the one that worked.

Slave families were another gray area in the system. According to the law, slaves could not legally marry, but they certainly did pair off (complete with rituals) and raise children. Owners learned quickly that slaves committed to family relationships were less likely to run away (Genovese, 1974); this same lesson had been learned by masters of indentured servants who initially preferred young single men but later encouraged servant marriages as a means of achieving more agreeable servant behavior (Salinger, 1987). Marriages and families among unfree labor make good economic sense. Although slave families were broken up, many owners sold unattached slaves first because they were the least stable within the system. Efforts were sometimes made to sell family members to nearby plantations, coupled with liberal visitation rights (Boles, 1983). When families were broken up, it was more common to sell children away from their parents than to break up "marriages"; even then, the greatest attack on the slave family came not from one owner's preference but from the breakup of estates upon an owner's death (Fogel and Engerman, 1974; Genovese, 1974).

In spite of powerlessness, slave families showed remarkable resilience. Men and women found ways to maintain some semblance of family life, often with separate quarters on both large and small farms. Perhaps the most interesting aspect of the slave family was the elastic nature of roles necessary to get basic family work accomplished. Care of small children, for example, could not be permanently assigned to only parents because they had no control over their time. Instead, slave family childcare often fell to older children, other relatives, and fictive kin—unrelated slaves who were honorary family members with all the responsibility that entails (Boles, 1983). When we later look at contemporary African American families, many of these features are found intact and for exactly the same reasons.

While our knowledge of slave families is sketchy at best, filled in largely by tracing sales records, our knowledge of both forcible and consensual sexual relations across the race barrier is even better hidden. The only records kept by slaveowners concern their strictness with white overseers should they molest female slaves. When it happened, they were fired and records were kept. But when an owner's son or the owner himself was involved, the only record is in the offspring of such matches. Fogel and Engerman (1974) argue that such children made up less than 2 percent of all slave children born. Given their overall perspective on slavery, that can probably be taken as a minimal estimate. If one takes into account the powerlessness of slaves, slaveowner desires to minimize such occurrences, and the low status of nineteenth century women of any color, one has to assume somewhat larger numbers. Observers who point out the disruptiveness of such behavior within the slave community may have a valid point with regard to large plantations, but recall the large number of small slaveholdings; throughout the South, rural isolation might cover much. It is in such a circumstance that the reading "Celia, A Slave" takes place. As twentieth century African American writer James Baldwin observes in a comment directed toward any white male, "You're not worried about me marrying *your* daughter. You're worried about me marrying your *wife's* daughter. I've been marrying your daughter since the days of slavery" (quoted in Genovese, 1974:414).

African Culture in the United States. The very process of the slave trade assaulted African cultures beyond any forces at a slaveholder's disposal. The process of capture, imprisonment, shipping, and subsequent work in a strange world had to have been psychologically devastating. In addition, the relative youth of most slaves shipped in the Atlantic trade meant they would not have encountered many elements of African cultures (Boles, 1983). In addition, slavery in the United States tended to erode African cultures because of the relatively small numbers of new Africans in the slave population versus the increasing native-born slave population. Unlike the Caribbean and South America, where new slaves were constantly arriving and providing continual cultural injections into the slave community, time worked to eliminate some African customs and beliefs. Even in contemporary times, the influence of African culture is much more evident south of the United States (Genovese, 1974).

In spite of obstacles, elements of African culture still survive in the United States. Although African religions were far more important in the Caribbean and South America, they also continued in the United States. Part of that strength is no doubt due to the movement of slaves and former slaves within the New World. Hoodoo spread from New Orleans throughout the South and into northern African American communities (Voeks, 1993). African medical treatments, which rely heavily on vegetable-based potions, continued in the United States. One slave in South Carolina was awarded both freedom and a salary in 1760 for contributing a cure for rattlesnake bite (Voeks, 1993). We also know something of the spread of such practices through recent archeological work in unearthing old slave quarters (Samford, 1996).

Slaves brought African stories and folktales with them; many of their elements seem more appropriate to the slave situation than to a free population. In particular, the clever trickster is a staple character in African stories (Watkins, 1994). We see him in African American folktales through Br'er Fox and Br'er Rabbit, the latter being a direct transport of Shulo the Hare from Africa (Wood, 1991). The trickster uses his brain rather than his brawn when he finds himself in a threatening situation. Slaves were much like Br'er Fox in being weaker than their adversaries; any success would have to come through cleverness and manipulation. People in continuing powerlessness develop such skills by necessity.

Slavery and Rebellion. Slaveowners in the American South lived in constant fear of slave revolts, with good reason. Earlier we looked at various ways slaves slowed production and manipulated their owners, but we only briefly touched upon more serious measures such as arson or poisoning. Arson and poisoning are usually the acts of an individual; a revolt implies the involvement of more people and some organization. Many slave revolts occurred throughout the history of North American slavery, with little resulting change to the system and only a few did much damage to either people or property. But the revolts were both large and frequent enough to serve as a constant reminder to slaveowners of the dangers built into their labor system. And if owners and slaves chose to look beyond their own realm to the West Indies (where slaves more greatly outnumbered whites), they could view some very successful revolts indeed. In this section of the chapter, we look at both slaves and their owners, watching each attempt to maximize power.

Slave revolts began in the New World almost as soon as slavery itself. In 1526, a very small Spanish colony in what is now South Carolina was formed with 500 Spaniards and 100 slaves. Disease, Native Americans, and poor planning doomed the colony. When several of the slaves joined with local Native Americans and fled the colony, the Spaniards left soon thereafter. This event introduced an often repeated pattern in which slaves formed alliances with Native Americans against Europeans (Aptheker, 1963).

As North America became both British and slave, continuing revolts occurred. Consider the following examples:

1709—Surry, James City, and Isle of Wight Counties, Virginia: A planned rebellion involving both Native American and African slaves was discovered and its leaders imprisoned (Aptheker, 1963; Boles, 1983).

1712—New York City: Slaves set fire to a building and shot whites who attempted to extinguish it. Piersen (1996) notes nine dead and 7 wounded. Twenty-seven slaves were condemned to death; six were later pardoned. The Governor described the executions: ". . . some were burnt others hanged, one broke on the wheele, and one hung a live in chains in the town, so that there has been the most exemplary punishment inflicted that could be possibly thought of . . ." (quoted in Aptheker, 1963:173).

1739—Stono, South Carolina: Twenty slaves killed two guards, stole weapons, and headed for Florida (then Spanish). Along the way, they acquired seventy-five to eighty more slaves and proceeded to kill all whites they met. Probably twenty-five whites died along with fifty slaves (Aptheker, 1963; Boles, 1983).

1741—New York City: A similar revolt to the 1712 revolt, this time resulted in 175 arrests, 14 slaves burned alive at the stake, 18 hanged, and 72 deported (Aptheker, 1963; Piersen, 1996).

1756–1853—Florida: Florida had long been a southern refuge for runaway slaves, both because it was Spanish and because alliances could be formed with the Seminole people who were native to the region. Many communities formed in which both groups lived. This situation ultimately brought out federal troops who finally were able to stop the ongoing rebellion (Carew, 1992; Heidler, 1993; Landers, 1993).

1800—Henrico County, Virginia (near Richmond): Slave Gabriel Prosser and other slave leaders hoped to take control of Richmond. Bad weather and betrayals combined to prevent their success. Prosser and 15 other slaves were hanged (Aptheker, 1963; Berlin, 1974; Forbes, 1990).

1811—St. John the Baptist Parish, Louisiana: Between 300 to 500 slaves involved in rural takeover. The revolt was put down by the militia (Berlin, 1974; Genovese, 1974).

1822—Charleston, South Carolina: Free black Denmark Vesey (he had purchased his own freedom in 1800) organized many slaves over a wide area, but the plot was betrayed; 131 people were arrested and 49 were condemned to death. Some pardons followed as some of those condemned slaves were sold to other areas, but 37 were ultimately put to death (Aptheker, 1963; Forbes, 1990; Genovese, 1974; Wood, 1991).

1831—Southampton County, Virginia: Nat Turner, a slave preacher who believed the Bible prophesied a slave revolt, planned to bring together slaves and march to the Great Dismal Swamp—a traditional area of refuge on the Virginia-North Carolina border where slaves could hide almost indefinitely. He planned to kill all whites they met en route. With a group of about 70 slaves, Turner killed 57 whites before being defeated by the militia. Turner was put to death; many followers were sold (Aptheker, 1963; Forbes, 1990; Genovese, 1974; Roper and Brockington, 1984).

Although these by no means exhaust the list of slave revolts, they provide some idea as to their scope, over both time and geography. The Turner revolt probably caused slaveowners the most fear, not only because of the large number of white victims but also because of the organizational skill behind it. Many feared it was the beginning of a general revolt throughout the South and new restrictions were put in place affecting all blacks, both free and slave (Berlin, 1974).

Restrictions on slaves, both old and new, were often more focused on preventing revolts than on the day-to-day problems of getting work done. Slaveowners approached the possibility of slave revolts both practically and ideologically, attempting to limit slaves in both opportunity and in will. The ban against slave literacy, for example, was designed primarily to limit opportunity by preventing communication among slaves and between slaves and northern antislavery propaganda. Religion provided a more ideological approach to the problem (Crowther, 1994). Whereas early Christian missionaries among the slaves were opposed to slavery, by the mid-nineteenth century most Christian faiths in the South had come to defend slavery; in addition, slaves were encouraged to focus on rewards in the next life (Wood, 1991).

A look at old slave spirituals suggests that Christianity was successful in focusing slave attention on the next world. Song lyrics about "one more ribber to cross" while one is "bound for the Promised Land" might indicate such thinking, but as Wood (1991) suggests, they are more likely references to the possibility of freedom in this life rather than salvation in the hereafter. The chariot in "Swing Low, Sweet Chariot" could well be a reference to the underground railroad—the system of safe houses where fugitive slaves could take refuge on their way to the north. Slave heroes in the Bible were Moses, Daniel, Elijah, Jonah, David, and Jesus, all of whom struggled against oppression and great odds. A spiritual about Samson contained the repetitive phrase, "If I had my way, I'd tear this building down." And the titles of other spirituals—"Let My People Go," or "Steal Away to Jesus"—carry interesting double meanings. Both Gabriel Prosser and Nat Turner felt a religious calling in leading their respective revolts (Wood, 1991). African American churches went on to become political as well as religious organizations, representing the needs of the African American community (Forbes, 1990). Christianity can be a very revolutionary religion indeed.

A combination of slave revolt fears, the dropping profitability of slavery in some areas of the South, and concerns about race relations down the historical road led some Southerners to believe in African colonization as a solution in the nineteenth century. The American Colonization Movement began in 1816. The idea was that American slaves could be returned to Africa and resettled. The fact that almost all American slaves were native-born Americans by the nineteenth century was not a problem to most whites. Ultimately from 1820 to 1899, 15,386 former slaves did leave to an area in West Africa renamed Liberia; in the earlier years, it was often the price they paid for emancipation (Forbes, 1990; Genovese, 1974).

Free Blacks in the United States: 1700–1845 From their introduction to the colonies in 1619, free blacks lived in a world of their own, restricted by the European American dominant group. They occupied a totally separate but constantly changing status, holding some rights of whites but facing many of the restrictions placed on slaves. Because their changing status was greatly affected by changes in slavery and white attitudes toward it, it makes sense to tell their story here.

Free blacks made up approximately 10 percent of the total African American population in the colonies and later the United States from the 1700s through the Civil War. This percentage varied, of course, from region to region. In 1800, for example, 56.7 percent of northern African Americans were free compared to only 6.7 percent in the South (Berlin, 1974). But the 6.7 percent is also important; until World War I, 90 percent of all African Americans lived in the South. By 1860, there will be almost 500,000 free blacks in the United States, more than half of them living in the South (Berlin, 1974; Genovese, 1974). They tended overwhelmingly to be mulattos. As the only colony to take such a census, Maryland in 1755 described 80 percent of its 1,800 free blacks as of mixed blood (Berlin, 1974).

In the midseventeenth century in Virginia, free blacks were probably in their strongest position until the end of slavery; however, Virginia slavery was unorganized and operated on a small scale. In addition, free blacks in the region were a very small minority. When slavery headed toward formal legal status in the 1660s, the rights of the free black population began to drop as well. Rights removed included holding political office, voting, serving in the militia, moving freely from colony to colony, testifying against whites in court, freedom from legal debt servitude, marriage with whites, and having sexual relations with whites (Berlin, 1974). As always, there was some variation from colony to colony (later state to state). For example, Delaware and Louisiana both allowed free blacks to testify against whites, Georgia and Florida required free blacks to have white legal guardians, and while Virginia eliminated intermarriage in 1705, North Carolina waited until 1838 (Genovese, 1974). When free blacks were useful, however, they were treated better. In 1765, Georgia was threatened by both Native Americans and the Spanish from Florida. Free blacks were invited to immigrate from other colonies, with the hope that a larger population would provide a stronger defense; they faced restrictions only on voting and serving in the colonial assembly (Berlin, 1974).

Some very important restrictions not on the law books involved employment. Although slaves were allowed to work in a wide variety of jobs, free blacks were more clearly job competition as free labor and were banned from many trades. One of the ironies of the Old South was that slave artisans, if they became free, often had to drop into the unskilled labor pool as a free worker (Berlin, 1974; Boles, 1983). Interestingly, with all these restrictions, free blacks never lost the right anywhere to own slaves themselves.

The free black population grew dramatically during the first years of the new republic at the end of the eighteenth century. On a legal level, northern states began to consider eliminating slavery altogether. A major stumbling block, of course, was that the largest slaveowners also tended to be wealthy and powerful people. Plans were constructed that either phased out slavery slowly and/or compensated slaveowners in some ways for their lost property. Not surprisingly, the states with the largest numbers of slaves instituted the longest periods of time for slaves to work their way to gradual emancipation (Piersen, 1996). On a more personal level, many individual slaveowners patriotically chose to free their slaves during this same time (Saillant, 1995). Comparing figures from 1790, 1800, and 1810, the percentage of northern African Americans who were free grew from 40.2 percent to 56.7 percent to 74 percent, respectively; in the South, the growth over those same years was 4.7 percent to 6.7 percent to 8.5 percent, respectively (Berlin, 1974). Since the abolition of slavery was not a factor in the South, these figures suggest the importance of increasing slave manumission.

Although Southern manumission after the Revolution was notable, it occurred mostly in the Chesapeake region where slavery was becoming less cost effective with every passing year (Berlin, 1974). Nevertheless, any manumission was too much for many Southerners. Virginia, for example, prohibited private acts of manumission in 1723, repealed that law in 1782 with proper revolutionary zeal, and then prohibited it again in the 1800s. In addition, Virginia also prohibited the entry of free blacks in 1793 and required newly manumitted slaves to leave the state in 1806 (Berlin, 1974). Such restrictions spread throughout the South.

Free blacks faced other forms of harassment in the South. Most were required to carry their freedom papers with them at all times and be willing to prove their free status upon being asked by a white person. (This law was primarily to maintain control over traveling slaves, some of whom might not be traveling with consent.) They also could be sold into servitude by courts if they could not pay fines; this practice no longer applied to indigent whites. In 1822, Virginia undertook an economy move with its prisons by selling free black convicts into slavery, replacing a state cost with a profit (Berlin, 1974). In addition to legal harassment, free blacks also had to fear white kidnappers who could capture them, transport them, and sell them into slavery elsewhere. If there is any light side to all this, it is the story of a pair of con artists, one Irish and the other black, who made a practice of selling the black member, arranging his escape that night, and reselling him in the next town (Berlin, 1974).

The numbers of free blacks in the United States grew through the combined occurrences of the 1792 slave revolt on the West Indies island of Saint-Domingue and the acquisition of Louisiana. On Saint-Domingue, many black and white slaveowners had to leave following the successful revolt (Blouet, 1990). Most entered the United States, with the majority moving to Louisiana. Even before this, however, Louisiana had a tradition of considerable intermarriage and the provision of civil rights for its free black population. While the rest of the South saw this increase of free blacks as threatening, many of them became slaveowners once again and were some of the strongest defenders of the institution.

It also should be noted, however, that not all free black slaveowners were in it for the money. It was fairly common for a free black person to have enslaved family relations—the easiest way to reunite the family was to purchase them. After that, they were often safer as slaves than as free blacks. If someone kidnapped a free black, white people did not care. But if someone kidnapped a slave, they were taking private property and threatening the system, even if they stole from a free black slaveowner.

Changes in the Southern Economy and Slavery Events in the early 1800s fundamentally changed slavery and American history. Slavery became profitable. Cotton was known to grow well in the Lower South (all states from South Carolina west to Texas), and the invention of the cotton gin in 1793 made large scale cotton production possible. Simultaneously, worldwide demand for cotton was steadily rising. For the first time, circumstances made the cash crop–gang labor combination of slavery work economically. In 1790, the total United States cotton output was 3,000 bales. This grew to 178,000 bales in 1810, 732,000 bales in 1830, and to 4,500,000 bales by 1860 (Fogel and Engerman, 1974). At this same point in time, slavery was becoming dramatically less profitable in the Upper South as the numbers of tobacco farmers dwindled because of low tobacco prices. But slavery was never as profitable with tobacco as it was with cotton. This economic change attracted many European American settlers into the Lower South and, in turn, brought many slaves (Figure 4.3).

In no other time in the history of American slavery was the slave family this threatened. The end of the U.S. slave trade in 1808 coupled with the growing demand for slaves in the Lower South forced the value of slaves sharply upward. Because most slaves were still in the Upper South and increasingly less needed, the demand for high-priced slaves in the Lower South brought about a major movement of slaves (Genovese and Fox-Genovese, 1979). Fogel and Engerman (1974) argue that 84 percent of slaves who made this move accompanied their owners who also were moving. Other historians suggest a little less togetherness probably occurred (Genovese and Fox-Genovese, 1979). However the upheaval occurred, it led to the large plantations that remained fairly stable until the Civil War.

This change in the pattern and economics of slavery is one of the primary reasons for slavery's existence, but why did slavery persist? From the slaveowner's standpoint, was slavery really worth the effort and the aftermath? Did slavery develop simply because of the overflow of labor from the West Indies where it clearly was extremely profitable? How big a factor was status in the desire of wealthy individuals to acquire a large number of slaves and the status that came with such ownership?

Within the United States, the North ironically benefited greatly from both the slave trade and its products; cotton, in particular, enriched the North because it could be transported to Europe for profit in its raw state or turned into manufactured goods in the United States and then transported. Slavery was also clearly profitable to the

FIGURE 4.3   MAP OF THE "COTTON KINGDOM"

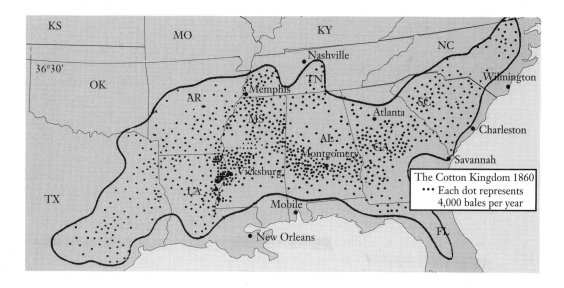

The Cotton Kingdom 1860
••• Each dot represents
4,000 bales per year

cotton growers themselves. Although slavery cost the nation the worst war in its history, and slaveowners their slaves, it provided wealth for some members of the South.

Native American Conflict and Relocation

The cotton boom that attracted both European American and slave populations to the Lower South brought only problems for the Native Americans who lived in those areas. The conflict over land that had characterized relations between Europeans and Native Americans on the Atlantic coast simply moved inland. The early 1800s produced large scale wars between European Americans and eastern Native Americans. These years also produced much of the legal basis that formalized relations between the U.S. government and its native peoples. The wars accomplished what they were intended to—the extermination of obstacles to American growth and expansion. The wars will recur in the last half of this same century with Native American peoples in the western territories. However, the legal battles begun led in directions no one could have foreseen at the time.

Conflict in the East: A Question of Land

As we saw earlier, the 1700s was a time of increasing land shortage along the Atlantic seaboard. Indentured servants arriving during that century did not have the economic opportunities their predecessors found in the 1600s. Likewise, more affluent entrepreneurs saw even better opportunities to the west, particularly in cotton. The movement of European Americans, new immigrants, and slaves toward the west created fundamental conflicts with native peoples. The new migration created battles over the land itself along with disagreements about how land should be used—a Native American hunting society was bound to have problems with a European American agricultural society. In addition, there was the gulf in cultural conceptions of land ownership. To Europeans, land was property; one could own this much of it or that much of it, here or there. To Native Americans, land was a communal resource, defended as property only when other Native American peoples infringed on one group's territory. From a European standpoint, land that belongs to everyone essentially belongs to no one and is therefore ripe for the taking.

The late 1600s and 1700s lessened the potential of war by the extreme drop in the Native American population. Disease, especially small pox, created huge losses for the eastern native peoples. Looking at the southern Native Americans, the years between 1685 and 1730 reduced the Cherokee from 32,000 to 10,500 and the Choctaws from 28,000 to 11,200; from 1685 to 1790, the overall drop in southeastern Native American peoples was from almost 200,000 to 56,000 (Rogers, 1988; Wood, 1989). When American heroes such as Andrew Jackson, Davy Crockett, and William Henry Harrison made their early nineteenth century reputations fighting southeastern native peoples, the wars were postponed until the odds were in their favor. In Ohio territory, a coalition of Algonquian-speaking nations was defeated in 1794; in Indiana territory, Tecumseh's coalition was defeated in 1811; to the south, General Jackson defeated the Creeks and Cherokees (Strong and Van Winkle, 1993).

Relocation and Recognition: The Birth of a Policy

In the late 1820s two political events related to Native Americans were on a collision course. The first was an effort by the Cherokee to use their knowledge of the American legal system to limit the powers of Georgia to take their land. This involved a direct challenge to the question of federal rights vs. states' rights. The second was a growing demand within Congress to relocate eastern tribes to the west, thereby opening up the eastern United States to European expansion.

The Cherokee went to court with an eye toward reaching the U.S. Supreme Court. They not only wanted their land in Georgia, but also they wanted to establish legal rights for themselves and other Native American groups that would apply to future issues. The U.S. Supreme Court under Chief Justice John Marshall was still in the process of defining itself within the rights granted it by the U.S. Constitution. The Constitution gave it a role in the federal government, but the extent of its powers had yet to be determined. In 1831, the case of *Cherokee Nation v. Georgia* reached that court followed by the 1832 case of *Worcester v. Georgia*. As we will soon see, the rul-

This painting depicts the forced migration of Native Americans along the Trail of Tears.

ings in these two cases had little immediate result because President Andrew Jackson completely ignored them, but they will be important later for all Native Americans.

The following principles regarding Native American rights flowed from these two Supreme Court decisions: (1) Native Americans are political entities separate from (and not to be ruled by) state or local government. (2) Native Americans are to be legal wards of the federal government who will hold their land in trust for them. (3) Native Americans are to be treated as *domestic dependent nations*, not allowed to have a foreign policy but allowed to interact with the federal government *as nations*; Native Americans will have no relationship with the federal government *as individuals* (Strong and Van Winkle, 1993). Emphasis (indicated by italic text) has been added in order to call attention to some brand new legal statuses. A domestic dependent nation had not existed before. What exactly was it? This would be determined down the road, but it clearly gave Native Americans a status that no other group had ever had. Equally important was the ruling that Native Americans would deal with the federal government as tribal units rather than as individuals. This means, for example, that treaties created between individual tribes and the federal government may be enforceable many years later as tribes (as opposed to members of tribes) sue the federal government for lost land (Norgren, 1994). Stay tuned for gambling casinos and land claims.

While the Supreme Court was busy with legalities, Congress was occupied with passing the Indian Removal Act of 1830. Although it barely passed Congress, this law gave President Jackson the control he desired to forcibly move all eastern tribes to territory west of the Mississippi River (Norgren, 1994). Civilized or not, eastern tribes

FIGURE 4.4   MAP TRACING THE REMOVAL OF THE EASTERN
NATIVE AMERICANS, 1840

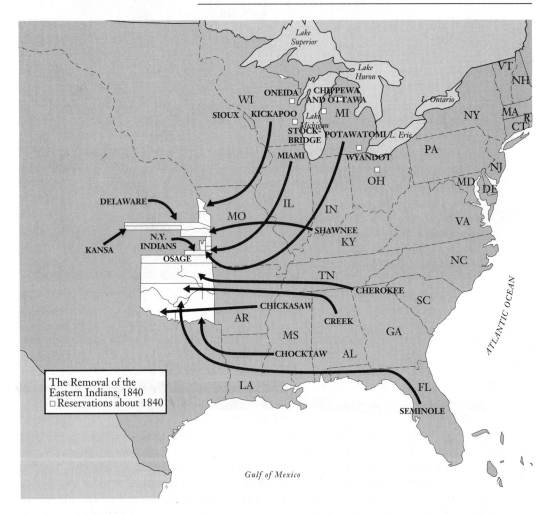

were obligated to relocate to the newly formed Indian Territory located in what is now eastern Oklahoma. Between 1830 and 1843, over 100,000 Native Americans from eastern tribes were removed (force marched) to Indian Territory; 12,000 died during the process (Spicer, 1980) (Figure 4.4).

In addition to having Native Americans removed, many European Americans also had visions of acculturating these people to American culture and Christianity. The American Board of Commissioners of Foreign Missions was founded in 1810 with the goal of establishing boarding schools for Native American children. Similar organizations with similar goals included the Missionary Society of the American Methodist

Episcopal Church (founded 1820) and the Presbyterian Board of Foreign Missions (founded 1837). Their techniques and goals are neatly summed up in the following comment from one of these missionaries:

> "This [acculturating Native Americans] can only be effectually accomplished by taking them away from the demoralizing and enervating atmosphere of camp life and Res[ervation] surroundings and Concomitants." (quoted in Devens, 1992:223).

Boarding schools with their assimilationist goals for Native Americans were common well into the twentieth century. With the aid of such religious organizations, governmental policy by the mid-1800s was becoming very clear: place Native Americans on reservations and convert their children to American culture. Before this policy changes, even fewer Native Americans will be alive in the United States.

Summary

The end of the fifteenth century brought Europeans to the New World—North and South America. While some, such as the Puritans in New England, had largely religious motivations, most Europeans came to the New World with hopes of economic gain. These latter explorers came primarily from Spain, England, France, Holland and Portugal. All but the last formed colonies in North America.

The Native American peoples already in the New World lived in relatively small groups, were not technologically advanced, and were very susceptible to many contagious European diseases. As a result, many millions died in the first century of contact. They were never a reasonable possibility as a large scale labor force for the incoming Europeans. The New World provided these explorers with much land and many resources but a severe labor shortage.

European's seeking wealth from agriculture were dependent on finding a profitable cash crop. Unlike subsistence farming, which generally provided for a wide variety of family needs, the principle behind a cash crop was to grow large amounts of one crop destined for export to Europe. But cash crops required large labor forces. With so much free land available, new arrivals from Europe generally preferred to work on their own rather than under the employ of someone else. The only solution for solving the labor shortage problem in the New World was some form of unfree labor.

Early English entrepreneurs in Virginia and areas around the Chesapeake Bay looked first to indentured servants to solve their labor shortage in growing tobacco. Indentured servants were poor Europeans who desired immigration to the New World but who could not afford the price of passage. They would therefore indenture themselves as laborers, selling several years of their future labor for compensation paid at the contract signing to ensure the cost of transportation. Such servants found their way to the middle colonies of England (Virginia and Pennsylvania primarily) and to English colonies in the West Indies.

The fastest growing and most prosperous of the seventeenth century English colonies was Virginia; the eighteenth century witnessed the growth of Pennsylvania, Georgia, and South Carolina. European immigration—both indentured and free—continued with significant numbers of Scots-Irish and Germans joining English immigrants in the trek to the New World. Some of these non-English immigrants attempted to retain their original cultures upon arrival, but most gradually gave way to the domination of English culture in the colonies. By the time of the revolution in 1776, many of these groups would see the war as an English affair and have minimal involvement.

Most colonies in the West Indies were using slave labor by the early 1600s. Some slaves were Native American, especially in Spanish colonies, but most slaves were imported from Africa. As the ultimate form of unfree labor, slavery was particularly profitable with the major cash crop of the West Indies—sugar. Because Africa had a long tradition of marketing slaves, both internally and to the Middle East, the New World slave trade was easily added. The African slave trade to the New World grew throughout the 1600s but would not reach its peak until the late 1700s. All of the European countries with colonies in the New World also became involved in the slave trade.

Slaves were slowly introduced to the English mainland colonies during the 1600s, first in the Chesapeake and later to other colonies. Slavery was officially recognized on the mainland in 1660. Slaves were used in the southern colonies primarily as agricultural workers, but they were also in demand in the northern colonies as personal servants.

By the early 1700s, indentured servitude had largely been replaced in the southern colonies and the West Indies. Demand continued for indentured labor in the middle colonies, however, as large numbers of German indentured laborers entered throughout the eighteenth century. In general, that labor system was coming to an end as economic conditions improved in Europe, free laborers in the colonies increased, and slavery increased. The biggest change came in the early 1800s when both England and the newly formed United States abolished the slave trade.

The New World slave trade grew enormously during the 1700s and was dominated by the English and French. Another eight million Africans were imported after 1700, most to the West Indies and South America. As the trade grew, it came to be dominated by fewer and larger enterprises; the profits were high but so also was the initial capital investment. While the Atlantic trade dominated Africa's slave trade during the 1700s, both the internal and Islamic slave trades continued even after the end of the Atlantic trade, with slaves still being bought and sold into the twentieth century.

The New World demand for slaves was largely determined by the export crops currently in production. Throughout the 1700s, the sugar crop in the West Indies created the largest demand for slave labor. By the 1800s, the cotton crop in the United States created more demand. With the end of the slave trade in 1808, this demand was met largely internally through the natural increase of the slave population. Still, demand in the United States forced slave prices up.

Slavery in the United States was very different at different times and in different places. In the agricultural realm, it ranged from very small operations with very few slaves to the large plantations with many slaves. Slaves also were found in urban areas, working as domestic servants, craftsmen, artisans, and even industrial workers. Of all these situations, the most organized and systematized were the plantations, which often included slave specialization by occupation as well as the use of slaves as drivers, occupying supervisory roles. Laws prohibited the teaching of slaves, designed to limit their communication among themselves as well as their understanding of the outside world. Slaveowners lived in constant fear of slave revolts. While such revolts occurred regularly, they were unable to make a permanent dent in the system.

At the same time most Africans lived in slavery in the United States, approximately 10 percent of that population was free, living in both the North and South. As free workers, they faced very different legal restrictions than did slaves. In particular, their employment was restricted to occupations that would not compete greatly with white workers.

While slavery grew and the United States came into existence, the Native American population diminished, largely because of disease. By the late eighteenth and early nineteenth centuries, European American expansion created conflicts with Native Americans over land. Following numerous wars and a few Supreme Court cases, the Indian Removal Act of 1830 gave President Andrew Jackson the authority to move the eastern Native American peoples to the western territories, primarily to what would become Oklahoma. Native Americans came to be recognized by the Courts as "domestic dependent nations," which would forever place them in a unique relation to the federal government.

Chapter 4 Reading
Celia, a Slave

Melton McLaurin

Melton McLaurin tells the following story of a young female slave named Celia who had the misfortune to be purchased by a man named Robert Newsom. His purchase was apparently sexually motivated, because he regularly raped Celia. Finally, she could endure no more and murdered him, burning his body. After this crime (to which she freely admitted), the courts took over. The question: Does a man have a right to rape his property? Read on.

On his return to Callaway County, Newsom raped Celia, and by that act at once established and defined the nature of the relationship between the master and his newly acquired slave. The emotional response of the master and his slave to this violent act lie[s] outside the methods of historical inquiry. Nevertheless, the historical record can be used to draw some reasonable conclusions. It is possible, even probable, that Newsom felt no remorse for his act. We know that his rape of Celia was no isolated incident, the act of a demented individual, an event which, had it been immediately discovered, would have raised a storm of moral outrage among white southerners, including the residents of Callaway County. Rather, recent historical scholarship has confirmed abolitionist charges that slave women were frequently abused by white men. One historian writing on the significance of rape by whites as a determinant of black female behavior has observed that "virtually every known nineteenth-century female slave narrative contains a reference to, at some juncture, the ever present threat and reality of rape." Others have shown not only that female slaves were frequently raped by masters, but that white southerners were aware that the sexual abuse of female slaves was widespread. Indeed, the practice of white male slaveholders using female slaves for their sexual gratification had its defenders, though the practice was never condoned by public opinion. Thus while it it is impossible to know the thoughts of Robert Newsom at the time he raped his newly purchased slave, it is entirely possible that he felt that his ownership of the young Celia entitled him to use her for his sexual pleasure. Whatever Newsom's thoughts about the matter, the sexual nature of the relationship between master and slave, once established, would never change.

Celia's probable emotional response to Newsom's attack upon her is suggested both by research carried out with modern victims of rape and by recent historical scholarship. The historical record indicates that this was very likely Celia's initial sexual experience. For a number of reasons, including the slaves' own standards of morality, sexual activity among female slaves under fifteen years of age was uncommon. Regardless of her previous sexual experience however, Celia's rape by her new master would have been a psychologically devastating experience, one which would have had a profound effect upon her. Modern research indicates that rape victims experience a variety of responses: fear, rage, an overpowering sense of violation, sometimes helplessness, and a loss of self-esteem. The evidence suggests that while the victim's response is determined by her unique circumstances, most victims go through several stages before coming to terms with the fact of the rape and restructuring their lives. Celia, however, had no opportunity to come to terms with a single incidence of rape, or to restructure her life. Life for Celia would entail continual sexual exploitation by her master.

———————

At some time before 1855, Celia became romantically involved with another of Newsom's slaves, a man named George. Little is known of the relationship, although Celia's trial record suggests that it developed toward the end of her stay on the Newsom farm. It must have been an intense relationship, for by early 1855 George had begun "staying" with Celia in her cabin. Whether his stays were for more than a night, and whether they were known to Newsom, we can only conjecture. What is known is that during this period Newsom continued to visit Celia regularly in her cabin and to have sexual intercourse with her. The very existence of this triangle suggests that Celia's relationship to George had only recently begun, for it is extremely unlikely that such a triangular relationship could have long endured without Newsom's knowledge. That Newsom had separated Celia from the male slaves on the farm and placed her in a special cabin close to his home strongly suggests that he was not prepared to tolerate her involvement with the male slaves. So, too, does George's insistence that Celia break off her relationship with Newsom.

Sometime in the winter of 1855, probably in late February or early March, an event occurred that changed forever the human relationships on the Newsom farm. Celia again conceived, this time without certain knowledge of the father, knowing only that the child she carried had been sired either by her master, Robert Newsom, or by her fellow bondsman George. Testimony given at Celia's trial suggests that Celia's pregnancy had placed an emotional strain upon George that he could not accept. Celia was his lover—he perhaps regarded her as his wife—yet he could not protect her from the sexual advances of the man who owned them both. At this time George faced a dilemma imposed by his own sense of masculinity and his inability to alter the behavior of his master. It was a dilemma common among male slaves, one that scholars agree was extremely detrimental to their status within the slave community and family. To have confronted Newsom directly at this stage to demand that he cease his sexual exploitation of Celia would have been an act that could have cost him his life. While some black males possessed the courage to take such risks, most, understandably, were unprepared to do so. So George made a demand of the most vulnerable member of the triangle. According to testimony at Celia's trial, George informed Celia that "he would have nothing more to do with her if she did not quit the old man." The question of how she was to accomplish this, given the nature of her relationship to Newsom, George apparently left unanswered.

George's ultimatum placed Celia in a quandary that exemplifed perfectly the vulnerability of female slaves to sexual exploitation by males within the owner's household. Faced with sexual exploitation by a white male, especially an owner, a female slave had few options but to submit. Of course, physical resistance was possible, and slave women in similar circumstances did on occasion resort to physical attacks upon the males who threatened them, some resorting to murder. Such resistance, however, always carried the possibility of physical retaliation, perhaps even death. If a slave woman had children, as Celia did, physical resistance could also lead to retribution against her children, including their sale. Understandably, under such circumstances only the most determined indomitable female slaves resisted their white exploiters physically. On large plantations, or in towns, a female slave might seek the protection of an individual whose power the master could not ignore–a white townsman of some prominence opposed to the sexual exploitation of slaves, for example. On larger plantations, the interceding individual could even be black, perhaps a black woman who had some standing within the plantation slave community. Celia, however, like many slaves, lacked the support of an organized slave community. She might have considered fleeing, one of the most prevalent forms of resistance among slaves, on plantations or on family farms, whether urban or rural. Here, again, she was handicapped by the presence of her children. In addition, there is no evidence to suggest that Celia had any contacts with the world beyond the Newsom farm, any relationships with persons, slave or free, who might aid her in an escape attempt. Finally, George's ultimatum clearly indicated that he would take no action to protect Celia from the advances of her master. The lone female slave on a family farm, lacking the support of a large slave community, cut off from the possible allies that might have been found in an urban setting, burdened with the responsibility for her two children and unable to depend upon the protection of her slave lover, Celia was forced to confront her dilemma alone.

———

Desperately seeking some means of complying with George's ultimatum, Celia threatened to hurt her master if he made further sexual demands of her. Newsom's response to Celia's threats is not recorded, but nothing in any of the testimony given at Celia's trial indicates that he took seriously threats from a slave girl who had served as his mistress for almost five years. Celia, on the other hand, was determined to break off the sexual relationship with her master, even if it meant acting upon her threats. After her confrontation with Newsom, Celia obtained a large stick, which she placed in the corner of her cabin upon her return. Should Newsom come to her cabin that night, as he said he would, Celia was prepared to resort to a physical attack to repel his advances.

As evening approached on Saturday, June 23, the members of the Newsom family prepared for bed. Since David and his bride had moved to their own home in April, only Mary, Virginia, and Virginia's children remained in the household with Newsom. Virginia last saw her father at twilight, sitting by his bedroom window reading. She had probably entered the bedroom to place her youngest son, Billy, who slept with his grandfather, in bed. Then Mary, Virginia, and the other children, including James Coffee, who slept in the room with his mother, retired for the night, leaving Newsom alone with his book and young Billy. It was the last time any member of the family saw him.

Later that night, at approximately ten o'clock, after the other family members were asleep, Newsom left his bedroom and walked the sixty or so paces to Celia's cabin. He entered the cabin, which was illuminated by the light from a small fire Celia had started in the fireplace. We know that Celia's children were in the cabin with her, but because of the late hour it is likely they were asleep. Precisely what occurred in the few minutes after Newsom entered the cabin is unclear. It is reasonable to assume that Newsom, as was his custom, went to the cabin solely for the purpose of having sexual intercourse with Celia. Since Celia had warned her master not to come to her cabin again for that purpose, and had threatened to hurt him if he did so, it is also reasonable to assume that a confrontation occurred. We do know from testimony at

Celia's trial that the two exchanged words. Whatever else was said, we know that Newsom demanded that Celia continue to have sex with him and that Celia refused.

As Newsom approached her, Celia retreated before him into a corner of the house, with Newsom positioned between her and the fireplace, his form silhouetted against the flames. Celia reached into the corner and retrieved the large stick that she had placed there earlier in the afternoon precisely for the purpose of defending herself should Newsom ignore her warnings. As the old man continued to advance, with one hand she raised the stick, "about as large as the upper part of a Windsor chair, but not so long," and brought it down against the head of her master. Dazed by the blow, Newsom "sunk down on a stool or towards the floor," groaning and throwing up his hands as if to catch Celia. Afraid that an angered Newsom would harm her, Celia raised the club with both hands and once again brought it crashing down on Newsom's skull. With the second blow the old man fell dead, to the floor.

Frightened by what she had done, Celia for a moment could only stare at the prone figure of her master. Then she bent down to examine him, "to see whether he was dead." Her examination revealed that she had killed Newsom, and momentarily she panicked. She realized her danger, feared that Newsom's body would be discovered and she would be hung. She had to think, to regain her composure. With Newsom's corpse sprawled upon her floor, it was imperative that she discover a method of disposing of the body, one which would draw no suspicion toward her, which would not reveal that Newsom had visited her cabin. She sat for an hour or more, watching the still figure on her cabin floor, perhaps checking to see that her children, after all the commotion, were still sleeping.

Finally she hit upon a plan, one both simple and ironic. She decided to burn Newsom's body in her fireplace. With his body consumed by flames, there would be nothing left to connect her with her master's disappearance, nothing to indicate that he had come to her that night. Celia took the stick with which she had killed Newsom and laid it in the fire. She stepped outside and collected staves intended for hogsheads, which were stacked near the cabin. Returning, she built a roaring fire, then doubled up the body of Newsom and pushed it into the flames. Through the night she tended the fire as it consumed the mortal remains of her former master. When the flames had disposed of the body, she picked the remaining bones from the ashes, crushing the smaller ones against the hearth stones with a rock and throwing the crushed particles back into the fireplace. The larger bones, those she could not crush, she placed "under the hearth, and under the floor between a sleeper and the fireplace." When the ashes cooled, she worked in the dark to remove some of the ashes, which she carried out into the yard just before daybreak. Then she went to bed.

———

[Celia was subsequently arrested and brought to trial for murder. Slaves did have a right to self-defense, but Judge Hall, who presided in this trial, prohibited such a plea in Celia's case. Celia's defense therefore approached their task in a wholly novel way for their time. According to Missouri law, both the prosecution and defense could request that the judge deliver particular instructions to juries. Getting the judge on your side in this matter could make or break a case. Celia's defense attorneys spelled out their defense in their requested instructions.]

The remaining six requested instructions contained the heart of the defense's case. Each of the six attempted to establish that Celia had the legal right to use force to repel her master's sexual advances. The defense's decision to make this, rather than self-defense, the basis for its case was dictated by Judge Hall's rulings on testimony given during the trial. By sustaining the prosecution's objections to defense questions about Newsom's possible threats against Celia's life, Judge Hall had removed all grounds for a plea of self-defense. The defense had been unable to obtain direct testimony from Celia about a perceived threat upon her life, for under Missouri law, as was the case in most southern states, a slave could not testify against a white person, even one deceased. Judge Hall's refusal to allow any reference to supposed threats on Celia's life, which were all ultimately based on statements made by Celia, was thus technically correct, though it was a severe blow to the defense, for self-defense was the sole legal argument extended by southern courts to slaves accused of capital crimes.

The remaining instructions requested by the defense represented a bold and imaginative response to its inability to plead Celia innocent on grounds of self-defense. Three of the instructions requested specifically called for Celia to be acquitted if the jury found that she had acted to protect herself from an "imminent danger of forced sexual intercourse." One of the three anticipated the jury's possible negative reaction to Celia's having a sexual relationship with both George and Newsom. The defense wanted the jury instructed that prior sexual conduct on Celia's part did not confer upon her master an absolute right to sexual relations with her. "Although the jury may believe from the evidence," the requested instruction read, "that Newsom and another had had sexual intercourse with Celia prior to the time of the said alleged killing, yet if they further believe from the testimony, that the said Newsom at the time of said killing, attempted to compel her against her will to have sexual intercourse with him, they will not find her guilty of murder in the first degree...."

Having presented its argument that, regardless of her past sexual activities, Celia should be acquitted if the jury found from the evidence that she had killed Newsom in an effort to prevent him from having sex with her, the defense proceeded to build its case that Missouri law allowed such a defense. At this point the defense's arguments began to threaten the very foundations of the institution of slavery. Celia, the defense insisted, even though a slave, was entitled by law to use deadly force to protect her honor. Section 28 of the second article of the Missouri statutes of 1845, the legal code in effect at the time of the trial, made it a crime "to take any woman unlawfully against her will and by force, menace or duress, compel her to be defiled." Those convicted of such an act were to be sentenced to three to five years in prison. The defense's requested instructions argued that "the use of a master's authority to compel a slave to be by him defiled is using force, menace and duress, within the meaning of the 28th section of the 2nd article of the Missouri Statutes for 1845." Finally, the defense requested that the jury be instructed that "the words 'any woman' in the first clause of the 28th section of the second article of the laws of Missouri for 1845, concerning crimes and punishments embrace slave women, as well as free white women." The contention that the term "any woman" included slave women was crucial for the defense, for, according to Missouri law, even first degree murder was justifiable if committed in resisting a person attempting to commit a felony upon the resisting individual.

The defense's contention that Newsom's death was justifiable homicide, that even a slave woman could resist unwanted sexual advances with deadly force, and that the sexual demands of even a master could be legitimately resisted by his human property was as bold as it was brilliant. It certainly was not the jury instructions that would have been expected from competent defense counsel selected to provide a merely adequate defense. With its claim that Celia had the legal right to protect her honor, defense counsel raised a multitude of legal questions about ownership of the reproductive capabilities of a female slave. If, for example, a slave could resist her master's advances, had she also the right to refuse a male partner her master selected for her? The issue of who controlled sexual access to female slaves held tremendous economic, as well as social significance, for the reproductive capabilities of female slaves were clearly viewed by slaveholders as an economic asset over which they had control. This innovative and daring strategy indicated that counsel for the defense was determined to employ every conceivable legal argument and device to see Celia acquitted.

The prosecution objected to each of the defenses requested instructions employing the justifiable homicide argument, forcing Judge William Hall to choose between the instructions proposed by the defense and those proposed by the state. The prosecution's requested instructions which, like its case, simply ignored motive, claimed instead that "the defendant had no right to kill Newsom because he came to her cabin and was talking to her about having intercourse with her or any thing else." In yet another of the prosecution's suggested instructions, the contention was put more forcefully: "If Newsom was in the habit of having intercourse with the defendant who was his slave and went to her cabin on the night he was killed to have intercourse with her or for any other purpose," Celia had no right to take his life and should be convicted. Finally, the prosecution requested that the jury be instructed that "there is no evidence before the jury that [Celia] was acting in self defense." The jury was to consider neither the defendant's motives nor her intentions. If the jury determined that the evidence and testimony presented proved that Celia, for any reason, had killed Newsom as charged, it was to return a verdict of guilty. Thus the instructions requested by the prosecution, were, in effect, a recognition of a master's right to demand sexual favors of his slaves.

The defense objected to each of the prosecution's proposed instructions, with one exception, a statement to the jury that it was to convict Celia unless the evidence showed "to the reasonable satisfaction of the jury that she was guilty of a lesser crime or acted in self defense." The inclusion of this item in the instructions requested by the prosecution was merely an acknowledgment that the law allowed a slave to protect her life, even against threats from a master. However, the prosecution had also specifically requested that the jury be instructed to find that no evidence indicated that Celia's life was threatened by Newsom. Since the prosecution objected to the crucial instructions requested by the defense, and the defense objected to requested instructions central to the prosecution's position, it fell to Judge William Hall to decide which instructions would be given. On his decision rested Celia's fate.

Because of the tenacity and creativity of the defense counsel he had appointed, Judge William Hall now faced a personal choice with inescapable moral implications. Whether the moral implications of the decision he faced troubled him is impossible to know. That he understood the moral implications of his decision, however, seems evident, for the simple reason that Celia's counsel chose to advance an essentially moral defense. Whether he found the choice easy or difficult, the judge came down squarely on the side of the prosecution. Over the objections and exceptions of the defense, which again indicated how seriously the defense approached this case, he delivered to the jury every instruction requested by the prosecution. He also sustained the prosecution's objections to all but three of the instructions requested by the defense, once again over the objections and exceptions of defense counsel. None of the three defense instructions the judge delivered to the jury introduced the central issue of motive. Rather, the defense instructions Hall read to the jury dealt with the technicalities of reasonable doubt and the weight to be given Celia's confession.

Judge Hall's denial of the defense's instructions to acquit Celia because of Newsom's sexual assault was practically a foregone conclusion. The trial testimony of both prosecution and defense witnesses established beyond dispute that Newsom had been in the habit of demanding sexual favors of Celia, and that she had resisted those demands over a period of several months. Testimony from both state and defense witnesses also established beyond reasonable doubt that Newsom went to Celia's cabin on the night of his death with the intention of again forcing her to have intercourse with him. Yet the defense's contention that a slave woman had the right to use force to prevent rape, especially rape by her master, was highly novel. In Missouri, sexual assault on a slave woman by white males was considered trespass, not rape, and an owner could hardly be charged with trespassing upon his own property. Missouri's failure to make the rape of a slave a crime was hardly unique. Rather, throughout the antebellum South, as historians from Ulrich Phillips to Eugene Genovese have observed, the law did not recognize the rape of black women. As Genovese has bluntly put it: "Rape meant, by definition, rape of white women, for no such crime as rape of a black woman existed at law."

The arguments advanced by the defense in Celia's case, on the other hand, posed an immediate threat, one of enormous magnitude to slaveholders. Had it been accepted, not only would it have struck a devastating blow to the authority of slaveowners, it also would have challenged one of the fundamental, if unspoken, premises of a patriarchal slaveholding society. The sexual politics of slavery presented an exact paradigm of the power relationships within the larger society. Black female slaves were essentially powerless in a slave society, unable to legally protect themselves from the physical assaults of either white or black males. White males, at the opposite extreme, were all powerful, with practically unlimited access to black females. The sexual politics of slavery in the antebellum South are perhaps most clearly revealed by the fact that recorded cases of rape of female slaves are virtually nonexistent. Black males were forbidden access to white females, and those charged with raping white females were either executed, or, as in Missouri, castrated, and sometimes lynched. Although on occasion a male slave was charged with raping a female slave, such cases were extremely rare, and convictions even rarer. Indeed, conviction was impossible since slave women were not protected from rape by law, no matter the color of her attacker. In Mississippi, for example, the state supreme court ruled in 1859 that the common law did not protect slaves, and that since no statute made slave rape a crime, a male slave so charged must be released.

While male slaves seldom were charged with raping female slaves, criminal charges against white men for raping slaves simply were not lodged in the antebellum

South. A search of Helen Caterall's compendium of slave cases of the American South reveals not a single case in which a white male was charged with raping a slave. While acknowledging that slave women were used by masters for sexual favors, state studies of slavery including Missouri's, fail to record charges against whites for rape of a female slave. Of course, the lack of such charges merely reflects that the law provided no protection to slave women against rape. If the courts would not convict black males of raping slaves, then such a charge against a white male was ludicrous. Thus, in the antebellum South the rape of slave women by white men, if not expected, was condoned by the law or, more precisely by the lack of it.

Significantly, the reasons put forward to explain this failing are less than convincing and underscore the significance of Jameson's defense strategy. The law failed to do so, historian Philip Schwarz contends, because to be efficacious, a law against the rape of female slaves by white men meant that the slave would have to testify. Allowing such testimony was unthinkable for white Virginians, the author claims, noting that Virginia judges even refused to hear cases "against any white man for violating common law when the victim was a slave." This questionable assumption avoids the possibility that the rape of slave women could have been made a criminal act, if only to discourage the practice, and assumes that convictions could only be obtained upon the direct testimony of the victim. Indeed, it could be argued that such legislation would have protected the slaveholders' female slaves from potential harm or injury inflicted by a rape committed by a white man. Such legislation also could have been offered in evidence of the masters' concern for the well-being and safety of their slaves. The absence of laws punishing any white men for raping slave women, since such legislation easily could have been framed so as to exclude masters, indicates that far more than the fear of slave testimony lay behind the failure of southern states to pass such legislation. Rather, such legislation was never passed because it would have threatened the master's absolute power over his female slaves.

In an ironic twist, Schwarz's study, while acknowledging the vulnerability of slave women to sexual assault by white men, sees the lack of criminal penalties for such acts primarily as a deterrent to masters who "would or could have ever taken the slightest legal—and therefore socially effective and significant—action to defend enslaved women against sexual assault by other white men." In other words, since no law made slave rape by white men a crime, it was impossible for whites so inclined to press charges for the rape of a slave. In yet another ironic twist, this same study states that "the almost complete inability of all female slaves to prevent being raped by white men had some influence on the conviction of slaves for rape of white women." The white women that black men raped, the work contends, were not women of position and wealth, nor were they women in male-headed households. They were instead, like slave women, those white women in the society who were basically unprotected and powerless. Thus even from the perspective of some recent scholars, slave women as victims of rape are seen as essentially insignificant, which was precisely the problem that John Jameson and the defense faced in their efforts to save Celia's life in a Missouri courtroom in 1855.

While the lack of criminal cases against white men charged with raping female slaves is an inevitable result of the fact that such rapes were not considered crimes, this void in the criminal record hardly means that such rapes did not occur. The literature on slavery makes it abundantly clear that white men regularly abused female slaves sexually, indeed, deemed sexual access their right. Sexual abuse of female slaves was a prominent theme in abolitionist propaganda precisely because it was an emotionally explosive charge that slavery's foes could document. Slave narratives by both men and women, were filled with references to sexual demands placed upon female slaves. Planters frequently warned their overseers about clandestine relationships with slave women, for planters were acutely aware of the tendency of overseers to use their power to demand sexual favors and of the potential problems such actions could provoke. The oral histories obtained from former slaves by interviewers employed by the Works Progress Administration during the New Deal contain frequent enough references to the sexual abuse of slave women to indicate that it was not, as southerners claimed, an infrequent occurrence, engaged in only by members of the lower class. The continuing significance of the issue of sexual abuse is evident in the works of some recent historians of slavery. Although the subject is no longer ignored by men, it is explored most effectively in the work of women scholars, one of whom estimates that as many as one in five female slaves experienced sexual exploitation. It also remains a common theme in the works of black creative writers, male and female, a haunting theme of tremendous emotional power, expressed sometimes as sorrow, sometimes as rage. Indeed, since the civil rights movement of the sixties, the sexual abuse of black women by white males has emerged as one of the dominant themes in the works of historians, sociologists, and creative writers.

The sexual vulnerability of female slaves, however, was not simply a metaphor that forcefully conveyed the power of slaveholding men. It was a reality of life under slavery, a feature of the routine operations of a system that regarded humans as property to be used for whatever purpose their owners might wish. It was also a practice that held the potential for economic gain for the master who abused his slave, because the children produced by such unions also became the property of the father. Nor is it accidental that white women were among the most vocal southern critics of the practice. White male access to slave women both threatened the stability of the white family and emphasized the fact that in many respects married white women were little more than the property of their husbands. Even when operating within the private sphere assigned to them by southern society, according to a recent study, white women "belonged within families and households under the governance and protection of their men." In fact, one of the essential legal differences between slave and free women was that free women were protected from sexual assault by law. Although once married white women had no legal recourse against unwanted advances from their husbands, they remained protected from other men.

Acceptance of the defense's argument that slave women were protected by law from sexual exploitation by white men, including their masters, would have granted slave women legal equality with white women in an area of social activity that, more than any other, symbolized class relationships within the South's slaveholding society. In one sense it would have given more protection to slave than to free women, because it would have allowed slave women the right to resist the sexual advances of any man, including masters and husbands, since slave marriages were not afforded legal status. It would have checked the master's absolute power over female slaves to some degree, perhaps not so much as the fear of physical attack by a black mate, but certainly more than fear of social opprobrium. Since slaves could not testify against whites, it would have encouraged slave women, such as Celia, to seek aid from sympathetic whites, aid that could be provided within the law. It would have recognized the humanity of the slave, granted slaves not just a right to life, which the law recognized, but a right to human dignity. White southerners sought to deny slaves precisely this prerogative because of the moral dilemmas inherent within the system. Although the slaves' humanity could never be completely denied, it had to be minimized for the institution of slavery to function. It was in an effort to deny slaves dignity, to deny their humanity, that the law sought to categorize, to define slaves as something other than human, a separate category of being. To have done otherwise, to have recognized by law the basic humanity of slaves, would have created even greater tensions within the society and posed additional moral dilemmas.

The arguments of the defense threatened not only the social assumptions under which slavery operated but the economics of slavery as well. The fertility of slave women was of obvious economic value, since their offspring became assets of the mother's master. Although scholars contend over the degree to which owners interfered in the sex lives of their slaves to insure high fertility rates, that masters were concerned with fertility rates is beyond dispute. By granting slave women the legal right to use force to repel unwanted sexual advances, the defense's instructions would have interfered to some degree with what owners saw as a property right. Such instructions, for example, would have prohibited an owner from arranging marriages between slaves on a plantation, a practice common in the slaveholding states. They inevitably would have had a direct impact upon the economics of slave fertility, one detrimental to the economic interest of slaveholders.

Perhaps most significant, the defense's instructions challenged the role of the white man as the protector of women within southern society. Indeed, it was the duty of the white man to protect all of those in his charge, especially white women. Another generally held expectation within the society was that the slaveholder be responsible for and behave morally toward his human property. While the southern male slaveholder was not a patriarch in the Roman sense, holding the power of life or death over slaves and family members, he was clearly a paternalistic figure, one responsible for those in his family, with slaves seen as part of an extended household. The planter patriarch faced few legal restrictions of his power. Those that existed were difficult to enforce, and thus seldom were. Instead, the society trusted a code of personal honor to restrain the power of the male head of household. Men were to act honorably toward those in their charge and toward those in the charge of other men The defense's instructions challenged the southern concept of male honor, a crucial element of the South's social system. If slaves could not rely on the protection of their masters, could masters be trusted to protect others in their charge? Specifically, the requested instructions posed the question that if slave women could not be trusted to the protection of their white male masters, and in fact required the protection of the law against their masters, did not white women stand in essentially the same relationship to their husbands?

———

[Celia's conviction was appealed to the Missouri Supreme Court. The ruling came down on December 14, 1855.]

"Upon an examination of the record and proceedings of the Circuit Court of Callaway County in the above case, it is thought proper to refuse the prayer of the petitioner:—there being seen upon inspection of the record aforesaid no probable cause for such appeal; nor so much doubt as to render it expedient to take the judgement of the Supreme Court thereon. It is therefore ordered by the Court, that an order for the stay of the execution in this case be refused." Whether this was the opinion of the full court, or that of a majority composed of Scott and Ryland, is unknown. Perhaps Leonard had argued for Celia, perhaps not. Whatever Leonard's personal stance, it did not save Celia. The court agreed with the trial judge. The moral issues in the case of Celia, a Slave, were judged irrelevant. There was no doubt that she had killed Newsom; the guilty verdict was valid and the death sentence deserved. The court's ruling was received and filed by George Bartey, clerk of the Callaway Circuit Court on December 18, 1855, three days before Celia's scheduled execution.

With the Supreme Court's ruling, all appeals were exhausted. Jameson, Boulware, and Kouns lacked any legal means to prevent Celia's execution. This time, no one in Fulton or Callaway County undertook heroic measures to keep Celia from the gallows. Defense counsel, having waged a prolonged and tenacious fight to save Celia, accepted the inevitable, as did those parties unknown who had engineered her "escape" from jail only a month earlier. Whatever Jameson's private thoughts about the impending execution of the nineteen-year-old slave girl, there is no evidence that he, Kouns, or Boulware took further actions to halt Celia's execution. The evidence clearly indicates that Jameson, Kouns, and Boulware saw Celia's conviction and sentence as travesties of Justice. Yet neither they nor others in the community who opposed Celia's execution staged a public protest or took additional extralegal measures to save her. This time the sentence of the court would be imposed, the laws of the state of Missouri upheld, the ultimate moral dilemma posed by Celia's case confronted.

During the night of December 20, the eve of her newly designated date of execution, Celia was once again interrogated in her cell. Her interrogators are unknown but probably included officers of the court. Her final interrogation was recorded by a writer for the Fulton Telegraph, once again the issue was whether she had help in the slaying of Robert Newsom. For the last time Celia denied that "anyone assisted her, or aided or abetted her in any way." She once more related her account of Newsom's death, claiming that she had first struck Newsom without intending to kill him. But, she continued, "as soon as I struck him the Devil got into me, and I struck him with the stick until he was dead, and then rolled him in the fire and burnt him up." This confession on the eve of her execution is probably accurate. Even Celia's insistence that the devil prompted her to strike the blows that killed Newsom seems an apt description of the rage she must have felt on that night. The reporter, however, does not mention Newsom's sexual exploitation of Celia, nor the children that resulted from his abuse.

On the following day, Celia was marched to the gallows. At 2:30 on a Friday afternoon, the trap was sprung and Celia fell to her death. The names of those who participated in or witnessed her death are not recorded, but given the time of execution, it is likely that many of Fulton and Callaway County's citizens stood at the foot of the gallows. One of the witnesses was the unidentified Telegraph reporter, in all probability the man who edited the paper throughout the decade of the 1850s, John B. Williams. With unintentional irony, that witness precisely characterized Celia's death: "Thus closed one of the most horrible tragedies ever enacted in our county."

5

Growth in Diversity and Size: 1845–1880

Although 1845 was a year of minority group stability, it marked the eve of change. By that year, western Native Americans occupied their traditional lands (relatively unbothered to date) and had been joined by relocated eastern native peoples. It was a year of relative truce; the eastern wars between Native American and European Americans had ended while the western wars had yet to begin. African Americans were held firmly in the grip of slavery, then close to its most profitable and most extensive period in American history. Those making money from slavery would not give up without a fight.

The remainder of the population was European American, a blend of northwestern European immigrants from the British Isles, Germany, and Holland, with a sprinkling of other countries from that general area. Between 1836 and 1853, 80 percent of all immigrants to the United States came from Britain, Ireland, and Germany (Cohn, 1995). In retrospect, most of those European Americans, almost all of whom were Protestant, seem to be pretty much alike. Northwestern Europe was growing industrially, which meant its people were used to cities, education, geographical mobility, and other characteristics accompanying such economic change. Yet despite these similarities, they still saw important differences within their group. Germans and Scots-Irish, for example, were still viewed as lower in status than those of English ancestry. Germans avoided English culture if possible, although they would soon turn toward assimilation (Cazden, 1987; Hoerig, 1994). But division among Protestant European Americans soon changed when more immigrant diversity arrived from Europe after 1880.

The coming European diversity had already achieved a beachhead on American shores. Throughout the early 1800s, immigrants began arriving from the southern,

Catholic regions of Ireland and from the Jewish communities in and around what is now Germany. In both cases, their numbers were still relatively small, and the immigrants were barely noticed.

Of the two new religious groups, Catholics generally drew greater concerns. The then tiny Jewish community in the United States had yet to penetrate Protestant consciousness; Sephardic Jews and the first of the German Jews lived and worked in general peace and quiet. But Protestant America was definitely Catholic conscious and concerned about Catholic growth. Few Jews ever lived in the British Isles, but Protestant conflict with the Catholic Irish was a centuries old tradition. The first Catholic Irish to arrive were noticed, and the onslaught to follow drew even greater attention.

The period between 1845 and 1880 produced many interesting changes in American race and ethnic relations. The European American community became much more diverse, especially religiously. The Civil War greatly divided European Americans in the course of ending slavery, and slavery's end moved the 90 percent of African Americans that had been slaves into a period of many freedoms and liberties in the South known as Reconstruction.

Meanwhile, the American West was being transformed. The growing eastern population was attracted by open land along with the lure of gold in 1849 California, including a new group—the Chinese in the 1850s, bringing the first Asians to the United States. All of this expansion occurred in the backyards of the western native peoples who found themselves and their way of life very much an obstruction to newcomers. By 1880, these native peoples had been defeated, African Americans had entered a new stage almost as horrible as slavery, and the door was in the process of being slammed on future Chinese immigration.

The Arrival of the Catholic Irish

Over 1 million Catholic Irish arrived in the United States between 1845 and 1849; in the century between 1820 and 1920, a total of 5 million joined American society (McCaffrey, 1992b). They faced more discrimination than had any other European immigrant group up until that time. For a generation or so, no ethnic groups were below the Irish economically, except racial minorities.

Famine and Migration: The First Urban Minority in the United States

The population of Ireland grew rapidly in the late 1700s and early 1800s. It jumped from 2.3 million in the mid-1700s to 6.8 million in 1821 (Miller, 1985). By 1841, the population reached 8.2 million. The huge growth was largely spurred by a tradition of early marriage and large families among the Irish; even high infant mortality rates did not prevent the growth (McCaffrey, 1976). But an equally important player in this demographic drama was the potato.

Long a New World staple, the potato is truly a miraculous vegetable for the days before refrigeration and canning. It is one of the few vegetables that lasts quite a while after harvest with only minimal care. Equally important, you can grow quite a few on a small piece of ground. In short, the potato was an ideal peasant crop in the eighteenth and nineteenth centuries. When Irish peasants turned to the potato on what little land was left to them, the improved nutrition allowed for greater population growth. Unfortunately, it also placed too many of their survival eggs in one basket. This total dependence on one crop made the potato famine of the late 1840s one of the world's great tragedies.

Although famines were not unknown in Ireland, some famines in the 1820s having already encouraged some Irish to emigrate, the famine of the 1840s was beyond anyone's imagination (Mageean, 1984). A potato fungus (*phytophthora infestans*) destroyed 40 percent of the 1845 Irish potato crop and virtually all of the 1846 crop (Newsinger, 1996). The famine lasted five years. Between 1846 and 1847, 500,000 Irish were evicted from their homes (Miller, 1985). Death estimates from starvation and starvation-related disease range from 1 million to 1.5 million (Guinnane, 1994; Miller, 1985; Newsinger, 1996). Approximately 1.8 million Irish immigrated to the United States and Canada (Miller, 1985; Newsinger, 1996). Taken together, these figures could run as high as one-third of the total Irish population.

The very poorest of the Irish could not afford the £2 to £5 required for passage to the United States. The areas with the highest death rates during the famine had the lowest emigration rates. Those who did leave Ireland during the famine were more likely to have slightly more money, more education, and to emigrate as families (McCaffrey, 1976; Miller, 1985). However the Irish emigrated and regardless of their destinations, most viewed their migration more as exile than as choice (Miller et al., 1980).

Life and Work in the United States   The United States European American population was as unprepared for the Irish arrival as the Irish were for the United States. To Protestant Americans, their shores were being stormed by a massive Catholic attack. And even though these new immigrants were slightly better off and more educated than those who remained in Ireland, they seemed woefully low on both scales to Americans. Never before had any one ethnic group entered the United States in such large numbers so quickly. We should remember, however, that such an immigration would have been impossible much earlier when crossing the ocean was both more expensive and took longer. As time went by, it became cheaper and therefore easier for future groups to make the same move.

To Irish immigrants, the United States was a hostile and frightening improvement on what they had left. Comfort is relative, after all. Protestant employers had already learned from both early Irish Catholic immigrants and northern free blacks that job discrimination produced cheap labor. Individual members from both minority groups, prevented altogether from competing for most jobs, had to stand in line for whatever was left, regardless of pay. The digging of the Erie Canal, completed in 1825, owed much to the Irish laborers who had so little choice. The process continued with the famine immigrants who were left with the worst, most dangerous, and most poorly paid employment available (McCaffrey, 1976, 1992b). Those who did not arrive in either New York City or Boston docked in the port of New Orleans. In the South, Irish laborers were often hired for work considered too dangerous for slaves. Slaves, after all, were valuable.

IRISH INDUSTRIES.

INFERNAL MACHINE MANUFACTURING.

THE IRISH VOTE MANUFACTURERS.

THE LANDLORD KILLING INDUSTRY.

A NEVER-FAILING IRISH INDUSTRY.

These 1881 cartoons represent the prejudicial attitudes toward Irish Americans that were widely held during that time.

In the North, where most Catholic Irish settled, men were most likely to be found working on canals, railroads, building construction, and as dock laborers. As just one example of how their desperation and powerlessness were manipulated, numerous Irish immigrant men responded to advertised wages for railroad work in upstate New York. When twice the number of needed workers arrived, the contractors cut the advertised wages and received aid from the state militia in forcing the laborers to work for the reduced wages (Miller, 1985).

By 1870, some Irish had improved their employment status to working as bricklayers, carpenters, miners, factory workers, building contractors, and small proprietors (Emmons, 1989; Mitchell, 1988). Still, employment discrimination continued into the next generation. Among native-born Irish toward the late nineteenth century, 40 percent continued working as unskilled laborers or domestics (Miller, 1985).

Irish immigrants faced other difficult circumstances beyond employment. Determined to avoid rural areas and farmwork after their experience in Ireland, they became urban dwellers in the United States. Even the earlier and more adventuresome Irish who headed to the West sought growing cities like Denver and San Francisco (Brundage, 1992; McCaffrey, 1976; Siegel, 1975; Wyman, 1984). Although the Irish wished to live together as most new immigrants do, housing segregation offered few alternatives. In New York City, the first Irish neighborhood was lower Manhattan (Dolan, 1972). Such poor neighborhoods spawned diseases such as tuberculosis, small pox, and Asiatic cholera; mortality rates from typically less lethal ailments were

Poverty forced many immigrants to live in primitive shanty towns, such as this one in New York City.

higher because of the poor environment (McCaffrey, 1976; O'Connor, 1979). Those neighborhoods almost certainly played a role in the disproportional Irish involvement with crime (McCaffrey, 1992a); in the 1850s, criminal convictions in New York of the Irish foreign-born were five times higher than rates for Germans, either native-born or foreign-born (Miller, 1985).

Although American employers benefited from inexpensive labor plus temporary freedom from labor organization, many Protestant Americans saw no benefits whatsoever to the newcomers. The Irish were simply living examples of the Catholic menace of which they had long been warned (Knobel, 1986). In 1854, the Order of the Star Spangled Banner (a secret anti-Catholic organization) was transformed into the American Party, which ran candidates for public office. The American Party came to be known more popularly as the Know-Nothing Party (based on the answer members provided when asked about their party). The party sought limitations on immigration, a longer naturalization process, and the restriction of political offices to the native-born (McCaffrey, 1992b). Although they did well in the 1854 and 1855 elections, their popularity and influence decreased thereafter. Nevertheless, the existence and success of the Know-Nothings foreshadowed future reactions to newer and seemingly more threatening immigrants.

The Know-Nothing attack was promptly followed by the Civil War. From the Irish point of view, this war was being fought to free from slavery one of the few ethnic groups in the United States worse off than they were. If successful, these newly freed slaves would arrive in the free labor market, seeking the same jobs as the Irish. The Irish already knew about American race relations from their contact with free blacks with whom they were currently competing in the North. To make matters even more unfair, wealthy Protestant Americans could purchase a poor man to fight in their place. Many of these poor men were Irish. In short, the Irish were being asked to fight a war and receive unusually high casualties for a cause that opposed their personal interests. This was the context of the July, 1863 riot in New York City, begun by poor Irish in response to the new draft law for the Union Army. The riot lasted five days during which eleven African Americans were killed along with a Native American mistaken for an African American. In addition, an African American orphanage was burned to the ground. The army was finally called in, resulting in 1200 more deaths. Ultimately, the riot resulted in $3 million worth of property damage (McCaffrey, 1976). In spite of this, 200,000 Irish wound up fighting in the Civil War, the vast majority for the Union, generally being assigned the most dangerous combat positions (Miller, 1985).

Urban Ethnic Politics and the Irish

A prominent Irish American sociologist and politician—Daniel Patrick Moynihan— once commented that the Irish seem to have little talent as entrepreneurs or intellectuals but top all other ethnic groups when it comes to organizational skills (Moynihan, 1964). The Irish involvement with many forms of politics in Ireland made them natural players in the wide-open arena of American style politics (Brundage, 1992). For all the forms of discrimination they faced upon arrival, one right they were granted was the right to vote. Their concentration in urban centers made city politics an obvious choice for them—they could use their sheer numbers to dominate. By 1870, the Irish comprised between 15 to 20 percent of the population in America's 50 largest cities (Daniels, 1991; Erie, 1988). Perhaps more important, they learned early how to register virtually the entire Irish community and get out the vote on election day (Erie, 1988; O'Connor, 1995).

As American city politics went in the nineteenth century, political power could be obtained and maintained in many creative ways. Most large eastern cities—particularly New York City—were already governed by political machines. A **political machine** (or party machine) organizes the electorate to maintain political control, election after election. This domination allows the machine to reward those voters with patronage jobs with the city while simultaneously shifting public funds into the pockets of the machine leaders (O'Connor, 1979). A machine leader, for example, might own a construction company that would be picked regularly for city construction projects.

Once a machine attains control of a city, its leaders can use that power to maintain its control. Patronage jobs obviously keep the voters happy, particularly if other kinds of jobs are proving hard to find. But even less ethical (and clearly illegal) techniques are available. For example, votes can be bought, naturalization procedures can be speeded up, imaginary voters can be registered from tombstones in the cemetery, voters can be harassed if there is doubt as to how they will vote, and ballot boxes can be tampered with. You can either add votes for your candidate (i.e., cemetery registration) or destroy ballots cast for the opposition. If you control the city, you control the police and they can take care of such matters for you. In the New York City 1886 mayor's race, the story goes that ballots cast for the machine's opposition could be seen floating down the Hudson River for days after the election (Erie, 1988).

In the late nineteenth century, the political machine in New York City was called Tammany Hall. It has the distinction of being America's most powerful and longest lasting political machine. Its name came from a private social club where machine leaders would gather to socialize, drink whiskey, smoke cigars, and plot political strategy. This is the origin of the famous "smoke-filled room." The machine leaders never ran for office themselves. They simply picked stand-ins, saw that they were elected, and then told them what to do.

When the Irish were new to city politics in the 1860s, they were only lieutenants of Tammany, then controlled by William Marcy Tweed—a powerful man of Scots-Irish background who would move from Tammany to prison in 1871 after being caught for a little political corruption (McCaffrey, 1992b). He was followed by "Honest" John Kelly in 1874. By 1892, 61 percent of Tammany was Irish (Erie, 1988). The Catholic Irish would control Tammany until 1933. The dividends for the Irish community were almost immediate. For workers who remembered the famine in Ireland, safe and secure city employment was a wonderful plum. City jobs in New York City went almost exclusively to the Irish population, moving them one more rung up the economic ladder. The image of an Irish cop on the beat in New York City from the old days is not just an idle stereotype. Most New York City police officers were indeed Irish. In the nation's fourteen largest cities between 1870 and 1900, Irish participation in public sector jobs rose from 11 to 30 percent (Erie, 1988).

Jewish Immigration from Germany

The term *German Jew* can be confusing because it was first used before Germany became unified into the country it is today. In the nineteenth century, central Europe contained many small, independent states, unified only by elements of culture. Of these elements, perhaps the most important was the common use of the German language. Thus, German Jews are probably best thought of as European Jews whose dominant language was German (Cohen, 1984). Keep in mind, however, that these people came to the United States from Poland, Bohemia, Alsace, and even as far east as Russia and Lithuania (Diner, 1993).

German Jewish immigration was stimulated greatly by early nineteenth century European changes in their legal status. European powers had become concerned by the American and French Revolutions and too much talk about freedom. The Congress of Vienna (1814–1815) sought to return Europe to the days before the French Revolution. This resulted in severely limiting German Jews' civil rights, restricting them in housing and employment. As their economic opportunities dwindled, so too did their standing in the eyes of most non-Jewish Germans. Jews came to be viewed as an alien race as German patriotism grew in fervor (Cohen, 1984; Goldstein, 1992). These "push" factors accelerated movement to the United States in the 1820s, but Jewish participation in revolutionary movements in 1830 and 1848 (both of which failed) made Germany even less hospitable (Barkai, 1994; Goldberg, 1975). The move was on.

German, Jewish, and American: The Growing American Jewish Community

By 1840, 15,000 Jews lived in the United States, the vast majority of them German Jews. Ten years later, their numbers had risen to 50,000. By 1860, 150,000 Jews called America home, and indeed, it was home. Whereas non-Jewish Germans often returned to Germany, German Jews had one of the lowest return rates of any immigrant group to the United States (Cohen, 1984). This should not be surprising, given the cultural and legal attacks they would have faced had they returned. Altogether, in the years between 1820 and 1880, approximately 150,000 to 200,000 German Jews immigrated to the United States (Diner, 1993; Cohen, 1984). To put this in perspective, that same time period brought 10 million immigrants total to the United States, including 3 million non-Jewish Germans (both Catholic and Protestant), 2.7 million Irish (mostly Catholic), 950,000 English, and 400,000 Scandinavians (Cohen, 1984). In terms of numbers, the American Jewish population was not all that noticeable in the nineteenth century. In addition, however, German Jews acted in such a way as to make themselves even less noticeable.

One of the ways German Jews decreased their visibility was to spread throughout many regions of the United States. In 1820, the only Jewish congregations were in New York City, Philadelphia, Richmond, Charleston, Savannah, and Newport, all founded and controlled by Sephardic Jews (Diner, 1993). In the early 1800s, Charleston had the largest Jewish population of any American city (Faber, 1993). By 1877, Jewish congregations were in thirty-seven of the thirty-eight states (Cohen, 1984). Although much of this dispersion was intentional, there were also many reasons for a new immigrant group to want to live in ethnic communities. With German Jews, two notable concentrations stand out. First, those who moved inland from the Atlantic coast concentrated in the triangle that connects St. Louis, Cincinnati, and Milwaukee. Not only did moving inland decrease visibility, but also that particular area contained many non-Jewish German immigrants with whom they felt culturally comfortable. Second, wherever they settled, German Jews were unabashedly urban. Much like the Irish and other European immigrants to come, cities were the magnet.

German Jews could be found in virtually all cities, large and small, in nineteenth century America, but not always in the same concentrations. Providence, Rhode Island, for example, contained about 500 German Jews in 1877 when the city's population stood at 100,000—a small percentage considering Jews made up 0.5 percent of the U.S. population (Goldstein, 1992). By contrast, 60,000 Jews were in New York City in 1870, representing 6 percent of that city's population. Their highest concentration was in the growing city of San Francisco where they comprised 10 percent of the population. When these extremes are averaged, German Jews usually accounted for 2 to 3 percent of the people in most nineteenth century American cities (Rosenwaike, 1994). By 1877, two-thirds of all Jews lived in only five states—New York, California, Pennsylvania, Ohio and Illinois (Goldstein, 1992).

German Jews and American Law   Although they left many legal restrictions in Germany, German Jews were not totally free of them in the United States. A few laws directly affected Jews: some were enacted specifically as anti-Jewish legislation, whereas others were simply reflections of the overwhelmingly Protestant domination of the new country. Anti-Jewish legislation can be seen clearly in laws prohibiting Jews from holding public office. Examples of the indirectly anti-Jewish laws and practices include mandatory Sunday closing laws for businesses along with the distinctive Protestant basis underlying the growing public school movement in the United States. Although German Jews preferred not to attract too much attention to themselves, they also did not believe in accepting such restrictions without fighting back.

Many states initially constructed constitutions prohibiting Jews from holding public office. Maryland, for example, passed an 1826 law popularly known as the "Jew Bill" which prohibited Jews from holding state office, becoming lawyers, and serving as officers in the state militia (Eitches, 1971). By the time most German Jews were arriving in the United States, only New Hampshire and North Carolina still had such laws on the books. With a constitution similar to Maryland's, North Carolina's would not allow any person to hold state office "who shall deny the being of God or the truth of the Protestant religion of the Divine Authority, either of the Old or New Testament, or who shall hold religious principles incompatible with the freedom and safety of the State" (quoted in Cohen, 1984:76).

Although such laws were more anti-anyone-who-is-not-Protestant than anti-Jewish, Jews often found themselves singled out for attack when they complained. In 1844, for example, Governor James Hammond of South Carolina asked all citizens to observe Thanksgiving by uniting in prayers "to God their Creator, and his Son Jesus Christ, the Redeemer of the world." When 100 Jewish residents of Charleston protested, the Governor attacked them, arguing that Jews had inherited "the same scorn for Jesus Christ which instigated their ancestors to crucify him" (quoted in Cohen, 1984:74–75). Probably the clearest attack on Jews came from then General Ulysses S. Grant who issued Order No. 11 in December of 1862. It expelled all Jews from the military zone under his command (Mississippi, Kentucky, and Tennessee).

Sunday closing laws created specific problems for Jews. The Jewish Sabbath begins at sundown Friday night and ends at sundown Saturday night. Some religious Jews are (and were) very restricted in what they can do during this time. Doing business is (and was) definitely prohibited. If a community required all businesses to close on the Christian Sabbath—Sunday—religious Jewish businessmen would lose a potential day of work. Some battled against these laws, but it was definitely an uphill fight against a large Christian majority. As one example, Solomon Benjamin, a Jewish merchant in Charleston, was arrested for selling a pair of gloves on Sunday. The city magistrate supported Benjamin's right to conduct business on Sunday, but the finding was overturned by the state court. Their grounds were that Christianity and not Judaism was the source of all morality (Cohen, 1984).

Religion and Assimilation in American Life Outside legal arenas, German Jews had concerns about other relationships in their adopted country. On the one hand, they had to be concerned about social relationships with their Christian neighbors. Jews had spent almost 2000 years away from their ancient homeland of Palestine, living everywhere as a minority amidst Christian and Islamic dominant groups. They appreciated the need to minimize potential hostility toward them. A second relationship, related somewhat to the first, was their own connection to their religion—Judaism. Many German Jews had drifted away from some or all of the laws and teachings of Judaism. For those who still wished to practice their religion, one option was to remake it as they were remaking other parts of their lives in the United States. And if they were to change Judaism, one possibility was to Americanize it.

The Know Nothing movement was directed at Jews as well as Catholics; it was generally opposed to almost anyone immigrating at mid-century. But, as we have already seen, there were only about 200,000 nineteenth century Jewish arrivals compared with 2.7 million Irish. With the emphasis of anti-immigrant hatred directed elsewhere, Jews recognized the possibility of blending in with American life. Their effort to spread out geographically was part of that strategy. They also adapted to the region in which they lived. Southern Jews were staunch defenders of slavery and the Confederacy, while northern Jews fought loyally for the Union (Greenberg, 1993). Unlike the Irish, they specifically avoided becoming associated with any one political party so as not to appear a threat. But even so, they faced many forms of discrimination based on their ethnicity. In particular, they found themselves excluded from many American private clubs and organizations—you could not be both a Jew and a Mason.

Social ostracism in American society started the Jews on a voyage of creating parallel social structures (see Berger, 1983). Whenever one institution in American society closed its doors to Jews, they would build a Jewish version with hinges well oiled for Jewish entry. When American fraternal organizations chose not to accept German Jews, their response was to start their own fraternal organization—the B'nai B'rith [Son of the Covenant] in October, 1843. Although the name of the organization and organizational titles were in Hebrew, its structure and activities looked for all that like the Elks, Masons, or Oddfellows. Numerous lodges were located throughout the country along with mandatory secret rituals, handshakes, degrees, and so on. When the YMCA was imported from England in 1851, Jews were not welcome. It was, after all, the Young Men's *Christian* Association. The Young Men's Hebrew Association was formed in 1874.

The Jewish Reform Society was organized in Chicago in June, 1858, under the leadership of Rabbi Bernhard Felsenthal. This new branch of Judaism was not totally new—its roots were in Germany back in the early 1800s—but the need for it in the United States was similar to the motivation for its origin in Germany. German Jews wished to modernize their religion for themselves while also making its practice more in line with dominant Christian religion. Reform Judaism spread rapidly throughout American Jewish congregations in the late 1800s, appearing in somewhat different forms here or there. Among the changes were the introduction of organ music, the seating of men and women together for services or in choirs, the elimination of prayer shawls, the use of English instead of Hebrew or German, and even occasional sermons delivered by the rabbi (Cohen, 1984; Goldstein, 1992). Other practices, such as the strict dietary laws involved in keeping kosher, were left largely to individual choice. As Jewish leader Issac Mayer Wise described the Reform movement, "Whatever makes us ridiculous before the world . . . may safely be and should be abolished" (quoted in Cohen, 1984:163).

As with all immigrant ethnic groups, no single description can capture the variety of experiences that occurred. For every tight knit German Jewish community, there were many Jews who lived with German Christians or elsewhere. Although many carefully married within their group, others were open to marrying non-Jews, especially on the frontier where wives were hard to find (Ehrlich, 1988; Faber, 1993). But American society would be making it increasingly difficult for Jews to forget their origins. The beginning of serious anti-Jewish persecution might be symbolically dated from June, 1877, when Joseph Seligman, a prominent Jewish banker and personal friend of President Grant's, was refused a room at the Grand Union Hotel at Saratoga. It would not be long before other resort hotels would flagrantly advertise, "No dogs. No Jews. No consumptives [tuberculosis sufferers]."

An Economic Strategy That Worked

One of the more popular trades within the European Jewish community was tailoring. Almost everyone needed this service, and tailoring was a highly mobile business if one needed to move on in the face of governmental repression. Many German Jews came to America with skills in the "needle trades" and as peddlers. As America changed in the mid-1800s, both of these trades provided great opportunities of growth for their practitioners.

Tailoring was transformed by the invention of the sewing machine in 1846. Commercially produced cloth had long been manufactured by the New England textile mills. The sewing machine offered the possibility of producing quantities of ready-to-wear clothes in yet other factories. All one needed was a large building, many hundreds of sewing machines, and many hundreds of workers. And best of all, such places did not already exist. Jews could start them without competing against Christian businessmen. The growth of this industry was phenomenal. By 1888, there were 241 clothing manufacturers in New York City, 234 of them owned by German Jews with an annual business of $55 million; outside of New York, Jews owned 75 percent of all clothing factories (Cohen, 1984; Howe, 1976). Some, such as Hart, Schaffner & Marx, became American landmarks. Along with this growth came business for Jewish bankers. Because Jewish businessmen often had difficulty getting business loans, it was necessary to find more understanding investors (Decker, 1979). Once again, discrimination forced the Jewish population to turn to each other for support.

Peddlers also found opportunity in the industrialization of America. As the population grew, it no longer made sense to drive a peddler's wagon from town to town. Jews became America's shopkeepers, starting with small dry goods stores and developing them into the huge department stores that populate the American landscape today. The list of success is almost endless: the Straus family (Macy's and Abraham & Straus), Julius Rosenwald (Sears, Roebuck, which pioneered the mail order business), the Rich family (Rich's in Atlanta), Levi Strauss (a dry goods store owner in Sacramento, California, well known for his blue denim pants), and the Gimbel family (Gimbel's, of course). We also should include other Jewish-owned department stores such as Neiman-Marcus in Dallas, Cohen Brothers in Jacksonville, Garfinckel's in Washington, Thalhimer's in Richmond, Sakowitz's in Houston, Goldwater's in Phoenix, and Goldsmith's in Memphis (Cohen, 1984; Daniels, 1991; Goldstein, 1992; Kaganoff, 1976; Laermans, 1993; Whitfield, 1982). Clearly geography was not an obstacle and neither were Christian businessmen because none of them owned department stores.

The War with Mexico: New Land and New People

By the mid-1800s, many U.S. citizens had become downright full of themselves. John L. O'Sullivan, editor of the *Democratic Review*, wrote in 1845 that it was America's "manifest destiny to overspread the continent allotted by Providence for the free development of our yearly multiplying millions" (quoted in McCaffrey, 1992). The term *manifest destiny* lived to become famous (or perhaps infamous); Americans became convinced that God was truly on their side in amassing as much real estate as possible by any means necessary. Steady population growth was pushing the country relentlessly westward as newcomers searched for fresh opportunities. This overflow ultimately bumped into northern Mexico.

The first encounter occurred in what is now Texas in the 1830s. The next decade would stir longings for all of northern Mexico—the region that is now the entire southwestern United States. For the most part, the land was there for the taking. Mexico had been relatively unsuccessful its attempts to encourage its population to fill the most northern frontiers of its territory (Benson, 1987). Most of the population lived south of the great desert that makes up today's northern Mexico. There were 60,000 settlers in what is now northern New Mexico, but relatively few Mexicans were in the other territory. The United States declared war on Mexico in 1846 (Figure 5.1).

After the War: Under American Control

The Mexican War formally ended with the signing of the Treaty of Guadalupe Hidalgo in 1848 (Griswold del Castillo, 1990). It contained, among other things, a guarantee that the U.S. government would protect all Spanish land rights. However, problems arose because of different social, legal, and surveying traditions between the two cultures. A major problem arose as a result of community grants and the

FIGURE 5.1 THE WAR WITH MEXICO, 1846–1848

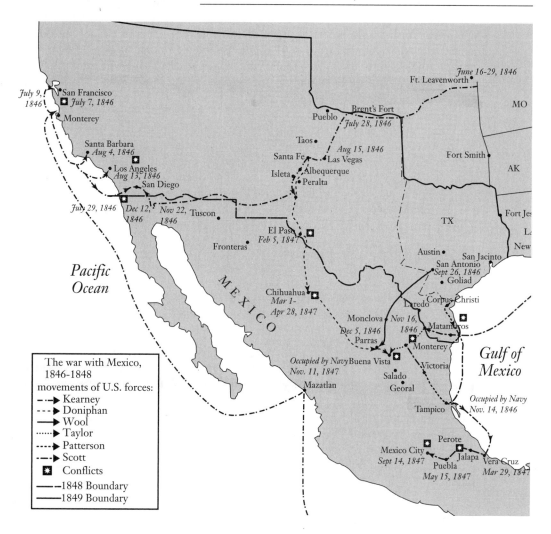

The war with Mexico,
1846-1848
movements of U.S. forces:
- –·–▶ Kearney
- ––––▶ Doniphan
- ──▶ Wool
- ······▶ Taylor
- ·––·▶ Patterson
- ·–·–▶ Scott
- ✳ Conflicts
- – – 1848 Boundary
- ── 1849 Boundary

ownership of common land held in those grants; in American law, land is individually owned, and there was no provision for community ownership in established legal tradition. (Knowlton, 1967; Van Ness and Van Ness, 1980). Another problem resulted from unclear titles and boundaries in which rocks and trees might mark property divisions (Knowlton, 1967). Finally, the surveying techniques that were employed in the United States involved neat rectangular lines that were incompatible with Spanish land lines. As a result, re-surveying property was virtually impossible (Westphall, 1965).

The Treaty of Guadalupe Hidalgo promised to protect property rights, but the Congress of the United States required that all such rights be proved to its satisfaction through legal procedures. The Land Law of 1851 set up such procedures, which required Spanish grant holders to prove their ownership. Later, the Court of Private Land Claims (in operation from 1891 to 1904) served a similar function. Over a half century, Anglos came to control four-fifths of former Spanish grant areas (Barrera, 1979). During its tenure, the Court of Private Land Claims confirmed 2,051,526 acres while rejecting claims to 33,439,493 acres (Barrera, 1979; Westphall, 1965). By 1904, the federal government had acquired control of 52 million acres in New Mexico, 9 million of which ended up as national forest, while the remainder came under European American control in ranching and mining (Swadesh, 1974).

Growing industrial ties to this area altered land use and resulted in the movement of the indigenous minority into wage labor. The Spanish in New Mexico were systematically and relatively rapidly displaced from their land. This not only shortened their list of economic options but also more drastically altered their culture because of the rapid transformations that occurred in their lives. The incoming dominant group did not have other arable land available to them nor did they have other ethnic groups available for wage labor. Thus, the Spanish had few choices in where they worked and what they did.

Opening the West

Cattle and ore in the new Southwest of the United States were strong magnets for population movements, but other areas of the West had similar attractions. Probably the most dramatic was the discovery of gold in California, leading to the massive gold rush of 1849. It began a population growth in California that has yet to drop below flood stage. But gold, silver, and other minerals were found throughout the West. The late nineteenth century West attracted immigrants from the eastern United States along with new immigrants from all over the world. The western economy boomed as old trails became railroads, and Native Americans were once again caught in the middle.

The "Indian Wars": Conflict in the West

By 1845, life had already changed for many Native Americans in the West. Some eastern peoples had already been moved to the new Indian Territory in what is now eastern Oklahoma. Other eastern tribes moved west of their own accord to escape the ever-increasing European movement into their lands. These new tribes created competition for the northern Plains Indians who by then had adapted their cultures to the horseback hunting of the huge buffalo herds. In the Southwest, native peoples in today's Arizona and New Mexico were on the eve of joining the United States as spoils of the Mexican War, and perhaps the worst fate awaited the California native peoples who found themselves between a miner and his gold.

Before the great central plains began to interest European Americans in the East, there was a strong desire to improve transportation to the Southwest and California. In the nineteenth century, better land transportation meant railroads. Thus, in June, 1861, two railroad companies—the Central Pacific and Union Pacific—were given the opportunity to connect California to the East. The Central Pacific began in Sacramento, laying track to the east, while the Union Pacific began in Omaha and laid track to the west. The two lines met in Promontory Point, Utah in May, 1869. Both lines were dependent on immigrant labor: the Union Pacific tapped immigrant groups in the East, particularly the Catholic Irish, while the Central Pacific tapped the new immigrants in California—the Chinese. The combination of cheap labor, the granting to the companies of real estate near their lines, and a monopoly over transportation thereafter created huge wealth for the railroads. In addition, it also increased the attraction of the West to settlers while simultaneously making the central plains more accessible.

It makes little sense to talk about federal Indian policy on the frontier because although there *was* one, the federal government was busy fighting a civil war and laws in the western territories were very flexible. The federal government favored the reservation solution, moving Native Americans from desirable land to less desirable land. Territorial governments just wanted the native peoples gone with little concern about where they wound up. They were encouraged in this goal by territorial business interests and supported by military might. Military support might include the U.S. Cavalry or territorial militias. The level of communication and conspiracy between the federal and territorial governments can only be estimated through historical hindsight, measured usually through recurrent tactics that produced similar results (Brown, 1970; Lewis, 1973). One such tactic was to provoke Native Americans to violence that would then provide a justification for major military intervention. The provocation was necessary because Native Americans generally saw little reason to start wars they were doomed to lose.

Certainly the worst example of such provocations occurred at Sand Creek, Colorado in 1864. Governor John Evans of the Territory of Colorado had a history of advocating wars against the Native Americans in his territory (Lewis, 1973). The massacre at Sand Creek occurred when the 3rd Colorado Volunteers led by Col. John Chivington attacked a camp of the Southern Cheyenne and Arapaho on November 27. At the time of the attack, most of the young men were away from the camp hunting, leaving about 600 people who were largely women, children, and the elderly. The 3rd Colorado Volunteers rode into the camp, killing and mutilating the inhabitants. Pregnant women were cut open and unborn children shot. Genitals were cut off. One militia member claimed he planned to use an Indian scrotum as a tobacco pouch. Even when the attack began, Indians did not resist because they believed themselves to be under a peace agreement with the military. When the shooting ended, 105 Native American women and children were dead along with 28 men (Brown, 1970; Lewis, 1973). This massacre led to reprisals and ultimately to a major war that ended with the Colorado native peoples moving north, opening up Colorado for development.

While Sand Creek is probably the most barbaric example from the war in the northern plains, the Nez Perces story is more typical. The Union Pacific wished to run track through Nez Perces land, and the federal government supported the desire, ordering the tribe to move. When one group of Nez Perces refused, the Cavalry was called in. The ensuing battles led to the 1,000-mile retreat of the Nez Perces into Canada with their chief Tooyalakekt [Chief Joseph] before they were finally captured. While the more peaceful Nez Perces remained on their reservation in Idaho, the rebellious band was moved to Fort Leavenworth in Kansas and later to Indian Territory in Oklahoma (Spicer, 1980).

Over time, these patterns were repeated. An alliance of several tribes did manage one victory in 1876 at the Little Big Horn, but it was far too little and far too late. Virtually all the fighting occurred between the 1860s and 1890, and many more movies were later made than battles had been fought. By the end, the native peoples of the plains wound up in federal custody on reservations. Promised government food was often sidetracked into the black market before it arrived. Hunting tribes often found themselves on land too dry to grow much and told to become farmers or cattle ranchers. Some Native American religions, such as the Sun Dance of the Sioux, were outlawed. Reservations produced massive cultural disruption, deadly boredom, and widespread starvation.

In the midst of all this, a very interesting religion formed and spread among the plains tribes. Known as the Ghost Dance, it originated with the Tovusidokado, a group of the Northern Paiute living in Nevada. Its leader was a man named Wovoka (a.k.a. Jack Wilson) who had had some contact with Christianity in his youth. The Ghost Dance borrowed the idea of the Apocalypse from Christianity, predicting that a day would come soon when all dead Indians would return to life (as would the buffalo) and all Europeans would vanish. The religion also contained an important dance ritual from whence its name came. Although the Ghost Dance was a religion and not a political movement, its prophecies clearly had political import. Its rapid spread from reservation to reservation was unnerving to the federal government.

As the Ghost Dance spread, some tribes added the ritual of the ghost shirt, believed to have magical properties that would stop a European bullet. This belief along with the religion would end at Wounded Knee Creek in what is now South Dakota. A small band of Oglala Sioux were being disarmed before being removed to a reservation. It was four days after Christmas in 1890. The Cavalry now had new automatic weapons. The Sioux were wearing ghost shirts and some were dancing. At some point, a shot was fired and the Cavalry opened fire. At the end, 128 tribal members were dead along with some Cavalry troops, mostly the victims of errant fire from the new machine weapons. The Indian wars of the plains were over and so was the Ghost Dance (Brown, 1970; Capeci & Knight, 1990; Hittman, 1992; Spicer, 1980).

The year 1880 did not mark the end of the native peoples in the United States—they began to grow in numbers again early in the twentieth century—but it did mark the end of many thousands of individuals and quite a few tribes in their entirety. By 1900, the entire population of Native Americans fell to 237,196 people (Spicer, 1980). Recalling the various estimates of the size of the Native American population in 1492, this represents a catastrophic demographic event. Although disease accomplished much of the population decline without assistance, assistance always seemed to be available. Given the experiences of other minorities in the United States, one has to

The burial of the dead at Wounded Knee, South Dakota.

conclude that Native Americans were doomed simply by the impossibility of their being useful as cheap labor. As it was, they were just in the way.

The First Asian Immigration: The Coming of the Chinese

The Chinese who came to the United States did not arrive as indentured laborers, although some were of similar status, and although the newcomers statuses varied, most came for remarkably similar reasons and with very similar goals. Nineteenth century China was in political, social, and demographic turmoil. The population was rising rapidly within an environment of increasing food shortages and increasing warfare, both between regions and with the British over opium importation. In these circumstances, economic advancement was almost impossible, particularly for poorer Chinese (Leung, 1986; Pan, 1990).

Chinese immigrants began arriving in the United States around 1850, searching for wealth. The gold fields of California were a particularly strong draw. Almost all immigrants came from the Kwangtung province in southeastern China (Pan, 1990; Wong et al., 1990). They were young men, most of whom probably saw the journey as temporary; the Chinese word for *emigrant* is synonymous with the English word for *sojourner*. Nevertheless, intentions are often left unfulfilled; the United States was far from unique among destinations in becoming a permanent home for migrating Chinese.

Chinese Immigrants and the Economy   The most important Chinese organizations to develop in the United States evolved from regional differences in China. A regional association (hui kuan) contained immigrants who shared the same region in China and the same dialect (McClain, 1990; Wong et al., 1990). In San Francisco, many hui kuan joined together in the mid-1850s to form the Chinese Consolidated Benevolent Association (commonly called the Chinese Six Companies). This umbrella organization operated somewhat like the Irish Tammany Hall, except it had nothing to do with politics. (Chinese immigrants were prohibited by Congress in 1870 from becoming naturalized and could not vote.) Nevertheless, the Six Companies provided mutual aid, welfare, and general economic support (including business loans) within the Chinese immigrant community (Wong et al., 1990). This aid was essential because Chinese immigrants found either discrimination or indifference from European American sources.

The first destination for Chinese laborers was Hawaii where the large sugar plantations badly needed workers; the discovery of gold in California altered the final destination for many. Chinese immigrants to California developed a pattern of adaptation, connecting certain kinds of labor to their ethnic community establishing what would be called "Chinatown" (see Ichioka, 1979; Pan, 1990). The Chinese are often singled out for their distinctive adaptation, but little in what they did was new. No other single immigrant group contained all the particular characteristics of Chinese adaptation, but all the elements had been seen before.

The promise of wealth in the gold fields of California drew many Chinese immigrants during the 1850s.

A Chinatown is a tight knit ethnic community with a centralized power structure (the Six Companies). Most immigrant groups form such communities upon arrival anywhere; even if they are welcomed elsewhere by the dominant group, they cling together for much needed emotional and economic support, evidenced in a wide variety of ways. Ethnic communities need special services not available elsewhere, such as ethnic food or clothing. The ethnic community offers numerous opportunities for immigrants to start small businesses serving an ethnic need. When those businesses grow, they provide employment for other ethnic group members. Economic support may come in the form of labor contractors who know where the jobs are outside the ethnic community. The community also may provide needed muscle on occasion because immigrant groups often do not receive the same police and legal protection as do native-born people. They must learn to take care of themselves. If the Chinatown is distinctive, it is mostly because it lasts across many generations, often becoming a tourist attraction, whereas most ethnic communities disappear over time.

Chinese immigrants to the United States probably also needed Chinatown for another reason: more than other immigrant groups, they were almost entirely men. Although it is typical of most immigrant groups for men to immigrate first, women and other family members usually follow. Because the Chinese generally planned to return to China for marriage, women did not immigrate. In the early days, only the wealthiest Chinese merchants had wives. Most of the women who did immigrate were prostitutes, many locked into a form of semislavery from which they could not escape (Daniels, 1988; Korus, 1992). In 1870 San Francisco, it is estimated that 7 of every 10 Chinese women were prostitutes (Pan, 1990). In 1860, there were over 18 men for every woman; by 1880, there were 100,000 Chinese men living in the United States but only 4,779 women (Wong et al., 1990).

Chinatowns initially were incredibly poor slums where unattached males sought each other's company along with prostitutes, opium, and gambling. Traditional secret societies called "tongs" found new life in the New World to satisfy these needs. They became highly organized and territorial with different tongs supplying different Chinatowns. The four largest were the Hip Sing, the On Leong, the Bing Kung, and the Suey Sing. When disputes arose over boundaries, tong wars would break out (Daniels, 1988; Tracy, 1993).

Many Chinese laborers ventured into the outside economy, doing whatever wage labor was available. Since they faced considerable employment discrimination, they were forced to work for less, usually one-third less than European American workers (Daniels, 1988). They also did not strike—they could not afford to—leading them often into the role of strikebreakers when European American workers unionized and went out on strike. This made them increasingly popular with employers and increasingly unpopular with striking workers. But they had little choice regarding the type of work they did, which is typical for first-generation immigrant groups.

The two major employers of Chinese immigrant laborers outside of Chinatown were railroads and mining companies. The arrival of the Chinese during the decade before the building of the transcontinental railroad made them prime candidates for work on the Central Pacific, which began its project in Sacramento with California labor. During the eight years of its construction, between 12,000 and 14,000 Chinese were on the payroll (Pan, 1990). Since the Central Pacific had to cross two major mountain ranges, use of explosives was considerable and deaths among the labor force were many. No accurate records were kept of the Chinese workers who died while building that railroad, but it is estimated that 1,200 died on the job (Tsai, 1986). Their employer's attitude toward them can be seen clearly, however, in the photograph for which workers and management posed together when the two railroads finally met in Utah: Chinese laborers do not appear in the picture.

With the end of the railroad construction in 1869, many Chinese laborers moved into mining. In many cases, railroad companies owned the mines. The large number of mines in the Rocky Mountains moved Chinese semipermanently away from the West Coast and into new locations such as Montana, Colorado, Oregon, Idaho, and Wyoming. As the Chinese moved, so too did the Chinatowns. Long-term mining towns such as Butte, Montana, developed thriving Chinatowns, complete with tong wars (Daniels, 1988).

Within the Chinatowns, many Chinese specialized in occupations currently unfilled or work generally viewed as women's work; sometimes their choices were both. A stereotypical fixture of early Chinatown is the Chinese laundry. During the beginnings of the gold rush, laundry was one of the many services that California was not prepared to provide. Some miners actually sent their laundry to Hawaii by ship. The Chinese filled a clear need. Laundries could be set up with minimal capital and required mostly just hard work to be successful.

Over time, Chinese immigrants came to make up between 7 and 8 percent of California's population, but because they were almost all young men and all in the labor force, that percentage comprised 25 percent of the labor force. When it came to the dangerous work on the Central Pacific, their labor was welcomed. After completion of the railroad, their willingness to work for less made them a threat to other California workers. Thus, those Chinese who survived the perils of railroad construction did not have an easy time.

Anti-Chinese Attitudes, Violence, and Legislation   In the western United States, violence directed at Chinese immigrants grew in intensity from 1870 on (Daniels, 1978). As with many minorities—particularly groups labeled as *racial* groups—they could not count on protection from either the police or the courts. In 1871, a mob in Los Angeles attacked the Chinese residents of that then small city of 5,728. When the violence ended, 21 Chinese were dead. Some had been shot, others hanged, and still others burned to death (Daniels, 1988). Riots occurred in San Francisco in 1877, Denver in 1880, Rock Springs, Wyoming in 1885, and the Oregon Snake River Massacre in 1887; altogether, 55 such mob attacks were directed against the Chinese (Pann, 1990; Daniels, 1988).

The Rock Springs riot, described in Box 5.1, shows the interplay between violence and labor competition. It may be the best known of the many mob attacks on Chinese, but the Snake River occurrence is very similar. Once again, Chinese miners were the target. Thirty-one Chinese were murdered and mutilated by a European American mob. While the murderers were known—one even testified for the prosecution—no one was convicted of the crime. In a grisly similarity to the Sand Creek Massacre of the Cheyenne and Arapaho, a local family saved and used a Chinese skull as a sugar bowl (Stratton, 1983).

BOX 5.1: THE ROCK SPRINGS MASSACRE OF 1875

Of all the incidents of nineteenth-century violence against Chinese immigrants, the Rock Springs Massacre is probably the best known. It shows not only the anti-Chinese sentiment common among European Americans of the period, but also the root of that sentiment in labor competition. Daniels (1988; 61–62) offers the following description of what took place:

> . . . the Rock Springs Massacre was more cause specific than most of the anti-Asian racial violence. Rock Springs, Wyoming Territory, was a division point on the Union Pacific Railroad and also a coal-mining center. Chinese had been brought there in 1875 by the Union Pacific Railroad to serve as strikebreakers in a mining strike.[1] The railroad had, of course, employed thousands of Chinese in its construction, and it continued to employ Chinese in its many enterprises throughout the century.
>
> In the Fall of 1885 there seem to have been 331 Chinese and 150 Caucasians employed by the Union Pacific in mining in Rock Springs. A dispute on the morning of September 2, 1885, over who had the right to work a particularly desirable "room" in the mine—miners were paid by the ton so that where one worked in the mine was significant—led to an exchange in which two Chinese were badly beaten by white miners. The white miners, most of whom were members of the Knights of Labor, then walked out.
>
> The men gathered in saloons waiting for a 6:00 P.M. meeting "to settle the Chinese question." Around 2:00 P.M. all the groceries and saloons were persuaded to close—almost certainly by a Union Pacific official—and perhaps 150 men, about half of them armed with "Winchester rifles," headed for the Chinese Quarter. Shots were fired; many Chinese ran. As one eyewitness later described it, "the Chinamen were fleeing like a herd of hunted antelopes, making no resistance. Volley upon volley was fired after the fugitives. In a few minutes the hill east of the town was literally blue with hunted Chinamen." Others, as the coroner's jury later reported, "came to their death from exposure to fire" as much of the Chinese quarter was burned to the ground by the mob.
>
> The official toll was 28 Chinese dead and 15 wounded; property damage was officially assessed at $147,000. The total Chinese population of Rock Springs, which Larson and Crane estimate at between six and seven hundred, was driven away. A deposition collected by the Chinese government in mid-September and signed by 559 former Chinese residents of the town reported that: "while they knew that the white men entertained ill feelings toward them the Chinese did not take any precautions . . . inasmuch as at no time in the past had there been . . . fighting between the races."
>
> Although hundreds of persons must have known who was guilty, the miners had the kind of community consent that lynch mobs often enjoyed. The grand jury of Sweetwater County, clearly speaking for the white majority, indicted no one, claiming no cause for legal action:
>
> > *We have dilligently inquired into the occurrence at Rock Springs . . . and though we have examined a large number of witnesses, no one has been able to testify to a single criminal act committed by any known white person that day. . . We have also inquired into the causes. . . While we find no excuse for the crimes committed, there appears to be no doubt of the abuses existing that should have been promptly adjusted by the railroad company and its officers., If this had been done, the fair name of our territory would not have been stained by the terrible events of the 2d day of September.*
>
> The sixteen whites who had been arrested for riot participation had to be released. The Coal Department of the Union Pacific

THE ROCK SPRINGS MASSACRE OF 1875

apparently knew more than the grand jury; it discharged forty-five miners for participation in the riot. United States Army troops, requested by territorial Governor Francis E. Warren, escorted some Chinese back to Rock Springs on September 9, and the Union Pacific continued to employ some of them there until well into the twentieth century. Sentiment in the towns along the Union Pacific mainline in southern Wyoming, where most of the territory's 50,000 people lived, was virulently anti-Chinese according to the historians of the massacre. The Laramie *Boomerang*, although it "regretted" the riot, found extenuating circumstances, as did the Cheyenne *Tribune*; the Rock Springs *Independent* denounced the railroad's alleged intention of making a "Chinatown" out of Rock Springs and called for a rebellion against both the railroad and law and order: "Let the demand go up from one end of the Union Pacific to the other, THE CHINESE MUST GO."

1. This account of the Rock Springs massacre is drawn from Paul Crane and Alfred Larson, "The Chinese Massacre," *Annals of Wyoming* 12 (1940): 47–55, 152–160. Only the first portion was reprinted in Daniels, *Anti-Chinese Violence*, owing to a publisher's error.

Personal attacks against the Chinese paralleled less violent attacks in state and federal legislatures during this same time. Employers liked the Chinese, but virtually no one else seemed to. Politicians were responsive to the general population's dislike of the foreign culture and labor competition. The first blow hit during the passage of the Fourteenth Amendment in 1870. Just following the Civil War, Congress was busily ensuring the rights of newly freed African Americans with Constitutional adjustments; the original document limited naturalization to free whites only. Then the question of the Chinese was raised. At the same time Congress ensured that newly freed slaves could become American citizens, Chinese immigrants joined Native Americans in being denied that right. With so few native-born children in the Chinese community, they were now a minority of permanent aliens.

A more serious legal blow hit over a decade later with the passage of the Chinese Exclusion Act on May 6, 1882. For the first time, Congress passed a bill that limited immigration to the United States. It would be the first of many. In this bill, Chinese immigrants were singled out for immigration exclusion for a period of ten years. The only exceptions were for Chinese teachers, students, merchants, and tourists. These exceptions left loopholes that led to a tightening of the law in 1884. It is easy to see, however, that wealthier Chinese would be left with far more options in the United States than would poorer laborers. This law stayed in effect until the middle of World War II when the United States found itself thrust into an alliance with China, both countries having been attacked by the Japanese.

The End of Slavery

As we have seen, the years 1845 to 1880 brought tremendous change to the country. But no ethnic group discussed thus far traveled the ups and downs of African Americans during those years. We find them in 1845 as we left them, rooted in an increasingly profitable system of slavery and owned by men willing to fight to keep them there. The war that followed ended slavery and sped former slaves up a roller coaster of freedom and opportunity in just a matter of years. Try to imagine being a slave all your life and then successfully running for the state legislature only five years after you become free. However, just as that roller coaster reached its apex, it traveled downwards just as quickly, landing African Americans in circumstances remarkably similar to the slavery they just left. Eighteen-eighty was just the beginning of a great many difficult years for African Americans in the United States.

Reconstruction in the Old South

The U.S. Civil War lasted less than half a decade, but it cost more American lives than any war ever fought by Americans. It not only ended slavery but also transformed the entire republic, altering every corner of its territory. The years from 1845 until the start of the Civil War were years of constant turmoil. Every new piece of territory gained by the United States held the promise of rocking the balance of political power. There is no need here to retell the history of the Civil War; it is told elsewhere. We need to focus more on its importance for our story. Looking first to the South, how did slaves respond as free people after the end of the war? And how did the South manage economically with its labor force so massively transformed?

Labor in the South: The "Black Codes" and Sharecropping   The "black codes" in the postwar South were a series of laws designed to accomplish two goals. First, Southern planters had just lost their work force and wanted it back. Newly freed African Americans wanted to do anything *but* work with cotton in the gang labor they were used to. Second, European American Southerners in general wanted to retain the social dominance over African Americans they had enjoyed so long. They were not pleased, for example, to be greeted on the street by a former slave; slaves spoke only when spoken to. But of these two goals, the first was clearly the more important. The Southern economy was in a shambles after the war, with both land and cotton prices dropping rapidly. Most of the sugar cane production was gone. The South was once again facing a labor shortage.

Southern states began passing the black codes in the summer of 1865. They were all slightly different but shared a common goal of limiting occupational choices of newly freed slaves. One common feature was to declare vagrancy (having no visible means of support) a crime. Poor people could be forced into labor with jail as an alternative. Other limitations involved African Americans' rights to buy or rent land, own dogs or weapons, engage in skilled urban work, move freely about at all hours, and be "disobedient," "impudent," or "disrespectful" (Boles, 1983; Foner, 1988; Roark, 1977). In Florida, African Americans who broke labor contracts could be whipped, placed in a pillory, or sold for up to one year's labor (Foner, 1988). The intent of laws prohibiting African Americans to own dogs and weapons was to prevent them from augmenting their survival by hunting.

Most Northerners also wanted African Americans back in plantation work at this time. Congress created the Freedmen's Bureau in 1865, which was charged with caring for newly freed slaves. They served as a charitable organization, distributing goods donated by Northerners, and also fulfilling an important educational function. Many Northern women came to the South as teachers sponsored by the Bureau, teaching some 200,000 African Americans hungry for the education denied them for centuries (Foner, 1988). But the Bureau also encouraged the creation of labor contracts between freedmen and planters so as to keep the former employed and the overall economy functioning (Roark, 1977). African Americans would often have to contract for a growing season and in return, planters would have to provide food and housing. If this sounds a little like short-term indentured servitude, it should. But neither African Americans nor planters were happy with the outcome. Planters felt they were not getting the quality of work they got from slaves and the former slaves had other plans altogether.

The end of slavery brought various concerns, problems, and dreams to African Americans in the South. One immediate concern was to locate relatives sold away as slaves. Many black newspapers appeared, all of them filled with ads such as the following from the Nashville *Colored Tennessean*:

> During the year 1849, Thomas Sample carried away from this city, as his slaves, our daughter, Polly, and son . . . We will give $100 each for them to any person who will assist them . . . to get to Nashville, or get word to us of their whereabouts (quoted in Foner, 1988:84).

Former slaves also wanted new lifestyles in every sense of the word. In addition to education, families wanted to own their own land. If men had to engage in wage labor, they wanted their wives (now legally their wives) to remain at home as was proper for all nineteenth century married women. Women were interested in changing their rough, crude slave clothing for the more feminine fashions of the day (which, not surprisingly, brought ridicule from white women). In short, African Americans wanted to live just like European Americans, and gang labor in the cotton fields did not fit those dreams.

Land ownership was not to be for most African Americans. By 1880, only 20 percent of African American farmers in the South worked their own land, and their farms tended to be considerably smaller than those owned by white farmers (Boles, 1983). Meanwhile, planters had experimented with alternative labor solutions. One Georgia planter brought 100 German immigrants to the South in 1866 with disastrous consequences. Another Mississippi planter enticed some Italian immigrants into farm work in 1885. Yet another dreamed of using some of the Chinese indentured labor then bound for the Caribbean to solve labor shortages there; however, this solution never came to pass (Roark, 1977). If Southern farm work was going to get done, African Americans would have to be the ones who did it.

A compromise between planter needs and African American dreams was sharecropping (see Crofts, 1995; Royce, 1993). In this system, a contract was created between a landless laborer and a landowner. The laborer would farm the land and then share the crop at harvest. Typically, if the landowner supplied seed, farm animals, and other materials, the laborer would receive one-third of the sale; if the laborer supplied materials, the split would be equal (Foner, 1988; Roark, 1977). The drawback for the laborer was that he took all the risks. If the landowner supplied materials, this was sometimes done on credit (even more risk for the sharecropper) with interest rates occasionally reaching as high as 60 percent (Boles, 1983).

An omen of what was to come for African Americans in the South occurred in Memphis, Tennessee in early May of 1866. Two horse-drawn hacks collided (accidents happened even then) and the police intervened. One driver was white and the other black. The black driver was arrested. A group of African American Civil War veterans witnessed the scene and became involved. Another group of European Americans formed and the riot was on, lasting three days and claiming 48 lives, all but two of them black. In addition, five black women were raped and hundreds of buildings were burned. The Memphis Police and Fire Departments, both dominated by Irish Catholics, generally supported the white rioters. Once again, we see the threat slavery's end caused poor immigrants who found themselves in competition. Free blacks continued to be labor market competition for all poor people, regardless of color, throughout the late 1800s to the present day (Foner, 1988).

Revenge from the North: Republicans and Reconstruction   Events described in the previous section were watched closely by influential people in the North. In particular, the Radical Republicans (so-called because they favored major transformations in the South) struck back (Mantell, 1973). The Thirty-ninth Congress met in December, 1865 to begin work. The Civil Rights Act of 1865 was followed quickly by the Fourteenth Amendment to the Constitution, designed to protect the rights of African Americans in the South. Both the Thirteenth (abolishing slavery) and the Fourteenth Amendments contained a section empowering Congress to enforce each amendment, marking a clear change in the degree to which the federal government could interact with local affairs throughout the republic. Southern politicians found loopholes in these laws (the "black codes" specified "vagrants"—not "black vagrants"—as lawbreakers) plus most Southern states refused to ratify the Fourteenth Amendment. In addition, practically no freedmen participated in the Southern elections of 1866.

When the second session of the Thirty-ninth Congress began, they were prepared to act. The Reconstruction Act of 1867 transformed the eleven states of the old Confederacy into five military districts. The South would finally become formally occupied military territory. All states were ordered to convene to create new state constitutions protecting the rights of African American citizens (Roark, 1977). This occurred between 1867 and 1869. Finally, Congress passed the Fifteenth Amendment to the Constitution, guaranteeing the rights of all American citizens to vote (although still excluding Native Americans, Asian immigrants, and women). The Fifteenth Amendment was ratified in 1870. All of this had an impact in the South, not only because of a federal military presence but also because of political changes in the region (Mantell, 1973). The Republican Party came to power in the South, and it would be almost a century before they did again, but they made the most of their first opportunity.

The Southern Republican Party was composed of (1) Northerners who moved to the South to invest (called *carpetbaggers* by most Southerners), (2) native white

Southerners who were traditionally opposed to secession in the first place (termed *scalawags* by most Southerners), and (3) free African American men. Party members held power in just about that order, with African Americans having the least power. Nevertheless, members of all three groups dominated Southern elections in the late 1860s, particularly local and state elections. The previously mentioned new state constitutions were written by many of these men. Many Southern police departments gained sizable numbers of African American members: the police departments of Montgomery and Vicksburg became half African American (Foner, 1988). African Americans also served as city magistrates, justices of the peace, and on juries. The late 1860s brought Southern African Americans into a brief moment in the sun.

Beginning in the early 1870s, the federal government slowly began to withdraw its active support from Reconstruction. The Ku Klux Klan also became active throughout the South during this same time. Composed of poor white laborers, planters, merchants, lawyers and ministers, the Klan made special targets of the Reconstruction Southern politicians, murdering white as well as black legislators. Reconstruction finally came to a fairly clear end with the close election of Republican Rutherford B. Hayes as President over Democrat Samuel J. Tilden. In the midst of the conflict over vote fraud, particularly in the South, Democrats called off their attack in what is generally termed the Compromise of 1877. Hayes took office but the federal government ceased its interference in Southern affairs (Foner, 1988; Lay, 1993; Mantell, 1973). The South was back in the same hands that controlled it before the Civil War.

African Americans in the North

While Northern Radical Republicans were busy seeking revenge against the South, they generally neglected to remove racism from their own backyard. Free African Americans had lived in the North since the arrival of the very first Africans via the slave trade. But from the beginning, they had faced discrimination Northern style (DuBois, 1969 [1901]). Most occupations, particularly skilled crafts, were closed to them. This continued because virtually all unions were white only (Pleck, 1979; Whatley, 1993). The results of this discrimination can be seen clearly in an 1871 survey of New York City's African American population. That survey found 400 waiters, 500 longshoremen, 2 physicians, and a handful of skilled craftsmen (Foner, 1988). When African Americans found themselves in competition with newly arrived Irish Catholics in the North, most employers barred African Americans and switched to an Irish workforce. These practices removed African American workers from the majority of construction and service occupations (Pleck, 1979). When they were brought in as strikebreakers, this only increased the conflict between African Americans and the Irish (Maloney, 1995).

Although workplace discrimination is probably the most destructive to a minority group, many other forms of Northern discrimination thrived as well. Oregon would not allow African American ownership of real estate. California, Illinois, Ohio, Indiana, and Iowa would not permit their testimony in court. Ohio required that African Americans post a $500 bond to guarantee "good behavior" while Illinois prevented them from entering the state. Most Northern states maintained segregated public transportation. Only five states, all in New England, permitted African Americans to vote before the Civil War (Foner, 1988; Pleck, 1979). During the Civil War, many of these restrictions were lifted. Nevertheless, it took the Fifteenth Amendment in 1870 to open the voting booths to African American men in New York, Ohio, Indiana, Illinois, Michigan, and Iowa.

In spite of this discrimination, free blacks were more than willing to fight for the Union in the Civil War. Northern whites, however, shared many stereotypes with white Southerners, believing African Americans to be inherently cowardly. It took the battles of African American leaders such as Frederick Douglass to obtain a trial of African American soldiers. The first into battle were members of the Massachusetts 54th led by European American Colonel Robert Gould Shaw. (White officers led nonwhite units of the American military until after World War II.) The heroic efforts of this unit and their high number of casualties opened the military to other African Americans, many of whom enlisted from the border states. At first, African American soldiers were offered less pay than European American soldiers. African American soldiers refused to accept any pay until the military equalized the rates. (Congress voted them past compensation in 1864 for the pay they refused.) In all, 180,000 African Americans fought for the Union in the Civil War; 37,000 of them did not return (Foner; 1988; Pleck, 1979).

Many former slaves headed to Northern cities following the Civil War, often aided by the Freedmen's Bureau in their relocation. The majority of those who moved were mulattos from the Upper South, many of whom had been free blacks before the war. The previously described Northern workplace discrimination continued even as voting booths and public transportation opened, so the move certainly did not solve all African American problems. In addition, some degree of tension existed between Northern African Americans native to the North and the new immigrants. They typically lived in separate neighborhoods and married within their own groups in the nineteenth century. African Americans native to New York went even further, starting a fraternal organization called the Sons of New York in 1884, which was not open to Southern African American immigrants. They apparently wanted no one to mistake them for the newcomers (Pleck, 1979).

The end of the Civil War clearly brought some civil rights to Northern African Americans. Still, the North continued to be home to only 10 percent of the African American population. The occupational opportunities there remained limited. In addition, many newly freed African Americans in the South either feared the cultural change involved in such a move or, more likely, simply could not afford it. In any case, 1880 found the North with a relatively small number of African Americans. That circumstance continued until the early nineteenth century when the labor needs of World War I changed Northern cities forever.

Summary

The year 1845 was a pivotal one in America's ethnic history. It began with the potato famine in Ireland which brought over 1 million Catholic Irish to the United States in less than half a decade. They became the first sizable Catholic ethnic group in the country and the first urban immigrant group, settling almost entirely in the cities of the northeastern United States. Initially, they faced considerable employment and housing discrimination, which forced them into low-paying, unskilled labor (and put them in direct competition with free African American labor). Ultimately, however, their race provided advantages. Perhaps most significant among these advantages was the right to vote—a right which led to the formation of urban political machines and a monopolization of patronage jobs.

The midnineteenth century was also the peak of German Jewish immigration to the United States, resulting in a nationwide community of some 200,000. Facing many discriminatory policies in Europe, these new immigrants created the first significant Jewish presence in the United States. Like the Irish, they settled largely in urban areas but spread out throughout the cities of the time, which decreased their overall visibility among Christian Americans. Many concentrated efforts in the newly forming clothing manufacturing industry—an economic strategy that minimized their competition with other ethnic groups. Others specialized in retail business, laying the foundations for what would become the major department stores in the United States. While they faced some discrimination directed specifically at them, they were more concerned with the taken-for-granted Protestant basis of the public schools and many laws.

These new European immigrant groups sparked anti-immigrant attitudes within the United States. In particular, the Know Nothing movement singled out both Catholics and Jews as targets. The movement successfully ran candidates for public office doing especially well in the mid-1850s. Irish immigrants gained power through the political machines and influence through the rapidly growing American Catholic Church. Germans Jews created organizations of their own—the B'nai B'rith was the best known—to protect their interests. They also were active in numerous legal challenges to laws they felt to be discriminatory toward non-Christians.

The late 1840s enlarged the United States through land annexed following the Mexican War. Most of the current southwestern United States was obtained by force from Mexico. The largest settlement of Mexicans in the area was in what is now northern New Mexico, mostly living in subsistence agriculture. As European Americans entered the new territory, they found land suitable for large scale cattle farming and mining. The coming of the railroads made both more appealing. During the thirty years following the end of the Mexican War, most of the land in northern New Mexico was taken from the Mexican population by European Americans through a combination of cultural differences in ownership, legal procedures, land taxes, and fraud. The now landless population was then forced into wage labor for European American enterprises.

As the new Southwest developed, so too did the remainder of the West. The discovery of California gold in 1849 coupled with mineral discoveries elsewhere spurred European American movement into the Western territories and created conflict over land with the Native American peoples of the area. The federal government and territorial political leaders began a process designed to move all western Native American peoples to assigned reservations where they essentially became wards of the federal government. By 1880, this process was largely completed, leaving Native Americans close to the point of extinction.

The first Chinese immigrants arrived in the United States around 1850. They settled almost exclusively on the west coast of the United States. Many apparently planned to return to China after saving their earnings in the United States; very few women immigrated. Like German Jewish immigrants, ethnic organizations were formed to ease the settlement of new arrivals, find employment, and provide loans for business enterprises. Unlike German Jews, many Chinese immigrants clustered together in tight knit ethnic communities that provided immediate employment. Following the Civil War and the continued Western development of the period, many Chinese laborers worked in mining and railroad construction. Paralleling the anti-immigrant Know Nothings of the East, anti-immigrant groups formed in the West in opposition to the Chinese, whose immigration was halted by law in 1882.

Following the defeat of the South in the Civil War, white southerners initially retained some political control. Efforts were made to control the newly free African American labor force through various laws known as the "black codes." Radical Republicans in the North soon took control of the federal government, which led to the period know as Reconstruction. During Reconstruction, former slaves were given many rights, including the right to vote and hold office. African Americans, Northern European Americans, and sympathetic Southern European Americans controlled the South as former slaves resisted the agricultural work of the past. As Reconstruction ended in the 1880s, political control in the South returned to more traditional hands.

Chapter 5 Reading
Boy, You Better Learn How
to Count Your Money

Aaron Thompson

Aaron Thompson is an African American sociologist whose research, in part, focuses on educational attainment among African Americans. He wrote the following article shortly before receiving his doctorate. Such times of transition are often times to reflect. In the first section of this article, Thompson describes his early childhood, focusing on the importance of his family in influencing him to work hard and to be successful in a newly integrated school system that was often less than hospitable. "Little Aaron" responded to those influences and succeeded, but his story is not the story of all African Americans in the American educational institution. In the second part of the article, Thompson examines some of the differences between black and white families, tracing those differences to the history of discrimination, particularly job discrimination, blacks faced. His overriding message is clear: Social institutions are connected to one another in significant ways. Here we see how the economic institution affects the family, which, in turn, influences the success of children in the educational institution.

The closer I got to the pinnacle of educational achievements, the Ph.D., the more I heard the remark, "Your family must really be proud of you." That remark always brought me pause because the word "pride" did not adequately accommodate the range of feelings that accompanied my approach to the Ph.D. You see, this guy had extremely humble roots, and it is precisely the family institution that made my academic success possible.

Two adjectives best describe my childhood days: poverty and discipline. Love played a major role in our household, but at times other things needed to come first, such as food. Being African American and living in southeastern Kentucky, the heart of Appalachia, did not provide for the grandest of living styles. Even though my father worked twelve hours a day in the coal mines, he earned only enough pay to supply staples for the table; our family also worked as tenant farmers to have enough vegetables for my mother to can for the winter and to provide a roof over our heads.

My father, Aaron Senior, was born in October 1901 in Clay County, Kentucky, with just enough African ancestry to be considered black. My mother, on the other hand, was a direct descendent of slaves and moved with her parents from the deep south at the age of seventeen. My father lived in an all-black coal mining camp, into which my mother and her family moved in January 1938. It wasn't long before the dark, beautiful lady met the tall, light-skinned, handsome man and married him. At age thirty-seven this was my father's first marriage, and he was marrying a women half his age. I remember asking him why he waited so long and why he chose to marry my mother. He answered, "How many eligible black women who are not your kin do you know?" I said, "Very few" and he replied, "There were even less then."

My father always seemed an extremely logical person; he was also quiet, reserved, and somewhat shy. I'm not sure my mother ever really appreciated these attributes in him. The one thing I remember the most about my father was his ability not to let anything or anyone antagonize him to the point of interrupting him in the pursuit of his goals. He often tried to pass this attitude along to his children. He would say to me, "Son, you will have opportunities that I never had. Many people, white and black alike, will tell you that you are no good and that education can never help you. Don't listen to them because soon they will not be able to keep you from getting an education like they did me. Just remember, when you do get that education, you'll never have to go in those coal mines and have them break your back. You can choose what you want to do, and then you can be a free man."

In early adolescence, I did not truly understand what he meant, but now I believe that I can finally grasp what he was trying to tell me. My father lived through a time when freedom was something he dreamed his children might enjoy someday, because before the civil rights movement succeeded in changing the laws African Americans were considerably limited in educational opportunities, job opportunities, and much else in what is definitely a racist society. My father remained illiterate because he was not allowed to attend public schools in eastern Kentucky. The eight brothers and sisters who preceded me also had many barriers to attaining a higher level of education, and many did not exceed the level of my father.

In the early 1960s my brother, my sister, and I were integrated into the white public schools. Since there were so few blacks in our small community, we three seemed to get the brunt of the aggression that so many whites felt toward a race they considered inferior. Physical violence and constant verbal harassment caused many other blacks to forgo their education and opt for jobs in the coal mines at an early age. But my father remained constant in his advice to me: "It doesn't matter if they call you nigger; it doesn't matter if they hate you with all their might; they probably will never know how to be your friend anyway. What really matters is that you are the hope of this family to achieve something better than breaking your back in these coal mines. If it gets too unbearable, get one of them alone and beat the shit out of him, but don't you ever let them beat you by walking out on your education."

My father died at the age of 77, after I had finished my undergraduate degree. He was truly proud that I had made it beyond the coal mines. Now, my mother is another story. You see my mother's method for motivating me was vastly different from my father's. My father was a calm, wise man who conscientiously searched for the right things to say and do. My mother, on the other hand, had no verbal or physical reticence. I can still hear her voice, "Boy, I'm gonna get your ass when we get home." My mother was fire and brimstone, and now in her 70s, her ardor has cooled very little. Mother was the true academic in our house, with an eighth grade education, which made her truly proud. An eighth grade education represented quite an achievement in the 1930s to a southern black woman.

My mother would tell me stories of how blacks were treated in the "old South" and what to watch out for with the white man. She would say, "They don't want you to have an education because then you would have a way to get money; they definitely don't want you to have money." This attitude was part of my mother's philosophy, and it colored mine, since she started my childhood education early. In my preschool years she taught me writing, reading, and, most importantly, how to count money. By the time I was 4 years old, I could walk to the little country store to buy small amounts of groceries, and I knew to the penny how much change was due me. This feat was not so much a credit to my learning abilities but rather a credit to my mother's belief that people will take your money if you don't watch them closely.

When I was ready to start school, the representative from the three-room, all-black school came by our house to talk to my mother. I was five years old and listened closely to the conversation. The representative said, "Mrs. Thompson, we would like to start Aaron in school early because we think that in the next several years the schools will be integrated, and we want him to be able to have enough skills to perform with the little white children."

I vividly remember my mother's classic response to this person. She said, "Shit, my son can read and write better than you can and will someday make a whole lot more money than you. You don't need to get him ready, just don't hold him back." So I started school that year and from then on "Little Aaron" could do no wrong academically in his mother's mind. If I ever had any problems, it was someone else's fault because she made sure I did my homework, attended school every day, and did everything the "No-Goods" down the hollow didn't do.

When I graduated from college, my mother was present and very happy. She often said, "He made it through school without our help. We had no money to help him, but he worked hard and made it. Reckon how much money he's going to make with this degree." My mother definitely appreciates my educational achievements, and the more she can empirically measure it monetarily, the prouder she becomes. She has always been my biggest cheerleader, and my achievements could not have happened without those cheers.

These circumstances help explain why I have trouble using adjectives like "proud" to express how my family truly felt about my education. I believe that most families are proud of their own, but it takes a truly special family to overcome such monumental social obstacles and continue to see hope in the future of their offspring. I look at the negative conditions that my brothers and sisters lived through, knowing how my mother and father wanted opportunities for them, and I understand the frustrations and sadness my folks experienced when the opportunities were not there and some did not achieve. But I lived in my own little world when my parents were giving me those motivational encouragements.

Many black families throughout the United States face seemingly insurmountable obstacles, and the future seems to be a shadow rather than a reality. Many live in conditions of poverty and many in one-parent households. But as my mother always said, "If you listen to the morals that are being taught to you, throw out the ones with hatred, and just learn how to count your money; then you will do okay in life." My mother seems to have the perfect answer to many of the problems of our society in this one statement. But as we all know, when many of us were growing up, there was very little money to count. Today, with both mother and father doing paid labor, it is harder for parents to give the direct attention to children that my mother showered on me. This issue could become problematic since socializing our children to understand that education is the important route to success usually starts in the family at an early age. Economics can be the culprit in the successes and failures of our children in this society. With education, success is not assured, but without education, failure seems imminent. Being poor, black, and Appalachian did not offer me great odds for success, but constant reminders from my parents that I was a good and valuable person helped me to see beyond my deterrents to the true importance of education. My parents, who could never provide me with monetary wealth, truly made me proud of them by giving me the gift of insight and an aspiration for achievement.

Insight and knowledge are the paths for success in people of all races and classes. The family is where these paths begin. The black family is a family of strengths and diversity. Growth and success can take place within our institution regardless of the structural make up. To make this growth and success possible, we need to know the truth of our history, a knowledge of the structure, the courage to seek assistance, and the strength to lend assistance.

To be located in the professional ranks inside most organizations, candidates need to have attained certain levels of academic credentials, at least a four-year degree. Many other organizations require a higher level of academic achievement. For example, a Ph.D. represents the minimum requirement to be a professor in a university. Black Ph.D.'s are at their lowest level since 1975 (Bunzel, 1990:46). African Americans are finding it difficult getting into major colleges and graduating from them (Thompson & Luhman, 1997). In 1984, 617,000 blacks were enrolled in four-year colleges and universities. By 1986 this figure had dropped by 2,000 and continues to drop. Blacks are the only demographic category affected to this degree (Bunzel, 1990:50). By 1990, approximately 29 percent of European Americans had completed four years or more of college compared to approximately 16 percent of African Americans (Pinkney, 1993).

When African Americans are admitted to elite universities, issues of reverse discrimination seem to loom in the background (Bunzel, 1989). Even if blacks are strong academically, research shows that because they are more disadvantaged economically, there is a greater chance of their not pursuing and continuing an education. This is in contrast to white students who might have lower academic skills but have the finances to pursue college (DeMott, 1991). Elite universities are having trouble in both gaining and retaining African American students (Thompson & Luhman, in review; Exum, 1983). Many of these problems' roots lie in the legacy of slavery and the power of racism; the impediments to gaining these necessary skills are rooted in the history of black America (Dill, 1979).

Black Americans and their families have faced segregation, discrimination, and inequalities throughout the history of industrial America. When compared to whites, blacks were more often faced with discriminatory laws, individually and in the family structure. Under slave law, black women, black men, and their children were the property of slave owners. Although during the slave period there were many freed married blacks, family units under slavery existed at the slavemaster's discretion. People could marry, but property could not, and slaves were considered property. Although many slaves defied this law and were married within their own community, slave owners could destroy this bond at any time they saw a need to do so by merely selling one or both of the partners to different slave owners.

After slavery, whites created formal and informal laws for the domination of black labor, a labor they once owned. These "Jim Crow" laws were enacted after Reconstruction.

These laws as much as anything else fostered an ideology of blacks as subordinate and whites as superordinate. These laws also contributed to a division of labor by sexes in the black family, as well as placing barriers to the formation of black two-parent households. For example, if a black woman married a black man, then the property she owned would go to him. Since the laws stated that property could only be owned by males, women did not relish the idea of working to give property to a male, so many decided to remain unmarried. If the black husband did not have a job, then the state could take his property. Of course, there was a good chance that the black male would not have a job, so many marriages did not take place. Thus, with the barriers to the black family being an intact family unit, there was a greater chance for poverty in the black community. Black women faced a dismal prospect for survival above the poverty level because they needed to find a job that could support them and, in many cases, their children. The state made laws saying that if black parents could not afford to care for their children, then the children could be apprenticed out as free labor. When girls were apprenticed out, most went to white households as domestic help. When boys were apprenticed out, they went as outside manual laborers such as blacksmiths. These divisions reflected a wider labor market distinction between men and women as well as the distinctions made in the African American community (Boris & Bardaglio, 1987).

Historically, there is a difference between the family structures of black Americans and white Americans. The work roles inside and outside the households seem to be one of the major differences. American plantation slavery did not make a distinction between the work performed by black men and that done by black women. Both worked in the fields and both worked within the household doing domestic labor. Gender role expectations were very different for black and white women. Black women were not seen as weak: in fact, they were seen as being able to work in the fields, have a baby in the evening, and cook breakfast the next day. (White women, on the other hand, were viewed as weaker than black women, unable to deal with the normal stresses of the day-to-day activities of the plantation. A woman's duties centered on pleasing her husband, whatever his wishes might be.) Black men also experienced different gender role expectations than did white men. Under slavery, the black male understood that both he and his family (whatever family could exist at this point) were at the service of the white family.

In the late 1800s when there was a need for more females in the work force, laws were loosened to include this need. These laws had a significant effect on the white family but very little effect on the black family. Later, when black family members moved into industrialization, they went into the paid labor market at a different pace and level than the white family. Black women most often were paid less than black men or white women, and they always maintained jobs in the paid labor market as servants, seamstresses, laundresses, and other domestic positions. Black women were not allowed to serve as salesclerks, cashiers, bookkeepers, etc., which were jobs filled by many white women in the labor market. In 1900, black women constituted approximately 20 percent of the female population and were 23 percent of the servant population. By 1920 they were 40 percent of the servant population. As the twentieth century got older, the black female servant proportionally grew compared to other demographic populations (Kessler-Harris, 1981:83).

Black men who had job skills in many cases could not practice those skills. For example, blacks were not allowed to join many of the trade unions in the South, where most blacks lived. The United Mine Workers Union in the South used blacks as strikebreakers but had many problems getting black members accepted as regular union members. Thus, in many cases blacks who worked as miners remained outside the union, with inadequate pay compared to the white union members (Gutman, 1975). The black family, because of a history of discriminatory laws undermining family structure, did not have a support system going into the paid labor market. Often there were no husbands in the family. Black women could not depend economically on their men because many did not have husbands or their husbands faced an economic market that discriminated against them. This lack of labor force participation by the black male led the black female to see him as a liability to her and her children, which further undermined the black family structure. To survive in the labor market, black women would accept any job to support their family, but the jobs that were available were the traditionally "female-specific" domestic jobs, such as house cleaners, cooks, nannies, etc. The few jobs that were available to black men were also jobs that were of a domestic nature. These jobs tended to pay less than jobs that were reserved for the white male.

Early in industrialization, a family wage system was enacted. A family wage system is one that is designed to pay enough money to the male in the paid labor force to support him and his family. This system allowed the female to stay in the household and the male to stay in the paid labor force with the title of "head of household." Although laws stated that men were heads of households, black men could not assert themselves as the undeniable heads of their households if they did not have the economic ability to back their claims. Thus, a pattern of single female-headed households started in the black family. Black men clearly did not and could not make a family wage for their family, and so black women continued to work. Since black men did not have the political or economic power structure on their side to help keep their families intact, the patriarchal father was not as dominant in the black household as he was in the white family.

White women and black women shared the burden of being forced to be in domestic positions in the home, and when they had to get paying jobs, they were forced to occupy sex-typed jobs in the labor market. The difference here is that the family wage white women depended upon was considerably higher than the wage black women enjoyed or expected. Without a doubt, black women from the beginning have not been able to depend on a constant family wage; thus, they never have.

Black women have headed their households for most of this century and have been accustomed to accepting all kinds of jobs throughout their lives to support their families. Black men are still experiencing unemployment and underemployment, and when they do get jobs, the majority of jobs are in the secondary labor market or in work that many white men would not accept. Black female-headed households comprise approximately 54 percent of all black families with children (U.S. Bureau of Census, 1994d). This percentage is almost as high as the total black male paid labor force participation. Though it is harder for a black woman to obtain enough education to increase her chances in the labor market, she still surpasses the black male in gaining these necessary resources. With the black male's inability to break the barriers of institutional racism, the ideal of a dual-career black family as the norm is not in the foreseeable future.

Women as a whole are getting more education, and dual-earner marriages are the norm in America now instead of single-earner marriages. Children expect to see their mothers as well as their fathers working outside the household and supporting the family financially. This change will likely bring about a change in the structure of the family. Hopefully, more egalitarian conditions for males and females will emerge. However, the black family, in general, still is not financially stable when compared to the white family. Black women are not experiencing the same level of new-found freedom in the paid labor market that white women are beginning to find. Black men are still underemployed or unemployed when compared to their white counterparts. Until black workers reach a point in our society where they are operating on the same footing as the white workers (equal education, equal employment, equal pay), blacks will be hard pressed to move into the twenty-first century with an egalitarian balance in the family and work.

In conclusion, labor market participation, low wages for both sexes, and discriminatory laws have affected the black family structure, producing the large number of households headed by a female. The family is the primary institution for socialization in our society, and this is where we should start looking for answers and providing solutions. Although I have no instant solutions to any problems suggested in this paper, as a sociologist I do believe there are some directions we can follow, and they can be stated in three simple steps: (1) We should teach our children the importance of education for the sake of knowledge as well as for economic survival. (2) We should set forth a pattern of appreciating cultural and economic diversity (understanding that race and class are social mechanisms for prejudice and discrimination). (3) We should teach our children to look beyond the limitations that society might have placed on them so as to build on steps one and two.

References

Boris, Eileen, and Peter Bardaglio. 1987. "Gender, Race, and Class." Pp. 132–152 in *Families and Work*, edited by Naomi Gerstel and Harriet E. Gross. Philadelphia: Temple University Press.

Bunzel, John H. 1989. "Affirmative Action Must Not Result in Lower Standards or Discrimination Against the Most Competent Students." *The Chronicle of Higher Education*, 1 March, B1(2).

_____. 1990. "Minority Faculty Hiring." American Scholar 59(Winter):39–52.

Dill, Bonnie T. 1979. "The Dialectics of Black Womanhood." *Journal of Women in Culture and Society*, 31(4):543–555.

Exum, William H. 1983. "Climbing the Crystal Stair: Values, Affirmative Action, and Minority Faculty." *Social Problems* 30(4):383–399.

Gutman, Herbert G. 1975. *Work, Culture & Society*. New York: Vintage.

Kessler-Harris, Alice. 1981. *Women Have Always Worked*. Old Westbury, New York: The Feminist Press.

Pinkney, Alphonso. *Black Americans (4th edition)*. Englewood Cliffs, N.J.: Prentice Hall, 1993.

Thompson, Aaron, and Reid Luhman. "Familial Predictors of Educational Attainment: Regional and Racial Variations." Pp. 63–88 in *Race, Ethnicity, and Multiculturalism: Policy and Practice*, edited by Peter M. Hall. New York: Garland Publishing, Inc., 1997.

6

The "New Immigrants" and the Old Minorities: 1880–1965

In 1880 many ethnic stories were either in limbo or transition; it was a time of endings and beginnings of unfolding action:

- Wars between Native Americans and European Americans were virtually over, with less than a decade to go before reservations came to dominate the scene. Native American people and their cultures were extremely weak and their land was available for development. Almost all lived in the West.

- African Americans were looking into an uncertain future as Reconstruction came to a close. Slavery had been over for fifteen years, but political freedom and economic opportunity were on the downswing in both the North and the South.

- Hispanic people in the Southwest had lost their influence and their economic clout. Facing racism and living in poorly paid wage labor, no better opportunities appeared on the horizon.

- Chinese immigrants were the first group to be singled out for immigration restriction. Those already here faced employment discrimination, racism, and violence. In addition, their communities were declining in size, with the drop in immigration and virtually no offspring.

- German Jews were settled throughout the United States and had achieved considerable economic success, particularly in clothing and retail sales. They were, however, facing increasing discrimination and were on their guard.

X The Catholic Irish had survived the worst thirty years they would ever spend in the United States. Although not as economically successful as Jewish Americans, they were developing some political clout, finding better jobs, and receiving some respect.

X The dominant group in the United States—people whose ancestors came from Protestant, northwestern European countries—was thriving. The country now stretched from ocean to ocean, and the economy was growing in leaps and bounds. This was a good time to build great wealth, but it would take a large labor force to keep the factories humming.

In 1880, the United States was settling into a general pattern that would continue, more or less, throughout the twentieth century. European Americans had the most rights and opportunities, with Protestants enjoying more of each than Catholics or Jews. Still, being "white" was becoming increasingly important as the number of racially defined minorities grew. These racial groups—African Americans, Native Americans, Asian Americans, and Hispanic Americans—were all either excluded from the labor force or highly restricted in the kinds of work they could do. Often located in rural areas where opportunity was further diminished, their employment options were dismal. As America became an industrial power, the dominant group much preferred European labor when possible for the growing number of manufacturing jobs. The few years between 1880 and 1914 brought millions of new Catholic and Jewish Europeans along with some Asians to the United States to fill that need. New York City alone grew from 1.5 million people in 1870 to 5 million in 1915, when 45 percent of its population was foreign-born (Friedman-Kasaba, 1996). The sheer size of this migration made this period unique in world history.

New Immigrants from Europe

The United States received much of its modern character from European immigrants who arrived between 1880 and 1914. Their numbers alone—26 million new Americans—would leave a permanent mark. Although immigrants continued to come from Germany (both Protestant and Catholic), Scandinavia, and the British Isles, southeastern Europe was the source of most (Figure 6.1). People from eastern Germany, Poland, Russia, Italy, Greece, Hungary—all the Slavic countries–arrived in New York harbor looking for opportunity, fleeing oppression, or both. Space constraints prohibit doing even minimal justice to this mass of humanity. Rather than trying to tell too many stories, we focus on two groups—Italians and eastern European Jews—to provide a flavor of the diversity of these new immigrants and the time in history they arrived. These two groups were selected for the size of their immigration plus their cultural differences from mainstream America in 1880.

FIGURE 6.1 EUROPEAN IMMIGRATION TO THE UNITED STATES, 1820–1930

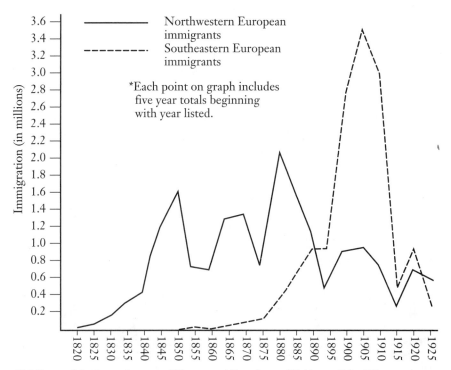

U.S. Bureau of the Census, *International Migration and Naturalization* (Washington, DC, 1970).

The Italians

The first Italians to enter the United States were men, either single or traveling without their wives and children. Typically, they were of working age, unskilled (44 percent), and illiterate (Brown, 1989; Cohen M, 1992). The primary goal for most was to earn money in the new country and return home with it as they had in the past (Cohen M, 1992). Now that the Atlantic crossing was quicker, safer, and cheaper with the coming of the steamship, it was possible for these men to go back and forth several times before finally bringing their families over. In many cases, wives and children would remain in Italy for decades before being permanently reunited with their husbands and fathers. By 1900, the Italian community in the United States was still only 25 percent women (Friedman-Kasaba, 1996).

As with most European immigrants from the mid-1800s on, most Italians became urbanites in the United States. By the 1880s, almost all European immigrants were being routed through the immigration processing center on Ellis Island in the New York harbor. After admission, New York City was their first experience of America. One could find work there and move into a growing Italian ethnic community where life seemed less foreign. Still, it did not take Italian immigrants long to learn of opportunities elsewhere. Even though the main ethnic community remained in New York City for years, Italians appeared all over the United States in the early twentieth century with sizable communities throughout the eastern seaboard, in New Orleans (200,000 by 1920) and even rural Louisiana, in Texas, and the West (particularly in California) (Baiamonte, 1989; Balboni, 1991a; Cotts and London, 1994; De Marco, 1981; Garroni, 1991).

As the Italian ethnic community ("Little Italy") grew in New York City and men were joined by their families, economic survival became a family affair. Men continued in the jobs traditionally open to them as unskilled labor in construction, on railroads, and in factories. Many found work through Italian labor contractors (*padroni*) who would supply groups of immigrant workers to American employers (Henderson, 1976). Because they were limited to the lowest paying jobs, it was necessary for the rest of the family to work as well. Sons tended to work at jobs similar to their fathers and only in rare circumstances were they allowed to attend the public school system. Wives and daughters also worked, but different rules applied and different jobs were available.

Some of these different rules concerned marriage. Almost a cultural universal at the turn of the twentieth century, it was considered inappropriate for married women to work *publicly*. This meant that wage labor or domestic service was out of the question for married Italian women. They could, however, help out in a family business if

A group of Italian "spot" lodgers huddle in a Bayard street tenement.

there was one (considered part of their assigned nurturing family role), take in boarders (if they had the room), or take home handwork to do in the secrecy of their kitchens. This handwork often consisted of sewing or piecework, but more often, Italian women contributed to New York's booming business in artificial flowers. Married women and their daughters often spent evenings painstakingly putting together these ornaments to grace finer homes (Cohen M, 1992).

As for unmarried daughters, New York City offered many opportunities but none very appealing. Daughters were expected to work—if anyone in the family attended school, it would be sons—and they could work anywhere because they were unmarried. In 1905, 46 percent of all Italian women over the age of 16 were wage earners, twice the percentage for all American single women (Cohen M, 1992). Young Italian girls made up large portions of the labor pool in clothing, candy, and box factories. In fact, 78 percent of wage-earning Italian women in New York City worked in factories, more than any other ethnic group in that city (Friedman-Kasaba, 1996). After a day's work, they would pick up the makings of artificial flowers to take home. Unlike their brothers, they were expected to turn over all of their earnings to the family (although many found ways around this). Marriage had to be postponed as long as the family needed this income (Friedman-Kasaba, 1996).

Movement up the economic ladder was slow for Italian immigrants, whose high rates of illiteracy coupled with lack of opportunity to utilize American education kept skills low. In addition, what being unskilled did not prevent, discrimination did. In 1918, for example, when the American economy was short of labor, some job advertisements specified that they would accept "Italian or Colored," showing not only how desperate the employers were but the general ranking of the two ethnic groups on the employment ladder (Henderson, 1976). In 1917, Italians had the highest rate of child mortality of any group in New York City (perhaps related to the fluff and chemicals from the homework women did) (Cohen M, 1992). It is not altogether surprising that some Italian Americans turned to crime, but the stereotypes far outstrip the reality (Gambino, 1991). Recall that this phenomenon occurred with the Chinese. One of the ironies about stereotypical Italian criminal activity is that it really did not begin to occur until most Italian Americans started to assimilate to American culture.

New Jewish Immigrants

In 1880, of the 250,000 Jews who lived in the United States, the vast majority were German Jews. By 1924, approximately 4 million Jews lived in the United States, and almost all had ties to eastern Europe (Daniels, 1991). What had been a relatively small, largely assimilated, and geographically dispersed minority became a large, culturally distinct minority concentrated in New York City and other large eastern cities.

Eastern European Jewish cultural differences not only distinguished them from mainstream American life but also from German Jewish culture in the United States. Eastern European Jews brought traditional Judaism, not to be confused with Reform Judaism favored by most German Jews. Some of them also brought Hassidic Judaism—a new, somewhat mystical, and very strict offshoot of Judaism—that produced cloistered urban neighborhoods in New York and New Jersey. They brought the Yiddish language—a Jewish dialect based on German–which was heard on the streets of New York's Lower East Side and in its theaters, and read in its newspapers in the early twentieth century. They brought the socialist politics of Europe, which were later reflected in labor union activity. They also brought a belief in the foundation of an all-Jewish state in their traditional homeland of Palestine. Eastern European Jews were a diverse lot and their arrival was responsible for major changes in the way American society operated.

Coming to America: From Eastern Europe to the Lower East Side   Eastern European Jews flooded into the Lower East Side of New York City in the late 1800s and early 1900s, with approximately less than 30 percent venturing outside of New York (Daniels, 1991). By 1910, close to a million Jews were in New York City, and half of them lived in the 1.5 square miles of the Lower East Side (Daniels, 1991; Henderson, 1976). At the turn of the century, the population density in that area was 700 people per acre (Howe, 1976). Unlike Italian immigrants, these Jews were much more likely to immigrate as families. If parts of families had to remain behind for economic reasons, it was usually not for as long as with Italian immigrants. By 1900, the sex ratio in the Jewish community was even (Friedman-Kasaba, 1996).

Married women in the Jewish community faced the same limitations we have seen in other ethnic communities of the period. By 1911, only 1 percent worked outside the home (Weinberg, 1988). Many took in boarders and some did homework (although neither as frequently nor for as long as Italian housewives). By 1911, 56 percent of Jewish homes had at least one boarder (Weinberg, 1988). The best solution for most of these women, however, was some kind of home business. In this manner, business, cooking, and childcare could be combined efficiently. Of perhaps equal importance for the women involved, they were not cut off from the community, as the following description from Russian Jewish immigrant daughter Mary Antin illustrates:

> [B]ehind the store was the kitchen, where in the intervals of slack trade, she did her cooking and washing. Arlington Street customers were used to waiting while the storekeeper salted the soup or rescued a loaf from the oven (quoted in Friedman-Kasaba, 1996:129).

Once again, as long as a married woman remained at "home," work was acceptable. Fathers, sons, and daughters were more likely to be employed outside the home. Few children among early immigrants attended school out of economic necessity. The Jewish tradition of literacy and education did play a role, however, in that 54 percent of Jewish immigrants arriving between 1899 and 1910 were literate compared with 26 percent of Italian immigrants (Cohen M, 1992). (The rates for Jewish men were even higher because only they were entitled to a religious education in Russia.) When Jewish immigrant families could spare a child for education, they favored sons over daughters and younger children over older children (Cohen N, 1992; Friedman-Kasaba, 1996). The logic behind the second preference was that older children could earn more, while younger children would be quicker to learn English. Because the preferences varied from family to family, sometimes younger daughters were treated to American learning. By comparison with Italian families, twice as many Jewish sons finished high school in the first generation and three times as many in the second generation; Jewish girls were twice as likely to reach that level as their Italian counterparts (Cohen M, 1992).

Part of the Jewish success with education was due to the skills and traditions they brought with them, but much can be explained by their relative economic success. Children could be spared from the labor force. Although fathers and unmarried children worked in many of the same clothing factories as Italian children, Jewish men tended to have higher level (and better paid) positions, while Jewish women earned 25 percent more than Italian women (Cohen M, 1992). Much of this was no doubt due to their employers: virtually all of the factories were owned by German Jews. With the exception of many young Italian immigrant girls stationed at rows of sewing machines, the clothing business became Jewish from top to bottom. By 1914, Jews made up 70 percent of the entire New York City clothing business (Henderson, 1976). In 1900, 53 percent of Jewish men, 77 percent of Jewish women, and 59 percent of Jewish children were in the "needle trades" (Friedman-Kasaba, 1996). If we include all Jewish immigrants of the time, regardless of job or location, they averaged 14 to 20 percent more in salary than any other immigrant group. Perhaps more impressive, they reached the income of native-born Americans after less than five years in the economy (Chiswick, 1992).

Although this story is beginning to look like a pleasant respite from all the terrible immigrant tales encountered thus far, the Jewish picture is neither entirely rosy nor is it without contradictions. For example, although Jews had the lowest rates of child mortality in New York City (Italians had the highest), they also had one of the higher rates of abortions (Cohen M, 1992; Lindenthal, 1981; Weinberg, 1988). Jewish women wanted to keep their family size down for predominantly economic reasons. They preferred (and were first in line for) Margaret Sanger's birth control clinics and her pamphlet, "What Every Married Woman Should Know," published in Yiddish for those avid readers (Weinberg, 1988). Meanwhile, first-generation Jewish families suffered from high rates of abandonment by their husbands and/or fathers. The *Jewish Daily Forward*—a Yiddish language New York City newspaper—used to run a "Gallery of Missing Husbands" in which photographs of shirking males were published (Friedman-Kasaba, 1996). Part of the family breakups were caused by poverty, but some were related to the emotional problems of reuniting families separated for several years by the serial immigration process so common at the time (Weinberg, 1988).

From Sweatshops to Universities: Overcoming Anti-Semitism   The German Jewish–owned clothing factories in New York City were commonly called sweatshops, at least by the workers. Most factories contained several floors, each covered with rows of as many sewing machines as would fit. The machine operators were almost all teenage girls, most eastern European Jews or Italians. Their job was to produce ready-to-wear clothing. As hard as the work was, it was generally better employment than the available alternatives for immigrant girls. For Jewish girls, it had the added bonus of being a family affair: many of them had fathers working in the same business (Glenn, 1990).

Low wages, poor working conditions, and a tradition of labor activism imported from Europe made this largely Jewish workforce volatile (Brandes, 1976; Glanz,

1976). In 1909 and 1910, 30,000 garment workers went on strike for 13 weeks. Of these strikers, 21,000 were Jewish women and girls. They were joined by 2,000 young Italian women and some 6,000 men, mostly Jewish (Friedman-Kasaba, 1996; Henderson, 1976). Although many young Italian women worked in the clothing trade, they were generally poorer and less involved in the strike than Jewish women. This strike succeeded in achieving higher wages for garment workers and served to found the International Ladies Garment Workers Union. Although this effort was successful for the workers, union activity in the early twentieth century was a potentially dangerous undertaking. The government generally was highly supportive of manufacturers who wished to prevent unions. One garment worker union leader in Seattle–Rebecca August–found herself arrested and held in immigration jail for two months, charged with entering the country for immoral purposes. "Immoral purposes" meant you were connected with prostitution (Weinberg, 1988).

As newly arrived eastern European Jews attempted to better their workplace and further their education, both they and German Jews saw anti-Semitism rise in the United States. Perhaps ironically, wealthier Jews (mostly German Jews initially) faced a greater variety of discrimination than did poorer Jews, including continued discrimination in obtaining bank loans. In the early twentieth century, they faced exclusion from the better hotels and resorts as such practices became common. In one case, Nathan Straus—a co-owner of Macy's—was refused a room at a New Jersey hotel; he

Out of economic necessity, many children of immigrant families worked rather than attending school, such as this Jewish boy at work in New York's garment district.

later returned to build another hotel right next to the offending one, except his was built twice as large (Cohen, 1984). Other Jewish entrepreneurs continued this approach well into the twentieth century, creating vacation resorts that catered almost exclusively to Jews.

Discrimination also appeared in private schools, clubs, professional organizations, and housing. Real estate in upscale neighborhoods was commonly limited by restrictive covenants, much like their modern versions (e.g., restrictions regarding house design) except they included lists of racial and ethnic groups who could not buy the property even if the current owner wished to sell to them. Such restrictions were legal until declared unconstitutional by the Supreme Court in *Shelley v. Kraemer* in 1948. As for schools and professional organizations, only wealthier Jews were likely to encounter such problems; it would take a generation before eastern European Jews sought higher education and professional careers for themselves. On the east coast, they found themselves largely limited to state institutions of higher learning when the ivy-covered doors did not open.

Jews became included in a growing anti-immigration sentiment that stemmed from the massive overall turn-of-the-century immigration. Although these attitudes did not reach their peak until after World War I, the beginnings were already present. In 1907, Congress created the Dillingham Commission to study the question of immigration to the United States. Three years of study produced a forty-two volume report, published in 1910 and 1911. The Commission concluded that (1) humans come from a wide variety of races, which genetically determined inferiority and superiority; (2) recent immigrants to the United States (including all Jews) were all of the inferior type; and (3) the government should strongly consider restricting immigration (Carter et al., 1996; Cohen, 1984). Congress took all of this advice in 1924.

The Birth of Jim Crow: African Americans at the Turn of the Century

At the beginning of the 1880s, about 90 percent of African Americans were in the South where Reconstruction was over and new laws similar to the Black Codes were being instituted. As for the 10 percent of African Americans living in the North, those born in the North were having little interaction with the new southern immigrants from among former slaves. All of them, however, were facing discrimination northern style, where housing, jobs, and education were clearly limited. Being able to ride on an integrated trolley car is of little use if you cannot afford the ticket. Furthermore, they had to compete with millions of new European immigrants. And each of those new European immigrant groups quickly learned just what it meant to be black in American society.

Jim Crow in the South (and James Crow in the North?)

The term *Jim Crow* expresses the overall theme for this portion of African American history. Although the term itself is fairly meaningless (borrowed probably from the words of an old song), it has come to stand for a collection of discriminatory laws coupled with the lifestyle they enforced. It was a lifestyle of black poverty punctuated by violence and subservience to whites.

As for *James* Crow—the wordplay in this section's heading—the intent is to call attention to a more genteel form of discrimination that was almost as damaging. Although the North passed far fewer laws to control African Americans, the results were really not all that different from the perspective of a turn-of-the-century African American. They found themselves at the bottom of the economy because of lacking opportunities elsewhere. They occupied the worst housing in the urban north, received the poorest education (even among poorly educated immigrants in the next neighborhood over), and they too faced violence against which they had little defense.

Jim Crow Southern Style   The Jim Crow laws of the South are what commonly came to be known as the laws of racial segregation. From state laws to city ordinances, black and white southerners came to be separated in almost all aspects of life. Southern African Americans actually preferred some social separation. During Reconstruction, most chose to attend separate churches and social organizations; many even preferred separate schools. But the Jim Crow laws not only enforced separation but also imposed an ongoing inequality. Services provided for African Americans were never equal to those provided for European Americans. "Colored" waiting rooms at train stations were poorly heated and furnished, while "white" waiting rooms provided some creature comforts. "Colored" schools typically were short of books, desks, heat, and sometimes even protection from the elements. "White" schools, on the other hand, were seldom troubled by the wind and rain disrupting lessons for which students had necessary books and desks.

This system of racial separation finally reached the United States Supreme Court toward the end of the century. A southern African American was placed in a segregated railroad car even though he had paid full fare. He challenged the segregation law but was turned down in the landmark Supreme Court case of *Plessy v. Ferguson* (1896). The Court ruled that segregation did not violate the Constitutional protections for citizens as long as the segregated services, while separate, were equal. They were, of course, far from equal. It would be 1954 before the Supreme Court overturned *Plessy* by declaring "separate" as inherently unequal. In the meantime, all of the Jim Crow laws in the South were free to thrive with protection from the highest court in the land.

While the Jim Crow laws of segregation both limited African American opportunity and demeaned African Americans, additional laws provided greater control over them as a labor force. Probably the most significant of these were the laws that imposed peonage on African American laborers. *Peonage* describes a variety of enforced labor situations in which laborers cannot leave their labor. Obviously, slavery provided such an outcome, but the Thirteenth Amendment required a new approach. Given that most southern African Americans were involved in agriculture, many as sharecroppers, instituting peonage was as simple as requiring any sharecropper in debt to continue that sharecropping and/or to not leave the county while in debt. An indebted laborer was therefore forced to work and was not free to relocate. This system worked well as labor control because sharecroppers were often in debt because of dropping

BOX 6.1: PEONAGE IN PRACTICE: UNITED STATES V. EBERHART

The first case tried under the federal government's 1867 anti-peonage law did not occur until 1898. The following, quoted in Boehm (1990:215) comes from the original complaint lodged against William Eberhart for controlling his workers through peonage. The complaint, written to the U.S. Attorney for Eastern Georgia, was lodged by H. O. Johnson of Winterville, Georgia:

. . . a man William Eberhart by name who for years on years has conducted a reign of terror in this community, and whose outlawry the good citizens of this country feel they cannot submit to longer . . . On this man's farm negroes [sic] are held in bondage, whipped almost into insensibility and forced to sign long termed contracts, and the children bound to said Eberhart until 21 years of age. . . Charles Calloway, colored, whose time had expired with Eberhart was hand-cuffed and beat by said Eberhart and one Thomas Erwin until forced to make again a contract for three years. His wife Mary Calloway also signing, also made to bind their six children to said Eberhart until 21. On the same day one John Robinson, also colored . . . was beat almost to insensibility and made to sign a contract for three years. He was more courgeous than Calloway and refused to the last to bind his children.

. . . A negro [sic], Henry Thomas moved to your city, Atlanta, to evade their brutality, having been beat unmercifully by them here in Oglethorpe County. These scoundrels followed him up, and found him in Atlanta, hurried him out of the City, and in your own County of Fulton beat this boy almost to death. The brother of Henry Thomas was in the act of bringing suit if he really did not do so, when Eberhart's brother in law plead and prevailed upon two parties in this Town to run up to Atlanta and hush up matters . . . We demand that these scoundrels be brought to justice, nor am I and others here afraid to stand up to every word I utter here.

---

farm prices, high interest rates, and a susceptibility to being swindled in labor contracts in which the landowner kept the books (Daniel, 1972; Royce, 1993). Contracts could be manipulated easily because African Americans in the South at this time had illiteracy rates between 75 and 80 percent (Daniel, 1979).

Although the federal government had outlawed peonage in 1867 right after they outlawed slavery (Boehm, 1990), this was accomplished by a law rather than a Constitutional Amendment. In addition, peonage was easier to hide than systems of slavery, and by the 1870s the federal government began moving away from enforcing laws in the South. The very first prosecution under the antipeonage law occurred in 1898 with *United States v. Eberhart*. Box 6.1 provides some of the original complaints lodged against William Eberhart. Although this case is more extreme than most instances of peonage, it is interesting to note that Eberhart was initially convicted in a federal court, but later a Georgia appellate court overturned his conviction on the grounds that the 1867 federal law did not apply to Georgia (Boehm, 1990). One wonders where it might have applied—apparently not in much of the South. A federal investigator estimated in 1907 that one-third of the larger plantations in the cotton belt had peons working their fields (Daniel, 1979).

James Crow Northern Style   In 1880, northern African Americans were in a precarious position. The overall community was split socially and physically between northern-born African Americans and post–Civil War migrants. All of them had more political rights than post-Reconstruction southern African Americans enjoyed, but occupational and housing discrimination against African Americans were both firmly entrenched in the 1880s North. Poor housing led to high rates of disease and death—outhouses were often located far too close to water pumps—and because the housing was isolated from white residential areas it also produced educational segregation (Pleck, 1979). In addition, many institutions of higher learning in the North closed their doors to aspiring African American students (Dixon, 1973).

Northern violence against African Americans continued to follow the "race riot" model in which large numbers of African Americans and their property were attacked. The New York City "draft riot" of 1863 was a model to be repeated many times over. The individual or small group victimizing of southern lynching occurred but did not grow to southern dimensions. It was most common in Michigan, Illinois, Ohio, and Pennsylvania (Pleck, 1979).

If the differences between northern and southern discrimination could be summarized in any one way, we would have to argue that discrimination was less formal in the North. Although laws limited northern African Americans in housing, marriage, and a few other activities, the discrimination that did the most damage came from custom more than law (see Sickels, 1972). Occupational discrimination and union restrictions were simply agreed upon by those who held control. Everyone knew the rules, but they were not written down.

Employment and Unions in the North.   Before 1880, African Americans in the North were already losing the job competition battle with the immigrant Irish—the other major group on the lowest rung of the ladder. After 1880, the gap widened dramatically, in part, because of the political domination of the Irish as previously discussed. In New York City, Boston, and other large northern cities, patronage politics provided many public payroll jobs for the Irish that were closed to African Americans. In addition, the Irish were welcome in craft unions, whereas African Americans were not. In 1870, the per capita wealth of Boston African Americans was 40 percent of the per capita wealth of the Irish. By 1890, the Irish in Boston were three times more likely to own their own homes than African Americans; by 1900, they were four times more likely (Pleck, 1979).

The lack of unionization in the South often provided African Americans more job opportunities there. For example, steel mills in the North were unionized; those in the South were not and they employed African Americans. Since American employers had a general pattern of using racial minorities as strikebreakers to create additional conflict among workers, the South became a natural source of African American strikebreakers for northern industry. In the case of a steel mill strike in 1875, the strikebreakers were trained workers from the mills in Richmond, Virginia (Whatley, 1993). This practice was repeated later in the meat packing industry (Maloney, 1995), and some unions finally caught on. In 1881, the Amalgamated Association of Iron and Tin Workers voted to include African Americans to prevent their use as strikebreakers, but most unions continued to follow their policies of racial discrimination. Between 1880 and 1929, African Americans were used as strikebreakers in 141 instances—all but 12 occurred in the North (Whatley, 1993).

In other areas of the northern labor market, African Americans actually faced worse conditions with the turn of the century because of the large numbers of new immigrants arriving every day. As just one example, African Americans maintained a major presence in the North as barbers, comprising 17 percent of Boston's barbers in 1870 (where they were 2 percent of the population). Most African American barbers maintained white-only businesses because their customers objected to sharing scissors and razors used on African Americans. Nevertheless, those customers moved steadily to southeastern European immigrant barbers over the years, dropping African American participation in that Boston occupation to 5 percent by 1900 (Pleck, 1979).

One area of the economy where African Americans could find work was in domestic service. Live-in domestic work was not appealing to immigrants, with the exception of young, single Swedish women who had filled that occupational slot in Sweden (Lintelman, 1991). This predominantly female occupation could not compete with industrial employment—the wages were lower plus live-in status was disruptive of family life. Not surprisingly, the work fell to women who had few other options. In his study of Detroit in the late nineteenth century and early twentieth century, Leashore (1984) found that African Americans ranged from 30 to 40 percent of domestics, while making up only 2 to 3 percent of the city's population. Job discrimination also forced some African American men into this occupation; they comprised 20 to 30 percent of the domestic workforce. Although African Americans needed this work, the African American family paid a price for it.

The Urban African American Family in the North.　African American families in the late nineteenth century and early twentieth century were nearly twice as likely to be headed by women as immigrant families (Pleck, 1979). This distinctive quality of the African American family is still true today. Widows headed the majority of these families. At the end of the nineteenth century, African Americans had slightly higher rates of illegitimacy than European Americans, but a more common cause of female-headed households was separation and divorce. Unlike many immigrant families faced with similar hardships, however, African American women were twice as likely to leave their husbands as the other way around (Pleck, 1979). Although it is difficult today to piece together the exact order of events, African American women had more economic opportunities than men. In a period when a woman's income belonged to her husband, it is easy to see the motivation. Many northern African Americans settled this matter by "housekeeping"—an unofficial living arrangement in which a man and woman would live as husband and wife but without the legal complications.

Although many African American families were female headed, this is not to say that children were neglected. Those northern urban families are best viewed as creative attempts to deal with great hardships in the best possible way, even if the results were not always what family members would have preferred. One characteristic of the early African American family—a characteristic still true today—is that higher percentages of children were raised either by an "other relative" or a nonrelative than other American families (McDaniel, 1994). It was (and is) necessary to have aunts, uncles, grandparents, older children, and neighbors pitch in with child rearing so that parents could be free to earn a living.

Jim Crow and Lynching: The Power Behind the System

In spite of all the laws governing segregation, peonage, and the like, African Americans in the South also faced a great many social constraints backed up by informal violence. As we will see, this violence was in fact highly organized, even ritualized, but none of it was officially part of the legal system.

To most of us, lynching is regarded as the result of some angry individuals taking the law into their own hands. In the nineteenth century South, lynchings occurred so often they became almost ritualized. Many lynchings, for example, were planned so that members of a given community would know both the time and location in advance. The audience—generally a cross section of the community, complete with children—often arrived before the victim. The victim might or might not be tortured (a ritual option) before being hanged, but corpses were regularly mutilated afterwards. Sometimes formal photographs were taken with the audience in a group picture and

As can be seen in this tableau, lynchings were often considered community social events.

hanging bodies in the background. The audience usually also expected souvenirs of the lynching, such as bits of blood-stained clothing or pieces of bone (Buckser, 1992). All in all, lynching was clearly a community social event.

Between 1882 and 1930, 1,655 African Americans were lynched in the cotton states of South Carolina, Georgia, Alabama, and Mississippi (Tolnay and Beck, 1990). Lynchings in all eleven states of the old Confederacy during this period totaled 3,819 (Miller, 1991). Not all lynchings occurred in the South nor were all victims African American. In this same time period, 4,761 lynchings were recorded in all of the United States, and of those, African Americans were 71 percent of the victims (Miller, 1991). Nevertheless, these figures clearly show lynching to be largely a racial act and a southern event. (African Americans also dominated legal executions in the South during this period; the 1,299 African American individuals executed represented 90 percent of all executions (Tolnay and Beck, 1990).

Lynchings reached their high point in the South during the 1890s, with Alabama leading the nation with 177 lynching victims during that decade (Feldman, 1995). The peak year for all of the South was 1892 when 255 people were murdered (Miller, 1991). After 1900, the number of lynchings decreased, but the percentage of African American victims rose to 90 percent, up from 70 percent in the 1890s. The highest percentage was 98.7 percent in 1913. In addition, the lynchings themselves became more brutal in the twentieth century (Feldman, 1995; Miller, 1991).

One-third of all African American lynching victims were accused of raping white women; the remainder were accused of a wide variety of crimes (Alexander, 1992). This fixation on sexual crime, however, tells us more about the accusers than the accused. Hall (1993) argues that the rape component of lynching functioned as a kind of folk pornography in which sexual topics became legitimated. He also points out, however, that these charges were a way of instilling fear into southern white women in an effort to maintain their subordinate status. African American journalist Ida B. Wells made this same charge in the 1890s when she spearheaded an attack on lynching, arguing that lynching was sexist as well as racist (Alexander, 1992; Miller, 1991). In 1930, Jesse Daniel Ames—a white woman—founded the Association of Southern Women for the Prevention of Lynching. A suffragist, Ames argued that the sexual aspects of lynching served to maintain the weak image of women; men placed women on pedestals as an alternative to allowing them rights (Hall, 1993; Miller, 1991).

Sociological explanations for lynching generally have economic roots. Beck and Tolnay (1990) noted the relationship between lynching and the price of cotton—when cotton prices went down, taking all aspects of the southern economy with it, lynching increased. Many social scientists have used these and similar statistics to argue that lynching was a kind of displaced aggression in which frustrated whites took out their anger on the most helpless around them. But there may well have been a more rational side to lynching as well. African Americans were helpless, but there was still the need to force them into certain occupations. If they were allowed to be free wage workers, they could easily become economic competition for poorer white southerners.

Holmes (1969) noted the potential for African American economic competition in his study of Mississippi farm foreclosures. When European American farmers lost their land to banks, those institutions often hired African American labor to farm the newly acquired land, exacerbating racial conflict. But a broader perspective is called for. Lynching was most common in socially and economically depressed southern counties where poor European Americans had the most to fear from African American competition (Feldman, 1995). In addition, mob violence peaked at certain times during any given year, rising at times of high labor demand when the cotton crop needed attention (Beck and Tolnay, 1992).

Violence against African Americans at the turn of the century appears to have had both rational and irrational components, which dovetailed nicely from a European American perspective. An individual could vent anger without fear of retribution, while simultaneously supporting his economic interests. In the broader scope of dominant-minority relations, lacking protection from the state against random violence is one of the most coercive forces in controlling behavior. If your neighbor can murder you at any time without a penalty and if both you and your neighbor know this, you have little choice except to become very agreeable.

Moving North: African Americans Enter the Industrial Era

World War I began in Europe during August of 1914, approximately two years before the United States joined in militarily, but because war is generally good for business, the economy was affected instantly. U.S. urban industrial centers began booming, and even with the large immigrant population, began facing labor shortages. These shortages became acute when the United States entered the war, transforming potential laborers into soldiers. In desperation, manufacturers in the North looked to the surplus African American labor force in the South and opened their doors to employment. Labor agents were sent to the South, offering free transportation along with a strong sales pitch (Halpern, 1992). Good jobs with unheard of pay for African Americans proved a magnet too strong to resist. Between 1916 and 1920, over 500,000 southern African Americans migrated to the North, creating a major urban presence there for the first time (Carter et al., 1996). Labor shortages during World War II would result in a similar migration, with the only difference being that jobs were then available in West Coast cities—Los Angeles and the San Francisco Bay Area in particular—which led to the growth of African American communities there also.

The Growth of African American Urban Life in the North

Around 1900, most northern cities contained African Americans. The largest community was in New York City, where most lived on the West Side. Typically, however, none of these communities exceeded 1 or 2 percent of a given city's population. In many cities, African Americans might live in more than one area because modern racial housing segregation had yet to become as rigid as it would be in twenty years. In short, northern African Americans had a relatively low visibility, especially when compared with the southeastern European ethnic communities that were bursting at the seams.

The World War I migration boom drastically changed the African American population in almost every northern city, including Detroit; Chicago; St. Louis;

A black family arrives in Chicago from the rural South.

Cleveland; Washington, D.C.; Boston; and New York. Between 1910 and 1920, the African American population of Chicago grew from 44,000 to 110,000 as the meat packing industry boomed; in the same period, Cleveland grew from 8,000 to 34,000 (Halpern, 1992; Meier and Rudwick, 1976). These numbers still cannot compete with the 3.25 million Italians that arrived in the United States between 1901 and 1920. Italians clearly faced their share of prejudice and discrimination—they would even be classified as a racial minority by 1920—but the African American heritage of slavery in the United States made them a special target and the source of special fears for European Americans. As a general rule, as the population of African Americans grew in a city, the more segregated housing became and the more forms of discrimination would develop. In Boston and Chicago, for example, hotels and restaurants that had previously served African Americans ceased to do so after the migration (Meier and Rudwick, 1976).

Just as the war had created opportunity for African Americans, its end produced a great many problems. At first, the economy continued to be strong, and returning soldiers found employment. Nevertheless, the ever-present tension among races only increased as the economy began to turn down into the recession of 1920 and 1921. Northern urban race riots occurred in almost all cities during these few years (Carter et al., 1996), but in 1900, New York City would host that city's worst riot since the 1863 "draft riot." The area of the city known as "Hell's Kitchen" had become the crowded home of several different southeastern European immigrant groups. Violence broke out between them and the then relatively small African American community nearby.

The largest of the postwar riots, however, occurred in East St. Louis (1917) and Chicago (1919). The latter began on the afternoon of July 27 when a group of African Americans wanted to swim at a traditionally white beach. As fighting began, roaming groups began to form and mobs clashed. Five days of rioting led to 23 African American deaths, 15 European American deaths, 500 persons injured, and hundreds of homes burned. The extreme length of this riot is attributed to Irish street gangs associated with Ragen's Colts—an "athletic club" sponsored by Democratic alderman Frank Ragen who allegedly provided police protection for the gangs to continue their violence (Halpern, 1992). The early competition and violence between the Irish and African Americans was apparently still very much alive in Chicago.

New York, New York: Harlem Happens   Of all the northern cities that became home to migrating African Americans, New York City was a singular experience. By 1900, it already had a thriving African American community on the West Side. As increasing numbers of African Americans migrated to New York in the early years of the century, housing segregation became stronger, and the West Side community became overcrowded. Segregation almost always produces overcrowding, which is why it pushes rent and property prices up—slums do not provide cheap housing if their occupants have no alternative places to live. As overcrowding on the West Side became almost unbearable, however, some interesting changes occurred in a nearby section of the city known as Harlem.

Harlem had once been a quite fashionable section of New York City and was still a thriving, all white neighborhood in the first years of the twentieth century. Rumors of a new subway line had encouraged real estate speculators to build a great many apartment houses between 1898 and 1904. The unbuilt subway coupled with overbuilding led to dropping rental prices and empty units, putting European American real estate speculators in a bind. Meanwhile, African American real estate speculators were watching with interest. One of them—John M. Royall—proposed that Harlem be opened to African Americans, whereby all involved could become rich. Other companies, such as the Afro-American Realty Company, also became involved. Within a few short years, Harlem was largely black and many real estate speculators (both black and white) were largely wealthy (Anderson, 1981).

Like the West Side, Harlem residents represented the entire African American community, from rich to poor, and because most wealthy African Americans gravitated toward New York City, Harlem became a very interesting and solid community indeed. The more interesting Harlem became, the more African Americans were attracted to it, particularly those whose interests could not be developed elsewhere. If you were African American and also an artist, philosopher, political activist, writer, or musician, Harlem in the 1920s was the place for you. This period in Harlem's history

Duke Ellington

is generally called the Harlem Renaissance because the unique social environment it offered allowed African American cultural life to thrive and become transformed. If modern urban black culture has a birthplace, it was surely Harlem.

African Americans and the Arts: The Beginnings of Northern Black Ethnicity
As African Americans became free to express religion, art, and emotions, a new sense of identity formed. Even under slavery, African Americans had the opportunity, although clearly controlled, for religious and artistic expression. Slaves produced music and dance both for themselves and at the orders of their masters. Following slavery, entertaining whites was one of the more (and one of the few) lucrative options open to African Americans. The end of the nineteenth century produced a wide variety of African American composers, performers, and writers, many of whom can truly be described as groundbreaking artists. African Americans have been behind most innovations in American popular music for the last century. Although whites stereotypically attributed such talent to blacks having "natural rhythm," African American musical success is much better explained by the lack of opportunity elsewhere. If the only good paying job around is playing the piano, it stands to reason that a great many people would try out their fingers on the ivories.

The syncopation behind turn-of-the-century ragtime music ultimately transformed popular music. African American composers such as Scott Joplin wrote such hits as "The Maple Leaf Rag" and "The Entertainer," bringing him a fine income from sheet music sales alone. By the 1920s, stride piano and the blues had evolved into modern jazz—America's true musical contribution to the world. Even rock and roll would not exist but for these African American musical roots.

African American performers in the nineteenth century had to work to build a niche in the entertainment world. Since they performed for white audiences, they had to provide what the audiences wanted. At first, this was the minstrel show. Minstrel shows were combinations of music, jokes, and short skits, all of which caricatured African Americans as stupid and lazy. The performers traditionally had been white performers in black face (usually burnt cork rubbed on) with large red lips painted on. Costumes were almost clownlike. Early African American performers got their start basically making fun of themselves.

The early African American team of Williams and Walker (Bert Williams and George Walker from Antigua and Kansas, respectively) originally had to bill themselves as "Two Real Coons" in New York City of 1896 to distinguish themselves from white performers in black face. Some performers, such as Ernest Hogan from Kentucky, were also composers. Hogan's most popular composition—"All Coons Look Alike to Me"—indicates how far these performers had to go to please white audiences. All of these early performers were fully aware of what they were doing and had hopes of moving their performances steadily away from the degrading minstrel format, however, even in the nineteenth century, many performers felt Hogan had gone "too far" with his popular song (Anderson, 1981).

By the early 1900s, New York City's Broadway had become the home to musical productions in which producers, directors, composers, and performers were all African American; only the audience was white, segregated by the African American producers who knew that whites would not attend if they had to sit with blacks. Artists such as Bob Cole, James Weldon Johnson, J. Rosamond Johnson, Noble Sisle, and Eubie Blake created distinctive theater. Was their impact lasting? If you can hum a few bars of "I'm Just Wild About Harry," then you can credit Eubie Blake for the song's longevity. As he once observed, that song kept his wife from having to scrub white folks' floors all her life.

New York's connection to the early development of African American art can in part be attributed to the vitality of its African American community (Baker, 1987; Huggins, 1971). The aforementioned artists were joined by writers and poets such as Paul Laurence Dunbar; Langston Hughes; Countee Cullen; and later, Richard Wright. Yet more musicians and performers who appeared during the Harlem Renaissance were Paul Robeson (also a fine actor), Bessie Smith, Ethel Waters, Bill (Bojangles) Robinson (Harlem's unofficial mayor), Fletcher Henderson, Duke Ellington, Louis Armstrong, Cab Calloway, and Fats Waller. In the early 1930s, Ella Fitzgerald and Billie Holiday would debut. Prohibition during the 1920s coupled with many nightclubs featuring such exceptional talent filled those club tables with white music lovers who could hear early jazz only there. The most famous was probably the Cotton Club, featuring Duke Ellington and his orchestra. As with African American Broadway productions, most clubs would not seat blacks (Anderson, 1981).

Black Politics Comes of Age: From the Courts to the Streets

While African Americans were expressing their developing sense of ethnicity in art, they also began expressing it politically (a very fine line separates the two). Frederick Douglass had been an impressive leader in the last half of the nineteenth century, but he had little company. A much milder new voice appeared at the 1895 Cotton States and International Exposition in Atlanta in the person of Booker T. Washington. In the speech he delivered there, he called upon African Americans to improve themselves through hard work and education. He was confident that white acceptance would follow. He went on to found Tuskegee Institute in Alabama, which focused on the vocational training he believed was appropriate for most African Americans. He remained a conservative voice in black politics, but the most important changes in African American life occurred through political activities in the North.

The Battle with Jim Crow Begins An alternative to Washington was African American sociologist W.E.B. DuBois, who had received his Ph.D. from Harvard. DuBois' ongoing study of race relations in the United States convinced him that whites would never find the acceptance of blacks to be in their best interests. If blacks were to make political and social gains in American society, they would have to do it themselves through an organized concerted effort. Along with similarly minded young intellectuals, he founded the Niagara Movement in 1905, which initially focused on developing African American politics.

Two new African American political organizations quickly followed the Niagara movement. In 1909, the National Association for the Advancement of Colored People (NAACP) formed, absorbing the members of the old Niagara Movement along with many other people. The NAACP would come to focus on political rights for African Americans, spending a great amount of its time and resources in courtrooms. Initially, they were aided by many politically conscious European American lawyers. The other organization was the National Urban League, founded in 1912. The primary focus of this organization soon came to be the informal job discrimination that was so effective at keeping northern African Americans away from advancement.

The NAACP legal staff argued numerous civil rights cases. In 1935, the NAACP created the position of special counsel, first filled by Howard University's Charles H. Houston. Houston's student, Thurgood Marshall, would later serve as special council and go on to become the first African American to serve on the United States Supreme Court. The NAACP special counsel and his legal staff filled the courts with attacks on *Plessy v. Ferguson*, showing that almost all separate facilities for blacks were very far from equal to those provided for whites. Most of these cases involved school segregation in the South. Years of such victories ultimately resulted in accomplishing the supreme goal—getting *Plessy* overturned in 1954 (Tushnet, 1987).

Other battles against discrimination carried great symbolic weight. African American contralto Marian Anderson was scheduled to sing in Washington D.C.'s Constitution Hall in 1939. The Daughters of the American Revolution (DAR) canceled the concert because of Marian Anderson's race. Activist First Lady Eleanor Roosevelt promptly resigned her own membership in the DAR and helped arrange for Anderson to give an outdoor concert from the steps of the Lincoln Memorial on Easter Sunday of 1939. Seventy-five thousand people attended.

Yet another battle was fought over the segregated United States military. World War II brought this segregation into the spotlight. African Americans were placed in racially segregated units under white officers. Even the military blood banks were segregated, lest a white soldier receive Negro blood. Ironically, an African American physician from Howard University—Dr. Charles Drew—invented the process of storing blood plasma. Segregated troops survived the war but not the administration of President Harry Truman following the war. In 1948, he desegregated the United States military by executive order. White officers protested, arguing that white soldiers would refuse to work with black soldiers or serve under black officers. Truman concluded that they would learn.

The Civil Rights Movement and Black Power   The desegregation of the American military was a significant step forward for African Americans, but American society continued to be largely segregated in both the North and South into the 1950s. The Civil Rights Movement was responsible for bringing about major changes in segregation, especially in the South, in the ten short years between 1955 and 1965.

Political efforts among African Americans in the South had been ongoing since the end of World War II, but they gained national attention in 1955 through the bus

Civil rights demonstrations were often met with resistance by local law enforcement, as in this photo of police turning water hoses on demonstrators in Birmingham, Alabama.

boycott in Montgomery, Alabama where African Americans were required to sit at the back of public city buses and to offer their seats to European Americans should the bus become crowded. On December 1, 1955, a young African American woman named Rosa Parks refused to give her seat to a European American male passenger and was arrested. She was a young political activist who intentionally broke the law, hoping to bring together the local African American community in protest. Although many other areas of inequality existed in Montgomery (e.g., in 1955, 70.2 percent of African Americans were engaged in unskilled (and poorly paid) labor compared with 18.5 percent of the European American population [Morris, 1984]), the seating on public transportation was such a clear form of discrimination that it served to be a better focal point on which to begin community protests. Montgomery's African Americans initially viewed the protests as local, but a young, African American Baptist minister at the Dexter Avenue Baptist Church in Montgomery soon changed that.

African American political leaders in Montgomery were searching for a new protest leader, preferably one who lacked long-term ties to the community who might be able to act more freely. They found that person in Martin Luther King, Jr., a minister still in his twenties. King stepped to the forefront as the community began what would be a year-long boycott of the Montgomery buses. King's frequent speeches of encouragement to the African American community helped maintain their enthusiasm while also attracting national attention. A charismatic speaker who presented himself with pride and dignity, King was the kind of leader who could speak to his own community while also touching many members of the European American community. When federal judges ordered the Montgomery buses to desegregate in late 1956, the Civil Rights Movement was on its way.

By January of 1957, the Southern Christian Leadership Conference (SCLC) was formed, with King as leader. As the scope of change broadened, movement tactics changed. Boycotts came to be replaced by civil disobedience—the intentional breaking of state and local laws as a means of generating both publicity and court tests of law constitutionality. By the late 1950s and early 1960s, organized groups of movement followers were forcibly desegregating restaurants, interstate bus travel, and other segregated areas of the South. In addition, local court injunctions against demonstrations and marches were ignored. An immediate goal of such lawbreaking was to stimulate local law enforcement personnel into using violence while arresting demonstrators. Demonstrators meanwhile remained nonviolent, creating a vivid contrast that, with luck, would appear on national television news programs that evening.

The tactic of civil disobedience worked better on some occasions than others. King's most famous success occurred in Birmingham, Alabama in the spring of 1963 when the local Commissioner of Public Safety—Eugene "Bull" Connor—played into the hands of demonstrators by attacking them with police dogs, cattle prods, and fire hoses. As public opinion moved toward the movement (and as federal courts offered increasing support), many of King's goals came to fruition with the Civil Rights Act of 1964 and the Voting Rights Act of 1965.

The SCLC was the most prominent organization of the Civil Rights Movement, but it was not the only one. Many younger African American activists were more attracted to the Student Nonviolent Coordinating Committee (SNCC), formed in

1960. SNCC specialized more in lower profile, grass roots organizing throughout the South and often found itself in conflict with the SCLC. By the mid-1960s, politics within the African American community became more diverse as northern, urban African Americans joined the scene, and movement tactics and goals came to be questioned. King seemed committed to a goal of assimilation for African Americans, but the growth of the new "Black Power" movement around the country carried a new message focused on bringing power into black communities and removing whites from movement organizations. Former SNCC leaders, such as Stokely Carmichael, were joined by young northern activists and "black power" became the battle cry for the last half of the decade. Assimilation moved aside for cultural pluralism.

The Anti-Immigrant Twenties

One might assume that the political changes that began in the African American community during the 1920s would have riveted European American attention in that direction. Strangely enough, African Americans were really not seen as all that much of a threat during that decade; the threat would not be perceived until the beginnings of the Civil Rights Movement in the 1950s. During the 1920s, Protestant America was much more focused on Asians, Catholics, and Jews, all of whom were current and potential future immigrants. An interesting collection of historical, economic, and scientific circumstances seemed to work almost in unison to maximize and rationalize their fears and prejudices. The result was the virtual end of legal American immigration for many years.

Anti-Immigrant Attitudes

Recall first that anti-immigrant feeling was alive and well before the outbreak of World War I. Restrictions on Asian immigration already had a thirty-year tradition. In addition, the flood of most Catholic and Jewish immigrants from Europe reached its highest level in the years just before the war. Congress had already launched a large scale study of immigrants in America, reflecting that concern. World War I is probably best viewed as a time of economic prosperity coupled with distracted attention. Both are clearly understandable, but the war's end removed the distraction and, within a few years, the prosperity. We have already seen how the recession of the early 1920s stimulated race riots in the United States; that same recession helped spur anti-immigrant attitudes.

Conveniently, the Ku Klux Klan had just been revived by William J. Simmons, who organized the new Klan in Atlanta on October 16, 1915. The original Klan formed in reaction to Reconstruction; the new Klan really did not have an agenda, but the original organization had recently been glamorized by the recent Hollywood film *Birth of a Nation.* Enough members were attracted to encourage Simmons, who had organized several fraternal organizations for a profit (among other things, he sold insurance to new members). By 1920, Simmons was joined in leading the Klan by two publicity agents. The Klan moved North to where anti-immigrant attitudes were

The schoolchildren's morning ritual of reciting the Pledge of Allegiance was originally an attempt to reinforce new immigrants' loyalty to the United States.

already brewing, and membership grew rapidly. Adapting to the scenery, the 1920s Klan practically ignored African Americans and Jews, spewing most of its hate on Catholics. Simmons later sold his interest in the organization for $90,000, and the Klan went on to be a major political force in 1923, primarily in Indiana and Ohio. By 1925, it was largely disbanded, but in the meantime, it successfully supported many political candidates who favored immigration restriction (Higham, 1981; Smith, 1978).

A book by Madison Grant, *The Passing of the Great Race*, was published in 1916. Grant postulated that Europeans belonged to three different races: moving more or less from the northwest to the southeast, one would find Nordics, Alpines, and Mediterraneans (Grant, 1921). As with the Klan, 1916 was not a prime year for these ideas, but a new edition of the book in 1921 was the right idea at the right time. Here clearly was evidence that southeastern European immigrants were lowering the racial stock of the United States. A few years before, the field of psychology produced its first intelligence (I.Q.) test, which was supposed to measure intelligence rather than knowledge. Amazingly, northwestern Europeans (the very people who made up the test) obtained the highest scores on this new test (Higham, 1981).

Immigration Restriction: The National Origins Act

With all of these beliefs, fears, and prejudices thriving in the 1920s, it is small wonder that Congress acted as it did. Having considered various kinds of immigration control over the years, the first major offering was the Immigration Act of 1921, which imposed a literacy test on immigrants (designed to keep out unskilled labor) along with a quota for every country. The law created an elaborate mathematical formula, which allowed every country a number of yearly immigrants equal to 3 percent of that nations population as counted by the 1910 census.

In 1924, Congress passed the Johnson-Reed Act (commonly called the National Origins Act), which would govern immigration to the United States until the mid-1960s. Modeled on the 1921 act, the National Origins Act changed the percentage from three to two and the census year from 1910 to 1890. These changes, especially the latter, served to undercut almost all further immigration from southeastern Europe. The law also tightened further Asian immigration. The result was that northwestern European countries were allotted 85 percent of the total 150,000 immigrants to be allowed in during any given year (Carter et al., 1996; Higham, 1981).

Native Americans: New Citizens and Old Struggles

Native Americans entered the twentieth century with very little going well for them. The fighting was over, but the peace was almost as hard. They lived on reservations as dependents of the federal government, unlike any ethnic group in the history of the United States. They also related to that government as groups (or tribes) rather than as individuals, again unlike any other ethnic group. And finally, they were not citizens of the United States.

Native American Cultures and Reservations

As we saw earlier, the earliest Native American reservations were augmented by boarding schools for children. These schools were originally run by religious organizations who had the conversion and assimilation of the children as their goal. As a general rule, the schools succeeded in damaging Native American cultures and were unsuccessful in Americanizing the children (Boseker, 1994; Rader, 1991). These schools were later placed under the control of the federal government's Bureau of Indian Affairs (BIA), which ran them from the 1880s until the 1930s (Hendrick, 1976; Liska, 1994). The BIA maintained both day schools and boarding schools, but both were geared toward the Americanization of the students. Starting in the 1930s, many states began incorporating Native Americans within the public school system. However, direct Native American control over the education of their children would have to wait until the 1970s.

The BIA was responsible for running reservation life until the 1930s, after which it played a strong role in conjunction with tribal governments. Ultimate authority was often blurred. The "dependent nation" status of Native Americans carried over into

A Navajo family sits in their hogan on the tribe's reservation in Northeastern Arizona.

BIA activities in that reservations were often viewed as being held in trust for the people who resided there. This unique relation with the federal government left a great many questions unanswered or unclear. The history of laws and Supreme Court cases involving Native American sovereignty (or its lack) is therefore of critical interest.

Native Americans v. the Federal Government: Laws and Legal Battles

The unique legal relation between Native Americans and the federal government began with the Cherokee court cases of the 1830s, but those decisions left many questions unanswered. The Indian Removal Act of 1830 set a precedent that "domestic dependent nations" should be relocated at will by the federal government whenever it thought best; thus, reservations were born. However, many other questions would arise over the coming years.

Going back a little into the late nineteenth century to pick up this story, we find the decision of ex parte Crow Dog (1883), which gave Native Americans on reservations the right to deal with reservation crimes independent of state or federal courts. Congress responded in 1885 with the Major Crimes Act, which removed reservation crimes such as murder from Native American jurisdiction. The law was followed two years later by an extremely important attempt by Congress to forever alter Native Americans and reservations—the Dawes Act (or General Allotment Act) of 1887. The goal of this law was to divide reservation land into individually owned parcels (allotments). The new owners of this private property could then do anything with their new possession, such as leasing or selling it to European Americans (Stuart, 1977). They also could hold their land and become independent ranchers or farmers; some historians have argued that a middle-class of Native Americans (with cultural traditions intact) and new Indian leadership might have developed from this law (Barsh, 1987). To most observers, however, the main intent of the law appeared to be opening up reservation land to European American speculators and ending the Native American tribal relationship with the federal government; the Curtis Act of 1898 continued in this tradition (Holm, 1979). After all, how far could Native Americans take this freedom, since they were still not U.S. citizens in 1887?

The citizenship issue was sticky after passage of the Fourteenth Amendment to the Constitution, just as it was for Asians on the West Coast. The intent of the amendment was to make citizens of former slaves, but the wording seemed to include quite a few other people. In the court case of *Elk v. Wilkins* (1884), the Supreme Court found that Native Americans were not granted citizenship under the amendment. Native Americans were likened to the children of foreign ambassadors who, although born here, would not be citizens (Stidham and Carp, 1995). It took Congress until 1924 to pass the Indian Citizenship Act, which finally granted citizenship status to all Native Americans and settled the issue. It gave Native Americans state, federal, *and* tribal rights, making them unique among American citizens.

The Dawes Act of 1887 lasted until 1934 when it was overturned by an equally important act of Congress—the Wheeler-Howard Act (Indian Reorganization Act) of 1934. Passed as part of Roosevelt's New Deal legislation, the goal of the Indian Reorganization Act was to turn the clock back to pre-Dawes by reaffirming tribal integrity and sovereignty. Land ownership once again became communal, and tribes would have the right to create business corporations if they so chose. Equally important, the federal government supported the building of democratic political institutions within tribes (Strong and Van Winkle, 1993; Stuart, 1977). This maintenance of tribal integrity was important down the road when tribes learned to take the federal government to court seeking the restoration of land granted to them in past treaties but subsequently taken away.

A final episode in this period of history began just before World War II. Because Native Americans were now citizens, they were subject to the draft under Selective Service. Many Native Americans objected to draft registration on the grounds that it violated tribal sovereignty. The courts sided with the Selective Service. The issue may well have been pointless, however. With the attack on Pearl Harbor, many Native Americans responded with patriotism. Some tribes even independently declared war on Japan and Germany. By the war's end, 44,500 Native Americans (10 percent of the Native American population) had served in the armed forces (Franco, 1990).

Immigration from Mexico

No racial or ethnic groups in the United States show the connection between minority status and the economy as clearly as slaves, who were brought here for the sole purpose of making money. However, if another ethnic group comes close to showing this stark relation, it would have to be Mexican immigrants, who came to the United States to make money. When they were welcome in the United States, their welcome was purely economic—inexpensive labor was needed. Today, Mexican Americans are part of the American fabric. Between 1900 and 1990, 2.5 million Mexicans legally crossed the border, complete with documents. Undocumented Mexicans are by definition uncounted, but they came in similar numbers. All these immigrants would become a large and influential minority group—economics is now only part of their story, but in the early twentieth century, everything connected with Mexican immigration grew from economic motives.

Immigration and Economics: The Push and Pull Factors of Cheap Labor

The story of Mexican immigration does not begin until a little before 1900, and even then it started slowly. Border officials counted around 50,000 crossings in the first decade of the century, those immigrants presumably joining the 103,393 Mexican citizens in the United States counted by the 1900 census. By 1910, the census counted 219,802 Mexican citizens plus another 107,866 American-born children who were automatically U.S. citizens (Cardoso, 1980).

At the same time in the United States, the economy was gearing up for World War I. We have already seen the impact of that economic upturn on African American laborers in the South. The western United States also was facing a labor shortage, exacerbated by the many restrictions on Asian immigration. The mining and railroad companies and farmers of the Southwest were as happy to employ Mexican immigrants as those Mexicans were happy to be employed. Between 1914 and 1920, it is estimated that over 1 million Mexicans crossed the border to work in the United States; the Mexican government estimates that number at 2 million (Cardoso, 1980; Meier and Ribera, 1993).

Up until 1920, most Mexican immigrants sought work in Texas. The need for their labor continued into the 1920s with increased demand from California agriculture. If you are wondering how these immigrants got beyond the first immigration restriction act of 1921 (which required literacy), Congress conveniently passed an exemption law that applied only to Mexican laborers, permitting the flow of immigrants to continue. Still, immigrants tended to be individuals away from families. A study done during the 1920s showed that Mexican workers in the United States sent a total of $58 million to Mexico in postal money orders alone (Cardoso, 1980). At that point in time, it was clear that almost all Mexicans in the United States saw their stay as a sojourn, lasting only long enough to earn needed funds. Many crossed the border numerous times, coming and going as they needed work.

The 1920s also produced various employment control techniques with regard to Mexicans that clearly benefited American employers. As with African American workers in the South, Mexicans made good strikebreakers when shipped around the country. The 1923 strike at the Bethlehem Steel plant in Bethlehem, Pennsylvania received Mexican workers from Texas. In that same year, a strike at the National Tube Company (part of U.S. Steel) in Lorain, Ohio acquired 1500 Mexican workers from Texas.

Many of the workers in both situations did not know they were being shipped in as strikebreakers (Meier and Ribera, 1993). When Mexicans were not needed either in the fields or as strikebreakers, Americans used the technique of repatriation.

Repatriation is a euphemism for mass deportation. Just as Mexican workers could be shipped around the country in groups, they could also be shipped to Mexico in the same manner (Guerin-Gonzales and Story, 1995). This occurred on a small scale during short recessions of the 1920s, but the depression of the 1930s initiated large scale repatriation. The 1930 census counted 639,000 Mexican citizens living in the United States; by 1940, that figure was down to 377,000. The Mexican government estimates 458,000 Mexican citizens returned to Mexico between 1929 and 1937 (Meier and Ribera, 1993). It is impossible to determine just how many of the returnees left voluntarily. Although jobs were difficult to find during the depression, signs also began to appear with messages such as "Only White Labor Employed" and "No Niggers, Mexicans, or Dogs." Still probably close to 170,000 were repatriated. Los Angeles county supported the use of this system as follows: 6,000 unemployed Mexicans would cost the county $425,000 in welfare but only $77,000 in transportation costs back to Mexico (Cardoso, 1980). The fact that some American citizens were undoubtedly among the repatriated was not a European American concern.

The Mexican American: A New Ethnic Group

In spite of these comings and goings, a stable population of Mexican Americans did begin to develop. Some families did immigrate together and some families were created in the United States. In 1928, the League of United Latin American Citizens (LULAC) was formed in Texas, with the goal of protecting the rights of Mexican Americans. As a means to achieving this, it recommended assimilation to Mexican Americans, with a particular emphasis on learning English. These are clearly the goals of people who expect to stay.

✗ Chinese Americans

Chinese Americans became more Americanized in the first half of the twentieth century, stemming in part from concurrent political changes in China. The formation of the new Republic of China included encouragement to Chinese all around the world to modernize. Chinese Americans could now cut off their queues and adopt Western hairstyles. Others eliminated ancestor worship, many second-generation Chinese explored the Christian religions of their adopted country. A Chinese boy scout troop was formed in San Francisco in 1914 followed by a Chinese YMCA in 1916. The newly formed Chinese American Citizens Alliance was an educationally oriented organization that encouraged greater Chinese assimilation to the United States (Tsai, 1986).

In spite of modernization, however, most Chinese still lived in Chinatowns in a narrower range of occupations over time. As mines and railroads either closed or looked elsewhere for labor, more and more Chinese moved into laundry or restaurant work. When the depression of the 1930s occurred, Chinatowns were especially hard hit because of the interlocking nature of the business interests.

Probably the most significant change in America's Chinatowns was the slow growth of the female population. Even with the difficulties of immigration, women comprised 25 percent of the Chinese population in the United States by 1940, up from 6.5 percent in 1910 (Tsai, 1986). As with Chinese men, these women also came to embrace more Western ideas. In addition to other factors that influenced men, they also had the role model of Madame Chiang Kai-shek—the very well-known wife of China's leader. Her visit to the United States in 1943 during World War II made a major impression on the Chinese American community (Chan, 1991).

World War II also produced an extremely important change in the status of Chinese Americans that would come to affect both their lives and future immigration. Although the Chinese were still aliens ineligible for citizenship according to the 1870 law, the alliance between China and the United States in opposition to Japan made this status embarrassing to the United States. On December 17, 1943, President Roosevelt signed an act, commonly called the Magnuson Bill, which allowed naturalization to Chinese nationals and eliminated the Chinese Exclusion Act of 1882. Followed by the War Bride Act of 1945, many doors, which had previously been ajar at best, opened for Chinese immigrant women.

The Arrival of the Japanese

Emigration from Japan became legal in 1885. Their first destination would be the sugar fields of Hawaii—30,000 Japanese made that journey from 1885 to 1894, coming as contract laborers much like the Chinese. Another 127,000 would come from 1894 to 1907 (Ichioka, 1980). Most of these early immigrants were young rural men from agricultural backgrounds whose education level tended to be lower than the average Japanese (Ichioka, 1980; Spickard, 1996). The average term of a labor contract was three years, during which time the laborer was essentially the property of the planter (Ichioka, 1983).

Japanese women, most of whom came after 1900, later joined the men in the sugar cane fields (Tamura, 1995). Young, single Japanese women had traditionally been part of the labor force in Japan, particularly in industrial labor where they comprised 68 percent of the workforce; thus, labor overseas for single women was not as radical as it might seem (Von Hassell, 1993), and they soon became 20 percent of the emigrants (Spickard, 1996). Many of these women continued to work in the cane fields. Others, especially those who wound up on the mainland, became prostitutes in the male-dominated society of Japanese immigrants. As is true for most illegal activities, we do not know as much about this as we would like. Some of these women used prostitution as a temporary occupation before moving into better circumstances; others appear to have been forced into the work against their wills (Spickard, 1996; Warren, 1989).

Beginning in 1890, significant Japanese immigration to the United States began, with the majority entering through the ports of Seattle and San Francisco. In the last decade of the nineteenth century, 25,942 Japanese would be admitted to the United States; in the first decade of the twentieth century, another 129,797 would be admitted (U.S. Immigration and Naturalization Service, 1993). Their numbers were not huge, but they concentrated almost entirely in the three Pacific Coast states where anti-Asian sentiment was already strong.

The first jobs open to the early Japanese immigrants were in farming and on the railroads, many being available because of the recent shutdown of Chinese immigration. In the early 1900s, Japanese labor moved into other forms of wage labor including sawmills and salmon canneries (Azuma, 1994). Many opened small businesses such as restaurants, barbershops, watch making shops, and laundries (Bonacich and Modell, 1980; Yamato, 1994). More important, they began moving from farm labor to farm operation, either through purchase or leasing. They tended to specialize in labor-intensive produce such as berries or vegetables, many of which were later sold at farmer's public markets directly to the public. In the 1914 Seattle public market, for example, Japanese farmers rented 300 of the 400 stalls (Daniels, 1988). The trend here is clear—get out of wage labor and into running your own business. Not only was wage labor poorly paid in general, but also the Japanese immigrants took an extra cut because of racism. While Italian American railroad workers made between $1.45 and $1.65 an hour in 1910, Japanese workers earned between $1.20 and $1.40. Japanese carpenters in the sawmills made between $1.65 and $2.00 a day, while their European

Japanese "picture brides" arriving in San Francisco, sent for by Japanese men working in the United States.

American coworkers earned between $2.75 and $3.50 (Azuma, 1994). Yet even though these new Japanese entrepreneurs were not in direct competition with European American businessmen (who did not, for example, grow strawberries), their success only increased the prejudice and discrimination they faced.

Anti-Japanese Attitudes

All in all, European Americans on the West Coast did not see much worth in their Japanese neighbors. Much of the hatred came from organized labor, which viewed any new cheap labor as a threat. As one labor organizer commented in 1900:

> Chinatown with its reeking filth and dirt, its gambling dens and obscene slave pens, its coolie labor . . . is a menace to the community; but the sniveling Japanese, who swarms along the streets and cringingly offers his paltry services for a suit of clothes and a front seat in our public schools, is a far greater danger to the laboring portion of society than all the opium-soaked pigtails who have ever blotted the fair name of this beautiful city (quoted in Yamato, 1994:35).

The school reference above concerned the practice of some older Japanese who wished to learn English at the public schools. They were often placed in classes with younger children. Politician Grover Johnson observed:

> I am responsible to the mothers and fathers of Sacramento County who have their little daughters sitting side by side in the school rooms with matured Japs, with their base minds, their lascivious thoughts, multiplied by their race and strengthened by their mode of life . . . I have seen Japanese twenty-five years old sitting in the seats next to the pure maids of California . . . I shudder to think of such a condition (quoted in Spickard, 1996:29).

You might notice a similarity with the earlier justifications for lynching African Americans, using the defense of women to feed prejudice.

If the previous two expressions of hatred do not seem cut from the same cloth, they are not. Most anti-Japanese feeling did not stem from a clear economic motivation (McClatchy, 1978). As we have seen, labor competition from the Japanese became less of an issue, especially as the Japanese moved from wage labor into business and owning property. These businesses, for the most part, thrived in independent economic niches, much like the German Jews with their clothing factories.

The Early Growth of the Japanese American Community

By 1920, California was becoming the most popular home for the Japanese American community. In 1900, only 40 percent of all Japanese Americans lived in that state, but by 1920, almost 70 percent did (Daniels, 1988). Although the largest urban concentration was the Japanese community in Los Angeles ("Little Tokyo"), Japanese Americans resisted the urbanizing trends of the twentieth century, continuing to focus their efforts on farm ownership (Yamato, 1994). In spite of the various West Coast laws prohibiting such acquisitions (nine states had done so by 1925), the percentage of the Japanese American labor force as farm owners increased from 18 percent to 39 percent in the few short years between 1915 and 1924 (Spickard, 1996; Suzuki, 1995). By 1940, 6,000 of the 125,000 total West Coast Japanese Americans would be farm owners, controlling a total of 250,000 acres worth $72.6 million (Daniels, 1988, 1991).

The real work of assimilation occurred within the Japanese American family. Families with Issei (first-generation) parents were characterized by strong leadership. Fathers were the ultimate authorities, but mothers maintained considerable informal authority. Mothers' attitudes were particularly important because women were responsible for early childhood socialization. In a typical immigrant family, children were treated quite leniently for the first several years of life but were then expected to follow quite strict guidelines. In short, these children (the Nisei) were expected to retain the Japanese tradition of hard work in the face of adversity but to apply that attitude to becoming as American as possible as quickly as possible. Some mothers intentionally withheld aspects of Japanese culture from their children so as to speed up this process (Von Hassell, 1993). As one Nisei woman described it, "It's a wonder we weren't all schizos. Our parents were always telling us to be 'good Japanese.' Then they'd turn right around and tell us to be 'good Americans'"(quoted in Spickard, 1996:80–81).

The first rule for a Nisei child was to excel in the public school system. For the most part, they attended integrated schools where they competed directly with European American children. Nisei children excelled in the classroom and on the playing

With the Japanese invasion of Pearl Harbor, West Coast Japanese-Americans were forcibly evacuated from their homes into guarded camps in the interior of the country.

field. They achieved grades well above the norm for California students and attended for more years. In 1940, 58 percent of Nisei men over the age of 25 held high school diplomas compared with 46 percent for comparable European American men (Spickard, 1996). This gap only increased with time. But Issei parents did not totally neglect Japanese education for their children. Japanese language schools were organized for Nisei to attend after their day at the public school. In addition, some Nisei were selected to attend school in Japan for a few years, rounding out their educational experience. Almost half of the West Coast Nisei took advantage of this opportunity (Spickard, 1996). Generally, parents encouraged boys more than girls in the educational arena. In spite of this, Nisei girls learned to identify with their female teachers and achieved well on their own (Tamura, 1995).

Just When Things Were Looking Up: Pearl Harbor and the Relocation Camps

The Japanese attack on the United States Naval Fleet at Pearl Harbor, Hawaii, seems to have caught civilian Americans as unprepared as the Navy. By 1941, the Japanese American community on the West Coast had become quite Americanized in both culture and attitudes, but the people were both unknown to and feared by their European American neighbors. Rumors spread rapidly about an impending Japanese invasion of the West Coast in which Japanese Americans would aid the invaders. Public opinion moved quickly to strongly support the removal of all Japanese Americans from coastal areas. Ironically, more Japanese Americans were then living on the Hawaiian Islands than on the West Coast, but no such rumors spread there; the Japanese American community in Hawaii was much more integrated with other ethnic groups and, as a result, was not feared.

President Franklin Roosevelt responded quickly to this pressure from the West, signing Executive Order 9066 on February 19, 1942. With the stroke of one man's pen—this was *not* a law passed by Congress—all Americans of Japanese ancestry were thereby ordered removed to inland prison camps (Daniels, 1975). This order resulted ultimately in the rounding up of 120,313 people, two-thirds of whom were American citizens, and their being incarcerated without due process of law. The United States Supreme Court soon declared this flagrantly unconstitutional order to indeed be constitutional. West Coast Issei and Nisei were given notice in early 1942 that they were to prepare to be evacuated, bringing with them only what they could carry (Daniels, 1988; Nakanishi, 1993).

In most cases having only a matter of days, Japanese Americans faced many economic and logistic problems. What could be done with farms or businesses? Lucky individuals found European Americans willing to look after things. Mary Tsukamoto describes her family's good fortune in having a European American neighbor not only willing to keep up their farm but also to pay taxes on it for them while they were absent (Tsukamoto and Pinkerton, 1988). Less fortunate individuals attempted to lease homes or businesses; many returned to find them poorly treated or looted. Personal possessions were often sold in yard sales at which European Americans rejoiced in finding such good bargains (Box 6.2). In short, relocation placed an incredible economic hardship on the Japanese American community (Spickard, 1996).

BOX 6.2: JAPANESE AMERICAN RELOCATION YARD SALES

This extract from Paul Spickard's work (1996:104) gives us some idea of the anger and frustration many Japanese Americans must have experienced at the injustice of their forced incarceration during World War II.

The secondhand dealers had been prowling around for weeks, like wolves, offering humiliating prices for goods and furniture they knew many of us would have to sell sooner or later. Mama had . . . one fine old set of china, blue and white porcelain, almost translucent. . . .

One of the dealers offered her fifteen dollars for it. She said it was a full setting for twelve and worth at least two hundred. He said fifteen was his top price. Mama started to quiver. . . .She didn't say another word. She just glared at this man, all the rage and frustration channeled at him through her eyes.

He watched her for a moment and said he was sure he couldn't pay more than seventeen fifty for that china. She reached into the red velvet case, took out a dinner plate and hurled it at the floor right in front of his feet.

The man leaped back shouting, "Hey! Hey, don't do that! Those are valuable dishes!"

Mama took out another dinner plate and hurled it at the floor. Then another and another, never moving, never opening her mouth, just quivering and glaring at the retreating dealer, with tears streaming down her cheeks. He finally turned and scuttled out the door, heading for the next house. When he was gone she stood there smashing cups and bowls and platters until the whole set lay in scattered blue and white fragments across the wooden floor.

Meanwhile, the War Relocation Authority (WRA) was created to form and run the relocation camps. As the camps filled, the United States government began to consider the possibility of an all-Japanese American unit in the army. Because the United States military was still racially segregated in World War II, this unit would consist of only Japanese American troops led by European American officers. On February 1, 1943, Secretary of War Stimson announced the formation of the 442nd Regimental Combat Team. The Selective Service System originally had classified the Nisei as 4-C (the same status as enemy aliens), but their classification was changed to 1-A, making them available for the draft. Hawaiian Japanese Americans comprised the majority of this new unit, but Nisei in the camps willing to sign loyalty oaths also were eligible. They were trained and sent to Europe to fight while their families remained in the relocation camps.

The 442nd would ultimately see 18,000 Nisei men in its service, most of them volunteers. By the end of the war, they had received 9,486 casualties and been awarded 18,143 individual decorations, including one Congressional Medal of Honor (29 total were awarded during World War II), 47 Distinguished Service Crosses, 350 Silver Stars, and 3,600 purple hearts. They became the most decorated unit in American military history (Daniels, 1988; Menton, 1994). One of these soldiers, Daniel Inouye of Hawaii (who later became a United States senator), first returned to the United States via ship to the port of San Francisco. He had lost an arm in Italy. Wearing his uniform with its one empty sleeve and covered with medals, he was unable to find a barber willing to cut his hair.

While the 442nd was fighting in Europe, the WRA was slowly trying to empty the camps. Japanese Americans who could find work (again, away from the coast) were allowed to leave. After World War II, most returned to the West Coast and attempted to pick up where they had left off. They once again became quite successful. Efforts to obtain some form of apology from the federal government began. In 1976, President Gerald Ford rescinded Executive Order 9066. On August 10, 1988, President Ronald Reagan signed the Civil Liberties Act of 1988, which provided for a payment of $20,000 to each surviving Japanese American who had spent time behind barbed wire. When the checks were finally mailed in October of 1990, 60,000 such people were still alive (Nakanishi, 1993; Spickard, 1996).

And What Happened to European Americans?

By 1950, the term *European American* was definitely becoming a reality for the people so labeled. Although they would generally use the much more simple term *white* to describe themselves, European ancestry distinctions were fast dying in the United States. They soon married across religious boundaries to further blur distinctions.

The interesting thing about white America is that people of color noticed the tie long before it became meaningful among white people themselves. From a slave's point of view, an Irish Catholic was little different from an English Protestant. The latter might have had more money in his pocket, but they were both white and both free; in a pinch, they would stick together. Hispanics use the term *Anglo* as a catch-all term for all white people. The term is obviously based on the English culture and ancestry of many Americans, but a Mexican American would apply it to a Jewish American with no qualms. Once again, the assumption made is that Christians and Jews will unite in their common race if threatened by Hispanics.

By 1950, diversity among the ethnic groups of color in the United States was still great. Among Asian Americans, there was very little connection between the Japanese American community and the Chinese American community. Mexican Americans had only begun to meet African Americans; neither group perceived it had anything in common with the other. And Native Americans still saw themselves in tribal identities rather than as Native Americans in relation to the rest of the country. In the mid-1960s, immigration laws changed once again, stimulating even more ethnic diversity among people of color. American ethnic relations will become even more complicated.

Summary

Over 4 million Italians immigrated to the United States between 1880 and 1914. Generally poor and uneducated, these immigrants initially concentrated in the northeastern United States, forming ethnic communities in the cities. In general, Italians moved into the more poorly paid, unskilled work available in those cities. To make ends meet, husbands, wives, and children were usually forced to work. This, of course, hindered education for children and extended the period before they would rise economically.

Between 1880 and 1924, 4 million Jews entered the United States from eastern Europe. As with Italian immigrants of the same period, they tended to cluster in ethnic communities in northeastern cities, particularly New York City. Also like Italians, survival was difficult and often required economic contributions by all family members. A significant difference between these two groups, however, was that Jews arrived with higher rates of literacy and a tradition of religious education that produced more immediate success with American public education.

These new Jewish immigrants had a complicated relationship with the German Jews already in the United States; although the new immigrants supplied inexpensive labor for German Jewish employers, they also increased the numbers and visibility of Jews in the United States, leading to significant levels of anti-Semitism.

With the end of Reconstruction in the South, Jim Crow began. "Jim Crow" is a slang term for a wide variety of laws designed to segregate African Americans in all walks of life while simultaneously controlling their participation in the labor force. Segregation in housing and schools served to eliminate opportunities for advancement; employment discrimination forced participation in agricultural roles for which African Americans were wanted. The most extreme example of this labor force control was the peonage laws, which tied laborers to the land by law. All of this discrimination was accompanied by lynching which, beyond the obvious terror inflicted, served to remind African Americans of the degree of their powerlessness.

African Americans in the North faced similar restrictions as those in the South but suffered more from custom than from law. Much employment was closed to them and unions tended not to admit African American members. This encouraged employers to use African Americans as strikebreakers, which added to conflict between the races. The surge of immigration from southeastern Europe added new job seekers to northern cities and increased the competition.

World War I created a situation in America's industrial north that results from almost all wars—a booming economy combined with a labor shortage. Thriving economies tend to exhaust labor forces anyway, but wars also put potential laborers in uniform, further decreasing their number. Northern employers turned to the South for a solution. The result was a northern migration of almost half a million African Americans seeking economic opportunity.

While all of the northeastern and north central industrial cities attracted African American migrants, New York City attracted the most. Because housing segregation was the norm in the North, these new migrants caused severe housing shortages. The result in New York City was the expansion of the African American community to Harlem. Even as African American employment dropped with the end of World War I, the new migrants remained in the North and were joined by yet more African Americans, many attracted specifically to Harlem. Harlem became the first center in the United States for African American artists, intellectuals, and political leaders who grew stronger and more productive as a result of their concentration.

The music of ragtime and jazz that entered the American mainstream at this time is probably best remembered, but significant changes began occurring in the realm of African American politics. Organizations such as the National Association for the Advancement of Colored People and the National Urban League were formed. Demands for racial equality in the economy and in politics became commonplace. In particular, the NAACP began a several decades long fight against the Jim Crow laws of segregation in the South.

The slumping economy of the 1920s following the massive immigration from southeastern Europe produced strong anti-immigrant sentiment in the United States. Racist notions appeared, which divided Europeans into different races depending on which region in Europe they originated from. This sentiment culminated in the passage of the National Origins Act of 1924, which essentially ended immigration to the United States by declaring all but northwestern Europeans unwanted.

By the 1920s, the reservation system for Native Americans was firmly entrenched. Government policies to date had been to attempt the Americanization of Native Americans whenever possible through boarding schools for children. Citizenship was granted to Native Americans in 1924. The Indian Reorganization Act of 1934 attempted to return some control over reservations to tribes and increased federal recognition of Native American sovereignty.

The early twentieth century was a time of political revolution and economic upheaval in Mexico. Coupled with a need for inexpensive labor in the Southwest, increasing numbers of Mexican immigrants began to enter the United States. Most were needed in agriculture and mining. Considered racially inferior by most European Americans, Mexicans faced employment discrimination, physical violence, and mass deportation during the Great Depression of the 1930s when their labor was not needed. Nevertheless, they clearly established themselves during this period as yet one more ethnic group in the United States.

Japanese immigrants came to Hawaii and later to the United States toward the end of the nineteenth century. They were in demand in the sugar cane fields of Hawaii; in the United States, they searched for a place in the economy. Facing the anti-Asian attitudes still prominent in the West, Japanese immigrants worked in a variety of low-level jobs including laundry work, railroad work, and farming. Farming turned out to provide the best future as Japanese immigrants sought to own small plots, which they planted with labor-intensive crops such as berries or vegetables. In the early 1900s, many women from Japan joined the Japanese immigrant bachelors. The parents of this next generation of Japanese Americans encouraged their children to become Americanized and yet to retain the basic Japanese value of hard work.

The 1920s and 1930s was a somewhat static time for Chinese Americans—their immigration had been cut off since 1882—but it was a time of growth for Japanese Americans. In spite of considerable efforts to minimize their economic success, the new generation of Japanese Americans proved increasingly successful at building a firm economic base for the Japanese American community. The Japanese attack on Pearl Harbor temporarily ended this success as anti-Japanese sentiment in the West of the mainland resulted in the imprisonment of all Americans of Japanese ancestry. Many lost their homes, businesses, and possessions during this incarceration.

By World War II, many of the differences once thought important among Americans of European ancestry were becoming less meaningful. As new generations from earlier European waves of immigrants became Americanized and intermarried, ethnic divisions appeared more and more strongly along racial lines. By 1950—only a few short decades beyond 1920s racist divisions among Europeans—America was moving increasingly toward boundaries that separated white Americans from all others.

Chapter 6 Reading 1
The Ethics of Living Jim Crow

Richard Wright

African American writer Richard Wright offers an inside view of the Jim Crow system of racial segregation that evolved in the American South following Reconstruction in the nineteenth century. Jim Crow truly came into its own, however, during the early twentieth century, affecting all aspects of life. In this essay, originally published in 1938, Wright shows how Jim Crow came to affect virtually all aspects of everyday life for African Americans.

My first lesson in how to live as a Negro came when I was quite small. We were living in Arkansas. Our house stood behind the railroad tracks. Its skimpy yard was paved with black cinders. Nothing green ever grew in that yard. The only touch of green we could see was far away, beyond the tracks, over where the white folks lived. But cinders were good enough for me and I never missed the green growing things. And anyhow cinders were fine weapons. You could always have a nice hot war with huge black cinders. All you had to do was crouch behind the brick pillars of a house with your hands full of gritty ammunition. And the first woolly black head you saw pop out from behind another row of pillars was your target. You tried your very best to knock it off. It was great fun.

I never fully realized the appalling disadvantages of a cinder environment till one day the gang to which I belonged found itself engaged in a war with the white boys who lived beyond the tracks. As usual we laid down our cinder barrage, thinking that this would wipe the white boys out. But they replied with a steady bombardment of broken bottles. We doubled our cinder barrage, but they hid behind trees, hedges, and the sloping embankments of their lawns. Having no such fortifications, we retreated to the brick pillars of our homes. During the retreat a broken milk bottle caught me behind the ear, opening a deep gash which bled profusely. The sight of blood pouring over my face completely demoralized our ranks. My fellow-combatants left me standing paralyzed in the center of the yard, and scurried for their homes. A kind neighbor saw me and rushed me to a doctor, who took three stitches in my neck.

I sat brooding on my front steps, nursing my wound and waiting for my mother to come from work. I felt that a grave injustice had been done me. It was all right to throw cinders. The greatest harm a cinder could do was leave a bruise. But broken bottles were dangerous; they left you cut, bleeding, and helpless.

When night fell, my mother came from the white folks' kitchen. I raced down the street to meet her. I could just feel in my bones that she would understand. I knew she would tell me exactly what to do next time. I grabbed her hand and babbled out the whole story. She examined my wound, then slapped me.

"How come yuh didn't hide?" she asked me. "How come yuh awways fightin'?"

I was outraged, and bawled. Between sobs I told her that I didn't have any trees or hedges to hide behind. There wasn't a thing I could have used as a trench. And you couldn't throw very far when you were hiding behind the brick pillars of a house. She grabbed a barrel stave, dragged me home, stripped me naked, and beat me till I had a fever of one hundred and two. She would smack my rump with the stave, and, while the skin was still smarting, impart to me gems of Jim Crow wisdom. I was never to throw cinders any more. I was never to fight any more wars. I was never, never, under any conditions, to fight white folks again. And they were absolutely right in clouting me with the broken milk bottle. Didn't I know she was working hard every day in the hot kitchens of the white folks to make money to take care of me? When was I ever going to learn to be a good boy? She couldn't be bothered with my fights. She finished by telling me that I ought to be thankful to God as long as I lived that they didn't kill me.

All that night I was delirious and could not sleep. Each time I closed my eyes I saw monstrous white faces suspended from the ceiling, leering at me.

From that time on, the charm of my cinder yard was gone. The green trees, the trimmed hedges, the cropped lawns grew very meaningful, became a symbol. Even today when I think of white folks, the hard, sharp outlines of white houses surrounded by trees, lawns and hedges are present somewhere in the background of my mind. Through the years they grew into an overreaching symbol of fear.

It was a long time before I came in close contact with white folks again. We moved from Arkansas to Mississippi. Here we had the good fortune not to live behind the railroad tracks, or close to white neighborhoods. We lived in the very heart of the local Black Belt. There were black churches and black preachers, there were black schools and black teachers; black groceries and black clerks. In fact, everything was so solidly black that for a long time I did not even think of white folks, save in remote and vague terms. But this could not last forever. As one grows older one eats more. One's clothing costs more. When I finished grammar school I had to go to work. My mother could no longer feed and clothe me on her cooking job.

There is but one place where a black boy who knows no trade can get a job, and that's where the houses and faces are white, where the trees, lawns, and hedges are green. My first job was with an optical company in Jackson, Mississippi. The morning I applied I stood straight and neat before the boss, answering all his questions with sharp yessirs and nosirs. I was very careful to pronounce my sirs distinctly, in order that he might know that I was polite, that I knew where I was, and that I knew he was a white man. I wanted that job badly.

He looked me over as though he were examining a prize poodle. He questioned me closely about my schooling, being particularly insistent about how much mathematics I had had. He seemed very pleased when I told him I had had two years of algebra.

"Boy, how would you like to try to learn something around here?" he asked me.

"I'd like it fine, sir," I said, happy. I had visions of "working my way up." Even Negroes have those visions.

"All right," he said. "Come on."

I followed him to the small factory.

"Pease," he said to a white man of about thirty-five, "this is Richard. He's going to work for us.

Pease looked at me and nodded.

I was then taken to a white boy of about seventeen.

"Morrie, this is Richard, who's going to work for us."

"Whut yuh sayin' there, boy!" Morrie boomed at me.

"Fine!" I answered.

The boss instructed these two to help me, teach me, give me jobs to do, and let me learn what I could in my spare time.

My wages were five dollars a week.

I worked hard, trying to please. For the first month I got along O.K. Both Pease and Morrie seemed to like me. But one thing was missing. And I kept thinking about it. I was not learning anything and nobody was volunteering to help me. Thinking they had forgotten that I was to learn something about the mechanics of grinding lenses, I asked Morrie one day to tell me about the work. He grew red.

"Whut yuh tryin' t' do, nigger, get smart?" he asked.

"Naw; I ain' tryin' t' git smart," I said.

"Well, don't if yuh know whut's good for yuh!"

I was puzzled. Maybe he just doesn't want to help me, I thought. I went to Pease.

"Say, are yuh crazy, you black bastard?" Pease asked me, his gray eyes growing hard.

I spoke out, reminding him that the boss had said I was to be given a chance to learn something.

"Nigger, you think you're white, don't you?"

"Naw, sir!"

"Well, you're acting mighty like it!"

"But, Mr. Pease, the boss said. . ."

Pease shook his fist in my face.

"This is a white man's work around here, and you better watch yourself!"

From then on they changed toward me. They said good-morning no more.

When I was just a bit slow in performing some duty, I was called a lazy black son-of-a-bitch.

Once I thought of reporting all this to the boss. But the mere idea of what would happen to me if Pease and Morrie should learn that I had "snitched" stopped me.

And after all the boss was a white man, too. What was the use?

The climax came at noon one summer day. Pease called me to his work-bench. To get to him I had to go between two narrow benches and stand with my back against a wall.

"Yes, sir," I said.

"Richard, I want to ask you something," Pease began pleasantly, not looking up from his work.

"Yes, sir," I said again.

Morrie came over, blocking the narrow passage between the benches. He folded his arms, staring at me solemnly

I looked from one to the other, sensing that something was coming.

"Yes, sir," I said for the third time

Pease looked up and spoke very slowly.

"Richard, Mr. Morrie here tells me you called me *Pease*."

I stiffened. A void seemed to open up in me. I knew this was the show-down.

He meant that I had failed to call him Mr. Pease. I looked at Morrie. He was gripping a steel bar in his hands. I opened my mouth to speak, to protest, to assure Pease that I had never called him simply *Pease*, and that I had never had any intentions of doing so, when Morrie grabbed me by the collar, ramming my head against the wall.

"Now, be careful, nigger!" snarled Morrie, baring his teeth. "I heard yuh call 'im Pease! 'N' if yuh say yuh didn't, yuh're callin' me a liar, see?" He waved the steel bar threateningly.

If I had said: No, sir Mr. Pease, I never called you *Pease*, I would have been automatically calling Morrie a liar. And if I had said: Yes, sir, Mr. Pease, I called you *Pease*, I would have been pleading guilty to having uttered the worst insult that a Negro can utter to a southern white man. I stood hesitating, trying to frame a neutral reply.

"Richard, I asked you a question!" said Pease. Anger was creeping into his voice.

"I don't remember calling you *Pease*, Mr. Pease," I said cautiously "And if I did, I sure didn't mean . . . "

"You black son-of-a-bitch! You called me Pease, then!" he spat, slapping me till I bent sideways over a bench. Morrie was on top of me, demanding:

"Didn't yuh call 'im *Pease*? If yuh say yuh didn't, I'll rip yo' gut string loose with this bar, yuh black granny dodger! Yuh can't call a white man a lie 'n' git erway with it, you black son-of-a-bitch!"

I wilted. I begged them not to bother me. I knew what they wanted. They wanted me to leave.

"I'll leave," I promised. "I'll leave right now."

They gave me a minute to get out of the factory. I was warned not to show up again, or tell the boss.

I went.

When I told the folks at home what had happened, they called me a fool. They told me that I must never again attempt to exceed my boundaries. When you are working for white folks, they said, you got to "stay in your place" if you want to keep working.

II

My Jim Crow education continued on my next job, which was portering in a clothing store. One morning, while polishing brass out front, the boss and his twenty-year-old son got out of their car and half dragged and half kicked a Negro woman into the store. A policeman standing at the corner looked on, twirling his night-stick. I watched out of the corner of my eye, never slackening the strokes of my chamois upon the brass. After a few minutes, I heard shrill screams coming from the rear of the store. Later the woman stumbled out, bleeding, crying, and holding her stomach. When she reached the end of the block, the policeman grabbed her and accused her of being drunk. Silently, I watched him throw her into a patrol wagon.

When I went to the rear of the store, the boss and his son were washing their hands at the sink. They were chuckling. The floor was bloody and strewn with wisps of hair and clothing. No doubt I must have appeared pretty shocked, for the boss slapped me reassuringly on the back.

"Boy, that's what we do to niggers when they don't want to pay their bills" he said, laughing.

His son looked at me and grinned.

"Here, hava cigarette," he said.

Not knowing what to do, I took it. He lit his and held the match for me. This was a gesture of kindness, indicating that even if they had beaten the poor old woman, they would not beat me if I knew enough to keep my mouth shut.

"Yes, sir," I said, and asked no questions.

After they had gone, I sat on the edge of a packing box and stared at the bloody floor till the cigarette went out.

That day at noon, while eating in a hamburger joint, I told my fellow Negro porters what had happened. No one seemed surprised. One fellow, after swallowing a huge bite, turned to me and asked:

"Huh! Is the all they did t' her?"

"Yeah. Wasn't tha' enough?" I asked.

"Shucks! Man, she's a lucky bitch!" he said, burying his lips deep into a juicy hamburger. "Hell, it's a wonder they didn't lay her when they got through."

III

I was learning fast, but not quite fast enough. One day, while I was delivering packages in the suburbs, my bicycle tire was punctured. I walked along the hot, dusty road, sweating and leading my bicycle by the handle-bars.

A car slowed at my side.

"What's the matter, boy?" a white man called.

I told him my bicycle was broken and I was walking back to town.

"That's too bad," he said, "Hop on the running board."

He stopped the car. I clutched hard at my bicycle with one hand and clung to the side of the car with the other.

"All set?"

"Yes, sir," I answered. The car started.

It was full of young white men. They were drinking. I watched the flask pass from mouth to mouth.

"Wanna drink, boy?" one asked.

I laughed as the wind whipped my face. Instinctively obeying the freshly planted precepts of my mother, I said "Oh, no!"

The words were hardly out of my mouth before I felt something hard and cold smash me between the eyes. It was an empty whisky bottle. I saw stars, and fell backwards from the speeding car into the dust of the road, my feet becoming entangled in the steel spokes of my bicycle. The white men piled out and stood over me.

"Nigger, ain' yuh learned no better sense'n tha' yet?" asked the man who hit me. "Ain't yuh learned t' say sir t' a white man yet?"

Dazed, I pulled to my feet. My elbows and legs were bleeding. Fists doubled, the white man advanced, kicking my bicycle out of the way.

"Aw, leave the bastard alone. He's got enough," said one.

They stood looking at me. I rubbed my shins, trying to stop the flow of blood. No doubt they felt a sort of contemptuous pity, for one asked:

"Yuh wanna ride t' town now, nigger? Yuh reckon yuh know enough t' ride now?"

"I wanna walk," I said, simply.

Maybe it sounded funny. They laughed.

"Well, walk, yuh black son-of-a-bitch!"

When they left they comforted me with:

"Nigger, yuh sho better be damn glad it wuz us yuh talked t' tha' way. Yuh're a lucky bastard, 'cause if yuh'd said tha' t' somebody else, yuh might've been a dead nigger now."

IV

Negroes who have lived South know the dread of being caught alone upon the streets in white neighborhoods after the sun has set. In such a simple situation as this the plight of the Negro in America is graphically symbolized. While white strangers may be in these neighborhoods trying to get home, they can pass unmolested. But the color of a Negro's skin makes him easily recognizable, makes him suspect, converts him into a defenseless target.

Late one Saturday night I made some deliveries in a white neighborhood. I was pedaling my bicycle back to the store as fast as I could, when a police car, swerving toward me, jammed me into the curbing.

"Get down and put up your hands!" the policemen ordered.

I did. They climbed out of the car, guns drawn, faces set, and advanced slowly.

"Keep still!" they ordered.

I reached my hands higher. They searched my pockets and packages. They seemed dissatisfied when they could find nothing incriminating. Finally, one of them said:

"Boy, tell your boss not to send you out in white neighborhoods after sundown."

As usual, I said:

"Yes, sir."

V

My next job was a hall-boy in a hotel. Here my Jim Crow education broadened and deepened. When the bell-boys were busy, I was often called to assist them. As many of the rooms in the hotel were occupied by prostitutes, I was constantly called to carry them liquor and cigarettes. These women were nude most of the time. They did not bother about clothing, even for bell-boys. When you went into their rooms, you were supposed to take their nakedness for granted, as though it startled you no more than a blue vase or a red rug. Your presence awoke in them no sense of shame, for you were not regarded as human. If they were alone, you could steal sidelong glimpses at them. But if they were receiving men, not a flicker of your eyelids could show. I remember one incident vividly. A new woman, a huge, snowy-skinned blonde, took a room on my floor. I was sent to wait upon her. She was in bed with a thick-set man; both were nude and uncovered. She said she wanted some liquor and slid out of bed and waddled across the floor to get her money from a dresser drawer. I watched her.

"Nigger, what in hell you looking at?" the white man asked me, raising himself upon his elbows.

"Nothing," I answered, looking miles deep into the blank wall of the room.

"Keep your eyes where they belong, if you want to be healthy!" he said.

"Yes, sir."

VI

One of the bell-boys I knew in this hotel was keeping steady company with one of the Negro maids. Out of a clear sky the police descended upon his home and arrested him, accusing him of bastardy. The poor boy swore he had had no intimate relations with the girl. Nevertheless, they forced him to marry her. When the child arrived, it was found to be much lighter in complexion than either of the two supposedly legal parents. The white men around the hotel made a great joke of it. They spread the rumor that some white cow must have scared the poor girl while she was carrying the baby. If you were in their presence when this explanation was offered, you were supposed to laugh.

VII

One of the bell-boys was caught in bed with a white prostitute. He was castrated and run out of town. Immediately after this all the bell-boys and hall-boys were called together and warned. We were given to understand that the boy who had been castrated was a "mighty, mighty lucky bastard." We were impressed with the fact that next time the management of the hotel would not be responsible for the lives of 'trouble-makin' niggers." We were silent.

VIII

One night, just as I was about to go home, I met one of the Negro maids. She lived in my direction, and we fell in to walk part of the way home together. As we passed the white night-watchman, he slapped the maid on her buttock. I turned around, amazed. The watchman looked at me with a long, hard, fixed-under stare. Suddenly he pulled his gun and asked:

"Nigger, don't yuh like it?"

I hesitated.

"I asked yuh don't yuh like it?" he asked again, stepping forward.

Yes, sir," I mumbled.

"Talk like it, then!"

"Oh, yes sir! I said with as much heartiness as I could muster.

Outside, I walked ahead of the girl, ashamed to face her. She caught up with me and said:

"Don't be a fool! Yuh couldn't help it!"

This watchman boasted of having killed two Negroes in self-defense.

Yet, in spite of all this, the life of the hotel ran with an amazing smoothness. It would have been impossible for a stranger to detect anything. The maids, the hall-boys, and the bell-boys were all smiles. They had to be.

IX

I had learned my Jim Crow lessons so thoroughly that I kept the hotel job till I left Jackson for Memphis. It so happened that while in Memphis I applied for a job at a branch of the optical company. I was hired. And for some reason, as long as I worked there, they never brought my past against me.

Here my Jim Crow education assumed quite a different form. It was no longer brutally cruel, but subtly cruel. Here I learned to lie, to steal, to dissemble. I learned to play that dual role which every Negro must play if he wants to eat and live.

For example, it was almost impossible to get a book to read. It was assumed that after a Negro had imbibed what scanty schooling the state furnished he had no further need for books. I was always borrowing books from men on the job. One day I mustered enough courage to ask one of the men to let me get books from the library in his name. Surprisingly, he consented. I cannot help but think that he consented because he was a Roman Catholic and felt a vague sympathy for Negroes, being himself an object of hatred. Armed with a library card, I obtained books in the following manner: I would write a note to the librarian, saying: "Please let this nigger boy have the following books." I would then sign it with the white man's name.

When I went to the library, I would stand at the desk, hat in hand, looking as unbookish as possible. When I received the books desired I would take them home. If the books listed in the note happened to be out, I would sneak into the lobby and forge a new one. I never took any chances guessing with the white librarian about what the fictitious white man would want to read. No doubt if any of the white patrons had suspected that some of the volumes they enjoyed had been in the home of a Negro, they would not have tolerated it for an instant.

The factory force of the optical company in Memphis was much larger than that in Jackson, and more urbanized. At least they liked to talk, and would engage the Negro help in conversation whenever possible. By this means I found that many subjects were taboo from the white man's point of view. Among the topics they did not like to discuss with Negroes were the following: American white women; the Ku Klux Klan; France, and how Negro soldiers fared while there; French women: Jack Johnson; the entire northern part of the United States; the Civil War; Abraham Lincoln; U. S. Grant; General Sherman; Catholics; the Pope; Jews; the Republican Party; slavery; social equality; Communism; Socialism; the 13th and 14th Amendments to the Constitution; or any topic calling for positive knowledge or manly self-assertion on the part of the Negro. The most accepted topics were sex and religion.

There were many times when I had to exercise a great deal of ingenuity to keep out of trouble. It is a southern custom that all men must take off their hats when they enter an elevator. And especially did this apply to us blacks with rigid force. One day I stepped into an elevator with my arms full of packages. I was forced to ride with my hat on. Two white men stared at me coldly. Then one of them very kindly lifted my hat and placed it upon my armful of packages. Now the most accepted response for a Negro to make under such circumstances is to look at the white man out of the corner of his eye and grin. To have said "Thank you!" would have made the white man *think* that you *thought* you were receiving from him a personal service. For such an act I have seen Negroes take a blow in the mouth. Finding the first alternative distasteful, and the second dangerous, I hit upon an acceptable course of action which fell safely between these two poles. I immediately no sooner than my hat was lifted— pretended that my packages were about to spill, and appeared deeply distressed with keeping them in my arms. In this fashion I evaded having to acknowledge his service, and, in spite of adverse circumstances, salvaged a slender shred of personal pride.

How do Negroes feel about the way they have to live? How do they discuss it when alone amongst themselves? I think this question can be answered in a single sentence. A friend of mine who ran an elevator once told me:

"Lawd, man! Ef it wuzn't fer them polices 'n' them ol' lynch-mobs, there wouldn't be nothin' but uproar down here!"

Chapter 6 Reading 2
We The People: A Story of Internment in America

Mary Tsukamoto and Elizabeth Pinkerston

This is the story of Mary Tsukamoto, a twenty-seven-year-old Nisei wife, her husband Al, her daughter Marielle, Al's sister Nami, and his mother and father. The year is 1942, and these Japanese Americans are living in Florin, California, just south of Sacramento. All but the grandparents are American citizens, yet all were relocated to internment camps for the duration of World War II because of their Japanese ancestry. They were first transferred to an assembly center in Fresno, California, and then to a more permanent residence in a WRA camp in Arkansas.

I felt old at twenty-seven as I looked out like a caged animal, captured and placed behind a barbed wire fence that seemed to stretch forever. The sound of the camp gates closing behind us sent a searing pain into my heart. I knew it would leave a scar that would stay with me forever. At that very moment my precious freedom was taken from me.

———————

The Fresno County fairgrounds was a somewhat logical place to set up a temporary relocation center for thousands of people. The fairgrounds' fences and gates had been easily converted to a guarded center. The site was far enough out of town that it did not create problems for the city; yet it was close enough that resources were available.

From the ominous watch an unmistakable uniformed figure looked down upon us frightened, confused evacuees as we entered the compound and looked for our new quarters. The glint of the rifle in the guard's hands sent terror through our hearts. We huddled in little groups, forlornly waiting to find out what this new life would be

Tired and numb from such an emotional day, our family stood in a long line waiting to be processed into the strange new place. Our identification was checked, and we were given room assignments. All the while, Al, Marielle and I were greeted and embraced by sisters and cousins who happily welcomed us. Our relatives had left Florin earlier and arrived at camp two days before us.

We were loaded onto trucks that rushed down the dusty roads between the tar-paper covered barrack buildings. A surprise row of fig trees with lush green leaves broke the monotony of the barracks. Inmates were seated under the welcome shade looking like a group of happy picnickers instead of the prisoners they actually were.

The truck stopped, and we were told to get off. Our familiar suitcases and bundles were thrown onto a huge pile of luggage. Part of the camp was fitted into the oval space where cars and horses had raced during the annual fair. Such festive occasions were in poignant contrast to what was taking place before us in the spring of 1942. Twenty neatly regimented and precisely measured long barracks stood, ten in a row, in each of the eleven sections. At one end were barrack buildings to be used as a mess hall, laundry, shower room, and two little huts for latrines.

I had never seen so many Japanese faces in such a small area. Some were already sun-burned; their skin and clothes were covered with dust. I could not believe how we were crowded together and how many of us there were!

As we were picking up our things from the pile of luggage, Marielle suddenly cried out, "Where's Obaachan? Where's Ojiichaan? Where's Namichan? They're not here!"

We searched frantically around Sections H, I, J, and K! Friends and relatives helped out, but Grandma and Grandpa, our two old ones, and Nami were nowhere to be found.

How could we have become separated? Each of us had our tags, #22076, attached to us since we had gone to the Elk Grove station early this morning. Moments ago, I had stood by Grandma and Grandpa as we waited in line to be processed into this chaotic place. How could they be lost?

I clung frantically to Marielle's hand, determined not to lose her too. Al and I had vowed to look after our three helpless ones to spare them undue stress. How could this have happened so soon after our arrival?

Al ran to the front office to find out which quarters they had been assigned. Marielle cried and I worried for what seemed hours. Finally, Al returned with our loved ones. Our family's anguish in the first hour of internment was over. Somehow, Grandma, Grandpa and Nami had been assigned to Section A with 24 single men who arrived with our group. Al found them on the opposite end of the fairgrounds. We quickly learned the consequences of not being absolutely careful. It would never happen again.

We dragged our luggage to the barrack as best we could—anxious to have some private space where the six of us could be by ourselves for the first time all day. We found, however, that the room assigned to us was already occupied. What a surprise! The Hideo Kadokawa family was in our tiny room. It was already dusk, so we decided to stay together for the night. The Kadokawas were cousins, so it seemed a shame to send them out. With nine cots, a baby crib, and a huge pile of luggage, there was not much room, but we managed to fit everyone in.

Each family was assigned a portion of the black tar-paper covered barracks. The floor was asphalt and still smelled strong, as it had been finished only recently. The penetrating heat of the Fresno sun turned the asphalt into a soft substance by midday. The floor sank when we stepped on it, and when we sat on our cots the legs penetrated the soft asphalt.

No one slept very much that first night. Searchlights blinked continuously in the darkness. The sounds of voices never stopped and kept everyone awake. One of the Kadokawa children became sick and cried with pain. She could not be quieted, and the commotion got everyone out of bed. Screams of other children reverberated to the roof and across the barrack to torment all five families assigned to our building.

In the morning the Kadokawas moved into a room of their own. We surveyed our quarters in Section H. For the six of us we were assigned a large room that measured 20 by 25 feet. The room was as large as our kitchen back home. Our six cots took up most of the space. The walls were of a cheap grade of knotty pine wood. I could see the roughly finished roof with its wooden beams and the tar paper covering on the outside. The tar was already beginning to drip onto the cots as the morning sun melted it.

The walls that separated us from the rooms of our neighbors were open at the top and unfinished. There were no ceilings. The partitions were so thin that even a whisper could be heard through the walls. Gaps between the boards and knotholes destroyed whatever privacy there was supposed to be.

Families eventually tried to create some measure of private dignity in whatever ways they could. Orange and apple crates from the garbage pile behind the mess hall were fashioned into tables and cupboards. Scraps of poles were turned into corner closets and covered with bedspreads. Wires were strung across the rooms and hung with sheets to create individual spaces.

Our basic needs at camp were provided by the military who had jurisdiction over us. Although they were accustomed to caring for adult males, camp authorities had no experience in dealing with women, the elderly, or children including teenagers and infants. This created innumerable problems, each of which we had to patiently resolve and maintain our emotional and mental stability.

The food was a problem for all of us. The mess hall was hotter than anywhere else in camp because they cooked for so many people. Eating in such a crowded, hot place with so many strangers, all in a hurry, was enough to ruin most appetites

Our first meal was hot stew. My plate took my appetite away: smelly mutton stew with jello melting into it, a distasteful mess. I tried to eat the buttered bread, but streams of perspiration ran down my face and would not stop. I gulped a salt tablet, which we were told to take regularly, but with my empty stomach, I was immediately nauseated. I hurried out of that noisy, crowded place that smelled so bad and rushed to our barrack. I was glad I had packed some soda crackers and jam to quench my hunger.

The bathrooms created another problem that we had to get used to. My first encounter left me smarting from the shameful insult to my modesty. I waited as long as I could before I finally went. The latrine was in an open public area with no privacy at all, and there was a long, long line queuing to the small shack. I heard water running, and at intervals, there was a great splashing sound like something heavy being knocked over. I tried to figure out what it was, but no bathroom I had ever seen made noises like that. When I got closer, I was shocked to find that the latrine's walls were only screens. I could see people sitting in a row. There was nothing private about it at all!

The door opened, and it was my turn to go in. I entered reluctantly with a very red face. Nothing in my life had prepared me to share my humanness with all these strangers. Five ladies sat back to back with no partitions, their tan bodies exposed, and almost touching each other. The banging came again just as I sat down. There was another great splash; then a large, wooden box spilled water that rushed through the trough beneath me. Raw sewage, putrid and unbearable, splashed out of the seat on the end, next to where I sat. The odor was overpowering. I thought surely I would faint. I ran out clutching my stomach and throwing up along the way, unable to cope with the stench. I never complained about the smell of the stew again.

Even the beds were hard to get used to. We were told to put straw in canvas bags for our mattresses, and a huge amount of straw was piled in the middle of an open area for this purpose. Most of us thought that the more we stuffed into the bags, the more padding would be provided. We were wrong, but we didn't find out until we kept rolling off our cots that first night. We couldn't sleep on top of the lumpy mounds. Every move we made was accompanied by the harsh rubbing sound of coarse straw as we tossed through the night. Away from the privacy and comfort of our homes, we new residents of this strange community tried to sleep, but most of us could not. Like the food and the latrines, it took us a while to adjust.

———

Wearily, we dragged our tired bodies to our new home—an "apartment" in Block 9, Barrack 8, Room E, 20 feet by 16 feet. It was smaller than our Fresno room, but what heaven! Al, Marielle and I had a room for ourselves! Grandma, Grandpa and Nami had the room adjoining ours. My parents, brother and sisters were near by, and the Ouchidas were also in our barrack. Nearly 350 of our Florin friends were near us in Block 9 including my sisters and their families. For this pleasure, we were very, very grateful.

Our room even had a wooden floor unlike the soft asphalt of Fresno's barrack. The walls and ceilings were well constructed of good wood. They were solidly built and did not have knotholes such as we had known in Fresno. We basked in the privacy, such as we had not known since leaving our Florin home five months ago.

The next morning we discovered the error of Jerome's excellent construction. Everything in the camp had been constructed for tiny people. We had wondered about the strange closet in our room; it was so close to the ground that Marielle could almost reach the rod. In the latrines, little, low toilets were set in rows. Laundry tubs were placed on stands as if they were built for kindergarten children. Our mess hall picnic tables were much too small.

A construction worker provided the answer in response to our puzzled looks. "Are you the people [we] are building this camp for?" he asked in surprise. He looked in amazement at young, Nisei teenagers who stood almost six feet tall.

"They told us little, brown people would be coming here," he said. "We were thinking they were like pygmies."

It was a silly mistake and an inconvenient one, but it provided us with a touch of humor that helped to brighten our arrival. Of course they had to adjust everything to fit our big bodies, and until this was done weeks later, most adults could not sit anywhere in comfort. But, it was still much better than Fresno.

In many ways our arrival, 17,000 of us at the two camps of Jerome and Rohwer (20 miles apart), was quite a cultural shock to the people who lived near the swamplands of southern Arkansas. Most residents of Arkansas had never seen people of Japanese ancestry before. No wonder there were those who were alarmed and appeared to be on the verge of panic. Rumors circulated through the small towns of an enemy invasion, which seemed strange for the "invaders" were locked up in concentration camps under armed guard. Our arrival, however, definitely altered life in the state of Arkansas, culturally as well as economically.

The sister states of Mississippi, Missouri, Louisiana, Oklahoma, and Texas were affected too. When 20,000 people come into an isolated area practically overnight, it is bound to affect almost everything. The same thing happened in all the areas where War Relocation Authority (WRA) camps were established. There were ten in all, and all were established in dismal, abandoned places far from the West Coast. Their colorful names disguise the sorrowful memories that are part of their story: Minidoka in Idaho; Amache Granada in Colorado; Heart Mountain, Wyoming; Topaz, Utah; Gila and Poston in Arizona; Manzanar and Tule Lake in California; and Rohwer and Jerome in Arkansas. Eventually 120,000 men, women, and children were confined behind the barbed wire fences of these ten camps.

We wondered how our loved ones in other camps were getting along and whether the problems we had were similar to theirs. There was little communication from camp to camp, so when letters were received, we eagerly gathered around to hear the news.

We waited many days for our freight to reach us at Jerome. There was a war going on, and we knew we were not a top priority, but it seemed to take such a long time. We were desperate for our bedding and clothing, especially when we started getting sick, one by one, many before the first day was over. A wave of dysentery hit the camp just as we were trying to get settled.

The sickness had disastrous effects because the camp facilities were not completed. I searched block after block to find a latrine ready for use and walked long distances to find hooked up showers. The few working latrines and showers had incredibly long lines. The people queuing up were just as upset and embarrassed as I was.

"If only the freight would get here," Al implored to the great spirit of slow trains. It was our chamber pot we needed so desperately!

By the time our crates arrived, our critical need was nearly over. Our illness was found to be gastroenteritis diarrhea caused by the water supply which had been exposed to contamination when new plumbing was installed. When they identified the source of the problem, we were warned not to drink the water unless it was boiled.

Our water source was 700-900 feet underground, and it was good safe drinking water except for this incident. It was strange water, very soft, and its sudsing factor was extremely high. The soapy feeling was impossible to rinse off. I had never seen so many suds; they just wouldn't go away.

Our bathrooms were such a big improvement over those at the assembly center. We were pleased with the gleaming white, porcelain, flush toilets. The only thing we didn't like was the long, open wall of toilets with no partitions. Some women took huge paper bags and wore them over their heads as they sat there so exposed. Their loud protesting finally convinced the camp directors to correct the situation. We scrounged for curtains to provide privacy, and finally, the bathrooms met the approval of the mothers and daughters among us.

Dreary, tiresome days fell into place week after week. We recovered from the stressful train ride and the attack of diarrhea that weakened so many of us. Al and I sat in our little room one afternoon and talked about the health of each of our loved ones. Though we worried about each of them, they all seemed to be better than they had been for weeks. Cautiously, we hoped that our lives might be free from worries for a while.

———

Life was certainly better than it had been at Fresno, but the barbed wire was still there to separate us from the outside world. Once inside the gate, no one could leave without a properly authorized pass. Sentries with guns were on the watch towers and always alert.

There was more privacy, but we still ate at picnic tables in large mess halls. Our buildings were farther apart, better constructed and much better insulated. That provided us with a sense of quiet that was good for body and soul.

Gradually as Al and I assessed our strengths, we realized how fortunate we were to have among us many who were so very talented. When I heard George Seno sing "Thanks Be To God," I felt as if the barrack's roof would lift off, as if God Himself had touched us and held us close, even though we felt so far from him in this strange camp city.

We were crowded into this concentration camp from which we could not escape. It was a time of humiliation and despair, but when I was asked to speak to the Christian Youth Fellowship, I pondered what to say. We hungered for spiritual food and we did not want to be here, but as Christians in search of God, we needed to come face to face with our Master, Lord Jesus the Christ. I decided to focus my presentation on living eternity day by day.

I spoke about pain, disappointment and degradation, and how they could be overcome with steadfast faith. Christ indeed could turn despair into triumphant joy. Hope and faith, edged with courage, could help us model His indomitable spirit day by day. This was my message to our youth, and I realized that first, I had to believe it myself. I needed to be the model they could see to live eternity here by taking one day at a time. What an awesome humbling responsibility. I faltered. But there could be hope even in this internment camp.

———

We had been deprived of our freedom, our liberty and our property, and no criminal charges had been filed against us. There had been no trials. Though the U.S. was at war with both Germany and Italy, only we citizens of Japanese ancestry had been subjected to this cruelty by our own government. Our only guilt was that we were of the Japanese race. Our Constitutional rights had been terribly violated, and there was nothing we could do about it.

Even the WRA was becoming the target of zealous racists as they attacked the program and claimed the government was pampering us;

"It is a waste of money."

"The government is too lenient with the evacuees."

"They are coddling to their wishes."

"The Japs are disloyal and dangerous."

"They should be kept confined."

Word reached us of this kind of talk, and I trembled as I wondered what else they could do to us.

The frenzy of injustice increased when a bill was introduced in the California legislature to deprive us Nisei of our citizenship. There were those who demanded our immediate departure out of the country. They wanted to send us all to Japan and make sure we would never be able to come back to California. Imagine! It was bad enough that we had been forced from our homes; now they wanted to send us even farther than Arkansas. Al and I just didn't know what to think.

In spite of all this talk, the administrators of the relocation camp and WRA officials showed sensitivity to our massive travail. They personally became involved to show their care and good will during our first Christmas away from home.

Huge Arkansas pine trees were delivered to each mess hall as gifts from the project administrators. The freshly cut giant trees reached up to the ceilings and filled the halls with the pungent fragrance of pine. The Arkansas pines reminded our Issei of the venerable pine tree of Japan, a classic symbol of enduring strength and noble dignity. Their eyes were bright as they enjoyed the nostalgia and stood a bit taller, determined to be noble and enduring themselves.

My eyes glistened with gratitude as I gazed at the symbols of hope brought into this barren place where Christmas joy seemed thousands of miles away. The life-giving green made my body bubble with the spirit of the Yuletide as I forgot for a while my weary soul.

Block residents were encouraged to think of ways to decorate the trees in the mess hall. Each day when we walked in, we saw new evidence of holiday decorations being placed about. There was such ingenuity displayed. Fruits and vegetables were turned into Santa Clauses with cotton whiskers. On the trees were festooned chains of jam jar rings, decorated cans and tiny boxes, berry chains, popcorn strings and dried fruit hung in interesting patterns. Paper wreaths, homemade bells, tinsels, crepe paper flowers and streamers were used as garlands to convert the mess halls into magic places fit for grand Christmas celebrations.

The banging of the dishpan sounded especially merry on this special day. Whispers had been heard for days of the holiday dinner that was being prepared. From our Block 9 Mess Hall, we smelled delicious smells all day. They kept us guessing joyously as we waited to be called.

Al and I gathered Marielle, Nami and our two old ones and together as a family we entered the Christmas dining hall. Before us were hundreds of colorful decorations that camouflaged the crude rafters overhead. Children were wide-eyed and smiling as they looked with wonder upon the transformed hall.

Marielle's eyes were filled with the wonderment of Christmas. I was grateful that my child could enjoy these simple pleasures. The smiling faces of the hard working kitchen crew—cooks, waiters, dishwashers—gave away their big secret. Their smiles were bigger than we had ever seen as they anticipated our astonishment.

Their surprise for us was a complete chicken dinner with all the trimmings. Our crew had scrimped and saved to make this dinner extra special, and that's exactly what it was for all of us. Fat chickens had been roasted for our table, chickens that had been raised by our fellow evacuees in Tule Lake. The variety of vegetables was from our own Victory Gardens. Tasty stuffing, cranberries and candied yams met our surprised gazes. For a few moments we forgot our loneliness and separation as we enjoyed this marvelous feast. We rejoiced, determined to bring the Christmas spirit to our miserable existence.

Each child under the age of fifteen had a gift. Some came with the names and addresses of their senders and messages of love and concern. They had been donated by sympathetic Caucasians from 30 states, gifts of friendship and gestures of the true Christmas spirit. Many were from Sunday Schools and churches.

We were comforted to know that we had not been forgotten and that so many people had sent greetings to us at this time. The Quakers of the American Friends' Church were on the top of the list of churches who remembered us and were foremost in their concerns and generous gifts. Christians and Buddhists alike, we celebrated together at this strange holiday time.

Messages such as this from an eight year old in Pennsylvania cheered us immeasurably:

"To a girl I have never met, but I know the receiver of this gift loves America just as much as I do, and I know you are true to the land both you and I love."

Gifts had also been purchased by camp officials. They had helped facilitate a grand shopping trip for the camp council representatives. Seichi Mikami headed a group of shoppers on a two-day buying spree to Little Rock. WRA Regional Director E.B. Whitake had authorized the expenditure of $1000 for gifts to supplement those sent from outside so that all the children would have gifts. The group also bought small gifts for the elderly Issei. The people the shoppers met in Little Rock were friendly and concerned. It made us feel happy about the good will in the spirit of the 1942 Christmas in Arkansas.

7

New Immigrants and Old Minorities: The Contemporary Playing Field

The beginning of the twentieth century is often thought of as the great age of immigration in the United States. By 1910, 14.7 percent of the entire U.S. population was foreign-born—the highest it had been since colonial days (U.S. Bureau of the Census, 1997b). The National Origins Act of 1924 put an effective stop to this influx, placing immigration quotas on every country. This law would remain in force for forty years, dropping the percentage of foreign-born in the United States by two-thirds between 1910 and 1970.

This chapter focuses on the newest Americans—those immigrants who arrived because of immigration law changes in the 1960s. The doors to America opened halfway, beginning a second great age of immigration. Unlike the wave from a century ago, which flowed largely from Europe, this new wave originated in Latin America and Asia (see Alba, 1999). The United States would once again face large numbers of people searching for ways to adapt while altering the ethnic landscape of the population. Not surprisingly, these changes also brought about the same anti-immigrant attitudes so prevalent in the early years of the twentieth century (Henry, 1999). Our current story of immigration is still in the process of unfolding.

BOX 7.1: WHAT'S IN A NAME

The United States Bureau of the Census and related governmental agencies will provide us with helpful information throughout this text. Because the topics at hand are race and ethnicity, it is important to understand the categories in use for sorting American citizens into this or that group. Different categories are used in different assessments, and to make matters worse, some categories allow the same people to be counted twice. What follows is a partial guide through this maze.

For many years, the bureau of the Census focused primarily on "white" people and "black" people as they counted. Native Americans and Asian Americans were either lost in the shuffle or counted as "other." The designation "Spanish surname," which neatly separated a Martin from a Martinez, was used for a time. Needless to say, it was of very little help.

Census categories are definitely more useful today, but they are still confusing at times. A new category—Hispanic—includes Mexican Americans, Cubans, Puerto Ricans, and all other peoples from the old Spanish colonies in the New World. Hispanics may be of any combination of European, African, or Native American ancestry. Hispanics, therefore, may be "black" or "white," according to common usage. Is a black American named Martinez treated differently than a black American named Johnson? In some cases, the answer is clearly *yes*. By the same token, a Martinez may have largely European ancestry. Will this Martinez be treated differently than a white American named Martin? Again, in some cases, *yes*. Remember that Hispanics may be of any "race."

"White" people are now also somewhat more confusing as a census category. The Census category of "white" may include Hispanics who are commonly labeled as "white" (or so label themselves). The Census category of "non-Hispanic white," eliminates this latter group and comes closer to the use of the term "European American" throughout this book.

"Black" Americans, to the Census, are either non-Hispanic Americans of African descent, or a larger group of people including Hispanics of African descent. At times, the Census will specify "black, non-Hispanic," but more typically we encounter simply "black," which includes those of Hispanic origin. Either Census term is considered the same as the use of "African American" in this book.

Native Americans appear in the Census category "American Indian and Alaskan Native," or "American Indian, Eskimo, and Aleut." Throughout this book, the term "Native American" has been used interchangeably to mean all the native peoples of the New World, or, in more contemporary usage, native peoples under the jurisdiction of the United States. The context should make the meaning clear. The latter definition of "Native American" matches the Census usage. Because there is great diversity among all these native peoples, however, it is always a pleasant surprise when the Census provides some breakdown within that overall category. This seldom happens, though, so we usually must work with the broad definitions we are provided.

Finally, the Census term "Asian and Pacific Islander" encompasses all people from southern and eastern Asia, including peoples from the southern Pacific islands. The category thus combines Samoans, Asian Indians, Vietnamese, Chinese, Koreans, Japanese, Filipinos, and many otehr peoples. This chapter makes the diversity within this category abundantly clear.

Hispaic is an ethnicity, not a Race.

The 1965 Immigration Act

In spite of what appears to be a forty year period of the status quo, immigration law did undergo some interesting changes via minor alterations between 1924 and 1965. Some alterations occurred largely for political reasons—granting rights of citizenship to Chinese immigrants during World War II, for example—but other changes were more significant overall. Immigration has always been linked with labor force demands. Immigration law changes in the 1940s and 1950s clearly showed that a closed-door policy did not allow fine tuning of the labor force during times of shortage.

The *bracero* program began quietly in 1942, receiving little notice because the United States had just entered a war that was then going badly. While the armed forces were in the rebuilding process, industrial production was going full swing and labor was needed. This labor shortage brought rural minorities and women into the labor force, but still more labor was needed, particularly in agriculture. The *bracero* program was designed to bring Mexican citizens into the United States on a temporary basis to work in agriculture. The program would remain operative until 1964.

A more global policy change in immigration law occurred with the passage of the Nationality Act of 1952 (the McCarran-Walter Act). This law left the quota system from 1924 intact but with three additions that would appear a decade later in much stronger form. First, the Nationality Act linked immigration preference with skills that were "urgently needed" in the United States. Immigrants with skills already in place could leapfrog over other applicants from their country. Second, it formalized preferences for immigrant relatives of U.S. citizens. Third, racial groups previously singled out as "ineligible for citizenship" were permitted to become naturalized citizens.

Basic Changes in the Law

The Immigration and Nationality Act of 1965 was the most significant immigration law since 1924; no immigration law passed since has come close to its importance. The cornerstones of the 1965 law were family reunification and an end to the racially and ethnically biased quota system. The law took shape with the addition of subsequent legislation, creating the following immigration preference system:

1. Restricted immigration would be limited to 270,000 individuals annually with no more than 20,000 individuals entering from any one country. Of that total, 80 percent would be limited to close relatives of citizens or residents of the United States, and 20 percent would be allocated on the basis of needed skills possessed by aspiring immigrants.

2. Unrestricted immigration would be granted to individuals in the following categories: (a) spouses, parents, and minor children of adult U.S. citizens and (b) refugees and asylees.

Interestingly, most governmental officials at the time had little idea of the impact this law would have on the United States, either in the overall size of immigration or its diversity. When Attorney General Robert Kennedy was asked how this law would affect Asian immigration, he replied, "it would be approximately 5,000, Mr. Chairman, after which immigration from that source would virtually disappear; 5,000 immigrants could come in the first year, but we do not expect that there would be any great influx after that" (quoted in Borjas, 1990:32). Kennedy was not alone. Most governmental officials believed that demand for immigration to the United States was relatively low.

The demand, however, was huge, particularly from Asians and Latin Americans. In addition, Congress did not seem to have a firm grasp on the nature of family structure—the more relatives that entered the United States, the more still other relatives became eligible. In most years, more immigrants entered under this unrestricted status than under restricted status. (Over time, the occupational preferences category decreased and was filled with increasing numbers of family relations in an effort to minimize total immigration numbers.) Congress thought it was half opening a door to a few newcomers. In fact, it was opening that door almost completely to a crowd of avid immigrants. Figure 7.1 shows the change in the percentage of foreign-born in the United States throughout the twentieth century. It also shows us the absolute change in the numbers of foreign-born in the United States. The impact of the 1965 law cannot be clearer.

FIGURE 7.1 FOREIGN-BORN POPULATION AND PERCENT OF U.S. POPULATION, 1850–1997.

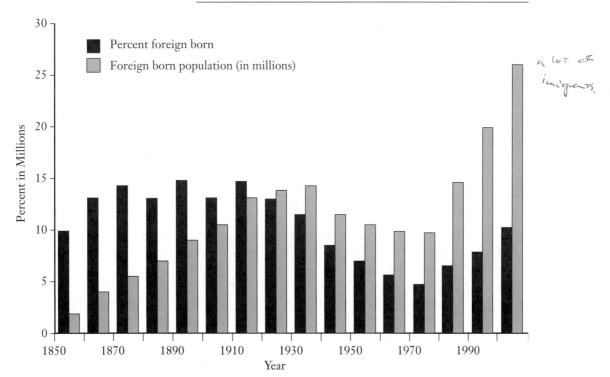

U.S. Bureau of the Census, 1999c.

Refugees and Asylees

Refugees and asylees comprise a fairly large proportion of unrestricted immigration. The United States has long admitted refugees and permitted asylum but only for certain kinds of refugees. Before 1980, refugees and asylees gained their status as immigrants only if they were fleeing a Communist country or a Communist-dominated area. Immigration policy was essentially an extension of foreign policy. The federal government often denied this status to individuals from non-Communist countries on the grounds that they sought to immigrate for economic reasons and were using fear of political oppression in the home country as an excuse. Because most countries with high levels of governmental oppression also offer fewer economic opportunities than the United States, prospective immigrants found such charges hard to refute.

The Refugee Act of 1980 redefined a refugee as an individual living outside his or her country of nationality because of fear of persecution on their return as a result of their race, religion, nationality, membership in a particular social group, or political opinion. The new law also attempted to place a cap on the total number of immigrants to be admitted under this status. This cap changes regularly, however, in response to changing world conditions. To date, most immigrants to the United States under this status have emigrated from Cuba and Vietnam. Unlike other immigrants, refugees and asylees receive federal governmental assistance, from finding housing and employment to receiving Medicaid. Between 1981 and 1987, approximately $600 million per year was spent on aid to refugees, amounting to about $7,000 per immigrant (Borjas, 1990).

The New Foreign-Born

Table 7.1 provides an overall view of the sources of recent immigration to the United States. One country alone—Mexico—is the birth country of over one fourth of all foreign-born. Other Spanish speaking countries from Latin America add another 24 percent. From across the Pacific, the many countries of Asia make up over 26 percent of all foreign-born. The Philippines regularly sends the most immigrants, but as Table 7.1 shows, Asian immigration is relatively spread out. The number from Europe may seem surprisingly large, but the impact on the United States has been more in numbers than in cultural diversity. Immigration was quite high between 1950 and 1970 from both the United Kingdom and Germany. Beyond that, there has been a steady stream of immigrants from virtually all European countries. The relatively smaller numbers from each country serve to lessen the imported cultural diversity. The greatest impact for the United States, both in numbers and in diversity, has come from Latinos and Asians.

These new immigrants settle in very definite geographic patterns. Some settle in certain areas based on earlier immigration to that area by their ethnic group. Such areas are typically urban but increasing numbers of new immigrants are settling in suburban areas (Alba et al., 1999). Others seek locations that are generally expanding economically and provide more opportunities. This combination has made California the most popular destination for both Latino and Asian immigrants. Fully 8 million foreign-born live in that state, making up 25 percent of its population. New York

TABLE 7.1	REGIONS AND TOP THIRTEEN COUNTRIES OF BIRTH OF THE FOREIGN BORN, 1997	
Country of Birth	**Number (in thousands)**	**Percent**
All Countries	**25,800**	**100.0**
Latin America	13,235	51.3
Mexico	7,017	27.2
Cuba	913	3.5
Dominican Republic	632	2.4
El Salvador	607	2.4
Asia	6,914	26.8
Philippines	1,132	4.4
China	1,107	4.3
Vietnam	770	3.0
India	748	2.9
Korea	591	2.3
Europe	4,360	16.9
Soviet Union	734	2.9
United Kingdom	606	2.3
Germany	578	2.2
North America	568	2.2
Canada	542	2.1
Other	748	2.9

U.S. Bureau of the Census, 1999c

contains 3.6 million foreign-born—19.6 percent of its population. Between them, California and New York contain almost half of the foreign-born in the United States (U.S. Bureau of the Census, 1999c). Beyond these two states, Hawaii is over 18 percent foreign-born, followed by Florida (16.4 percent) and New Jersey (15.4 percent). The impact of the new immigration is clearly focused in a few select regions of the United States.

Latino Immigration

Mexico and countries to the south have experienced a rocky twentieth century. In spite of many cultural differences among them, most are either newly or recently encountering industrialization. Their economic peripheral connection to the industrial

United States and other industrial countries has paralleled their generally weak political position on the world stage. The growing pains of new industrialization, producing urbanization, poverty, and political upheavals, created strong push factors that stimulated immigration beginning in the early 1900s. For Mexico alone, the twentieth century experience with U.S. immigration has been an on-again, off-again fluctuation; labor needs in the United States were constantly at war with the fear and hatred of foreigners. This fear and hatred seemed particularly active with relation to Latino foreigners.

Nowhere in the world does such a highly industrialized country share so long a border with an undeveloped country as that between the United States and Mexico. In a larger sense, the United States is equally close to many Latin American countries, and the 1965 Immigration Law also would make it accessible.

Puerto Rican Immigration   *not parts of U.S*

Puerto Rico remains in the circumstances once shared by Hawaii and Alaska—it was long a territory of the United States but never a state. In 1952, it became a semiautonomous commonwealth, having been acquired along with the Philippines and assorted other real estate following the Spanish American War at the end of the nineteenth century. Its previous Spanish ownership gave this small Caribbean island a definite Spanish cultural cast (including language). At the same time, Puerto Rico is both physically and socially removed from the other major sources of Hispanic immigration to the United States. Although it is convenient to categorize all the Spanish-speaking people of Latin America as Hispanics or Latinos, the differences among these groups are equally important. Puerto Ricans were distinctive when they first arrived in the United States and they have remained distinctive.

Unlike other Latin Americans, Puerto Ricans can enter the mainland at will because of the unique status of the island. They were unaffected by the Immigration Law of 1965, but the timing of their migration to the mainland is indicative of how other Latinos responded once the door was open. Very few Puerto Ricans migrated during the first half of the twentieth century. World War II brought American investment to the island, however, producing the modernization and industrialization that was spreading throughout all of Latin America at about that time. As traditional means of survival disappeared, Puerto Ricans were forced to look elsewhere for different kinds of work. Some looked to the capital city of Puerto Rico—San Juan—but many others turned to the large urban centers of the northeastern United States. The northern migration grew rapidly following World War II: from 1950 to 1960, 41,200 Puerto Ricans left for the mainland, dropping to 15,000 between 1960 and 1970 (Morales and Bonilla, 1995).

New York City became the primary destination for migrating Puerto Ricans. As with so many immigrant groups, a combination of housing discrimination and cultural ties created a distinctive ethnic community (Santiago and Galster, 1995). Although New York City had seen much cultural diversity over the years, Puerto Ricans introduced the first significant Spanish influence. The already well-established patterns of discrimination affecting Mexican immigrants and Mexican Americans were transferred to this new group in its urban setting. Changes in the postwar American economy created an additional hurdle: the strong manufacturing base that had been the bedrock of northern urban economies was on the way out. Puerto Ricans, along with many newly arrived African Americans, would soon find northeastern cities offering fewer and fewer employment opportunities.

The new urban environment and high unemployment rates had an immediate impact on the traditional Puerto Rican family. Women's roles, in particular, changed dramatically. For the smaller group of middle class Puerto Rican immigrants, women found greater opportunity and liberation in the United States; husbands, sometimes grudgingly, went along. Working class women faced difficult challenges, many attempting to maintain traditional roles while facing economic demands to join the labor force. Extended family relations strengthened in order to provide childcare for working mothers; in many cases, older children were thrust into this role (Toro-Morn, 1995).

As employment opportunities declined in the Northeast, Puerto Rican families increased in welfare participation and single-parent, female headship, particularly as unemployment rose for Puerto Rican men (Enchautegui, 1992; Melendez and Figueroa, 1992; Rodriguez, 1992; Santiago and Galster, 1995). During the 1970s, when other Hispanic groups were improving economically, Puerto Ricans' economic situation was in decline. Their percent below the poverty level during that period rose from 29 percent to 35 percent. In the early 1980s—a time of economic recession— Puerto Rican income declined 19 percent compared to 14 percent for African Americans, 9 percent for other Hispanics, and 5 percent for European Americans. By the late 1980s, Puerto Ricans were clearly worse off than they had been in the mid-1960s (Tienda and Jensen, 1988).

Much of declining Puerto Rican fortunes can be explained by geography. Northeastern and north central urban centers, which had provided so much opportunity for immigrants for a century, became traps for minorities in the 1970s and 1980s. When northern, urban Puerto Ricans are compared to those living in other parts of the United States, many differences stand out. Those who escaped the cities have higher rates of labor force participation, lower levels of welfare participation, and lower levels of female headship (Enchautegui, 1992). They also, however, have higher educational levels than those who remained. It is very possible that those who bring more to the labor market are the same people who move in search of better markets. Nevertheless, the picture of Puerto Ricans in northern cities continues to be bleak. There are few indications that increased education and migration will become realistic possibilities for most Puerto Ricans at the present time.

Cuban Immigration

As with Puerto Rican migration, Cuban immigration to the United States also was unaffected by the immigration law changes of 1965; they entered the United States under refugee status. They also shared a Spanish cultural and linguistic background as a former Spanish colony, but there ends the resemblance between Cubans and any other Latinos in the United States.

With the exception of some Central American immigrants, Cuban immigration was politically rather than economically motivated. They left Cuba as a direct result of the revolution in the late 1950s and the establishment of a Communist regime under Fidel Castro. Refugee status under immigration law in the United States has traditionally favored individuals fleeing Communism. The relocation of Cubans to southern Florida was a natural move because 34,000 Cubans already lived there by 1950, many having fled following previous revolutions on their home island. Their population in Florida would grow to 79,000 by 1960, 500,000 by 1970, and 831,000 by 1980 (Morales and Bonilla, 1995).

Political refugees are always different from economically motivated immigrants. As we have seen, they are more likely to be "pushed" than "pulled," and for obvious reasons, they also are far less likely to return. But political refugees fleeing Communist regimes tend to be distinctively professional and middle class because they typically lose the most through a political change to Communism. Southern Florida thus became home to a rapidly growing ethnic enclave—"Little Havana" inside Miami—populated largely by individuals with money and skills. As with many middle-class immigrants, those skills may not have been immediately culturally transferable; indeed, Caribbean immigrants from former British colonies tend to fare better in the United States in part because of cultural similarities (Kalmijn, 1996). But many intangibles come with middle-class status in any culture. A professional might turn to business, for example, and fare better because of an educated background. Thus, Little Havana turned into a thriving ethnic enclave quite quickly.

Ethnic enclaves tend to be most successful when four conditions coexist: (1) many group members have prior business experience; (2) sources of capital are available; (3) inexpensive labor is available; and (4) the market for those business enterprises is somewhat sheltered, providing somewhat captive customers who have little choice in where they patronize (Portes and Manning, 1986). All of these factors were present in Little Havana during the 1960s and 1970s. Cuban rates of self-employment rose higher than for European Americans during this period and greatly surpassed African Americans with whom these immigrants shared Miami (Pérez-Stable and Uriarte, 1995). The enclave protected Cubans from many of the changes in the manufacturing sector then occurring in Miami; African Americans had no such buffer and strained relations developed between these two groups.

Portes and Manning (1986) argue that enclave economies not only provide for the immediate needs of an immigrant group but also provide members with opportunities for future social mobility outside the enclave. Cubans with the highest levels of education did indeed become successful outside the enclave, particularly Cuban men who moved into professional occupations. Self-employed Cubans and women tended to fare better economically inside the enclave (Portes and Jensen, 1989). Such growing opportunities will no doubt diminish the enclave with future generations.

Ethnic enclaves are primarily economic units, but they have an important cultural aspect as well: the more economically successful the enclave, the more that group's culture will gain strength and become a potential pluralistic challenge to the dominant group. The Cuban enclave in southern Florida brought a Latin flavor to Miami while creating more trade in that area with other parts of Latin America (Levine, 1985). These changes stimulated European American backlash movements such as the

"English Only" movement, which seeks to have English declared the national language of the United States.

Mexican Immigration

Unlike the more recent Puerto Rican and Cuban immigration, Mexican immigration had been steady throughout the twentieth century. Political and economic disruption within Mexico combined to make the United States a popular destination. When Mexican immigration slowed or temporarily stopped during the century, it was a direct result of downturns in the American economy and a dropping need for inexpensive Mexican labor. The repatriation of Mexicans in the 1930s (see Chapter 6) was dwarfed by "Operation Wetback" of the 1950s recession. In 1954 alone, over 1 million Mexicans (including some American citizens) were forcibly returned to Mexico (Morales and Bonilla, 1995). When Mexican immigration was on the increase, the American economy was invariably booming and in need of labor. The flow of immigration from Mexico was controlled by the United States. At least, that was the case until 1965.

The special circumstances of Puerto Ricans and Cubans placed them outside the impact of the Immigration Law of 1965. For the rest of Latin America, and for Mexico in particular, the effect was immediate and striking. Mexico received the same 20,000 per year quota that other countries were granted, but the family reunification rules for unrestricted immigration opened much greater opportunity. The relatively

The immigration law of 1965 allowed previous Mexican-American immigrants to bring their families to join them in the United States.

large number of Mexicans and Mexican Americans already in the United States by 1965 were allowed by the law to bring their families to join them. Because many of the earlier Mexican immigrants were married men with wives and dependent children in Mexico, the potential for new immigration was substantial. During the 1970s, Mexican immigration was far above the numbers from any other country. The second largest source of 1970s immigrants—the Philippines—contributed just a little more than half the people coming from Mexico. As we saw in Table 7.1, no other country comes close to Mexico in its contemporary percentage of the foreign-born in the United States.

As the legal immigration from Mexico grew, so too did illegal immigration. In their study of the 1980 census, Warren and Passel (1987) estimate that 14.5 percent of the foreign-born in the United States were illegals—a total of 2,057,000 people. Of those, they estimate that 55 percent—1,131,000 people—were of Mexican origin. Those illegals of Mexican origin comprised almost 45 percent of the total foreign-born Mexican population in the United States at that time. When those numbers are added to the legal immigration rates from Mexico previously noted, Mexican dominance in American immigration becomes even more apparent.

An early United States response to illegal Mexican immigration was the Border Industrialization Program, begun in 1965, which resulted in the creation of the *maquiladoras*—the Mexican factories near the U.S. border that benefit from special trade and tax laws with the United States. An American automobile manufacturing company, for example, can build a factory in northern Mexico just as if they were building it in Texas: materials and manufactured goods come and go in the same way. The difference, however, is that Mexican labor is much less expensive than American labor. This program therefore encourages U.S. investment in Mexico while simultaneously encouraging Mexican citizens to stay in Mexico to work instead of crossing the border into the United States.

The *maquiladoras* have had an impact on Mexican immigration, particularly in decreasing the immigration rates of young women who are much desired as workers in the border industry. Still, many Mexicans prefer the United States for employment as the above immigration numbers suggest. But the American economy has changed considerably from the days when Mexican labor was found predominantly in agriculture, mining, and railroads. Latinos in the United States today are 90 percent urbanized, compared to 75 percent of the European American population (Morales and Bonilla, 1995). Modern Mexican immigrants are more likely to find urban employment in the industrial sector than native-born Mexican Americans or even recent immigrants from other countries (Morales and Ong, 1995).

Mexican immigrants not employed in the industrial sector are just as likely to be urban residents, employed in low-level service occupations. Women, in particular, are much in demand as domestic workers and childcare providers. In both Mexico and Central America, informal networks exist to induce women to emigrate (Repak, 1994). In the United States, yet more informal networks exist, functioning as an apprenticeship-type program in which newer immigrants work for earlier immigrants (Hondagneu-Sotelo, 1994). Sometimes new immigrants from one area find employment with immigrants from another area. In Los Angeles, for example, Asian-run businesses in the garment industry employ significant numbers of Latino workers (Light et al., 1999). These systems thus provide employment contacts for the newer immigrants and, of course, inexpensive labor for the earlier immigrants. Such systems are reminiscent of middleman minority theories in which immigrants turn to some form of self-employment for survival (see Spener and Bean, 1999). Housecleaning, like turn-of-the-century laundries, can operate with limited capital.

As the previous domestic work example suggests, Mexicans and Mexican Americans occupy very different places in the labor force. The critical factor in employment is the acquisition of American culture skills—native-born Mexican Americans are much more likely to have those skills in place. Both legal and illegal recent immigrants find themselves at a disadvantage in the job market, unable to compete with their co-ethnics. This observation also contains an implicit caution: any statistics about Mexican Americans includes *all* of these individuals. Thus, when Mexican Americans turn up with low overall levels of education, for example, appreciate that newer immigrants may be responsible for lowering those numbers.

Mexican Americans occupy many different communities in the United States, in large part because of their continual immigration. Unlike ethnic groups who had a surge of immigration over a twenty-year period, continued immigration separates an ethnic group into native-born and foreign-born; in addition, those native-born whose parents were also native-born are very different from those with shorter histories in the United States. The former are likely to be very Americanized, while the latter likely maintain strong elements of Mexican culture (see Pardo, 1991). As just one example of these many differences, native-born Mexican Americans are less likely to divorce as their educational attainment increases. One obvious advantage of increasing educational attainment is that it usually removes some economic strains from marriages. By contrast, the most poorly educated among foreign-born Mexicans in the United States have the lowest divorce rates. In their case, the strains of immigration and poverty status are filtered through the strong value of familism in Mexican culture (Bean et al., 1996).

Immigration from Mexico to the United States shows no signs of significant change. Push factors in Mexico continue to combine with economic pull factors in the United States. The family reunification priorities of the 1965 Immigration Law should only serve to open the doors more widely in coming decades. Combined with high birth rates within the United States, Mexican Americans should increase their percentage within the Latino population and help increase the overall percentage of Latinos within the United States.

New Asian Immigration

Although Mexicans benefited from the family reunification feature of the 1965 Immigration Law, those with needed (and largely professional) occupational skills had less motivation to emigrate. By contrast, both features of the 1965 law had a major impact on immigration from Asia. Asian ethnic groups already in the United States made extensive use of family reunification. Professionals from among those Asian ethnic groups not well established in the United States by 1965 turned to the occupational preferences of the law. Some of those professionals were trained in the United States and were already familiar with the language and culture; others came from countries in which that language and/or that culture were already established, particularly India and the Philippines. Between family reunification and occupational preferences, the 1965 Immigration Law produced a major increase in Asian immigration.

Table 7.2 follows Asian immigration since the 1965 law. Overall, Asian immigration reached 30 percent of all immigration in just six short years after the law changed. Since then, Asian immigration has hovered between 30 percent and 50 percent of all immigration. Although overall immigration has been relatively constant, variation by country has been extreme. Vietnamese immigration—unique among these countries because of its dependence on refugee status—began suddenly in 1978 with the end of the Vietnam War and has continued since. Japan is alone among these Asian countries in the lack of immigration interest among its citizens. Most other countries began a steady rise in immigration since the 1970s, with India being the last to join the movement. Overall, the Philippines stands out in both the magnitude and regularity of its immigration. We examine these individual differences by country in the following sections.

Before turning to differences among Asian ethnic groups, some similarities shared by many or most of these groups provide a useful starting point. First, Asians dominate the occupational preference immigration entry, showing particular strength in the fields of engineering and computer science (Kanjanapan, 1995). The percentage of total immigrants who enter under this status varies by country, but almost all make a noticeable contribution. Those groups whose cultural differences from the United States prevent their importation of occupational skills have shown unusually high rates of entrepreneurship, both in enclave economies and as middleman minorities.

Second, most Asian cultures place a high value on education, which continues after immigration. Asians with little English background have generally concentrated their efforts in mathematics and science where they are less disadvantaged in competition with native English speakers. This educational emphasis, which has been termed a "risk aversion strategy," clearly provides some immediate employment benefits, but it may create barriers for future advancement (Wong, 1995).

 —— Third, other Americans have saddled Asians with a "model minority" stereotype, which springs from the general "quietness" of Asian immigrants (i.e., low crime and public assistance rates) and their educational and occupational success. Although there is some truth to this stereotype, many Asian groups have higher-than-average poverty rates to accompany high-average income rates. This suggests that many immigrants from the same country of origin actually belong in two groups—those who have been fairly successful and those who have not. Considering that Asian professionals are paid less than European American professionals with the same training and that Asians are concentrated in some of the highest cost-of-living areas in the United States, they can hardly be considered problem-free minorities (Min, 1995b; Nishi, 1995; Tang, 1993; Zhou and Kamo, 1994).

Fourth, Asian arrivals have almost all settled in urban areas; their locations range from 85 percent to 95 percent urban compared to 60 percent for European Americans and 79 percent for African Americans (Min, 1995a). This is not too surprising in that almost all new immigrants to the United States from around the world head for the cities (Carlson, 1994). For some immigrants, this is simply a rational decision based on the location of employment opportunities. For others who may be joining families and/or co-ethnics, urban settlement may be automatic. While some Asians lived in rural America earlier in the twentieth century—most notably the Japanese and Filipinos—Asians have historically been urban dwellers. Those arriving after the 1965 law employed a wide variety of economic strategies, but all were urban.

Chinese

Chinese immigration history began in the midnineteenth century followed by many years of low immigration in the early twentieth century. Still, most of those early decades saw the arrival of 20,000 Chinese on average. During the 1960s, however, that average quintupled to over 100,000 as a direct result of the 1965 change in immigration law. Table 7.2 shows the steady growth that followed. No other Asian group in the United States has quite this pattern of old and new with so much time in between of virtually no immigration. To be of Chinese ancestry in the United States today might make you a foreign-born, new immigrant or it might make you a fifth-generation (or more) American. Not surprisingly, Chinese Americans today are far more than one ethnic group (Kwong, 1987).

Chinese immigrants and Chinese Americans are both largely urban, but their concentration is even more striking. Nineteenth-century Chinese were located exclusively in the western United States. By 1890, over 90 percent of all Chinese lived in the West. That percentage decreased during the early twentieth century as the New York City Chinatown grew. Post-1965 immigrants tended to follow this bicoastal arrangement. By 1990, 52 percent of all Chinese still lived in the West (43 percent in California) and another 27 percent lived in the Northeast. The percentages are almost identical for both new immigrants and Chinese Americans (Wong, 1995).

Native-born Chinese Americans have made impressive gains in the American economy and American education over the last century plus. Over 50 percent of this more acculturated group has completed four years of college—approximately double the comparable percentage among European Americans (Agbayani-Siewert and Revilla, 1995). Twenty-five percent are found in professional occupations and 18.2 percent in managerial positions; the comparable figures for European Americans are 14.4 percent and 12.8 percent, respectively (Wong, 1995). They also have a much lower representation in more poorly paid occupations in service, operatives, and labor and are less likely to be self-employed than European Americans. This last fact is of particular importance in that early Chinese immigrants in their enclave economies thrived on self-employment. Obviously the native-born Chinese American population has achieved in other areas.

The new Chinese immigrants offer a different and much more complex story that is all the more important because the new immigrants comprise about 70 percent of all ethnically Chinese in the United States (Min, 1995a); any statistics on Chinese Americans are dominated by these newer immigrants. The major complexity of this

| TABLE 7.2 | ASIAN IMMIGRATION BY COUNTRY OF BIRTH, 1960, 1965–1997. | | | | | | | | |

Year	All Countries	All Asian Countries	Percent Asian	China[a]	Japan	The Philippines	Korea	India	Vietnam
1960	265,398	23,864	9.0	3,681	5,471	2,954	1,507	391	—
1965	296,697	19,788	6.7	4,057	3,180	3,130	2,165	582	—
1966	323,040	39,878	12.3	13,736	3,394	6,093	2,492	2,458	275
1967	361,972	59,233	16.4	19,741	3,946	10,865	3,956	4,642	490
1968	454,448	57,229	12.6	12,738	3,613	16,731	3,811	4,682	590
1969	358,579	73,621	20.5	15,440	3,957	20,744	6,045	5,963	983
1970	373,326	92,816	24.9	14,093	4,485	31,203	9,314	10,114	1,450
1971	370,478	103,461	27.9	14,417	4,357	28,471	14,297	14,310	2,038
1972	384,685	121,058	31.5	17,339	4,757	29,376	18,876	16,926	3,412
1973	400,063	124,160	31.0	17,297	5,461	30,799	22,930	13,124	4,569
1974	394,861	130,662	33.1	18,056	4,860	32,857	28,028	12,779	3,192
1975	386,194	132,469	34.3	18,536	4,274	31,751	28,362	15,733	3,039
1976	398,613	149,881	37.6	18,823	4,285	37,281	30,803	17,487	3,048
1977	462,315	157,759	34.1	19,764	4,178	39,111	30,917	18,613	4,629
1978	601,442	249,776	41.5	21,315	4,010	37,216	29,288	20,753	88,543
1979	460,348	189,293	41.1	24,264	4,048	41,300	29,248	9,708	22,546
1980	530,639	236,097	44.5	27,651	4,225	42,316	32,320	22,607	43,483
1981	596,600	264,343	44.3	25,803	3,896	43,772	32,663	21,522	55,631
1982	594,131	313,291	52.7	36,984	3,903	45,102	31,724	21,738	72,553

1983	559,763	277,701	49.6	25,777	4,092	41,546	33,339	25,451	37,560
1984	543,903	256,273	47.1	23,363	4,043	42,768	33,042	24,964	37,236
1985	570,009	264,691	46.4	24,789	4,086	47,978	35,253	26,026	31,895
1986	601,708	268,248	44.6	25,106	3,959	52,558	35,776	26,227	29,993
1987	601,516	257,684	42.8	25,841	4,174	50,060	35,849	27,803	24,231
1988	643,025	264,465	41.1	28,717	4,512	50,697	34,703	26,268	25,789
1989	1,090,924	312,149	28.6	32,272	4,849	57,034	34,222	31,175	37,739
1990	1,536,438	338,581	22.0	31,815	5,734	63,756	32,301	30,667	48,792
1991	1,827,167	358,533	19.6	33,025	5,049	63,596	26,518	45,064	55,307
1992	810,635	348,553	43.0	38,735	10,975	59,179	18,983	34,629	77,728
1993	880,014	357,041	40.6	65,552	6,883	63,189	17,949	40,021	59,613
1994	804,416	292,589	36.4	53,985	6,088	53,535	16,011	34,921	41,345
1995	720,461	267,931	37.2	35,463	5,556	50,984	16,047	34,748	41,752
1996	915,900	307,807	33.6	41,728	6,617	55,876	18,185	44,859	42,067
1997	798,378	258,561	32.4	44,356	5,640	47,842	13,626	36,092	37,121

SOURCE: Immigration and Naturalization Service, 1993;1999.

a. Until 1981, immigrants from China included both immigrants from mainland China and those from Taiwan. Since 1982, immigrants from mainland China have been tabulated separately from those from Taiwan.

group stems from their extremes in educational and occupational skills. Some foreign-born Chinese in the United States occupy the highest levels of educational and occupational attainment; others find themselves barely scraping by, often working the very lowest jobs in Chinatowns and vulnerable to labor exploitation as a result of their low acculturation levels (Kwong, 1987; Lowe, 1992; Min, 1995b; Zhou and Logan, 1989).

The old Chinese immigration coupled with the complexity of the new immigrants produces three more or less distinct groups, occupying extremes in American society (Kitano and Daniels, 1988). Native-born Chinese Americans and foreign-born professional immigrants are responsible for the overall high averages for Chinese on income and education. Evidence is ample that both native-born and foreign-born Chinese ethnic professionals and managers face promotional discrimination, especially in the sciences and engineering, indicating they should probably rank higher occupationally than they do (Min, 1995b; Tang, 1993; Wong, 1995; Zhou and Kamo, 1994). At the other extreme, over 11 percent of all Chinese live below the poverty level compared to 8.5 percent of non-Hispanic whites (Baugher and Lamison-White, 1996; Min, 1995a).

The ethnic enclave of Chinatown persists, in part because many Chinese ethnic entrepreneurs still find it a lucrative environment for certain kinds of business enterprises with steady growth in import-export and the tourist industry. The continuing flow of inexpensive immigrant labor makes those businesses even more lucrative (Zhou and Logan, 1991). At the same time, many professional, middle-class Chinese ethnics are looking more toward the suburbs and away from the enclave. In some cases, this produces a decrease in Chinese culture while increasing American acculturation (see Tuan, 1999). Some suburbs, however, have become more like ethnic bedroom communities in which some Chinese culture remains while residents commute to work outside the community. The city of Monterey Park in California is now 60 percent Asian and almost all middle class (Wong, 1995). Many contemporary Chinese middle-class people are following a path originally traveled by Jewish Americans who desired a bicultural lifestyle in the United States.

Asian Indians

Asian Indians have been a distinctive immigrant group to the United States in the following ways:

1. Their immigration rates have always been relatively low, apparently guaranteeing that they will not soon be among the larger of Asian ethnic groups in the United States.

2. Their immigration "surge" was slow to gather steam, as Table 7.2 shows; their immigration began slowly after 1965 but has grown steadily throughout the 1990s.

3. As an immigrant group, they are singularly well educated and financially secure as compared to other Asian groups; unlike the Chinese, no sizable group of poor Indians accompany the better off.

4. They are particularly able to profit from their education in the United States because English is one of the official languages of India, spoken by all in the upper classes.

5. Their ability to compete on their own in the American economy has prevented any hint of an ethnic enclave economy.

6. They have not concentrated in the West as other Asian immigrant groups have.

As a group, Asian Indians are not problem free because they are still Asians in a non-Asian society. For Asians, however, they are in a position of strength.

Asian Indians have some of the highest educational levels among Asian immigrants, particularly earlier immigrants who clearly were the elite of Indian society. While immigrants to the United States from all countries of the world have between 20 and 25 percent rates of college completion, Asian Indian immigrant rates have ranged from 42 to 68 percent (Min, 1995a). By comparison, less than 30 percent of native-born European Americans have completed college (Day and Curry, 1996a). The educational gap between Asian Indian immigrants and native-born European Americans widens at the graduate level, where their percentages at that level of education are three times higher, rising to 6 times higher among the earliest (before 1979) immigrants (Sheth, 1995).

 These high levels of education are directly translatable into employment opportunities because of this group's widespread facility with English. Asian Indians have the highest median income among all Asian immigrant groups, without the downside we saw with the Chinese; Asian Indians have the same percentage living below the poverty level as native-born European Americans (Kitano and Daniels, 1988; Min, 1995a). In their study of Asian Indians in the United States, Rao and associates (1990) found family incomes ranging between $50,000 and $100,000. Not surprisingly, very few respondents needed or received economic help from family or friends, which is very unusual for any immigrant group.

With no need for a protective enclave economy, Asian Indians have moved into a wide variety of economic activities. Many are professionals in medicine and academia. Still more are self-employed, operating both small-scale and large-scale businesses. Asian Indians with less investment capital are coming to dominate the newsstand industry in New York City. Those in stronger positions have acquired larger businesses, particularly in hotel and motel operation (Kitano and Daniels, 1988; Sheth, 1995). These entrepreneurs are classic examples of the middleman minority, operating outside any discernible enclave and doing business with other ethnic groups. All of these factors suggest increasing rates of assimilation for future generations (Bacon, 1999).

The geographical dispersion of Asian Indians in the United States is remarkable when compared with other recent Asian immigrants. While others have concentrated in the West, Asian Indians have not. As of 1990, 23 percent lived in the West (almost all in California), 18 percent in the Midwest (especially Illinois), 35 percent in the Northeast (primarily New York and New Jersey), and 24 percent in the South (Sheth, 1995). Their predominant means of employment tend to concentrate them in urban

centers—largely true of all recent immigrants—but there are no regional requirements if one is a physician or a hotel owner. Because their numbers in the United
States are currently small anyway, this geographical dispersion tends to decrease their
visibility among non-Asian Americans.

Filipinos

The Philippines have had a long and not necessarily pleasant experience with the
Western World. They were a Spanish colony for years until acquired by the United
States following victory in the Spanish-American War of 1898. The Spanish influence
made them a largely Catholic people, and the American influence made the English
language commonplace. In addition, Filipino children were force-fed much American
history as part of their normal schooling (Agbayani-Siewert and Revilla, 1995). As a
result, immigration to the United States for Filipinos is neither the cultural shock nor
the economic setback that it is for many immigrants. Much like Asian Indians, their
imported skills have been immediately marketable.

Filipinos have a long and varied immigration history to the United States. As a
U.S. territory, they were allowed easy entrance until the passage of the Tydings-
McDuffie Act in 1934. Final independence for the Philippines in 1946 placed Filipinos in the same status with other Asians thereafter. In the early 1900s, young Filipino men entered the United States to pursue higher education. Between 1910 and
1938, an estimated 14,000 Filipino men enrolled in American colleges and universities (Kitano and Daniels, 1988). Many returned home but a fair number remained. Yet
other Filipinos immigrated to Hawaii and the West of the United States as agricultural workers. In Hawaii, sugar plantations owners actively sought Filipino workers
for their cheap labor. The plantations tended to have *haole* managers (pronounced
"howlie"—a Hawaiian term for European Americans), Japanese skilled workers, and
Filipinos as unskilled workers, producing a convenient local system of ethnic stratification (Espiritu, 1996). On the West Coast of the United States, many Filipino immigrants in the 1930s found themselves working for Japanese run agricultural
enterprises. Unlike the Japanese, few Filipino agricultural workers would advance to
farm ownership (Kitano and Daniels, 1988).

These immigration patterns changed drastically with the passage of the 1965 Immigration Law. Some Filipino immigration fell under the family reunification sections
of the law. The extremely high proportions of men within the Filipino American population produced a highly understandable enthusiasm for bringing Filipino brides to
the United States. More important initially, however, was the occupational skills entrance status. Highly educated Filipinos were thereby provided a means to increase
their success in the more lucrative climate of the United States.

Before 1965, Filipinos in agricultural work were poorly prepared and lacked resources for educational attainment. Educated Filipinos in the United States had faced
employment discrimination when seeking higher status jobs. (This, of course, is why so
many of the earlier student immigrants returned to the Philippines.) There were understandably few incentives for education among the earlier immigrants. The post-1965
wave of Filipino immigrants was educated, however, and faced the decreasing overt

discrimination characteristic of the 1960s. Filipinos were unique among almost all immigrant groups in that the *foreign-born were better educated* than the native-born. In 1990, 42.3 percent of foreign-born Filipinos in the United States had completed four years of college compared to 22.3 percent of native-born (Agbayani-Siewert and Revilla, 1995). Only Asian Indians had similar proportions and the differences have been declining among them with post-1985 (and less educated) immigrants (Sheth, 1995). In terms of education and occupational skills, Filipino Americans are a highly diverse group.

In addition to the diversity of the U.S. Filipino population resulting from the many years of immigration, regional diversity within the Philippines also plays a role. The Philippines are a collection of islands inhabited by many distinct ethnic groups and different languages. Filipinos often had their first taste of national consciousness thrust upon them when they immigrated. To European Americans, Filipino regional differences were unimportant and they were all treated as one group. Nevertheless, those regional differences clearly prevented the formation of ethnic enclaves as a cultural/economic strategy (Agbayani-Siewert and Revilla, 1995; Kitano and Daniels, 1988). For those Filipino immigrants with educational and occupational skills, the challenge of "making it" in the United States was similar to that faced by Asian Indians. The best strategy was to make it on one's own. They wound up with a median family income almost as high as Asian Indians and even a lower percentage living below the poverty level. As with so many Asian immigrants, the locations in which they chose to settle were Hawaii and the West on the mainland (Min, 1995a).

Koreans

By 1964, only 15,050 Korean immigrants had found their way to the United States (Min, 1995c). Until that time, the United States was primarily a destination for Korean students seeking education. The occupational preferences of the 1965 Immigration law would change that.

Korean immigration began slowly after 1965, starting at about 2,500 annually in 1966 and reaching 10,000 annually by 1970. By 1972, however, annual Korean immigration numbers were near 20,000; they have continued to hover between 25,000 and 35,000 yearly since, dropping off only in the 1990s (see Table 7.2). Up until the mid-1970s, a high percentage of these immigrants were admitted under occupational preference status: in 1967, 38.1 percent of all Korean immigrants entered in that status; by 1972, that percentage had risen to 45 percent. It only began to drop in the late-1970s, averaging a little less than 10 percent since (Min, 1995c). This drop had much more to do with changes in U.S. governmental policy than with Korean credentials or their desire to immigrate.

Unlike Asian Indians and Filipinos, Korean immigrants brought neither familiarity with the English language nor American culture with them. As a result, white collar and professional Korean immigrants faced immediate downward mobility in occupational status upon arriving in the United States (Light and Rosenstein, 1995; Min, 1990). With personal resources, sometimes combined with "borrowed" resources from Korea, new immigrants looked to build an economic beachhead on American soil. As with so many such groups, their strategy was entrepreneurship.

Korean immigrants often opened small retail businesses, such as grocery stores, in economically depressed urban areas.

Whereas Chinese immigrants clustered in enclaves and Asian Indians struck out on their own, Korean immigrants have done a little of both. Sizable "Koreatowns" have grown in Los Angeles and Flushing, New York, with the former containing 35 percent of the Korean population in Los Angeles by 1980 (Min, 1990, 1995c). (California is home to 33 percent of the Korean population and New York contains another 12 percent.) Like most enclaves, these communities contain many elements of Korean culture in both commerce and lifestyle (see Park, 1999). Still, most Koreans live outside enclave economies. The most common economic pattern is the classic middleman minority solution. For Koreans, the dominant mode has been retail business—especially small groceries and liquor stores—which operate in low income and usually nonwhite areas of America's largest cities; by 1992, 1,800 Korean-owned green grocery stores were in the New York City metropolitan area, accounting for 60 percent of such stores (Kim and Hurh, 1993; Light and Rosenstein, 1995; Min, 1990, 1995c; Yoon, 1995).

Retail business in economically depressed urban areas has the advantage of facing little competition and requiring minimal start-up capital. Most such Korean entrepreneurs lower costs still further by employing family members (often later immigrants) at low wages. While these businesses have considerable economic interaction with non-Koreans in both wholesale and retail trade, social interaction is minimal. In particular, Koreans have faced considerable conflict in African American communities, many of whose members complain that the Korean businesses take from the communities without putting anything back; the stores do not even provide much needed employment (Chang, 1993; Jo, 1992).

Korean–African American conflict has been both nonviolent and violent. Major boycotts against Korean-owned businesses occurred throughout the 1980s and 1990s in New York City. More striking was the Los Angeles riot of 1992 following the acquittal of police officers charged with the beating of African American Rodney King. African Americans and Hispanics singled out Korean-owned stores in Los Angeles, looting and burning a total of 2,300 such businesses and causing damage estimated at $350 million (Min, 1995c).

The Korean economic strategy has produced obviously mixed results. Although Korean median income in the United States is well above African Americans and Hispanics, it is below European Americans and all other Asian groups with the exception of the Vietnamese. In addition, over twice as many Koreans (proportionately) live below the poverty level as European Americans—again, the highest figure for Asians with the exception of the Vietnamese (Min, 1995a). Their businesses require hard work, place them in physical danger, and leave them open to exploitation from wholesalers (Kim and Hurh, 1993; Min, 1990). The tight-knit families with which they arrived have experienced change through increased numbers of married women in the labor force—the growing independence of women has led to marital tensions—and the Americanization of children, some of whom have turned to gang activity (Min, 1995c). Still, the high levels of entrepreneurship among Koreans throughout the United States suggest future economic stability; traditional high values placed on education within the community also should produce future generations with marketable skills.

Southeast Asians   *poorest group.*

Table 7.2 provides statistics on Vietnamese immigration under refugee status following the end of the Vietnam War. However, these figures do not quite tell the whole story of that war's aftermath. While the Vietnamese were 78 percent of the immigrants from the old French colony of Indochina, another 16 percent arrived from Cambodia, and 6 percent from Laos (mostly Hmong) (Kitano and Daniels, 1988).

There were also differences among the Vietnamese. The earliest refugees tended to be professional and military people, whereas later immigrants were more mixed. Among the latter were individuals hoping to avoid the "re-education" camps of the new government, poorer people attempting to avoid a poor economy and famine, and ethnic Chinese. The ethnic Chinese in Vietnam had long been a separate ethnic group from the Vietnamese. Increased conflict between Vietnam and China following the war made emigration much more appealing.

Another class of refugee immigrants, much smaller than the rest, is composed of children born to Vietnamese women but fathered by American servicemen. These children are not accepted within Vietnamese society; they are termed *bui doi* (the "dust of life") by Vietnamese. The 1987 Amerasian Homecoming Act created special provisions for these children to enter the United States. Much of this story was dramatized in the Broadway musical production of *Miss Saigon*.

Refugee status provides a wholly different entry than any form of immigration status. The two most striking differences are (1) governmental (and private) assistance in resettling, and (2) immediate access to all forms of available public assistance. Refugees are never truly "on their own."

Resettling refugees usually falls under the responsibility of the federal government. Private organizations, especially religious groups, also have played a significant role in this process. For example, many of the Hmong (pronounced "Mung") people of Laos found themselves resettled in the Minneapolis/St. Paul area through the efforts of private religious organizations (McNall et al., 1994). Although the cold tundra of the American north central regions may seem an odd location for a tropical people, it was definitely in keeping with the federal government's desire to disperse the southeast Asian refugees throughout the United States; the government apparently wanted to avoid an ethnic enclave such as "Little Havana" in Miami. While some of these refugees did wind up in unusual locations for Asian immigrants, approximately 50 percent of Vietnamese, Cambodians, and Hmong settled in California (Rumbaut, 1995).

Immediate public assistance upon arrival for refugees clearly provides much needed short-term help. In particular, Medicaid is a source of free health care. The large percentages of many Asians in self-employment make health coverage problematic upon arrival because the most affordable private health coverage is group coverage provided through employers. On the other hand, public assistance has created some problems for southeast Asian refugees that other minorities face. Assistance is removed when recipients find employment. Typically, that employment does not include health coverage, leaving the more ambitious people worse off for their efforts.

Whatever the shortcomings of public assistance, it has been more necessary for southeast Asian refugees than for any other Asian group of immigrants. Vietnamese immigrants have the lowest median family income of any Asian group in the United States; more important, 23.8 percent of them live below the poverty level (Min, 1995a), which exceeds the poverty rate of Hispanics and approaches the poverty rate of African Americans. While some refugees came from a wealthy and professional background, the majority of southeast Asians came to the United States from rural, poor, and uneducated backgrounds. Some, such as the Hmong, lived on the fringes of society even in their native Laos. Acculturation to the United States has been difficult indeed for these people (see Box 7.2; Celano and Tyler, 1991; McNall et al., 1994).

The Immigration Reform and Control Act of 1986

Obviously, the U.S. Congress received a surprise in the response to its 1965 Immigration Law. Not only were legal immigrants arriving in numbers far beyond anyone's wildest expectations, but also illegal immigrants were on the increase. A glance at the first column in Table 7.2 shows the millions of legal immigrants that entered the United States in the two decades following the 1965 law; several million more entered illegally during that same time period.

The large numbers of total immigrants produced many of the same anti-immigration feelings that earlier immigration surges motivated. Two familiar complaints were (1) new immigrants are taking jobs away from native-born Americans, and (2) new immigrants are bringing very different cultural traditions and may not acculturate to mainstream American society. A new wrinkle in the complaints, often unvoiced, was that these new immigrants were not from Europe. The selective racism in

BOX 7.2: THE HMONG PERSPECTIVE ON AMERICAN LIFE

Because the Hmong people from Laos were on the losing side in the Vietnam War, the United States allowed them to enter under refugee status. However, the culture shock was greater for the Hmong than for most immigrants. Rumbaut (1985: 471–472) offers the following assessment on American life by a Hmong refugee:

Any jobs they have require a literate person to get. We have the arms and legs but we can't see what they see, because everything is connected to letters and numbers. . . .When we were in our country we never ask anybody for help like this, [but] in this country everything is money first. You go to the hospital is money, you get medicine is money, you die is also money and even the plot to bury you also requires money. These days I only live day by day and share the $594 for the six of us for the whole month. Some months I have to borrow money from friends or relatives to buy food for the family. I'm very worried that maybe

one day the welfare says you are no longer eligible for the program and at the same time the manager says that I need more money for the rent, then we will really starve. I've been trying very hard to learn English and at the same time looking for a job. No matter what kind of job, even the job to clean people's toilets; but still people don't even trust you or offer you such work. I'm looking at me that I'm not even worth as much as a dog's stool. Talking about this, I want to die right here so I won't see my future. . . .How am I goind to make my life better? To get a job, you have to have a car; to have a car you have to have money; and to have money you have to have a job, so what can you do? Language, jobs, money, living, and so on are always big problems to me and I don't think they can be solved in my generation. So I really don't know what to tell you. My life is only to live day by day until the last day I live, and maybe that is the time when my problems will be solved.

nineteenth century anti-immigration attitudes (such as against the Chinese) would now be more widely applied.

In spite of growing anti-immigration attitudes, Congress chose not to alter the qualifications for legal immigration as specified in the 1965 law. Family reunification would still predominate and legal immigration would still rise. As Table 7.2 shows, legal immigration increased in the late 1980s and early 1990s. Instead, Congress turned to illegal immigration, focusing particularly on illegal immigration from Mexico.

 Closing the Border? 1986 Changes in the Law

The Immigration Reform and Control Act of 1986 (IRCA) was designed to stem the rising tide of illegal immigrants from Mexico. It approached this goal with a variety of changes in both immigration and naturalization. The major changes created by IRCA include the following:

 1. IRCA created an amnesty program for illegal immigrants already in the United States. The amnesty program allowed current illegals to obtain legal resident alien status in the United States after a several year period, provided they demonstrated some English language proficiency at the end of that

period. Those eligible included all illegal immigrants who had lived in the United States continuously since January 1, 1982. The starting date of 1982 was designed to eliminate the most recent immigrants.

2. The Special Agricultural Workers program provided amnesty for illegal alien workers in perishable-crop agriculture if they had worked in the United States for at least ninety days during 1986. Mexican labor was obviously still needed in the fields.

3. The Replenishment Agricultural Workers program was created to allow still more Mexican laborers into the United States as needed for agricultural work. Those who participated in the program for three consecutive years would be eligible for resident alien status.

4. IRCA made it illegal for employers to knowingly hire illegal aliens. Fines were established for first offenders followed by larger fines for repeat offenders and the possibility of a six-month prison term. The teeth in this part of the law were obviously not very sharp but it was the first time Congress had turned its attention to American employers. Before IRCA, American employers could hire all the illegals they cared to. If the illegals were caught, they would be deported and the employer would then hire new illegals (or perhaps the same ones when they returned). Congress had clearly been

The Immigration Reform and Control Act of 1986 resulted in an increase in enforcement by the INS border patrol.

protecting American business interests, but IRCA was designed to diminish the "pull" that was bringing illegals across the border.

5. IRCA increased border enforcement. This aspect of the law has received increased attention since the law's passage, with increased numbers of INS officials and law enforcement officers assigned to the task.

Responses to IRCA

The first response to IRCA came from Hispanic Americans, especially Mexican Americans. They feared that IRCA would lead to employment discrimination against native-born Hispanic Americans along with legal immigrants from Mexico. The key term in the employer sanction section of the law was "knowingly." Would that mean that an employer might be held responsible for an illegal employee if that employee possessed forged papers? Might employers decide that all Hispanic applicants were therefore suspect and to be bypassed? As it turned out, the employer sanction part of the law was only loosely enforced. Employers did apparently decrease their reliance on illegal immigrants, but this affected neither the native-born nor legal immigrant (Donato and Massey, 1993).

A big question with regard to IRCA concerned the amnesty programs it offered. Would illegal immigrants trust the federal government enough to come forward? For the most part, they did. Approximately 3.1 million illegals applied for amnesty under the various programs. Between 70 and 80 percent of all applicants were of Mexican origin (Borjas, 1990).

The biggest question, of course, concerned the whole point of the law. Would it stem illegal immigration from Mexico? The number of apprehensions dropped somewhat in the two years following the law, but began to rise again in the late 1980s. During the 1990s, the numbers for apprehensions and for estimates of increases in the illegal immigrant population both declined. As of 1996, the U.S. Immigration and Naturalization Service (1997a) estimated a yearly growth in that population of 275,000. Woodrow and Passel (1990) argue that IRCA has decreased employment opportunities for undocumented workers in the United States. Jones (1995) attributes some of the decline in illegals to the success of the *maquiladoras* in Mexico, which have provided increased employment opportunities for those who stay home. The Immigration and Naturalization Service (1997b) attaches much importance to improved border enforcement in decreasing illegal immigration (see also Nuñez, 1992).

Estimates for the total population of illegals in the United States run at 5,000,000, more than half of whom are of Mexican origin. California is the home of 40 percent of all illegals, with Texas a distant second at 14 percent (U.S. Immigration and Naturalization Service, 2000).

New Competition on Old Turf

Although IRCA was directed almost exclusively at illegal crossings of the border between the United States and Mexico, all kinds of immigrants from all countries could be viewed as potential competitors. Fear of immigrant competition is as old as the history of the nation. In the past, fear of immigrant competition was often unwarranted. Japanese immigrants in the West, for example, tended to play a middleman minority role, specializing in employment that no one else was doing. Still, many European Americans saw them as a threat. In other cases, immigrants were indeed a threat but to other minorities—not to the dominant group. Irish Catholic immigrants in the midnineteenth century, for example, were a threat to (and were threatened by) free African Americans in New York City and Boston. What is the case with contemporary immigration? Are today's immigrants a threat and, if so, to whom?

There is no evidence that immigration, legal or illegal, damages mainstream American employment. Immigrants generally either become self-employed or work in the lower sector of the economy. If they are self-employed, immigrants tend to concentrate in economic areas that are either new (such as one would find in an ethnic enclave) or currently unoccupied (such as middleman retail businesses in central cities). The only exception to this generalization might be Asian Indians who have moved into hotel and motel ownership.

As for those employed in unskilled occupations, very few fill jobs sought after by native-born Americans of any race or ethnicity. Torres and Bonilla (1995) note that Puerto Ricans have suffered somewhat in New York City as a result of increasing immigration from Asia and the Dominican Republic. Pfeffer (1994) reports the replacement of African American agricultural workers in Pennsylvania by Cambodian immigrants, the latter being more appealing to employers because they were organized under immigrant labor contractors. The most common finding regarding immigrant competition by far, however, is that immigrants do not compete with the native-born, even those in minority status. They compete with each other (Borjas, 1990; Jaret, 1991; Reischauer, 1993; Sorenson and Bean, 1994).

A closer look at American employment shows some interesting relationships. Illegal immigrants earn 37 percent less than legal immigrants, but most of this difference is due to education and skill differences. When these factors are controlled, documentation does not, in fact, generally produce more poorly paid workers (Borjas, 1990). The real gap occurs between immigrants of any type and the native-born. The latter acquire numerous cultural skills as a birthright, qualifying them for employment beyond the reach of almost all immigrants. A more common picture of employment is a complementary relationship between the native-born and immigrants: the native-born may well be employed at midlevel or upper level jobs while immigrants handle the lower level jobs.

The complementary relationship between native-born traditional minorities and immigrants also can be seen on a more macro level. African Americans, for example, report higher fears of immigrant job competition than do many other groups. In reality, African Americans have higher employment rates in those cities with the highest proportions of immigrants (Jaret, 1991; see also Murdock et al., 1999). Immigrants do head for cities with expanding economies, but they also increase the demand for goods and services when they arrive. Their presence can serve to create employment for minorities already living there.

Borjas (1990) paints a less rosy but very similar picture. He calculates that a 10 percent increase in the number of immigrants decreases the earnings of the native-born by 0.2 percent. Their presence has no impact on labor participation or employment opportunities for the native-born. By contrast, he calculates that the same 10 percent increase in the number of immigrants decreases earnings *of other immigrants* by at least 2 percent. Again, immigrants do no damage in the labor market to the native-born.

The connection between employment needs and immigration law continues into the 1990s. The Immigration Act of 1990, for example, was designed to increase the percentage of immigrants admitted on the basis of occupational skills and to create new temporary forms of employment for aliens. The U.S. economy is not growing at the rate it was during the early 1900s, but immigration is still clearly perceived as necessary for that economy to run smoothly.

Summary

The National Origins Act of 1924 had effectively cut off immigration to the United States throughout much of the twentieth century. The first major change to this law occurred with the Immigration and Nationality Act of 1965, which removed the racial bias of the earlier law and permitted unrestricted immigration for the purposes of family reunification. Coupled with the restricted slots created by the new law for needed occupational skills, a flood of immigrants began to arrive, almost all from Asia and Latin America.

The majority of Latino immigrants came from Puerto Rico, Cuba, and Mexico. Puerto Ricans were allowed special access because of the legal relationship between Puerto Rico and the United States. Industrialization on the island following World War II stimulated considerable immigration, almost all to northeastern U.S. urban centers. Arriving just when those cites were in decline economically, Puerto Ricans never managed to find a solid foothold in American society. They are still among the poorest of the Hispanic groups in the United States.

Most Cubans arrived as refugees following the 1959 revolution led by Fidel Castro. Refugees receive more governmental assistance than other immigrants, but Cuban immigrants also needed less; they tended to come from the professional and business classes in Cuba—those most likely to lose under Communism. The vast majority of these refugees settled in Miami, Florida, creating a distinctive ethnic enclave. They lead most Hispanic Americans today in income, education, and professional/managerial employment.

Although the 1965 law change greatly increased the numbers of Mexican immigrants to the United States, their immigration history spans the entire twentieth century. Political and economic turmoil in Mexico early in the century spurred many Mexicans to seek employment in the United States. During times of labor shortages, they were more than welcome in lower level work such as in agriculture and railroads; however, during economic recessions and depressions, they were herded toward the border. The proximity of Mexico coupled with the length of the border has always made illegal entry possible. Today, Mexican Americans are the largest single group of Hispanics in the United States, moving increasingly to urban status as an ethnic group.

Unlike most Latinos, Asians have entered the United States under both the family reunification and occupational preference categories of the new law. In particular, India, Korea, and the Philippines have taken advantage of the occupational preferences category, placing many of the new immigrants in better economic positions than earlier immigrants from their respective countries. By contrast, Chinese immigrants were much more likely to arrive under family reunification. The poorest of the Asian immigrants were refugees from Vietnam following the end of the Vietnam War. In spite of governmental assistance, most Vietnamese groups today rank at the bottom of Asian ethnics in most income and educational rankings.

Most Asian immigrants settled in California, with New York receiving the next largest group. The only exception to this rule is the Asian Indian ethnic group. All Asian immigrants have become urban dwellers. Some, such as the Chinese, have turned toward enclave economies. Others, such as Koreans and Asian Indians, have become classic middleman minorities. Most Asian ethnic groups in the United States contain considerable internal variation with some very poor and others well off. This creates apparently high levels of success as measured by income and education, but also high percentages of immigrants live below the poverty level.

The increasing flow of illegal immigrants from Mexico led to the passage of the Immigration Reform and Control Act of 1986. The purpose of this law was to provide amnesty to long-term illegals already in the United States while decreasing the flow of new illegals. The latter goal was approached through instituting sanctions against American employers who knowingly hired illegals, the creation of new temporary worker programs for Mexicans, and increased border enforcement. Approximately 5 million illegals are currently in the United States, half of whom are of Mexican origin.

The 1986 law responded in part to growing fears among Americans that new immigrants were posing a competitive threat to the native-born. Most research on immigration—both legal and illegal—indicates that immigrants do little if any damage to native-born Americans in the job market, including American minorities. Immigrants are a competitive threat to one another, however, as they tend to all compete within the same labor markets.

Chapter 7 Reading
The Real Immigrant Story:
Making It Big in America

Denise M. Topolnicki and Jeanhee Kim

New immigration following the Immigration Law of 1965 brought both diversity and large numbers of foreign-born to the United States. Such immigration surges have always produced anti-immigration attitudes in American history; there is no reason why contemporary immigration should produce any different results. Topolnicki and Kim examine both the new immigrants and the opposition they face from native-born Americans.

A lot of the angry Californians who helped pass Proposition 187—to deny costly welfare, health care and public education to undocumented aliens—probably had someone like Hilda Pacheco in mind when they pulled the lever in November. Pacheco, 32, entered the state illegally from Mexico in 1978, never finished high school and is the single mother of two children. People like her, the argument goes, are draining $2.3 billion a year from California's strained health care, prison and education systems while also filling some of the relatively few jobs available in the state's recessionary economy. What's more, they lack the skills ever to contribute as much as they will take. People like her should go home.

Such anti-immigrant sentiments echo well beyond California's borders today. Arizona, California, Florida, New Jersey and Texas are suing the federal government for a collective $14 billion—the states' estimate of their outlay to support, educate, hospitalize and imprison illegal aliens. In Washington, the Commission on Immigration Reform, headed by former Democratic Rep. Barbara Jordan, urged Congress to create a national registry of legal workers, effective[ly] barring jobs from the estimated 200,000 to 300,000 undocumented immigrants each year. And the new Republican majority in Congress has gone further, threatening to cut off welfare benefits even for legal immigrants, except for refugees and those over age 75.

Before you reach your own conclusion about these initiatives, you may want to learn more about people like Hilda Pacheco. What you discover may not conform to the talk radio image of immigrants as leeches. Rather than being a drag on the economy, Pacheco—like most immigrants—is making it in America. She has never been on welfare, has attained legal status and has elevated herself from a subminimum-wage job at a hamburger stand 16 years ago to her current $50,000-a-year managerial position at a worker-training firm. "I'm sure that illegals pay more taxes than they get credit for," Pacheco says.

With immigrants entering the U.S. at a rate of 1 million a year, foreign-born residents—legal and illegal—now represent 8.5% of the U.S. population, nearly twice the percentage (4.7%) in 1970. And in California, where fully 40% of recent immigrants settle, 22% of the population was born outside the U.S. Still, foreign-born residents today make up a much smaller portion of the U.S. population than they did following the great wave of immigration at the start of the century, when foreign-born residents peaked at 15%.

People who criticize today's immigrants, however, contend that as a whole, the current newcomers are fundamentally different from the 13 million Eastern and Southern Europeans who immigrated to the U.S. in the first half of this century. They surely are different in at least one sense: Just 38% of today's arrivals are white, compared with 88% of those who came before 1960. Critics also argue that our high-tech economy now demands brains, not brawn, which means poorly educated and unsophisticated immigrants have little hope of following their predecessors into the middle class.

If you look at the research on immigrants, however, you'll find that much of the pessimism is unwarranted. Contrary to what many Americans believe:

The vast majority of today's immigrants—legal and illegal—are doing well, or at least striving to pave the way for their children to live better lives. Figures from the Census Bureau reveal that immigrants who arrived in the U.S. before 1980 actually boast higher average household incomes ($40,900) than all native-born Americans ($37,300).

Few immigrants come here to get on welfare. In reality, working-age, nonrefugee immigrants are less likely than their native-born counterparts to be on the dole.

Immigrant children aren't gobbling up precious educational dollars, either. In fact, only 4% of the $227 billion we spend to educate our children is spent educating legal immigrant children and just 2% is spent on the estimated 648,000 kids who are here illegally.

Immigrants are not long-term drains on our economy. Yes, the estimated 3.8 million illegal immigrants cost us about $2 billion a year, chiefly because many work in low-wage jobs and often don't pay income taxes. But over time, immigrants become productive. As a group, the foreign-born pay $25 billion to $30 billion a year more in taxes than they consume in government services, says the Urban Institute.

Like yesterday's immigrants, the newcomers choose America because it offers a chance to prosper. Jeffrey S. Passel, the Urban Institute's director of immigration research and policy, is optimistic about their prospects. "The very act of pulling up stakes and moving to a foreign country indicates that you have initiative and want to better yourself," he says.

The successes of today's immigrants hold lessons for us all, whether our ancestors came here on the Mayflower, in slave ships, on a turn-of-the-century steamer or on a jetliner.

Jobs, Not Welfare

Immigration's foes are fond of pointing out that 9% of immigrant households collect cash welfare benefits, compared with only 7% of households headed by native-born Americans. But that single statistic paints a misleading picture. Welfare use is high almost exclusively among legal refugees from war-torn or Communist countries, including Cambodia (50% of all households), Laos (46%), Vietnam (26%), the former Soviet Union (17%) and Cuba (16%). Unlike other immigrants, these favored refugees are immediately entitled to public assistance. As a result, 16% of the refugees, in contrast to only 3% of other immigrants who came here during the 1980s, get public aid.

The notion that illegal immigrants come to the U.S. to obtain welfare benefits is a myth. Illegals already are barred from all public assistance except for emergency medical care under Medicaid and the women, infants and children (WIC) nutrition program. Further, even a legal immigrant who goes on the dole during his first five years in the U.S. risks deportation. Though few actually get the boot, the law still acts as a deterrent because an immigrant on welfare would have difficulty getting the approvals necessary to sponsor relatives for residency in the U.S., which is a prime goal for many immigrants.

Thirty-two-year-old Iraji Khiar reflects the prevalent imminent attitude toward welfare. He fled war-torn Ethiopia in 1977 and spent the next 10 years with family friends in the Sudan before being sponsored for U.S. entry by a cousin who had come a few years earlier. But when Khiar arrived in San Diego in 1987, he couldn't locate his relative, and in order to survive, he accepted the Catholic Church's help in signing him up for welfare—for all of four weeks. At that point, Khiar refused further aid, insisting that he wanted to earn his keep "with my own sweat." He began working as a high school janitor at $7.75 an hour and attending classes toward an associate's degree in business administration from San Diego City College. He later went into the food business with another cousin and her brother. Today the trio typically work 141 hours a week at the Maryam Sambussa Factory, which bakes savory East African pastries, and the Sphinx International Restaurant, which serves up a multiethnic stew of East African and African-American foods. The Sphinx features African and American music—when it's not karaoke night.

The Dream is Alive

Academics have found that the longer immigrants are here, the more likely they are to have obtained two staples of the American dream: a home and their own business. For example, among immigrants who have lived here five to nine years, 44% own their own homes. That figure rises to 55% after at least 10 years.

Some scholars believe that immigrants eventually pull ahead of natives in the income race because their work habits aren't constricted by our notions of the typical eight-hour workday. Further, a willingness to strike out on their own has allowed many immigrants to earn more money sooner than they would have in the corporate world, given their often limited command of English. Overall, the same portion (7%) of immigrants as native-born Americans are self-employed, and both groups of entrepreneurs earn, on average, about $30,000 a year. Yet for some ethnic groups, self-employment rates are significantly higher, particularly for Koreans (18%) and Iranians (12%). Immigrants also are well represented in highly skilled professional and technical jobs. Two of every 10 U.S. physicians are foreign-born, for example, as is one in eight engineers.

Nevertheless, some immigration experts argue that immigrants who arrived here after 1980 will never do as well as natives because they're more likely than their predecessors to have come from Third World nations. Only time will tell whether recent immigrants' median household income of $31,100 will rise. Still, a closer look at the facts reveals that these newcomers aren't as disadvantaged as they first appear. Explains University of Texas sociologist Frank D. Bean: "To say that today's immigrants are of lower quality than their predecessors puts an unfair onus on them. They actually have more education than immigrants who came here 20 years ago." Indeed, between 1970 and 1990, the percentage of immigrants with college degrees climbed from 19% to 27%. Meanwhile, the portion of immigrants who dropped out of high school fell to 37% from 48%. (By comparison, 15% of native-born Americans are high school dropouts and 27% are college graduates.) Nearly half (47%) of African immigrants hold college degrees.

Even if you assume that most immigrants who lack college degrees will never earn much in today's demanding job market, it's wrong to presume that they won't become taxpayers or that their children will get stuck in low-wage jobs. As Michigan State sociologist Ruben G. Rumbaut, an expert on recent immigrants, reminds us: "At the turn of the century, many people argued that the U.S. was attracting immigrants who had little education and few job skills. But the fact that you came here as a peasant didn't mean that your children would forever be part of the unwashed underclass."

Immigrant Kids: Moving to the Head of the Class

The widely held belief that most immigrant kids demand to be taught in their native languages indefinitely is also dead wrong, as is the notion that we are spending a ton of money on bilingual education. Federal spending on bilingual education, adjusted for inflation, actually fell 48% during the 1980s, despite a 50% increase in the number of public school children with limited proficiency in English. In addition, studies show that English is the language of choice for the children of immigrants, no matter what their nationality. The experience of the Rev. Nancy C. Moore, senior pastor of Faith United Presbyterian Church in Los Angeles' predominantly Hispanic Highland Park neighborhood, is instructive. Since most of the 72 children who signed up for Sunday School two years ago were Hispanic, Mrs. Moore decided to assign two teachers to each classroom, one who spoke English, another who spoke Spanish. She dropped the plan, however, when she discovered that 69 of the kids already knew English and that the three who didn't wanted to be taught in their adopted language, not their parents' tongue.

There's also plenty of evidence that immigrants' children are performing well academically, despite poverty, poorly educated parents and discrimination—problems often associated with underachievement in native-born Americans. Even children who missed years of school while detained in refugee camps abroad do amazingly well. In one study, for example, University of Michigan researchers tracked 536 Vietnamese, Laotian and Chinese-Vietnamese children who attended public schools in low-income sections of Boston, Chicago, Houston, Orange County, Calif. and Seattle during the early 1980s. Most were B students, more than a quarter regularly got A's, and only 4% had grade point averages at or below C. They also did better than average on a standardized achievement test; in math, an impressive 27% ranked in the top 10% nationwide.

Why do these kids remind us of Horatio Alger rather than Bart Simpson? Because their parents preached a mantra that has served immigrants for generations: Control your destiny through education. The kids, in turn, relish the chance to learn; in their homelands, education is generally reserved for the wealthy. As a result, families gather around their kitchen tables on weeknights, with older children expected to assist younger siblings. The University of Michigan researchers found that, on average, immigrant grade school students studied two hours and five minutes a night, while high school kids hit the books three hours and 10 minutes. The typical American junior or senior high school student studies only an hour and a half per day. Unfortunately, other researchers have found that when immigrant kids' grades falter, it is often because of overassimilation into American culture. In other words, the longer they live here, the more television they watch and the less homework they do—results that reflect more poorly on us than them.

Another myth: Success is limited to Asian kids. A study of Salvadoran, Guatemalan and Nicaraguan illegals who attended overcrowded, violence-plagued schools in the San Francisco area found that they were the academic stars of otherwise dismal institutions. Although two-thirds of the 50 Central American students surveyed worked 15 to 30 hours a week to supplement their families' income, half made the honor roll.

The most astonishing achievements, however, belong to the Hmong, people who were subsistence farmers and CIA operatives in the mountains of northern Laos during the Vietnam War. Many adult Hmong are not literate even in their native language, and a disturbing three-quarters of their households are on welfare. Yet studies of Hmong schoolkids in San Diego and St. Paul conducted during the past four years reveal that they earn better grades than native-born white children. Ruben Rumbaut is still haunted by one San Diego teenager he interviewed a few years ago. The girl's mother had died giving birth to her eighth child; her father remarried and had six more children. In the U.S., the family of 16, joined by the girl's maternal grandparents, squeezed into two apartments. The girl was responsible for keeping house, so she usually couldn't start studying until midnight. Yet she scored 1216 on the SAT (the national average is 902). Muses Rumbaut: "Whenever I think of that girl, I know it's unwise to make pronouncements about the future success of immigrants' children simply by looking at aggregate census data on recent immigrants' education and income."

Yet despite immigrant accomplishments, some Americans seem determined to keep whispering: No matter what, they'll never be real Americans. They'll keep their strange customs, congregate in ethnic enclaves, and as their numbers and economic well-being increase, they will demand political power.

And if they do, well, they won't be very much unlike the largely unschooled, ragtag ethnic tribes that landed on our shores three or four generations ago and still insist on clinging to such rituals as polka dancing, playing boccie and marching in the St. Patrick's Day parade. Aren't we better off for having let them in?

III

Theories and Dilemmas: Modern Racial and Ethnic Relations in the United States

Modern racial and ethnic relations in the United States occur in a wide variety of contexts. Race and ethnicity are important factors in where you live, how well you live, the amount and the quality of education you receive, where you work, the state of your health, and how you interrelate with the law. Part III explores these and other questions about contemporary race and ethnicity.

The remaining four chapters of this book take a close and hard look at contemporary American society. Chapter 8 focuses on the educational institution of the United States, comparing how both new immigrants and old minorities fare when they enter its doors. Chapter 9 relates the educational picture to the structure of the American economy; educational credentials and skills obviously play a major role in the economic niches we all occupy. Chapter 10 brings all of this down to its most basic level, showing how minority status affects health and causes premature death. Finally, the legal system affects different racial and ethnic groups in distinctive ways as it defines some behaviors as legal and others as criminal; the relations between America's racial and ethnic groups and the legal system conclude this exploration in Chapter 11.

The contemporary look at the United States that follows continues the perspective that directs this book, viewing all groups in a comparative manner. Some readers may be primarily interested in African Americans while others might have a greater interest in the current status of Hispanic Americans. Still, no one group can be fully understood if its social and political environments are ignored. The comparative focus on contemporary relations in the remaining chapters provides a unique view of group interactions within the most important institutions of American society.

8

Education: The Allocation of Skills and Resources

Industrial societies and formal educational institutions develop in a parallel manner. Industrial societies cannot exist without formal, public, and usually mandatory education. The technological development characteristic of such societies requires an increasingly trained labor force if manufacturing is to be efficient. At early stages of industrial development, simple literacy may suffice for most employment; at later stages of development, the workforce needs to be much more highly skilled as the number of unskilled jobs decreases. Educational systems arise to fill these needs in the economy and people are generally given little choice about attending. From an individual standpoint, an education is a means to a good job with higher pay; from a societal standpoint, an educated workforce is essential. Similar to the situation of colonial days when labor was in short supply, a lack of trained workers on today's industrial market drives up the wages they might receive. The more workers are trained, the better for employers. In short, education and the economy cannot be separated.

Having said that, Chapters 8 and 9 do indeed separate them—the alternative would be one, very long chapter. Education comes first here simply because it fits the chronology of every individual's life experience—education leads to employment. But, as we will see, the education you receive has much to do with the economic position of your family. Examining the relationship of the two quickly becomes a chicken and egg question. Therefore, expect considerable overlap in these two chapters.

Just as the economy plays a major role in the education you receive, so also does racial or ethnic status. Part of that influence stems from ethnic stratification; if poor people generally get less or poorer education, then those racial and ethnic groups disproportionately poor will be disproportionately affected. But race and ethnicity have effects on education independent of social class. This chapter examines the interplay among education, social class, and race or ethnic status with an eye to the many interrelationships that exist. While proceeding with this examination, the impact of all this on the economic advancement of minorities is never far in the background.

Education in the United States: Racial and Ethnic Differences in Attainment

Education has a long and wandering history in the role it has played in American racial and ethnic relations. It was denied to slaves in an effort to minimize slave revolts but encouraged for European immigrants and Native American children in hopes it would "Americanize" them faster. At the same time, minorities who achieved too much with education were viewed as threats. Jewish immigrants in the Northeast and Japanese immigrants in the West both faced educational discrimination for moving too far, too quickly. In the 1980s and 1990s, Asian students—many immigrants or the children of immigrants—argued they were limited in their attempts to enroll in American colleges and universities in spite of their high grades and entrance exam scores (Nakanishi, 1989).

For most minorities, experiences with education have been less extreme and far from positive. As a general historical rule, poorer people and/or minorities have been offered poorer education, *and* they have received less of what was offered to them than those less poor. They were offered poorer education because it was less expensive for the state in which they lived; they received less of it because (1) they could not spare the time because they had to earn money, (2) schools were too far away, (3) they saw little value in it, or (4) they objected to what they were taught and/or the way they were treated. Some variation of these possibilities describes the twentieth century educational experiences of the three groups with the worst current educational attainment—African Americans, Native Americans, and Hispanics.

African Americans entered the twentieth century in segregated schools. In the South, such schools were written into law (*de jure* segregation); in the North and other regions, it was the outcome of housing segregation (*de facto* segregation). As late as 1940, expenditures for European American schools in the South were three times that of African American schools (Miller, 1995). Such schools were one of the many targets of the Civil Rights Movement of the late 1950s and early 1960s. Ironically, segregated schools had been declared unconstitutional back in 1954 when the Supreme Court reached a unanimous decision in *Brown v. Board of Education*. Still, it would take years before most schools desegregated. None of these changes, however, had any impact outside the South where *de facto* segregation continued. The level of African American educational segregation has not changed since the 1960s, more pronounced today in the North than in the South.

Native American education moved slowly from the boarding schools offered by the Bureau of Indian Affairs to institutions run by tribal governments. The boarding schools attempted to Americanize Native American children by changing everything from their style of clothing to their values. Regardless of who runs the schools, however, conflict between the values, knowledge, and behaviors of the Western style classroom and native cultures is inevitable. For Native Americans interested in Western education, difficulties await that are reflected in poorer attainment and lower achievement.

Hispanic education has been many different things for different Hispanics at different times. Mexican Americans in the early twentieth century found schools to be of poor quality and generally inaccessible to a largely rural population. This was especially true for families involved in migrant agricultural labor. Hispanics also faced the cultural and language barrier experienced by Native Americans in school. As Mexican Americans became more urban and were joined by Cubans and Puerto Ricans, many of the African American urban experiences with schools were duplicated, still complicated by language barriers. With many new Hispanic immigrants entering the United States daily, many of these problems have persisted. As of the 1990s, Hispanics along with African Americans and Native Americans still lag significantly behind the rest of the American population in educational attainment.

The Current State of Minority Attainment

An educational snapshot of American educational attainment at the end of the twentieth century depicts European Americans and Asian and Pacific Islanders as the clear success stories. Educational variation within each group is considerable, especially for the many ethnic groups categorized as Asian and Pacific Islanders, but overall records are positive. From the first grade through postsecondary education, students from these two census categories receive the highest grades, attain the highest test scores, are least likely to drop out along the way, and wind up with the most prestigious degrees. Generally, European Americans score slightly higher on tests of English language verbal abilities while Asian and Pacific Islanders excel in mathematics and science abilities; the careers ultimately dominated by the members of each category reflect that variation.

By contrast, African Americans, Native Americans, and Hispanics fall behind in all these forms of attainment. Members of these census categories tend to enter the primary grades with fewer necessary educational skills in place and proceed to drop further behind as the educational years pass by. We have seen gains in some areas since the 1970s for members of these categories, but in general, the picture has changed little. This lack of change is particularly striking considering that higher percentages of these students drop out of education during the secondary school years. By the time students are enrolled in colleges and universities, higher percentages of the European American and Asian and Pacific Islander populations are still active students when compared with the other three categories. Assuming that high dropout rates generally remove weaker students from the statistics, the persistent racial and ethnic gaps in attainment throughout postsecondary education become even more problematic.

One striking change in educational attainment since 1964 has been the relative gains of African Americans and Hispanics in high school graduation rates compared with European Americans. In 1964, 72 percent of European Americans (aged 25 to 29) attained a high school diploma or more compared with 45 percent of African Americans. By 1974, when Hispanics joined the census categories, the European

American figure had risen to over 83 percent compared with 68 percent for African Americans and 52 percent for Hispanics (Miller, 1995). Figure 8.1 shows that the high school diploma gap between European Americans and African Americans had virtually closed by 1998; Hispanics still fall noticeably behind but the language problems of so many new immigrants among Hispanics cloud this figure considerably. Some of these changes in high school attainment are due to the overall "education inflation" that accompanies industrial development; more highly skilled jobs require a more educated workforce and more years spent in school as time passes. Higher percentages of *all* racial and ethnic groups are attending longer, but the slow closing of the 1964 gap is the significant point here; we will return shortly to the other obvious gaps depicted in Figure 8.1.

Figure 8.2 shows one of the reasons why that gap closed somewhat. European American reading scores changed little between 1971 and 1996, but scores for both African Americans and Hispanics rose in relation. Scores for science and mathematics for these groups over these years have an almost identical pattern. In addition, that pattern is repeated across these three subject areas for 9-year-old students (Mullis et al., 1991). Figure 8.3 provides a more contemporary picture of these scores,

| FIGURE 8.1 | DIFFERENCES IN EDUCATIONAL ATTAINMENT BY RACE/ETHNICITY, AGED 25–29 YEARS, 1998 |

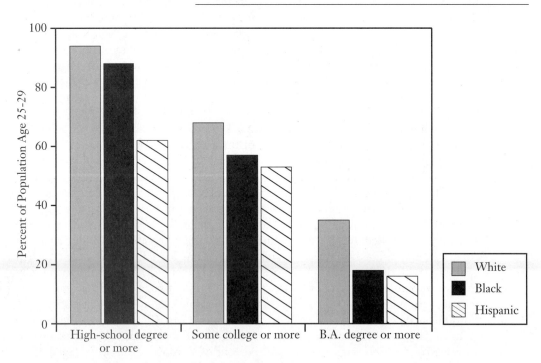

U.S. Department of Education, 1999

FIGURE 8.2 AVERAGE READING PROFICIENCY OF HIGH SCHOOL STUDENTS, AGE 17

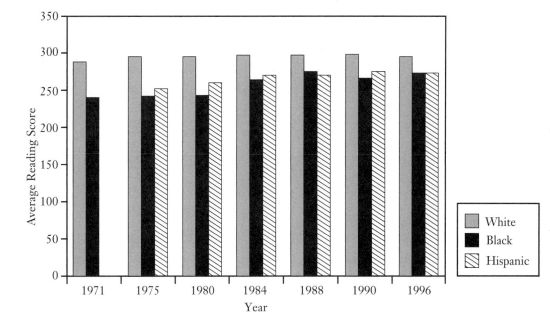

Mullis et al., 1991; U.S. Department of Education, 1999

FIGURE 8.3 AVERAGE READING (1998), MATHEMATICS (1996), AND SCIENCE (1994) PROFICIENCY OF HIGH SCHOOL SENIORS

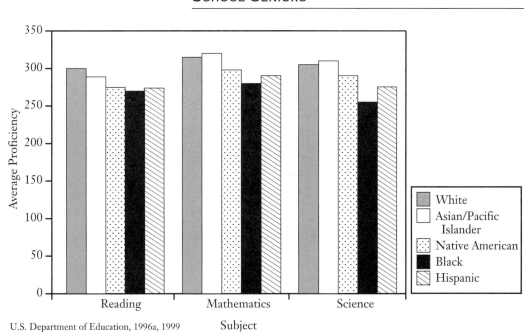

U.S. Department of Education, 1996a, 1999

showing comparisons on all three subject areas and with additional racial and ethnic variation. The European American and Asian and Pacific Islander dominance on verbal, science, and mathematics scores noted earlier is clear in this figure. African Americans, Native Americans, and Hispanics score lower in all areas but particularly so in science and mathematics—two subject areas that often lead to highly skilled and highly paid employment.

Before we move to a discussion of postsecondary education, dropout rates must be examined. The strongest predictor of high school dropout rates is family income (U.S. Department of Education, 1999). Other important causes exist as well, but most are related to family income. For racial and ethnic groups, ethnic stratification becomes a primary factor in explaining group differences because many minorities are disproportionately in lower income levels. Thus, just as those income levels have not changed all that much since 1970, we might expect to find similar results with dropout rates. In 1972, African Americans were 79 percent more likely to drop out than European Americans, and Hispanics exceeded that rate by 111 percent. By 1997, the African American rate dropped to 39 percent over the European American rate while the Hispanic rate rose to 164 percent (U.S. Department of Education, 1999). Again, the large Hispanic immigration during those twenty years is likely an important factor, but the gap is more persistent than those two dates might suggest. For example, as recently as 1991, the African American dropout rate was 88 percent higher than the rate for European Americans (U.S. Department of Education, 1999). Because dropout rates are related to the economy, we should not be surprised to find such yearly changes.

For high school graduates planning to attend postsecondary education, taking a standardized entrance examination is one of the first steps. Figures 8.4 and 8.5 provide Scholastic Aptitude Test (SAT) scores from 1976 to 1995 for selected racial and ethnic groups. (These comparisons stop in 1995 because SAT scores were rescaled after that year.) Note that we should be looking at better prepared and more motivated students than in previous statistics because these students paid to take this exam and presumably most will enroll at a college or university. The same relative pattern we saw earlier is still present. European American students taking the SAT have changed little in verbal or math skills in the two decades depicted here. The familiar contrast by subject matter between European American and Asian and Pacific Islander students also remains. And as with high school scores, the slow closing of the gap between European Americans and minorities remains; almost all groups in both subject areas improved over those twenty years, moving closer to the European American standard.

At the point of postsecondary education enrollment, the picture begins to change more noticeably. Figure 8.6 tracks the variation in college or university enrollment since 1972 among high school graduates. In comparison to the fairly stable white increase in postsecondary enrollment over the years, black and Hispanic enrollment has greatly fluctuated. Although both groups exceeded white enrollment briefly in the mid-1970s, the drop and increasing gap between them and whites are dramatic. Between 1995 and 1997, black college enrollment showed some improvement relative to white enrollment for the first time in a decade; by contrast, Hispanic college enrollment dropped relative to white enrollment during most of the 1990s, showing some

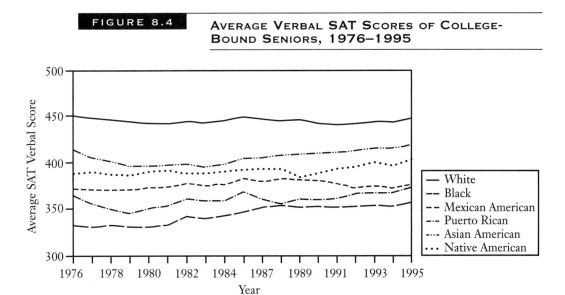

FIGURE 8.4 — AVERAGE VERBAL SAT SCORES OF COLLEGE-BOUND SENIORS, 1976–1995

U.S. Department of Education, 1996a

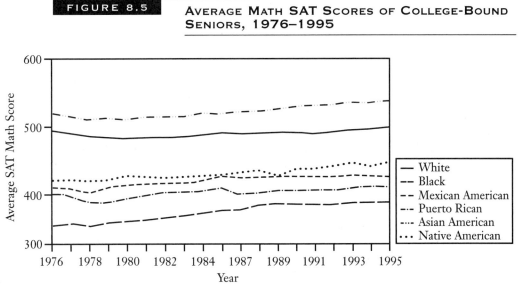

FIGURE 8.5 — AVERAGE MATH SAT SCORES OF COLLEGE-BOUND SENIORS, 1976–1995

U.S. Department of Education, 1996a

PERCENTAGE OF HIGH SCHOOL GRADUATES
ENROLLED IN COLLEGE, AGED 18–24 YEARS

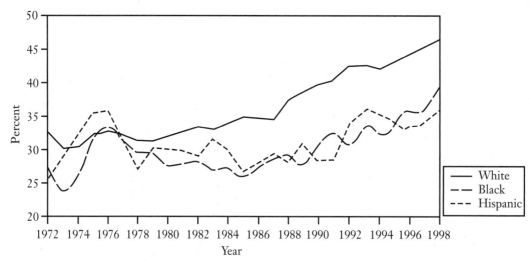

U.S. Department of Education, 1999

improvement toward the end of that decade. Nevertheless, the college enrollment story of the last few decades is in stark contrast to the situation in the mid-1970s.

Figure 8.7 provides a somewhat different picture of higher education in the United States. Instead of focusing on the percent of certain populations that enroll, Figure 8.7 displays the breakdown of students enrolled between 1976 and 1996. During these years, both African Americans and Native Americans maintained roughly the same percentage of the overall population of students enrolled in higher education. African Americans did show mild gains between 1986 and 1996. Hispanics, Asian and Pacific Islanders, and nonresident aliens, on the other hand, all increased noticeably. The only group to decrease in percentage was the white population.

If the rising percentage of the white population attending higher education in Figure 8.6 seems contradictory to the data in Figure 8.7, keep in mind that white Americans have a relatively low birth rate, augmented only slightly by immigration. This means that a smaller percentage of the white population falls in the traditional age group for higher education attendance. Therefore, a growing percentage of young whites enrolling can still produce a declining percentage of overall student population. By contrast, all other U.S. resident groups in Figure 8.7 have grown in population during this time period because of high birth rates, immigration, or both. We should expect to see relative increases among such groups, simply because of the demographics of their populations. When we *do not* see increases among groups with growing numbers of young people—African Americans and Native Americans, for example—some explanation is required.

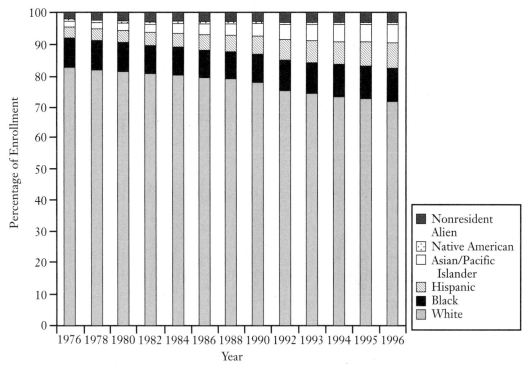

FIGURE 8.7 PERCENT OF TOTAL ENROLLMENT IN HIGHER EDUCATION, 1976–1996

U.S. Department of Education, 1999

Returning our attention briefly to Figure 8.1, the relative parity between European Americans and African Americans in attaining a high school diploma or more only serves to call attention to what happens next. The smaller percentage of African American high school graduates that enrolls in postsecondary education is reflected in Figure 8.1 in the middle of the graph. The drop between the two groups, even in attaining "some" college or more, is significant. Among students with a Bachelor of Arts (BA) degree, we find even greater gaps between groups. Hispanics are almost three times less likely to attain a BA degree than European Americans. African Americans come closer to European Americans but with a 15 percent completion rate compared with 26 percent for European Americans. We already noted that a minority drop occurred before enrollment. Obviously other factors operate after enrollment that differentially affect completion rates.

Figure 8.8 provides a more complete picture of what happens to different kinds of students who enroll in higher education. Of all students who enrolled in American institutions of higher learning in 1989 to 1990, whites and Asian and Pacific Islanders were equally likely to have earned a BA degree by the spring of 1994. Both groups also

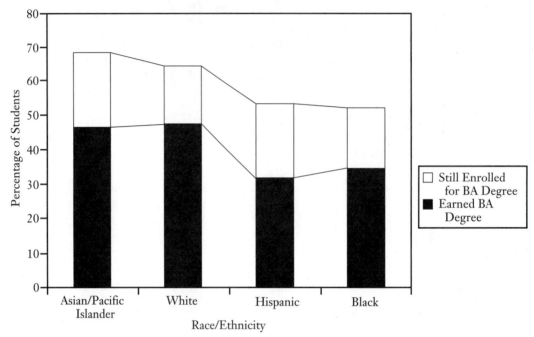

FIGURE 8.8   PROPORTION OF 1989–1990 COLLEGE ENTRANTS WHO RECEIVED OR CONTINUED PROGRESS TOWARD BA DEGREES IN SPRING, 1994

U.S. Department of Education, 1996a

had significant numbers of students who were still enrolled in pursuit of that degree. For Hispanic and African American students, however, a very different picture emerges: both groups contain fewer students with completed degrees. Since the percentage of unfinished students still working toward a degree is very similar for all four groups, Hispanic and African American students either have higher dropout rates than the other groups or take more time off from full-time student status. Of all American students who completed the BA degree as measured in 1993, white students spent an average of 6.24 years; for African American and Hispanic students, that figure was 7.19 years and 6.79 years, respectively (U.S. Bureau of the Census, 1999a). Even counting the late arrivals, the long-term outcome is that these minorities are not completing BA degrees at the same rates as European Americans and Asian and Pacific Islanders, and time has changed little. The gap in BA degree attainment rates by race/ethnicity has decreased very little since 1971 (U.S. Department of Education, 1999).

Education and the Economy: When Education Becomes Human Capital

This chapter began by emphasizing the many connections between formal education and the changing (and increased) skills required by a developing industrial economic system. The relationship is reciprocal in that new economic potential develops from increased skills and, in turn, requires even more skilled workers as that potential develops. Changes in one produce changes in the other. We look at this relationship much more thoroughly in Chapter 9, but for the present, note that contemporary American wealth is more often produced by knowledge than by owning steel mills. For example, very little manufacturing is involved in computer software. Once computer applications are developed, their manufacture involves little more than their reproduction on some kind of media (often a disk of some kind) so that they may be distributed. Bill Gates of Microsoft did not fill his pockets by manufacturing much of anything.

Because knowledge grows in importance as the economy changes, we need to alter our thinking about it. When knowledge can be turned almost directly into wealth, then it becomes a kind of capital in the economy just like a factory is capital. In this case, it is human capital—an attribute of a worker that can create other kinds of capital. Education can therefore be conceptualized as human capital that can be acquired by an individual and subsequently invested in the economy. Its growing economic importance makes the study of its accessibility vital to understanding social stratification and social mobility.

Education has one fundamental difference from other kinds of capital. Producing other kinds of capital always requires capital to start with. Before a steel mill can make its owner wealthy, he or she must first have capital to build the steel mill. It may be borrowed, but it still has to be there. By contrast, education is made available to the very poorest members of society. It is no more free than the steel mill, but the tax basis on which public education rides makes it a very inexpensive form of capital to acquire indeed. As we saw in the preceding section, however, those poorer members of society—the very people with the least access to other forms of capital—seem to find education equally difficult to acquire.

It is easy to speak of skills and education interchangeably. The two often do go together, but the relationship is more complicated than that. Some beginning skills (many acquired in the first few years of life) make the acquisition of formally presented educational skills easier. In addition, many skills can be acquired outside of formal educational institutions through experience or independent study. An individual who comes up with a grand idea that produces wealth probably does not care how the skill was acquired. Most individuals, however, must first convince an employer that their skills are worth hiring (and that a grand idea may be down the road). Because many skills are difficult to measure, most employers rely on educational credentials as a stand-in. The more formal such processes become, the more essential it becomes for individuals to gain access to formal education to acquire the official credentials (Collins, 1979). As we continue our focus on racial and ethnic diversity in the United States, we need to keep two related questions in mind: How much access do minority group members have to necessary skills in general and to formal educational credentials in particular?

Inequality of Educational Opportunity

Educational opportunity covers a wide range of topics. Who has more or less opportunity to acquire this capital? Do those with opportunity have the beginning skills, willingness to work, and acceptance of school structure and authority necessary for success? Answering these questions will lead us to the families and communities of those with educational opportunity.

Additional information is found in the schools themselves and in the wider society that funds and structures those schools. Do schools equally provide the necessary and appropriate training for the diversity of students they encounter? Can schools accommodate the needs of students with differing kinds of beginning skills, values, and attitudes? Do poor students have the same opportunity as rich students? Do nonwhites have the same opportunity as whites? And when all is said and done, do students who receive identical credentials actually receive the same human capital that will bring them equal returns on their investment in the economy?

The Funding of Education in the United States

Education in the United States is available both publicly and privately. The former is funded by a combination of tax dollars and, for most higher education, student fees. The emphasis of this chapter is almost entirely on public education. It not only serves more students overall, but also it serves the vast majority of poorer students—those individuals whose social mobility through educational credentials is under study here. The focus on this section is the relationship between how public education is funded and the quality of the educational experience provided.

Differences in school funding often mean substandard building or educational materials in poorer school districts.

Turning first to primary and secondary public education, public schools in the United States are funded through taxation. Although different kinds of taxes fund education, the most important of these is property tax. Typically, every school is assigned a physical school district (the neighborhoods from which it draws students) and its funding varies, depending on the value of that district's real estate. Hence, schools in wealthier districts receive more funding than those in poorer districts. For example, wealthy school districts in the United States spent 24 percent more per student than the poorest school districts in 1994 to 1995 (U.S. Department of Education, 1999).

How are different levels of funding reflected in educational processes? To start with, teachers from poorer districts earn less. In the 1993 to 1994 school year, teachers in high poverty level school districts earned 28 percent less than their counterparts in wealthier districts ($35,496 as compared with $45,547 in average earnings) (U.S. Department of Education, 1997). Those teachers are also a little less likely to have been trained in their teaching areas. Table 8.1 shows some of the variation in mathematics and science teachers across different types of school districts. Because teachers in these high poverty level schools report greater dissatisfaction with their income, it seems likely that those with specialized training in the sciences would gravitate toward districts with higher salaries (U.S. Department of Education, 1999).

In addition to hiring teachers, schools also fund educational programs and purchase equipment. In a 1993 to 1994 study of fourth grade students in public schools, those enrolled in low poverty level schools were more likely to have programs for the gifted and talented along with extended day programs. Low poverty level schools also were more likely to have computers and internet connections. Class size was the same in all schools, regardless of the poverty level of the district (U.S. Department of Education, 1997).

TABLE 8.1   PERCENTAGE OF PUBLIC HIGH SCHOOL MATHEMATICS AND SCIENCE STUDENTS TAUGHT BY TEACHERS WITH MAJOR OR MINOR IN CLASS SUBJECT, BY PERCENTAGE OF STUDENTS ELIGIBLE FOR FREE OR REDUCED-PRICE LUNCH: SCHOOL YEAR 1993–1994.

Subject	Percent of Students Eligible for Free or Reduced-Price Lunch			
	0–5	6–20	21–40	41 or more
Mathematics	83.3	79.7	76.1	74.1
Biology	80.9	73.5	75.7	72.7
Chemistry	77.2	64.8	72.1	62.9
Physics	42.8	50.5	39.3	29.3

U.S. Department of Education, 1997.

Separate But Unequal

Different levels of funding make schools different. But are there other ways that school variation affects educational opportunity? The variation in the poverty levels of school districts reflects the broader differentiation of American society. Social stratification in the United States separates people economically, but beyond that, they generally choose to separate themselves both socially and physically. Shacks and mansions are not built on the same streets. Housing segregation by social class ensures that school districts do not contain very much economic variation, and when ethnic stratification is figured in, housing segregation by social class carries a racial or ethnic dimension as well. Thus, if overt housing discrimination occurs that affects specific racial or ethnic groups, that dimension becomes all the more important. As schools vary by economic geography, therefore, they also vary by race and ethnicity.

In earlier chapters, we traced the roots of present day housing segregation in the United States while watching the urban centers grow with new immigrants and internal migrants. The racial and ethnic dimension of housing segregation was a standard component of discrimination, from the earliest days of the Republic throughout the twentieth century. Just as nineteenth century African Americans, Catholic Irish, and eastern European Jews faced restrictions on where they could live in American cities, twentieth century Puerto Rican immigrants and African American internal migrants faced similar limitations. As overt discrimination became illegal in housing segregation following the passage of the Civil Rights Act of 1964, that which remained went underground. More important, however, was the force of ethnic stratification in maintaining much of that segregation. Many middle-class minorities could find housing in middle-class neighborhoods, but the large numbers of poor minorities had little choice. Segregated housing continued and, in turn, segregated schools continued.

The mid twentieth century also brought other changes to the American economy that affected employment and housing. Although we take a detailed look at these changes in the next chapter, it is important for the present to recognize a few points. First, most central city manufacturing employment disappeared in the northeast and north central United States, either moving to the suburbs, the Sun Belt, or overseas. Second, most skilled, middle-class Americans followed those employment changes, moving to the new jobs. These two changes left those remaining in central cities increasingly poor, nonwhite, and unemployed as the suburbs grew increasingly middle class, white, and employed. As neighborhoods became more polarized, so did the schools that served them (Massey and Eggers, 1990; Rivkin, 1994).

If we define a segregated school as one that is between 50 and 100 percent minority, the vast majority of African Americans and Hispanics attend segregated schools. In Illinois 89% of African American students attend segregated schools. The percentage drops only a little for other states—86 percent in New York, 85 percent in Michigan, 80 percent in New Jersey, and 79 percent in California. Similarly, 86 percent of New York's Hispanic students attend segregated schools. In other states with large Hispanic populations, the percentages stay equally high—85 percent in Illinois, 84 percent in Texas, 84 percent in New Jersey, and 79 percent in California (Miller, 1995). Obviously, many nonwhite students attend many of those high poverty schools examined in the previous section. Districts with less money to spend have less education to give. Are districts with predominantly minority schools providing even less opportunity?

Bankston and Caldas (1996) studied 40,000 Louisiana high school students whose schools varied from being predominantly European American to having high concentrations of minority students. Their goal was to determine the effects, if any, of high minority concentrations (i.e., essentially segregated schools) on achievement test scores. The strongest predictor of high test scores was race: European American students tested better than African American students. The next strongest predictors were parental socioeconomic status (SES), in this case a mixture of parental educational levels and occupational status, and the level of school segregation. When they analyzed results for European American and African American students separately, however, SES and degree of school segregation explained only about 10 percent of the variation in test scores. School segregation had a negative result on test scores for both groups but was definitely not the most important factor.

Other research on school segregation has focused more on long-term effects for African Americans. Education in a desegregated school apparently predicts later attitudes and employment for African Americans, both related to their more positive associations with European Americans (Braddock et al., 1994; Dawkins and Braddock, 1994). In particular, African Americans from nonsegregated schools are more likely later to work in desegregated work places and to be less likely to use racial categories to explain or describe the behavior of coworkers (Braddock and McPartland, 1989). Desegregated schools, however, also can place minorities in a true numerical minority position, which can negatively affect their peer relationships when racial barriers to interaction persist (Cookson and Persell, 1991; Wilkinson, 1996).

While minority concentrated segregated schools are also the poorest schools with the poorest students, neither of which is positive for educational success, the individual impact of group separation is hard to gauge. One factor that does seem clear is that group separation is perceived very differently, depending on whose choice it is (Miron and Lauria, 1995). Whereas children have little choice in which school they attend, adults have more options. Many African American college bound students prefer historically black colleges and universities where minority students predominate. These institutions were originally developed to provide alternatives for African Americans who were denied entry into European American institutions. Today, they continue by choice and, in many ways, are more successful in educating African American students than predominantly European American institutions (Davis, 1994; Garibaldi, 1991).

What is Worth Learning? The Dominant Culture Dominates

In tracing the American story from colonial days to the present, we have followed the ongoing domination of basically English culture in the United States. This domination not only made Americans of English ancestry feel more comfortable (one's own ways are always the best), but it also provided advantages for their children and grandchildren who were culturally "one up" on everyone else. Cultural domination becomes both more important and clearer when institutions of formal education appear on the scene. English culture controls both the style and the content of American education. Success in school requires success with that style and content. Children either bring that culture to the classroom with them or adapt to it after they arrive. Not surprisingly, those already equipped with the basics do the best. As for those not equipped, some adapt better than others. This collection of knowledge and skills is often referred to as "cultural literacy" (Brayfield et al., 1990). In the next section, we examine many attempts to explain those varying levels of knowledge and skills, but that discussion will make more sense if we spend a moment here to examine the cultural foundations upon which all else rests.

In style, American education emphasizes individual achievement and competition. Students also quickly learn that their achievements are heightened by the failure of others. This approach is clearly in keeping with the overall cultural environment in the United States; structuring the classroom in any other way would be inappropriate given that education is supposed to prepare students for what lies ahead. But while individualism and competition are hallmarks of English culture, they often clash with the values brought to school by students with non-English culture backgrounds. In particular, many Native American and Hispanic cultures emphasize cooperation and a focus on the family. Considerable research has pointed out this fundamental culture clash that some students face in the American classroom (Deyhle and Margonis, 1995; Patthey-Chavez, 1993; Wojtkiewicz and Donato, 1995). On the other hand, many Asian cultures share this emphasis on cooperation; the relative success of Asian students in American education suggests that many factors play a role in cultural adaptation (Domino, 1992; Fejgin, 1995; Lee, 1994; McNall et al., 1994).

In content, American schools generally teach students "what they need to know" to succeed (although it is hard to see the memorizing of fifty state capitals as essential knowledge). As with style of presentation, content reflects a general western European outlook on the world, but that, of course, is also true of American society for which students are being prepared. There are elements of school content, however, that deserve a closer look. The point can probably most easily be made with a look at language.

American schools (with the exception of bilingual schools) use Standard American English—that dialect of English that is the "correct" or standard form in the United States at any point in time. Obviously, American schools need to teach Standard American English because students will need that skill when they seek employment. But here again is one of the skill differences that students bring with them to the classroom. Worse yet, it is probably the hardest to change because we do our fundamental language learning in the first few years of life. Students who speak any other variety of English (or another language altogether) will face more educational barriers than children who speak standard English. (See the reading by Murai following this chapter.) Because education comes through the medium of standard English and students are tested in it, attainment in most subject areas is closely tied to a student's particular language skills.

Predictors of Educational Attainment

One of the more famous research efforts in educational attainment was headed by sociologist James S. Coleman in the 1960s (Coleman et al., 1966). The overall goal of the research was to learn why some students met success at school while others failed. To their surprise, the researchers discovered that the schools themselves seemed to have little to do with who succeeded and who failed (although they were more important for minorities than for European Americans). Class size, teacher education, and the quality of facilities did not seem to matter much. The most important predictor of school success was the student's home: factors such as parents' education levels, family size, and educational resources in the home were apparently more important than what students encountered in the classroom. Also very important were student attitudes, both those of individual students and those of their peers. Positive attitudes toward education and high attainment expectations were related to achievement. (It also should be noted, however, that successful achievement both motivates and strengthens such attitudes.)

The Coleman Report (as this study came to be called) did not close the door on educational attainment research for several reasons: (1) With sociological research, it is very difficult to separate causes when they are all so interrelated. Parents with high educational attainment, for example, also have books in their homes and often successfully pass along positive attitudes toward education. Many types of statistical analyses have been developed that attempt to isolate effects, but outside of rank ordering the importance of predictors, such techniques cannot provide all the answers. Although we know a parent's educational attainment is a very important factor in how a child will do in school, statistics cannot tell us why. (2) The Coleman Report did raise many interesting questions. As previously noted, school characteristics are more important in explaining attainment for some racial and ethnic groups than for others. Although a school cannot change a child's home experience, apparently some schools can be more effective with poorly achieving minority students than others. The Coleman Report invited such exploration.

The Role of the School

The Coleman report lessened the assumed impact of the school on student achievement but, as we saw, did not discount it either. Perhaps more important, the study was limited to easily measured aspects of the school environment such as teacher qualifications or class sizes. It could not touch the many intangibles—both positive and negative—that can be so important in shaping the outcome of learning. This omission was of particular importance because student attitudes were indeed singled out by the study as significant factors. What role could the school play in shaping attitudes and improving the learning experience?

Teacher Attitudes   Just after the Coleman Report was published, a striking piece of research appeared that caught the attention of social scientists. It was clever both in its concept and its title—*Pygmalion in the Classroom*—but it raised more questions than it answered. Robert Rosenthal and Lenore Jacobson argued that children could be transformed as students if teachers could be influenced to alter their expectations of the children. In their study, they picked children at random and gave their names to teachers as students who were measured as "late bloomers"—students ready to improve greatly regardless of the level of their past achievement. The researchers reported that these children did indeed "bloom," simply because their teachers expected them to and treated them differently accordingly (Rosenthal and Jacobson, 1968).

What is the role of teachers' expectations? It is probably not as great as Rosenthal and Jacobson report. Many efforts have been made to repeat their experiment with varying and often contradictory results (see Boocock, 1978). Teachers are obviously important in the educational experience, but students would have to truly and uniformly care what their teachers thought of them to respond so dramatically. Nevertheless, teacher attitudes toward different kinds of students have been shown to matter significantly in student outcomes (Dusek, 1985). Teacher education programs warn prospective teachers of the dangers inherent in stereotyping children, but most people apply many stereotypes unconsciously. A teacher might consciously avoid stereotyping an African American student based on the student's race, but subsequently stereotype him or her on the basis of speaking nonstandard English. Most people in fact apply stereotypes regarding intelligence simply on the way words are pronounced (Luhman, 1990). In addition, stereotypes can help as well as hurt. Asian students often find themselves stereotyped as "model" students by their teachers, which can make their school work appear better than it actually might be (Lee, 1994).

Tracking and Special Education   Teachers' attitudes can impact students in more concrete ways. Some of the most important decisions made by high school counselors (with input from teachers) involve the assignment of students to different tracks and/or special education classes. Tracking refers to the separation of students into classes with different levels of difficulty for the same subject. A given high school might have three eleventh grade English classes, ranging from a high level of difficulty (designed to be college preparatory) to a low level of difficulty. The theory behind this practice is straightforward—teaching occurs more efficiently when most of the students in a given class are at approximately the same skill level.

Special education classes include a wide variety of programs, designed to meet the needs of a diverse student body. Congress mandated such programs through the passage of the Education of the Handicapped Act in 1975, which required "free and appropriate public education" for all children. Schools were then required to meet the special needs of students who were deaf, mentally retarded, learning disabled, speech impaired, or emotionally handicapped. Between 1977, when the act was implemented, and 1988, the percent of students in special education classes in the United States rose from 1.8 percent to 4.8 percent, with over 40 percent of those enrolled in classes for the learning disabled (U.S. Department of Education, 1990). Unlike deafness, where the categorization is clear, "learning disabled" is a judgment call and therefore more open to mistaken decisions.

Beyond the obvious uses, both tracking and special education also can serve to re-move students from the educational experiences they would need in postsecondary education should they desire to continue. A decision to place a student in a lower track or into special education also therefore implicitly decides that student's educational future. Any system of decision making can be in error at times; the question here is: Do high school counselors and school psychologists sometimes make systematic errors that affect particular groups of students?

One of the earlier studies of special education was Jane Mercer's (1973) study of Mexican American students in Riverside, California. She found them assigned to classes for the mentally retarded in greater numbers than their population in the school should warrant. She located adult Mexican Americans who had previously been assigned to those classes while students and discovered they were indistinguishable from any other adults. Upon checking the school's decision-making process, she discovered that Mexican American students' language skills in English were limited. This affected the ability of school personnel to properly assess intelligence because standardized tests were unable to bridge the cultural and linguistic gap.

Serwatka and others (1995) found that African American students were overrepresented in classes for the emotionally handicapped and underrepresented in classes for gifted students. This result was more likely to occur in schools with small percentages of African American students; as the percentage of minority students grew, the assignments to special education became more equitable. This finding is in keeping with the work of Harry and Anderson (1994) who found that African American males in particular tended to be placed in special education. Their explanation of this occurrence was that assignments were made by European American, middle-class women who did not understand the norms of interaction governing young, African American men. The implication of both studies is that better communication and understanding between students and school personnel improve the decision-making process.

Tracking focuses more directly on specific student achievement, grades, and test scores in the decision-making process. While high school counselors have the primary responsibility for track assignments, they also have input from teachers and sometimes from parents. Grades and test scores provide very concrete information, but counselors also typically take more impressionistic information into account. If a teacher, for example, "feels" that a student is brighter than grades suggest, a poorly achieving student might be tracked higher. This kind of "soft" data should definitely enter the picture because good teachers often have very accurate "feelings" about their students. On the other hand, it also opens the door to stereotypical "feelings" that might well be off the mark.

Kubitschek and Hallinan (1996) studied the tracking of ninth grade students in one high school. In that study, parental involvement was significant in only about 5 percent of tracking decision changes; counselors definitely maintained control. The researchers knew at the outset that boys were more highly tracked in math and the sciences while girls were overrepresented in higher track English courses. They also knew that African American students were overrepresented in lower track courses.

What they did not know was whether those students arrived in those classes through their own achievement or through the less clear impressions of school personnel.

They approached this question by comparing students at every track level with a focus on grades and test scores. Would girls, for example, have to get *higher* test scores than boys in math and science to receive high tracking? Would African Americans have to similarly overachieve to obtain high tracking? In English classes, as it turned out, gender was the only issue. It was easier for girls of any race to acquire top track classes than any boys. The exception was for poor females who were required to over-achieve along with boys to receive top tracking. Tracking in mathematics classes, on the other hand, was sensitive to both gender and race. European American males had the easiest route in while all other students were required to achieve higher grades and/or test scores to obtain the same result.

Educational researchers have come to view student progress with a "pipeline" metaphor. In many fields of study—most notably mathematics and science—class content is clearly linked; that is, if you do not master concepts taught in one of the earlier classes, you will be greatly handicapped in future classes to the point where you will probably give up in frustration. If students are to emerge from the end of the pipeline, we have to be sure they get into the pipeline to begin with and then maintain steady progress.

Pearson (1987) notes the dearth of African American students prepared for mathematics and science at the college level. The frustration dropout period for most was the tenth grade, but he attributed the problem to their not receiving the basics in elementary school instruction. Tracking becomes an issue in pipeline debates because it detours students and lowers their potential options in later years (Kershaw, 1992; Mehan, 1997; Wilson-Sadberry et al., 1991). Students who attempt to attend postsecondary education without this background face remedial classes at that level. There is a direct relationship between the number of remedial classes a college student is assigned and his or her retention in school: students taking three remedial courses as entering college students are half as likely to attain a BA degree as those taking none (U.S. Department of Education, 1996a).

Structure, Programs, and Role Models The Coleman Report minimized the overall effects of factors such as school curricula, special programs, and the like on achievement. That finding did not prevent schools from continuing to search for new ways to enhance the achievement of poorly prepared students. We look at some of the major efforts along those lines in the last section of this chapter where we encounter Head Start, busing, bilingual education, multicultural education, and other programs. Before leaving this section on the role of the school, however, it might be helpful to look at some of the less obvious aspects of school structure and programs that seem to have an impact on education.

Public schools have changed their staffing in many ways since 1975. Most notably, women and minorities are much more likely to be found as assistant principals, principals, and officials and administrators (U.S. Department of Education, 1998; U.S. Equal Employment Opportunity Commission, 1995). Those advancements are particularly striking for women whose percentages in those job categories increased between three and four times. By contrast, very little change has occurred among classroom teachers at either the elementary or secondary levels. The only racial or ethnic group to increase their numbers in the classroom between 1975 and 1995 was Hispanics. They increased their percentages in all elementary teaching jobs from 1.7 percent to 5 percent; percentages in secondary level teaching jobs increased from 1.5 percent to 3.6 percent. Outside of slight gains by Asian and Pacific Islanders (who had very few of those jobs in either year), African Americans and Native Americans fill the same percentages of teaching jobs in 1995 that they did in 1975. In addition, both groups hold lower percentages of those jobs than their percentage in the population (U.S. Equal Employment Opportunity Commission, 1995).

How important are role models for children in the classroom? They are obviously not critical for Asian students who do well in education with very few Asian teachers. But Asian students, like European American students, have access to highly educated role models elsewhere. African American, Hispanic, and Native American children, however, may have little personal experience with educated role models. Both the idea of being educated plus the educational content offered may well be strange notions indeed. In such circumstances, an increase in the number of minority teachers could only help, particularly in improving student motivation (see Davis and Jordan, 1994; Orlans, 1989). They also might serve students well in terms of improving lines of communication as discussed above with tracking and special education decisions.

School activities and procedures can influence student attainment and persistence. In particular, participation on school athletic teams has been shown to be a significant factor in lowering the dropout rate. Other activities, such as fine arts, also lower the dropout rate but not to the extent of athletics (McNeal, 1995). The degree to which a school emphasizes discipline is also a negative factor in attainment, particularly if the school imposes harsh penalties such as expulsion for a single incident.

BOX 8.1: EDUCATION AND ROLE MODELS: A PERSONAL NOTE

I cannot resist relating a personal experience involving minority students and role models. I worked with Hispanic fourth-grade children in the 1970s. Their small community was very poor, and many of them had difficulty with school subjects. While working with a boy who read at about the first-grade level, I tried to convince him that reading would get easier for him if he practiced and got better at it. It might even get so easy, I told him, that he might want to read just for fun. He looked at me in total disbelief. "But you're a teacher," he responded. Obviously, only a teacher could say something so stupid. He had learned that there were two kinds of adults in the world—teachers and normal people. Normal people who knew how to read did so only when they had to.

I was not a Hispanic role model for this boy, but all of his other teachers were not only Hispanic, but also from his community. Would he have reacted differently if one of them had made my comment? I doubt it. Teachers were teachers to him, regardless of their ethnicity. He was not going to buy an idea from a teacher until he saw some normal adult accept it first. Role models are important in the classroom, but they are even more important in the home and in the community.

Such penalties are more likely to be applied to poorer and minority students and accordingly have a negative effect on those groups (Davis and Jordan, 1994; Jordan et al., 1996; Noguera, 1995).

The Family and Community: Structure and Values

The Coleman Report directed attention toward family, community, and student attributes as important factors in predicting educational attainment. It is not surprising that those areas of research have been extensively mined over the years. Although this area of research covers considerable ground, it is not easily categorized into neat compartments; the overlap in the social factors examined is extensive. Research that examines basic family attributes (parent's educational attainment, parent's occupation, family income) can either focus on the human capital that children acquire or it can focus on educational attitudes that children acquire. Other family factors, such as being raised by a single mother, geographical mobility, and so on often are viewed as emotional disruptions for children rather than as related to human capital or attitude. In addition, when examining student attitudes, it is often unclear if they are family related, peer (or community) related, or both. This section begins with a focus on how children acquire basic human capital from family conditions and moves on to issues of values and attitudes.

Family socioeconomic status is almost always positively associated with a child's educational attainment, and this more often holds true for European American than African American families (Thompson and Luhman, 1997; Watts and Watts, 1991; Wilson-Sadberry et al., 1991). Recall that socioeconomic status is a composite measure of social class usually including parents' educational attainment and occupational prestige levels and family income (see Chapter 2). For many people, measures on all three of these dimensions is similar; a corporate lawyer, for example, would rate high on all dimensions. But many Americans, and especially minorities, vary from this assumption, as becomes apparent if you examine each part of SES separately for different groups. For example, the strongest family characteristic predictor of African American educational attainment is mother's education followed by father's education. Income and occupation are far less important; in many cases, they have no association with an African American child's educational attainment (Thompson and Luhman, 1997). However, it is wise to use care when interpreting research employing this much-used measure (see White, 1982). Income does play a role in the selection of higher education institutions, with poorer students attending less selective institutions regardless of their academic ability (Hearn, 1991) and often turning to two-year institutions (Wilson-Sadberry et al., 1991).

The higher numbers of single-parent families among African Americans and Puerto Ricans have led many researchers to examine the connection between family head and child's attainment. The findings are often contradictory. Watts and Watts (1991) conclude that father absence does not affect the child's attainment while Mulkey and associates (1992) and McLanahan and Sandefur (1994) find just the opposite. Marx and Crew (1993) find that grandparent-headed families (with no parents present) have only a minimal negative impact on a child's attainment. McLanahan and

Sandefur (1994) argue that the presence of a grandmother along with a parent in a home greatly increases dropout rates due to family disruption. The major educational attainment problems associated with single-parent homes are lack of income and other secondary causes (Milne et al., 1986), although other research indicates that single-parent status affects a child's education independently of income (McLanahan and Sandefur, 1994; Mulkey et al., 1992). When a single mother is employed, it generally has a negative impact on a child's attainment, except for African American families for whom a mother's employment has a positive impact (Milne et al., 1986). Family geographical mobility always has a negative impact on dropout rates (Jordan et al., 1996; McLanahan and Sandefur, 1994).

Research regarding family values, attitudes, and expectations has been more consistent in its findings. African American parents have equally high educational aspirations for their children as European American parents (Solorzano, 1992); African American students also equal European American students in educational aspirations, even surpassing them when social class is held constant (Solorzano, 1992; Wilson and Wilson, 1992). High aspirations do lead to higher educational achievement, but students tend to measure that success against other students with whom they identify (usually members of their own racial or ethnic group). Hence, high achievement within a low achieving group may produce somewhat limited attainment (Kao et al., 1996). Minority and/or poor students may have high aspirations but are generally realistic with educational expectations (McLanahan and Sandefur, 1994). The gap between aspirations and expectations can be a negative influence on attainment itself (Hanson, 1994).

Student attitudes toward schools and education can take an altogether different route, particularly when peer influences strengthen. The youth scene in the United States includes values and behaviors that often work at cross-purposes to educational attainment. As evidence of this influence, Wojtkiewicz and Donato (1995) find higher attainment among Mexican American students whose parents were foreign-born as opposed to students with native-born parents. These first-generation students in fact match European American students in high school and college graduation rates. The researchers explain part of this phenomenon by the pro-education values and behaviors that thrive in these families. This suggests that the Americanization of immigrants and educational attainment can be at odds.

Although minority students have high educational aspirations, other attitudes seem quite the opposite. To many African American and Hispanic youths (especially male), educational success is considered "acting white" (Clark, 1991; Fordham and Ogbu, 1986; Hood, 1992; Kao et al., 1996; Ogbu, 1988). This attitude—sometimes called an oppositional identity—stems from building minority pride through emphasizing contrasts with mainstream American society, which supports education; thus opposing education can be an act of pride. In addition, you really do have to "act white" to succeed in American education; speaking dialects such as Black English or "Spanglish" will not bring rewards on the verbal section of the SAT.

By contrast, most Asian students are apparently outside such cultural forces. Educational success does not bring charges of "acting white" for these students. Although educational diversity within this broad census category is considerable, many

European Americans tend to think of Asian students in "model minority" stereotypes (Lee, 1994). For many Asian groups, however, there is a tradition of cultural values attached to educational attainment. These values are reflected in the day-to-day lives of many Asian families where a general value on education is backed up by very concrete behaviors promoting homework, high test scores, and overall achievement (Fejgin, 1995; Hsia and Hirano-Nakanishi, 1989; Lee, 1994). For newer immigrants, difficulties with English promote an emphasis on science and mathematics, which is reflected in the test scores examined earlier (Tang, 1993). The large numbers of Asian educational successes from both native-born and foreign-born students does tend to validate the importance given to family structure and values by the Coleman Report.

Responses to Inequality of Educational Opportunity

More than three decades have elapsed since the publication of the Coleman Report. The report found largely segregated schools and many uneducated students throughout the country. Even though it placed much of the blame for educational failure outside the school, that failure could not be ignored. A companion "call to arms" was published in 1983 by the National Commission on Excellence in Education entitled *A Nation at Risk*. It observed that no country spent more money per student than the United States, yet American students typically placed out of the top twenty in major subjects when compared with students in other countries (National Commission on Excellence in Education, 1983). This outcome, the Commission felt, did not bode well for the continued economic and industrial growth of the United States. This section examines some of the federal, state, and local efforts to equalize the educational playing fields in the United States.

School Segregation and Housing Segregation

The Coleman Report certainly made the inequality in American education public. Segregated and poor nonwhite students monopolized some school districts, while middle-class European American students dominated others. The school segregation was *de facto* in that it was the result of housing segregation rather than law, and the housing segregation was partly the result of poverty and partly the result of a great many informal instances of discrimination. We explore all of this more fully in the next chapter, but however it occurred, it led to many kinds of educational inequality.

If your goal is to eliminate school segregation based on housing segregation, your choices are very limited. You can either move families to new neighborhoods (and new school districts), or transport students out of their districts into other school districts. The first choice would produce massive expense. The only option appeared to be moving the students on a daily basis. In 1971, the Supreme Court decided in *Swann v. Charlotte-Mecklenburg Board of Education* that segregated schools in Charlotte should be remedied by the busing of students.

Busing is the attempt to create student bodies with diversity in race, ethnicity, and social class through the movement of students within and across school district lines.

Busing within a single district may produce few desegregation results if all schools within the district contain mostly poor nonwhite students. In fact, many American cities today lack diversity because increasing numbers of middle-class families have moved from cities to suburbs with totally separate school districts. When busing plans cross district lines between city and suburb, the goals of diversity become possible to achieve.

Busing created considerable political conflict when it was first introduced. There is evidence that busing does help minority students achieve higher grades and test scores, when buses bring them to schools containing European American students from a higher socioeconomic background (Mahard and Crain, 1983). Still, results were uneven and all were dwarfed by the massive public outcry against the program. During the Carter administration of the late 1970s, the federal government generally backed busing solutions to segregation and was supported by the courts. Beginning in the 1980s with the first of two Reagan administrations, both the executive branch and the judicial branch of the federal government decreased federal support for busing.

Some school districts attempted to defuse the busing controversy by making the central city schools more attractive and calling them "magnet schools." Magnet schools are public schools that specialize in a particular area of the curriculum, providing students with focused (and presumably excellent) instruction in classes with other students who share their interests. A school might specialize in the humanities, science, the performing arts, or foreign language and culture. A school might even choose traditional studies, emphasizing the basics of education in a highly disciplined environment. By providing such a focus, the schools attract not only the best and most highly motivated students but also dedicated teachers to whom that educational environment is appealing. In general, magnet schools have been successful in attracting European American students back to the inner city. In some cases perhaps, they have been too successful because inner city nonwhites have had some difficulty in competing for enrollment space. In spite of their value, magnet schools are only a partial answer to the problems of unequal and segregated public schools in the United States because funding is inadequate to provide such facilities for all students.

In a famous example, Kansas City, Missouri attempted a magnet school solution to its segregated school system. When the new schools required new funds, a federal judge—Judge Russell G. Clark—ordered an increase in property tax to fund the new schools. In 1990, the Supreme Court overturned this lower court order in *Missouri v. Jenkins*. The decision ruled that a federal judge could not order a city to raise its property taxes when the citizens of that city voted them down. In short, school districts could be asked to desegregate, but they could not be ordered to pay for it. The federal courts were clearly moving away from busing as a solution (Persons, 1996; Rist, 1996). American schools remain about as segregated today as when Coleman reviewed them.

School-Based Responses

The federal government stepped directly into the classroom by requiring new programs. Probably the best known of these are the preschool compensatory educational programs (e.g., Head Start) and bilingual/multicultural programs that attempt to provide avenues to mainstream educational success for an increasingly diverse student population. There also have been efforts to create "new basics" curricula to give students a more solid and college appropriate background. Many "new basics" programs have been the mainstay of much private education in the United States for years. We conclude this section with a brief look at the differences between public and private education.

Compensatory Education   Students bring many aspirations, attitudes, and expectations with them to the school system, but they also vary in human capital. All students arrive with skills but some of those skills are much closer to classroom expectations than others. Compensatory education assumes that certain students come from cultural backgrounds that ill prepare them for the demands of the public school system; it is designed to fill in the gaps in that background so that those students are able to compete on an equal footing with other students. The best-known such program is Head Start, which attempts to prepare preschool children for the demands of first grade and beyond. Early studies on the efficacy of Head Start were somewhat inconclusive; it obviously helped students in the short run, but it was less clear how much it helped in the long run. Such questions are difficult to answer because so many factors affect school achievement. Nevertheless, some studies suggest that Head Start graduates do make long-term gains (Brown, 1985).

Bilingual and Multicultural Education   Although public schools generally have been under local control, the form of instruction and curriculum content, for the most part, are standardized, so that children throughout the country receive a remarkably similar experience. The schools, not surprisingly, reflect the predominantly English culture that characterizes American life, with an emphasis on English language and literature couched within a generally Eurocentric view of the world (Sankowski, 1996; Schiele, 1994; Seltzer et al., 1995). In the past, students who brought non-English or non-European cultures with them to school faced a large gap between who they were and what their teachers wanted them to become.

Traditionally, public schools have been less than tolerant about such student problems, sometimes punishing students for speaking languages other than English in class or even on the playground. Such treatment has helped eliminate certain non-English cultural traits on the American landscape, but it has not always been successful at replacing those traits with the skills promoted by the school curriculum (Lee, 1992). Minority children often fail to learn what schools teach, and a significant number drop out before receiving a diploma. An assumption made by many educators is that this lack of attention to the cultural diversity of students lessens their achievements through lowering their self-esteem (see Hymowitz, 1992; McCarthy, 1990;). Beginning in the 1960s, American schools have tried to respond to these shortcomings through bilingual schools and multicultural programs.

Schools providing bilingual education and/or multicultural programs vary both in form and in degree. Some provide additional course content designed to match the cultures children bring to school, while others devote a significant portion of the curriculum to non-English content and to classes taught in a language other than English. The official goal of such programs is to provide children with an easier route to traditional American educational objectives. For example, children who do not speak

Bilingual education programs help students achieve their educational objectives through strategic use of the student's native language.

English are first taught to read in their native language while simultaneously learning to speak English; once they have accomplished these goals, learning to read in English should be easier.

Bilingual education programs have been geared primarily to Spanish-speaking children in the Northeast, Florida, and the Southwest, and to Asian children who have immigrated to the United States over the last twenty-five years. The three states with the highest percentages of non-English speaking children are California (14.9 percent), Texas (11.3 percent), and New Mexico (8.5 percent) (U.S. Department of Education, 1996a). An unofficial goal of such programs, particularly with regard to Spanish-speaking populations, has been to maintain some of the elements of Spanish culture through the institutional support of the educational system. Currently, 31 percent of Hispanic students speak Spanish in the home, but much of that is maintained through continuing immigration (U.S. Department of Education, 1997). Although the programs and schools have been in place for a number of years, it is difficult to fully assess their success in achieving either of the goals mentioned (Spener, 1988; Stanton-Salazar and Dornbusch, 1995). Factors beyond school curriculum also have a major impact on students' attitudes toward school achievement and toward mainstream American culture. Under the circumstances, it is difficult either to fully credit such programs with success or to fully blame them for failure. Nevertheless, they do reflect a fundamental change (if they remain permanently in place) in how open the educational institution is to other cultures.

Do Private Schools Do a Better Job?   Over 83,000 public schools offer primary and secondary education in the United States; they educate over 88 percent of American children enrolled in schools. The remaining 12 percent primarily attend over 27,000 private schools, of which almost 20,000 are run by religious institutions. Half of those religious schools are run by the American Catholic Church, and half are organized by other religious institutions (U.S. Department of Education, 1999). Although at first glance it might appear that the relatively large number of private schools enrolling so few students would result in many fewer students per teacher, but class sizes in private schools are only slightly smaller than those in public schools. Private schools tend to be much smaller than public institutions and to have smaller staffs. But while the student-teacher ratios are similar, the results of student achievement vary considerably. Private schools are doing a better job.

A higher percentage of private school students attend college than do the graduates of public schools (Falsey and Heyns, 1984). A logical explanation for this difference might seem to be the difference in cost between the two forms of education. Since private education creates direct costs for parents, the students would be economically screened, producing far fewer lower class children in the private school classrooms. That economic explanation certainly accounts for some of the differences in educational achievement, but it does not explain all the differences. If we compare private and public school students whose parents are similar in social class and education, students in the private schools still achieve more than their matched comrades in the public schools (Coleman et al., 1982; Coleman and Hoffer, 1987). If we narrow that comparison down to just Catholic schools and just minority students, the private school again appears to be doing a superior job of both educating and motivating students (Greeley, 1982). As we have seen, there are no major differences in student-teacher ratios between public and private schools, but there are differences in other areas. Catholic schools tend to have both stricter discipline and higher educational standards for their students than do public schools; presumably, their students are willing and capable of responding.

Could public schools effectively raise both academic standards and standards for discipline to create an environment approximating that of Catholic schools? Other factors may be involved as well. Even inner city Catholic schools with predominantly nonwhite students may be able to create higher motivation because of their distinctiveness amidst a public school environment. Private schools also may be attracting students from those parents who are most committed to and supportive of their children's education. Nevertheless, the relative success of private schools suggests what is possible in education. Even though a child's socioeconomic background has a major impact on school success, educators are not completely helpless in countering that influence.

Summary

The institution of education becomes critical for industrial societies like the United States because of its intimate connection to the economy and employment. Most skilled work requires not only the skills and abilities available through education but also the credentials obtained there as well. Discrimination with regard to education therefore translates directly into employment discrimination.

Minority historical experience with education in the United States has been complex. Some immigrant groups—notably Asians, both historically and contemporarily—have excelled. This has often been explained by traditional Asian values attached to educational success and hard work, but in many cases, part of the explanation lies in previous educational successes. Notably, the ethnic groups in the contemporary United States with the poorest educational attainment are those groups who historically faced the most educational discrimination—Hispanics, Native Americans, and African Americans. All have advanced relative to European Americans in the last half of the twentieth century, but significant gaps remain.

One form of inequality built into public education in the United States is the way public funding is provided. Schools are funded through taxation, but most schools receive a large amount of funding from property taxes. This produces more funds for schools in wealthier districts. Poverty area schools in particular pay poorer salaries to teachers and are not able to purchase the same equipment that better funded schools enjoy.

To the extent that racial and ethnic differences follow neighborhood income differences, poorer schools tend to have higher percentages of minorities in their student bodies. In fact, the vast majority of America's minorities currently attend largely segregated schools. Beyond differences in funding, most research indicates that students in these schools experience long-term damage that carries over into adult life and the job market.

An additional factor in minority educational attainment concerns cultural clashes between schools and students. With some exceptions—bilingual and multicultural schools stand out—minority students bring different skills to school than do dominant group children. Many of these skills are comprised of "taken for granted" knowledge upon which schools expect to build. This places many minority children at an initial disadvantage that grows over time.

Attempts to explain unequal educational attainment have produced a great amount of research. Most of this research indicates that schools themselves are not the major factor in why some students succeed while others fail. When schools are important in these results, most findings have emphasized the importance of teacher expectations and differential treatment, the effects of tracking and special education in depriving many minority students of a necessary foundation, and the lack of role models among teachers and administrators.

Better predictors of educational attainment are generally found in families, communities, and student peer groups. Family socioeconomic status is usually an important predictor, the most important aspect being parents' educational attainment. Single-parent families tend to work against the educational attainment of children as does any kind of family disruption. Among peer groups, many minority youths have developed "oppositional" subcultures, which take pride in rejecting school values.

Efforts to equalize educational opportunity in the United States have been varied and not particularly successful. Segregated schools are a direct result of segregated housing, and both forms of segregation persist today. Busing students across district lines to desegregate schools created significant conflict; in spite of some successes for minority students, federal support has been lacking since the 1980s. An alternative approach was to transform central city schools into magnet schools, making them more attractive to higher socioeconomic students.

Within schools, efforts at compensatory education, such as Head Start, date back to the 1960s. These programs have not solved all problems but have generally been successful. Schools also have instituted bilingual education and multicultural education in an effort to make the atmosphere and curriculum match students better. Probably the best results have come from central city private schools, particularly Catholic schools, which have emphasized traditional studies and discipline.

Chapter 8 Reading
No Can Geeve Up:
Crossing Institutional Barriers

H.M. Murai

Institutional discrimination affects particular racial or ethnic groups not because of their group membership but because of skills or abilities that tend to distinguish them from dominant group members. When our institutions discriminate on the basis of such skills or abilities, particular racial or ethnic groups may be adversely affected just as surely as if their group had been singled out for unequal treatment. In the following article, H.M. Murai tells the story of his childhood and his experiences as a university freshman. Growing up in a distinctive Japanese American community in Hawaii, he differed in many ways from the dominant group that conceived and operates the University of Hawaii. Of those many differences, Murai focuses on his language. He grew up speaking a form of language known as a pidgin, which combines and simplifies elements of two or more other languages. A pidgin language accomplishes the task of communication (which is all a language really need do), but Murai found it less than acceptable to his freshman English teacher at the university; to Murai, freshman English might as well have been freshman Greek. His authorship of the following article is evidence that he succeeded, but many in his place would not have.

I grew up in a sugar plantation community near Hilo on the "Big Island" of Hawaii. In spite of its size, it is still considered to be one of the "outer islands" of Hawaii; i.e., not the center of industry and especially the tourist industry, which thrives on the island of Oahu and more recently, Maui and Kauai. Growing up in a plantation community one was considered to be "from the country(side)? People from the country are often considered lower class and lower in "class."

Plantation language, which was my first language, was closer to the pidgin spoken by the original immigrant workers, somewhere between a Creole and a dialect of English. You were likely to find interaction between ethnic groups in the plantation camps in spite of management's attempts to keep the groups separate and ununited. The integration of ethnic groups contributed to the perpetuation of the original pidgin created for communication across languages. On any given Sunday you would find Filipinos, Japanese, and Portuguese gathering at the local cockpit and billiard hall. Growing up in this environment made it difficult to identify the sources of words borrowed from Filipino, Japanese, Portuguese, Hawaiian, and so on in the local dialect.

I believe that the majority of the children coming from the camps were proud of their roots. Being proud of our backgrounds I believe further contributed to our maintenance of the language and culture of the camps. Life in the camp was rarely dull. One could always go "pick cane" from the cane fields or "catch cane" from the flumes delivering the cut sugar cane stalks to the mill. There was also "poking crayfish" and opu (mudfish) in the rivah (river), catching opal (river shrimp), picking wild guava and rose apple, or just following the rivah. In my case there was also body surfing with scraps of board at one of dah bes beaches in the Islands. During the summer months, baseball and the Japanese bon dances (commemorating the deceased) provided more weekend recreation and entertainment. Entrance into the middle and secondary schools located in the city provided for a broadening of one's cultural-linguistic circle; however, the foundation of the personalities of the kids from the camps was formed in the camps, and Hilo, Hawaii, is indeed a country town when compared to Honolulu.

Leaving the warmth and security of home for the first time to venture off to school in the big city, Honolulu, provided as much cultural-linguistic conflict as I could handle. The institutional barriers to being successful in college were many.

I had heard about how one-third of the freshman class failed English composition at the University of Hawaii and how failure in composition usually led to flunking out of school. English comp. was known as the screening class for those who could and those who couldn't. This was especially true of students with my background; i.e., students who were from the rural areas of the islands and who spoke "pidgin" English (actually a dialect of English). I was from one of the "outer islands" (islands other than Oahu, where Honolulu, the capital, is located). Adding to my fears was the self-doubt that haunted me throughout my first year—self-doubt that was set in motion by my almost total failure during my final two years in high school and by a counselor who thought I would be "wasting time and money" by pursuing a university degree. I am quite sure that my father had the same sense of doubt when he waved me off at the airport with the somewhat ambiguous, "No can come home" (may be interpreted as "You can't come home" or the intended, I believed, "If you can't make it, come home"). In fact, till this very moment it is still not clear to me how I was able to attend the university out of high school. I had taken college preparatory courses such as chemistry, geometry, and a second year of algebra, but I would have been the first to admit that I had no demonstrable proof that I had learned anything in these classes. Nearly straight D grades were an accurate reflection of my near total neglect of academics. Attitudinally, I learned that I was not capable of anything beyond the usual requirements for high school graduation. I definitely felt much more comfortable playing for extra spending money in the pool room or drinking beer "wid dah boys" than in any classroom.

Perhaps my greatest fear during the first semester was that someone would discover that I was mistakenly admitted to the university and that yes, my high school counselor was correct in suggesting that it was a waste of time and money for me to pursue a university education. This led to more than a bit of paranoia. I felt that I was being watched and could not afford to attract any attention through failure. Failure on any examination and definitely failure in any class I was sure would lead to immediate discovery. Visiting the office of any instructor was thus out of the question. My inability to speak "good English" would surely lead to my being spotted as an imposter. Much to my relief none of my professors required individual meetings. In fact, I do not recall any professor mentioning office hours, much less inviting students with problems to come in for consultation.

I can still recall, after 25 years, what my first semester freshman schedule was like. I know I felt compelled to enroll for 18 units, which was the maximum allowed. I thought (incorrectly) that I had to do this in order to graduate in four years, just as the catalog had illustrated. I recall speech class, where I was obviously in deep trouble trying to pronounce the "th," "ed," short "i" and "schwa" sounds and vainly attempting to speak "like one haole" (anglo). Western civilization class was at 7:30 A.M., Tuesday, Thursday, and Saturday and freshman English was Monday, Wednesday, and Friday at 2:00 P.M. Western civ. was a memorable class not simply because of the time and the 300 or so students packed in an auditorium, seat numbers assigned, but because of the difficulty I had in note taking. Note taking was a skill I had never acquired in high school. I assumed that my notes had to be accurate and thought that someone might ask to read them. I wrote in pencil, filling every line and margin as though I had spent my last penny on the notebook. I also erased words while desperately trying to record lectures verbatim. Someone forgot to tell me that I was only supposed to capture the main points (assuming I could determine what they were). Needless to say, when it came time to review my notes I had difficulty deciphering my own handwriting and could barely make sense of the little that I was able to decipher. I received a well-earned D in the first semester of Western civilization. But ah, English composition, there was the ultimate challenge.

The instructor for English 101 was, I learned later, a former high school English teacher. She had what may be referred to as a "no nonsense" approach with an emphasis on actually writing rather than grammar or the how-to of writing. In correcting papers, my papers at least, Mrs. C. tried to interpret what I, more frequently than not, was unable to clearly express. She was a risk taker. She was not afraid to substitute words or rewrite sentences in attempting to clarify my compositions. Throughout the first half of the semester, Mrs. C. came close to totally rewriting my compositions. More often than not, she was, as far as I could ascertain, on the money. I was impressed, amazed, astonished even, to see how clearly she could represent my deepest thoughts and feelings without ever consulting with me. Nevertheless, I managed to receive straight F's on my first six compositions. We were at mid-term and I thought, "You in real trouble bruddha. How you going do dis!" There were no remedial classes offered. It was the prototype of the "sink or swims," "give up or geev um!" (show what you can do) situation.

Receiving graded papers, due on Monday and returned on Friday, was the most demoralizing and embarrassing part of that first semester. In high school I could laugh at the fact that I did not try to do well and thus expected poor grades, but this was the big time and I was indeed trying my utmost to succeed. Thus when I received my straight F's, I simply slid my papers into my notebook so they would not be revealed to my classmates. I also avoided looking at any of my classmates who spoke aloud about their disappointment in receiving B's or C's and even A minuses! Even the gal with the friendly smile sitting next to me who did not say anything about the straight A's she was receiving was a threat to my ego. Surely these people realized that I did not belong in the same room, much less the same university. Surveying the class of twenty-five or so students, it was difficult not to conclude that I was the one of three who were destined for failure: "I just gotta show um (them)" (that I could do it).

As soon as we were given the topic for the next composition on Friday, I headed for the library. Thanks to the good advice of two older students, I had left time after each of my classes to review notes or conduct the research necessary to clarify points in doubt. Although it was extremely painful, and partly because it was required, I immediately began the rewrite of my paper. Because of the detailed correcting by Mrs. C., this was not a terribly difficult task. During the first few weeks, library work included looking up words that were part of the assigned topics. After the very first class, the first word I had to look up was "composition." I was not sure what was being required. I also looked up "plagiarize" (I knew the "Don't" part of the phrase). Although my dictionary never left my side, I frequently found it necessary to look up more detailed descriptions of words in the encyclopedia or unabridged dictionaries in the library. Then began the writing process, outlining, and starting the rough draft. Because the composition class was my last class on Friday, I could spend two to three hours in the library before heading back to my apartment for dinner. My goal was to write the exact number of words required for the composition.

The writing continued immediately after dinner and ended sometime around midnight. As reward for my hard work, I would often treat myself to a bowl of crisp kau chee mein (Chinese noodles with vegetables) and several cups of tea in a restaurant sparsely patronized mainly by people who had closed the local bars. This was a treasured time for meditation and contemplating my world view. Friday evening or early morning at the Golden Duck and martial arts practice during the week were the few breaks in my daily academic routine. I was always happy to get out of the studio apartment I shared with two working brothers. Although only fifteen minutes from Waikiki, I never spent time on the beach that first year in school. Saturdays were devoted to Western civilization, so the final draft on the composition was not completed until Sunday. Until my teacher hinted that it would not be a bad idea to either write the composition with a pen rather than a pencil or even type the final draft, I did not have to spend a lot of time producing the final draft. In spite of the usual ten to fifteen hours of writing spread over three days, I seemed to make little progress during the first six weeks. None of my classmates seemed to be experiencing any difficulty in the class.

I vividly recall standing outside the classroom a half hour before class began, listening to students who were actually completing their compositions before class! These were students who were receiving A's and B's. All I could think of was "You in beeg trouble bruddha!" There were many days of loneliness and discouragement during which times, for some reason, I refused to accept failure, thinking that all I had to do was work ten times as long and hard as my classmates, a hundred times if necessary. "No can geev up!" was always my final thought.

Then, alas, during the seventh week of class, I was asked to read my paper to the class, an honor given to the chosen few each week. I panicked and hesitated. I was also having serious problems in my speech class at that time. In addition, I had been turned down as a volunteer reader for a blind student because of my inability to read English fluently, and I was not at all confident about getting up in front of the class. The kind teacher obliged by reading my composition for me while I listened with ambivalent feelings of pride and shame: I made it through English composition with a C grade the first semester and a B for the second semester.

I often wonder how and why I persisted during that first semester, especially during the first half of that first semester. I had no one to turn to except myself and a sister-in-law who sometimes read my papers for obvious flaws. I can think of some obvious reasons, such as the fact that I was able to survive without a great deal of effort in my Spanish and psychology classes and of course, P.E. and the required ROTC. However, it was also clear to me that by any combination of grades that I could expect, failing English composition would definitely lead to disenrollment from the university. Reflecting back into that time, I can see numerous reasons for my survival.

I come from a family background where it was assumed that anything could be achieved through effort. I think I felt that failure in school was worse than death. "No can come home" was an ever-present thought throughout my four years as an undergraduate. Hereditary limitations in intellect were not considered important. My father, who [sic] I greatly admired, had himself risen from the ranks of a plantation laborer to a supervisorial position. During my junior year in high school he had been selected to join a team of engineers and horticulturists to help modernize sugar mills in a third-world country. Daddy, though extremely strict in discipline, always provided a loving arm around the shoulders in a timely manner, and obviously put the needs of his children over and above his personal needs. He had an eighth-grade education. My mother I admired for her determination to provide for the family through hard work and persistence under difficult economic, sociocultural, and oftentimes highly stressful psychological conditions. She was the model of the nurturing mother, always there in times of need and never punitive. She, too, put her children's needs ahead of her own. Relatives, especially Uncle and Aunty Y, Roy and Kay, also impressed upon me the importance to success of hard work and the willingness to suffer through hard times. These values were reinforced by participation in the martial arts from an early age and by Japanese samurai films that my aunt and uncle introduced to me early on.

Participation in sports must have had some influence on my attitude about persistence in the face of difficult odds. Although never an outstanding athlete, I had experienced some success in baseball, track, and football and even tried swimming and amateur boxing during my junior year when I was injured and unable to participate in football. I had the experience of competing under teachers and coaches who were, for better or for worse, sometimes sadistic in their training methods. Being able to run the last wind sprint and stand up for the final tackle drill and experiencing some success throughout all of the trials I am sure contributed somewhat to my belief that hard work could conquer all.

Although academic successes were few and far between throughout my early school years, I recall reminding myself that I had had some moments of success achieved through hard work. I had vague memories of being fairly successful in challenging social studies, science, and English classes during my junior high school days just prior to being initiated into one of the local gangs. I also had memories of being able to learn Spanish without too much effort and elementary algebra, which did not seem to come too easily for some of my classmates who I thought were higher achievers. It is funny how a few successful experiences can be so meaningful in times of doubt and carry you over hurdles.

I have always attributed some of my success to a cousin whose intelligence I respected although he had not completed his college work. He had told me that I would never fail if I attended classes faithfully and kept up with my daily work. For the first two years I took his words to heart, never missing a class even though I suffered through several bouts of illness during that time period. I could think of absolutely nothing that could keep me from attending classes, an extreme contrast with my attitude during my high school days when I missed more classes than I attended. This was all in spite of the fact that attendance was never applied as a factor in grading in any of my classes.

My financial situation must be included as a factor that encouraged me to persist and survive. It was my feeling that I would have no second chances in any of my classes; my financial situation, I believed, did not allow for second chances. Although my parents were able to help me with cost of room and board during my first year, I had saved from my part-time job in high school and from several summer jobs including work with a trucking company and in a fertilizer factory.

There is little doubt that the teacher of my composition class was of primary importance in helping me persist. Although the grades she gave me were not encouraging, the fact that she obviously took the time to correct my errors gave me hope that I could improve if I could learn to self-correct as she had been modeling for me. In fact, I think I spent time committing most of the corrections she modeled for me to long-term memory. I was determined never to make the same error more than once, and though I was not always successful, I knew that I was improving by my fourth or fifth failure. I wonder if Mrs. C knew how her comments and grading of my papers affected me. She gave me no indication that she did, yet she must have surely wondered how I ever managed to get into the university and why I kept coming back for more punishment.

Any one of the reasons discussed above could account for my persistence in English composition. I am of the opinion that, as with most of human behavior, only a complex combination of many reasons, including but not limited to the above, must be considered. Nevertheless, today, as I relate my experience to my students who seem to be in similar situations, I emphasize that one can only fail if one lacks the desire to succeed, the willingness to sacrifice for one's goals. No can geeve up.

9

The Economic Sphere: Job Competition and Discrimination in the United States

Your connection to the economy is your survival. Nothing else could be more important in the long run except your health, and as we will see in Chapter 10, even your health is affected by your place in the economy. Economic relations are central to understanding the racial and ethnic boundaries that remain so important over time. The very definition of a minority group springs from economic standing. This chapter focuses on the economic circumstances of contemporary racial and ethnic minorities in the United States. We first look at changes in the U.S. economy since the 1970s that have had a negative impact on many minorities in the job market. The poverty resulting from those changes has lessened the resources of those affected, creating additional problems for families and communities. This has made adapting to changing employment needs even more difficult. The final two sections of the chapter take us into the workplace for an examination of unemployment rates, hiring and promotion practices, and the impact of affirmative action programs.

Income and Employment in the United States

In 1997, the median income for white households was $38,972; this amount is 56 percent above the median income for African American households ($25,050) and 46 percent above the median income for Hispanic households ($26,628) (U.S. Bureau of the Census, 1999a). Table 9.1 turns our attention to the worst-case results of lower pay, showing the percentages of different groups that fell below the poverty level.

Adding one more group not listed in that table, 31.2 percent of Native Americans fell below the poverty level as of the 1990 census (U.S. Bureau of the Census, 1990). As is clear, African Americans, Native Americans, and Hispanic Americans are between three and four more times likely to be poor than European Americans. Asian and Pacific Islanders appear to be doing much better, but all of these numbers need closer inspection. As we saw in Chapter 7, the Asian and Pacific Islander and Hispanic origin categories include considerable variety.

Table 9.2 offers us a closer look at how many of the different Asian and Pacific Islander ethnic groups compare with each other and to the larger society based on data from the last available census. With the exception of Vietnamese immigrants, all are in an apparently better economic position than blacks and Hispanics. Still, as we saw in Chapter 7, figures concerning these groups can be misleading because of (1) their geographical concentration in high income/high cost of living regions of the United States, and (2) the tendency for Asian ethnic groups to contain more rich and more poor with fewer in the middle ranges. As you can see, Chinese Americans have a higher median family income than whites but also a noticeably higher percentage living in poverty.

Similarly, Table 9.3 provides a Hispanic group breakdown in terms of median family income and percentages in poverty. Clearly, Cuban Americans are more economically advantaged than any other Hispanic group by both measures. Just as clearly, Puerto Ricans are the most disadvantaged. Probably the most confusing category in this table is "Mexican Origin," which contains both native-born and foreign-born people. In spite of Hispanic diversity, however, note that all of these groups fare less well economically than whites in the United States.

These data provide a snapshot of American society, but a longer term view does not suggest very much change in the picture. In 1979, for example, black men's earnings were 27 percent lower than white men's; by 1993, that gap narrowed only slightly

TABLE 9.1	PERSONS IN POVERTY IN THE UNITED STATES, 1995.	
	Number	Percent
Total persons	36,425,000	13.8
Race		
White, non–Hispanic	16,267,000	8.5
Black	9,872,000	29.3
Asian and Pacific Islander	1,411,000	14.6
Hispanic Origin	8,574,000	30.3

Baugher & Lamison-White, 1996.

TABLE 9.2	MEDIAN FAMILY INCOME AND PERCENT OF FAMILIES BELOW THE POVERTY LINE FOR SELECTED ETHNIC AND RACIAL GROUPS, 1990.

Race/Ethnic Group	Median Family Income	Percent of Families Below the Poverty Level
White	$37,152	7.0
Black	$22,429	26.3
Hispanic	$25,064	22.3
Asian and Pacific Islander:	$41,251	11.6
Chinese	$41,316	11.1
Japanese	$51,550	3.4
Filipino	$46,698	5.2
Korean	$33,909	14.7
Asian Indian	$49,309	7.2
Vietnamese	$30,550	23.8

U.S. Bureau of the Census, 1993a.

TABLE 9.3	MEDIAN FAMILY INCOME AND PERCENT OF FAMILIES BELOW THE POVERTY LINE FOR HISPANIC ETHNIC GROUPS IN THE UNITED STATES, 1993.

	Mexican Origin	Puerto Rican Origin	Cuban Origin	Central and South American Origin	Other Hispanic
Median Family Income	$23,714	$20,301	$31,015	$23,649	$28,562
Percent of Families Below the Poverty Level	26.4	32.5	15.4	27	21.7

U.S. Bureau of the Census, 1994a.

to 26 percent. In those same two years, black women fell below white women by 8 percent and 10 percent, respectively (Bennett, 1995). Although parity in women's earnings is far greater across race and ethnicity than men's earnings, the important point here is that the two gaps are not changing.

Hispanic median family income in 1997 was 32 percent less than all white Americans (U.S. Bureau of the Census, 1999a). If we check the same two categories of people in 1980, Hispanic family income was 27 percent of the comparison group (U.S. Bureau of the Census, 1994a). We find a similar lack of change with poverty rates over the same period: Hispanic persons moved from 21.8 percent to 27.1 percent over that decade while whites began at 10.2 percent and ended at 11 percent (U.S. Bureau of the Census, 1999a).

Although income and poverty rate levels have been relatively stable, some interesting occupational changes have occurred at the top end of the economy—changes to which we return later in the chapter. However, in looking at the changes discussed here, keep in mind that African Americans represent approximately 13 percent of the population, Hispanics 8.9 percent, Asian and Pacific Islanders 2.9 percent, and Native Americans less than 1 percent (U.S. Bureau of the Census, 1999a). Between 1966 and 1994, African Americans increased their percentage among officials and managers in the economy from 0.9 percent to 5.3 percent; Hispanics rose from 0.6 percent to 3.5 percent; Asians and Pacific Islanders improved from 0.2 percent to 2.2 percent; and Native Americans moved from 0.1 percent to 0.4 percent. Overall, these four groups increased their participation in these occupational categories from 1.8 percent to 11.5 percent. Similar changes can be seen within the ranks of professionals, with participation increasing from 3.9 percent in 1966 to 14.7 percent in 1994 (U.S. Equal Employment Opportunity Commission, 1995). These numbers are not overwhelming, particularly with reference to the percentages of these groups in the overall population; in addition these upwardly mobile individuals do not receive as much income as white males in the same job categories. Nevertheless, there has been a change. The flip side of this coin, however, is that opportunities at the lower end of the job market have decreased, not only for minorities but also for all Americans. Growing income inequality in American society figures prominently in the pages to come.

This quick economic overview of contemporary American society does not tell us all we need to know, but it certainly makes two points quite clearly. First, the American stratification system is structured along racial and ethnic lines as discussed in Chapter 2. Second, the system's structure has proved to be considerably stable over time; that is, the economy can rise and fall with business cycles, but the gap between specific ethnic and racial groups is retained during those cycles. We saw in previous chapters how American society has been ethnically and racially stratified from its very beginnings. The primary purpose of this chapter is to explain how it has stayed that way in contemporary times.

American Industrial Restructuring

Economies in industrial societies are usually conceptualized as having three sectors. The primary sector of the economy refers to activities associated with the production of raw materials. These would include farming, fishing, mining, harvesting timber, and so on. The secondary sector of the economy refers to manufacturing activities that transform these raw materials into usable goods, such as food, clothing, fuel, construction, machines, and all consumer goods. The tertiary sector refers to the production and distribution of services—such activities as law making, house cleaning, teaching, running a grocery store, and frying potatoes at the local fast-food outlet.

The sector approach to economies allows us to talk about differences among countries as well as regional differences within countries. For example, preindustrial societies (such as some countries in Latin America) are often characterized as having one big primary sector but little else—they produce raw materials for industrial nations to use in manufacturing. Historically, the American colonies served this role. A major change occurred, however, when the North entered the secondary sector by turning to manufacturing while the South remained a supplier of raw materials. But economic change did not stop there. When we speak today about industrial restructuring, we are referring to a decline in manufacturing coupled with a rise in the service (or tertiary) sector. As we have seen, much greater wealth is generated in the secondary and tertiary sectors.

Changes in U.S. Manufacturing and Income Equality Since the 1970s

The United States has long been an industrial nation, but its industrial focus has changed at times. Just as it was once more profitable to import raw materials into New England than to create them, it would later become more profitable for the entire United States to move away from the primary sector to better develop its secondary and tertiary sectors. During the 1940s and 1950s, no nation could compete with the manufacturing might of the United States, which created great wealth and many jobs. This economic concentration gradually changed so that since the 1970s, it generally has been more profitable to import most manufactured goods into the United States than to make them here. Inexpensive labor, especially in newly industrialized countries around the world, made that change possible. As the secondary sector declined, the tertiary sector grew: every closing factory was replaced by the opening of a new law firm or investment brokerage house. Economic opportunities still existed, but there were *fewer* openings for the unskilled employee and *increasing* opportunities for skilled and educated workers (Caputo, 1995b; Wacquant and Wilson, 1993).

Manufacturing jobs are neither glamorous nor easy, but they do create wealth and allow young, unskilled workers entry into stable and adequately paid employment. U.S. factories and the workers who flocked to them created all the large American cities as people and jobs concentrated. They created opportunities for immigrant workers looking for a foothold, and these opportunities also were sought after by longer term American minorities who migrated to cities in hopes of greater advancement. In short, the secondary sector provided a solid middle to the American economy at the end of the nineteenth century and throughout most of the twentieth century. The growth of this sector began in the Northeast and spread to the north central United States, only more recently developing in the West and the South. However, just as those factories once provided the steam for the chilly northern "frost belt" of the United States, their increasing abandonment since the 1970s brought a new term into American use—the "rust belt."

Industrial restructuring, sometimes termed deindustrialization, is an apt description of the economic change that occurred in the United States during the last quarter of the twentieth century (Susser, 1996). Although the change was most pronounced in the northeast and north central regions, it also was evident to some degree in cities such as Los Angeles and Miami (Morales and Ong, 1995; Pérez-Stable and Uriarte, 1995). As a general rule, the restructuring involved several simultaneous changes. First, manufacturing jobs declined as a percentage of the overall U.S. workforce as more goods were imported. Second, the remaining northern manufacturing moved to the South and West in search of lower wages and fewer unions to maintain profits. Third, central city manufacturing jobs in all regions tended to move toward suburban areas. The second and third changes had drastic effects on residents of northern central cities, many of whom were minorities by the 1970s. Before we turn to their situation, however, we need to look at the decline in manufacturing jobs, which had a significant impact on all Americans, regardless of ethnicity or region. A decrease in middle level employment coupled with more high-end and low-end employment moved the American population toward greater economic inequality (Sheak and Dabelko, 1993).

Figure 9.1 divides the American population into five equal parts based on income, each part consisting of 20 percent of the total population. The first 20 percent represents the poorest 20 percent of the people and the last 20 percent reflects the richest. Those segments in between make up the other three categories. In a hypothetical society with total income equality, each 20 percent segment would receive a total of 20 percent of the income. Figure 2.1 in Chapter 2 presents the 1997 results from Figure 9.1, showing that the American stratification system is far from equal. By adding data from 1950 to 1997, we can see the results of industrial restructuring over time. Most of the population segments changed little during most of these years, at least up until 1970. The only pre-1970 changes of interest are a slight increase in percent of income for the very poorest and a slight decrease in percent of income for the very richest. From 1950 to 1970, the United States moved slowly toward greater income equality. Since 1970, however, the poorest 60 percent of the population has steadily received less of the available income while the top segments have increased their share. The richest 20 percent received close to half of all income in 1997 with the richest 5 percent enjoying 20.7 percent of all income.

Finally, we return to the poverty rate. Figure 9.2 displays the actual number of Americans living below the poverty level between 1959 and 1997 as well as changes in the rate of poverty over those years. Because the American population was growing during this time span, the actual number of poor people could be expected to rise at the same rate as overall population growth without any change in income inequality. As such, it is not all that useful a statistic, except for the 1960s when the population was rising but the number of poor people was dropping. This drop is of course reflected in the poverty rate decline during those years, which tells us the percentage of the population living in poverty at any given time. As with the earlier figures, Figure 9.2 reflects the late 1960s growing income equality and shows its impact on lowering poverty rates. Since 1974, however, that rate has been on the rise—understandably the result of even less income in the poorest sector of the population.

FIGURE 9.1

FAMILY INCOME AS PERCENTAGE OF INCOME IN THE UNITED STATES, 1950–1997.

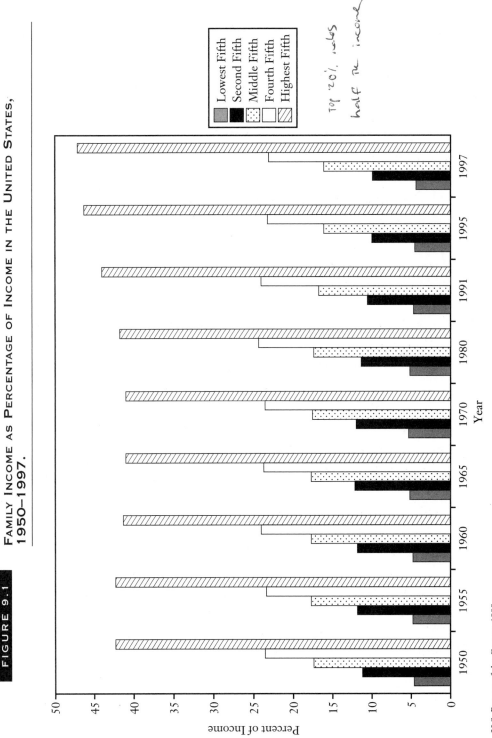

U.S. Bureau of the Census, 1999a.

NUMBER OF PERSONS IN POVERTY AND POVERTY RATES IN THE UNITED STATES, 1959–1997.

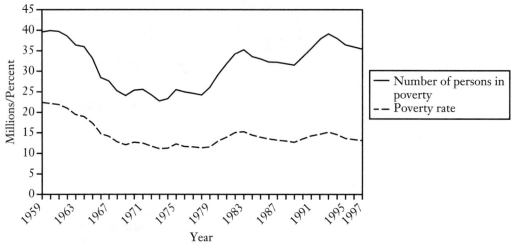

U.S. Bureau of the Census, 1999a.

FIGURE 9.3   NUMBER OF PERSONS BELOW THE POVERTY LEVEL BY YEAR.

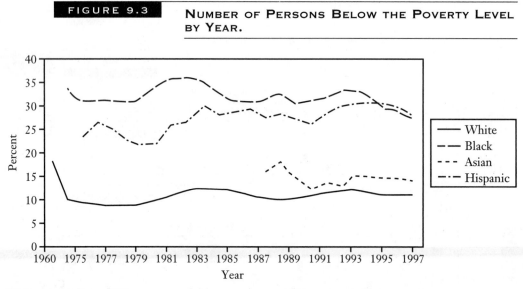

U.S. Bureau of the Census, 1999a.

Figure 9.3 shows how changes in the poverty rate since 1960 are distributed among different racial and ethnic groups in the United States. The mid-1970s—a period of low poverty—were generally good for all groups. Even for whites, no period since 1960 has produced a smaller percentage of that population below the poverty level. The period of the 1990s is often described as one of unprecedented economic growth, but it has not affected the lowest portion of the white population that greatly. By contrast, both black and Hispanic Americans seem to bear the brunt of economic cycles, being hit severely by the recession of the 1980s. While both groups have benefited from the late-1990s economic growth, their poverty rate runs two and one half times that of whites. Finally, the Asian poverty rate should be emphasized. Although much lower than for either blacks or Hispanics, it shows the dual character of the Asian American community in which individuals either fare better than the norm or noticeably below it.

The Two-Tiered Service Sector

The service (or tertiary) sector of an economy contains an extremely wide range of activities, differing in skill level, income level, and social prestige. Corporate lawyers and hotel housekeepers both belong in this category. Neither creates raw material or a product. Both are employed to provide an intangible service. They also have another connection: coal miners and machinists do not typically travel in the course of their employment, but lawyers do. More lawyers therefore create more hotels and more hotel housekeepers. As the high-end service sector grows, those so employed require many kinds of low-end services to aid them in producing their high-end service. Bankers require the help of clerks, and physicians could not function without the hospital housekeeping staff. When high-end service increases, low-end service increases.

The change from manufacturing to service is important to the overall economy because middle-tier jobs are missing. Compare a bank with a factory. A factory hires low-level service (janitors, etc.), middle-level operatives (the old "blue collar" jobs), and upper-level service (managers and executives). Both of the lower tiers requires fairly minimal skills, and the middle tier provides stable and well-paid employment. A bank, on the other hand, contains no level that compares with a factory's middle tier. Industrial restructuring produces an increasing polarization of the population into highly skilled, well-paid service workers and unskilled, poorly paid service workers.

This change in employment opportunity happened quickly and made a significant impact. Between 1960 and 1970 in eighteen older northern cities, 750,000 blue collar jobs disappeared, while professional, technical, and clerical jobs increased by 300,000 (Wilson, 1981). Between 1979 and 1989, 4 million operative and 1 million laborer jobs were lost to the economy, while professionals gained 1.4 million jobs, sales jobs increased 5.3 million, and manager positions increased by 250,000 (Fasenfest and Perrucci, 1994). The racial component of this change can be seen in the general occupational category of "sales," which contains both lower tier and higher tier positions. When European American workers shifted from "operatives" to "sales," they gained an average $7,000 in yearly income; when African American workers made the same change, their yearly income dropped an average of $2,000. This change in the overall structure of the American economy explains much of the increasing income inequality we examined in the last section.

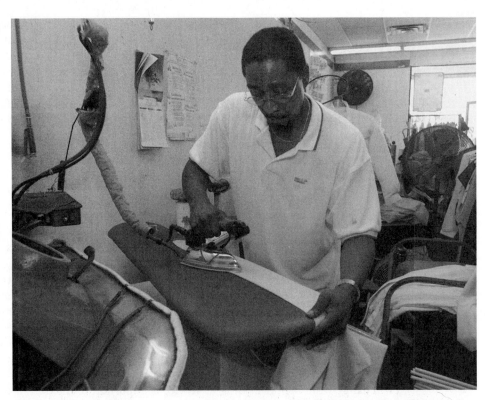

The "two-tier" service sector means that service occupations tend to be either low-level or high-level, with little middle ground.

Urban Unemployment: The Growth of the Underclass

The overall decrease in manufacturing and increase in service is the first of three changes involved with industrial restructuring. The second and third changes both involve the geographical locations of different jobs. Remaining manufacturing has moved either to suburbs or the Sun Belt, while service has increased in all areas. In particular, service has become a very important sector in the urban economies of the United States as factories have been replaced by office buildings. The purpose of this section is to examine employment geography. We need to know where different employment opportunities are located and where prospective employees live and how easy it might be for them to move should they live in the wrong place.

The Modern Urban Mismatch: Housing, Jobs, and Transportation

Between 1953 and 1986, New York, Baltimore, Boston, and Philadelphia lost 1.35 million jobs in manufacturing and the wholesale/retail trades; during those same years 420,000 new jobs appeared in Atlanta, Houston, Denver, and San Francisco (Kasarda, 1990). Those lost jobs came primarily from the central areas of those four northern cities. Although each suffered net losses of employment, their suburbs increased in employment opportunity in all sectors. The new jobs opening in those northern central cities were managerial, professional, and high-level technical jobs (Kasarda, 1993, 1995; Levy, 1987).

Just as urban centers grew in the nineteenth century as workers followed jobs, many late twentieth century northern urban workers moved to the suburbs and the Sun Belt. Those who moved were those who (1) had the financial resources to move, (2) were not tied to their current location in any significant way, and (3) already had the skills and/or credentials to make them eligible for higher level service occupations. Of these, the only vague condition is the second because "ties" to a community can come in many forms. Individuals can be tied by family obligations, dependent on others because of ill health, or a wide variety of factors (see Sandefur, 1986). Poorer people are more likely to have such connections and needs (which we discuss shortly), but they also lack the financial resources and needed skills to move (Murphy and Welch, 1993).

The upshot of all this is that central cities in the northern United States became largely poor and nonwhite when those who could leave did (see Jargowsky, 1996). Of European Americans who did not leave, there is evidence that downwardly mobile workers displaced African American workers in the declining manufacturing sector (Bates, 1995). In 1972, central cities housed one-third of the nation's poor; by 1985, that proportion rose to half. Two-thirds of this overall increase occurred in New York City, Chicago, Detroit, and Philadelphia. Of these new urban poor, less than 10 percent were European American and 70 percent were African American (Kasarda, 1992).

As we have seen, central cities still provide service employment. Cohn and Fossett (1996) point out that central city African Americans are every bit as close to employment opportunity as European Americans. But they also observe that the jobs available to such unskilled workers include "physical plant workers, drivers, security agents, food-service workers, and cleaning service workers" (Cohn and Fossett

1996:564). With few exceptions, these jobs are poorly paid, lack advancement potential, and many are short on fringe benefits such as health insurance. The only options for poor, unskilled, central city workers are either to increase their skill levels or to make manufacturing jobs accessible to them. Accessibility requires that people move closer to the jobs (now in the suburbs), the jobs move closer to the people (back into the central cities), or better public transportation be created.

Can the people move? As noted earlier, those with resources generally do. In a study of the San Francisco bay area, socioeconomic status was a strong predictor of residential relocation for African Americans, Hispanics, and Asians, although somewhat less of a factor for African Americans (Massey and Fong, 1990). This study and others suggest that more than just economics plays a role in the continued housing segregation of African American and Puerto Ricans (Massey and Eggers, 1990). Both groups face housing discrimination, either directly from renters or through the "steering" of minorities into minority neighborhoods by realtors (Massey et al., 1994; Turner et al., 1991; Yinger, 1991). Another important factor is the high level of female-headed, single-parent families in both of these groups, which tends to lower income and labor force participation (Massey et al., 1994; Rodriguez, 1992; Santiago and Galster, 1995; South and Crowder, 1997). There is growing evidence that African Americans are the least likely to move out of housing segregation and the most likely to move into it of all American minorities (Massey and Eggers, 1990; South and Crowder, 1997). Some efforts, such as Chicago's Gautreaux experiment, have attempted to break up the massive central city housing projects and relocate their residents to smaller and more suburban locations; the effects on education and employment, especially in the next generation, were clearly positive (Rosenbaum, 1991).

Could the jobs move? Location is a critical decision in any business and a wide variety of factors enter into the final equation. Important variables include property tax rates, the availability and stability of an adequate labor source, the availability of quality housing and schools, levels of crime, and many other similar concerns. Cities can manipulate tax rates to attract business, but they can do little about most of the other considerations. The federal government attempted to provide additional economic attractions for business relocation by establishing "enterprise zones" in central cities, but most businesses remained unattracted (see Fasenfest and Perrucci, 1994).

Could transportation be provided? Public transportation is once again appearing in major American cities as the American love affair with the automobile has cooled somewhat during commuter traffic jams. The primary factor in such a change, however, came from the numbers of high-level service workers whose jobs were in the cities but who preferred to live in the suburbs. Few American cities provide effective and low-cost transportation for central city residents who seek employment in outlying areas (Blackley, 1990). The general assumption is that automobiles will be involved at the suburban ends of such systems.

The picture painted in this section is not very promising. Even if the logistic problems could be solved, the issue of dwindling manufacturing jobs for largely unskilled workers remains. Although education is a necessary piece of this puzzle, the economic isolation described here has attracted major attention from social scientists in recent years. Much of that attention focuses on the idea of an American underclass.

The Urban Underclass

The idea of an "underclass" in an economy has been around for some time. It has been the object of considerable attention, however, since it was brought into the public forum by sociologist William Julius Wilson (1987; 1993). Wilson applies the term specifically to individuals who reside in central cities (especially those in the North) in neighborhoods of concentrated poverty in social isolation from other groups, who maintain marginal (or non-existent) ties to the labor force, and who share a culture (or "social milieu") that reinforces their labor force marginality (Wilson, 1993). Wilson's initial interest was to explain modern African American urban poverty. The term has since been widely applied to Puerto Ricans living in the Northeast (Tienda, 1993). Their urban similarity to African Americans provides a logical explanation because Puerto Ricans declined more economically during the 1970s and 1980s than did other Hispanics (Tienda and Jensen, 1988). The critical limitation that prevents expansion of the term to include all people in poverty is the variable of housing segregation; the underclass, by definition, are locked into urban neighborhoods where joblessness is endemic. As we have seen, there is racial and ethnic variation in the degree to which this occurs.

The idea of underclass raises several issues. One of the more explosive points is the assertion of a culture or milieu that supports a life away from the labor force. Elements of such a culture might include a willingness to live on public assistance, membership in the underground economy (notably trafficking in illegal drugs), high levels of gang membership and violence, and an acceptance of single-parent families with never-married mothers. Wilson perceives these behaviors and values as an adaptation to the hopelessness of ghetto poverty. On the other hand, none of them seems conducive to reattaching to the labor force should that become a future possibility.

The issue of a poverty culture dates back to the 1960s. Originally termed the "culture of poverty," the suggestion was made that poor people develop lifestyles (or a culture, if you will) that emphasize immediate gratification, fatalism, and a few other traits that would appear to help keep poor people poor (Lewis, 1965). If you do not sacrifice in the present (save your money, work for an education, and so on), how can you expect to reap rewards in the future? The counterargument then—similar to Wilson's now—is that sacrifice in the present is irrational if no reasonable possibilities exist for cashing in down the line (see also Morris, 1989; 1996). The dispute finally focused on the permanence of such cultural traits. Those who supported the "culture of poverty" in the 1960s argued that such traits were transferred from generation to generation, permanently keeping poor people in a lifestyle that would lead them straight into poverty. Wilson's point is that such behavior is a response to poverty (made more extreme by social isolation) that would vanish in the face of employment opportunities.

A second point of conflict with Wilson's theory concerns his perspective on middle-class African Americans. The social isolation of the underclass is intensified, he argues, by the departure of middle-class African Americans for nonsegregated housing. Individuals who could have been leaders and role models (termed "social buffers" by Wilson) are now absent from the central cities. Unlike Harlem in the 1920s, modern segregated neighborhoods do not have the social class diversity that more extreme housing segregation once created.

People are poor because jobs move!

The loss of manufacturing jobs from the Northeast to the Sun Belt has resulted in fewer opportunities for skilled employment.

Massey and Eggers (1990) acknowledge the high levels of residential segregation among African Americans. They argue, however, that the out-migration levels of middle-class African Americans have not been studied enough to show that trend to be a cause of segregation. Beyond that, they are unconvinced that the type of segregation plays a major role in economic advancement (Massey et al., 1994). For example, they point out that African American poverty is relatively low in the California cities of San Jose and Anaheim yet middle-class African Americans are more likely to live separately from poorer African Americans there than in Chicago (Massey and Eggers, 1990).

In spite of these disputes, there is no argument that African Americans do live in high levels of residential segregation in areas with few employment opportunities. If San Jose and Anaheim are taken into account, the critical factor may be the lack of employment rather than the levels of segregation or the presence of middle-class African Americans within those communities. Whichever theoretical perspective is more appealing, it means little for the urban poor in the North; Chicago is not likely to turn into Anaheim anytime soon.

Social Isolation and Gender Differences

Wilson notes that the social isolation of the underclass creates an atmosphere supportive of labor force marginality. Poverty affects men and women differently, so, as we might expect, the cultural response to persistent poverty is in many ways gender specific. In particular, we need to examine the reasons for high percentages of never-married single mothers in the African American and Puerto Rican communities. We also need to examine the system of public assistance that often supports these children, the impact of that assistance on urban communities, and the impact on children growing up in these circumstances. In addition, when men are not fathering these children, many are involved with gangs, illegal economic activities, and unprecedented levels of violence.

It is easy to see the direct connection between persistent poverty and adaptive cultural change with men. More so than women, men's sense of efficacy is closely tied to income (Downey and Moen, 1987). In 1970, 9 percent of young African American men between the ages of 20 and 24 were neither employed nor in school; by 1990, that figure rose to 28 percent (McLanahan and Sandefur, 1994). The male role as breadwinner is a dominant feature in American culture. Persistent poverty denies this role to men, forcing them to look elsewhere for self-esteem. Alternative sources can be found with gangs and street violence. Status in a gang or in the community at large can stem from public aggressive posturing and successful fighting (Miller, 1996). This self-esteem comes at a high price: the death rate from homicide of young African American men, ages 15 to 24, is approximately seven times greater than for young European American men (National Center for Health Statistics, 1997).

Another source of self-esteem is fathering children. Poverty prevents the more complete live-in husband/father role. A cultural "stand-in" for many young African American men is the act of impregnating women. Anderson (1993) observes that status increases according to the number of children one fathers and the number of different women involved. Because these young men cannot afford more lasting relationships with women, they tend to bond more tightly with male peers at ages when other young men are turning toward family relationships.

While young African American men are fathering children, young African American women are raising them. In 1993, 19.3 percent of European American children lived with one parent compared with 57 percent of African American children and 31.8 percent of Hispanic children (U.S. Bureau of the Census, 1994b; Bennett, 1995). Only about 3 percent of those single-parent families were male headed. The number of single parent families has been growing steadily for all racial and ethnic groups in the United States for many years. Much of this overall increase stems from rising divorce rates, but in the urban African American population in 1995, 70 percent of all mothers were never-married single mothers; by comparison, 21.2 percent of European American mothers that year had never been married (National Center for Health Statistics, 1997). While never-married single mothers typically have fewer resources on which to draw, all households headed by single mothers face more economic difficulties than two-parent households.

In 1993, 6.5 percent of married-couple households in the United States fell below the poverty level compared with 35.6 percent of households headed by single mothers (U.S. Bureau of the Census, 1995a). These percentages would be even farther apart if not for decreasing family size among female-headed families (Gottschalk and Danziger, 1993). For European American single mothers, only 25 percent fall below the poverty line. The rate drops even further—to 18.6 percent—for Asian and Pacific Islander single mothers (U.S. Bureau of the Census, 1995a). By contrast, almost 50 percent of African American and Hispanic households headed by single mothers were below the poverty level in 1993 (Bennett, 1995; U.S. Bureau of the Census, 1994a). Of the three major Hispanic groups in the United States, however, only Puerto Ricans have similar characteristics to African Americans with 40.5 percent of households headed by single mothers; both Mexican Americans and Cubans fall close to European American percentages (Moore, 1989; U.S. Bureau of the Census, 1994a). The similarity here between Puerto Ricans and African Americans is clearly in keeping with the underclass thesis.

The greatest concern with so many children being raised by single mothers is the economic handicap so many such children face. Lack of money obviously perpetuates all of the underclass elements we have explored. The only alternative for such families is public assistance. Such programs include Aid to Families with Dependent Children (AFDC), food stamps, Medicaid, subsidized housing or rent assistance, and Supplemental Security Income. Fourteen percent of Americans participated in at least one of these programs in 1993. The rate of participation rose to 28.9 percent for Hispanics, 35.5 percent for African Americans, and 42.9 percent for persons in female-headed families (U.S. Bureau of the Census, 1996d).

Taking all of these programs into account for 1993, only Alaska and Hawaii provide enough assistance for families to barely reach the poverty level. Few states reach 90 percent; for all fifty states, the median amount of assistance provided is 70 percent of the poverty level (Tonry, 1995). Since the programs are need based, any additional income earned by welfare recipients reduces their assistance. If public assistance leaves a single mother and her family at 70 percent of the poverty level, employment at the low end of the service sector would be pointless unless it made up all of that 70 percent plus quite a bit more because she may be facing daycare expenses and working a job with no health benefits. The only rational course of action, since families do not survive at 70 percent of the poverty level, is to work in the underground economy where earnings are not reported. As many researchers have noted, these programs that were designed as a safety net can easily become a dependency trap for the women they cover (McLanahan and Garfinkel, 1993; Leahy et al., 1995).

Although the greatest concern with single-parent families is economic, McLanahan and Sandefur (1994) explored additional problems for children in such environments that are independent of income. They focus on the effects of single parenthood and family disruption on levels of teenage pregnancy, high school dropout rates, general school performance, and being idle (i.e., not attending school or working). Family disruption refers to such changes as divorce, remarriage, cohabitation, death of a parent, or the addition of a relative to the household. As a general rule, their findings confirm that single parenthood and/or disruptions have negative effects on teenage behavior. Two somewhat unexpected findings—surprising and ironic, respectively—are that the death of a parent does relatively little damage to teenagers and a mother's welfare status has only a minimal effect on rates of teen pregnancy. Most important, however, is their conclusion that single parenthood has clear negative effects on teen behavior that are beyond the effects of income and residential change. They also note, interestingly, that all of these disruptions cause greater changes in European American teenagers than either African American or Hispanic because European American teens with single parents essentially lose the advantages they would normally have.

We have had an opportunity in previous chapters to follow the African American family over time, constantly struggling with adverse circumstances. We also have seen a large number of creative responses—especially by women—that made difficult family circumstances run somewhat more smoothly. In the midst of all these statistics on household composition, we should note one more: African American households have the highest rate (7.3 percent in 1993) of children being raised by neither parent; in 1980, this figure was at 12 percent (Bennett, 1995). Who are these nonparent parents? Many naturally are grandparents, uncles and aunts, and other relatives. Still others, however, are nonrelatives or "fictive kin." Fictive kin have a long tradition in the African American family. Sometimes nonrelatives informally adopt one another for life; at other times, the connection may be more temporary (see Randolph, 1995). In still other circumstances, the help may be in the form of more casual community networking (Stack, 1974). The important point is that people in difficulty are attempting to fill in the gaps that poverty leaves in familial relationships. These connections are also important for understanding lower levels of African American geographical mobility. If your family's life is dependent for survival on many other people, you are effectively tied to the neighborhood where that support exists.

The concern over single parenthood has a rocky history in the social sciences. Much of it began with the 1965 publication of *The Negro Family: A Case for National Action* by sociologist (later politician) Daniel Moynihan. Moynihan examined what were then thought to be high levels of single parenthood in the African American community and their connection with persistent poverty. Many African American leaders in the 1960s responded to the book as an attack, perceiving it to be the "culture of poverty" perspective that blamed poor people for being poor. Moynihan argued this was not his intent and that his focus was misinterpreted. But whether single parenthood is viewed as a cause of poverty or as an adaptation to poverty by women who do not wish to forego motherhood just because they are poor, it does tend now to be a roadblock to economic opportunity for mothers and children.

Race and Ethnicity at the Workplace

Thus far, our attention has been somewhat global, tracing the relationship between large scale economic change and American minorities. The interaction between changes in the job market and minority job-related skills clearly explains a lot about contemporary poverty in the United States. But what happens with those who do acquire job skills and enter (or attempt to enter) the labor force? This section examines the interplay between race and ethnicity and the occupational structure of American society.

The Structure of Employment and Unemployment

You can practically set your watch by the racial and ethnic gaps in United States unemployment rates. Unemployment rates rise and fall with the overall economy, of course, but the rates are always higher for racial and ethnic minorities. In comparing Hispanics to all non-Hispanics in 1983 (when unemployment was high), 16.5 percent of Hispanics were out of work compared with 10.6 percent of non-Hispanics. In a much better economic year—1990—Hispanics dropped to 8.2 percent while non-Hispanics enjoyed a 5.3 percent rate. By 1993, in spite of a strong economy, Hispanics had risen again to

11.9 percent while non-Hispanics were up to 7.1 percent (U.S. Bureau of the Census, 1995a). If you focus on the percentage *gap*, however, you will find that Hispanic unemployment rates outdistanced non-Hispanics in those three years by 36 percent, 35 percent, and 40 percent, respectively. Similarly, black unemployment has been approximately 50 percent higher than white employment, dating back to the 1950s, with an even greater gap for younger people (Bennett, 1995; Bowman, 1991b; Newman et al., 1978). Considering that unemployment rates do not count people who have given up searching for work, the race and ethnic gap is actually greater than it appears.

Tables 9.4 and 9.5 provide some information about the occupational structure in the United States by race and ethnicity. The census categories employed can be a

TABLE 9.4   PERCENTAGE DISTRIBUTION OF MALES, 16 YEARS AND OVER, IN SELECTED OCCUPATIONAL CATEGORIES BY RACE/ETHNICITY, 1993

Occupational Category*	Non-Hispanic White	Mexican Origin	Puerto Rican Origin	Cuban Origin	Central and South American Origin	Black	Asian and Pacific Islander
Managerial and professional speciality	29.2	8.7	15.5	20.3	15.3	14.7	37.8
Technical, sales, and administrative support	21.7	13.8	18.0	20.1	17.9	18.1	24.9
Service occupations	8.8	15.2	22.4	12.5	17.8	19.4	11.7
Farming, forestry, and fishing	3.7	11.9	1.8	2.7	4.3	3.2	2.0
Precision production, craft and repair	18.7	20.5	15.1	21.1	16.7	14.0	11.7
Operators, fabricators, and laborers	17.8	29.9	27.3	23.3	28.0	30.6	11.9

SOURCE: U.S. Bureau of the Census, 1994a; 1995c.
*Occupations are organized as follows:
Managerial and professional specialty occupations:
   Executive, administrative, and managerial occupations
   Professional specialty occupations
Technical, sales, and administrative support occupations:
   Technicians, and related support occupations
   Sales occupations
   Administrative support occupations, including clerical

Service occupations:
   Private household occupations
   Protective service occupations
   Service occupations, except protective and household
Farming, forestry, and fishing occupations
Precision production, craft, and repair occupations
Operators, fabricators, and laborers:
   Machine operators, assemblers, and inspectors
   Transportation and material moving occupations
   Handlers, equipment cleaners, helpers, and laborers

little misleading in some cases. For example, the category "technical, sales, and administrative support" combines well-paid, highly skilled occupations with many poorly paid, semiskilled occupations, and subcategories such as "sales" are very general indeed. Still, the dominance of non-Hispanic whites in the "managerial and professional specialty" categories stands out. Both European American men and women outdistance many minorities in those occupations. By contrast, both male and female minorities dominate the "service" and "operators, fabricators, and laborers" categories, which contain only the lower end of service occupations. These data offer some explanation for both the income differential for minority workers as well as the unemployment gaps: most minority workers are clustered in unskilled, high-turnover sectors of the economy.

TABLE 9.5	PERCENTAGE DISTRIBUTION OF FEMALES, 16 YEARS AND OVER, IN SELECTED OCCUPATIONAL CATEGORIES BY RACE/ETHNICITY, 1993

Occupational Category*	Non-Hispanic White	Mexican Origin	Puerto Rican Origin	Cuban Origin	Central and South American Origin	Black	Asian and Pacific Islander
Managerial and professional speciality	30.9	13.6	18.5	18.4	15.7	20.5	37.8
Technical, sales, and administrative support	43.9	40.7	48.4	49.0	31.3	37.8	36.6
Service occupations	16.0	24.9	19.9	20.1	31.6	27.5	12.6
Farming, forestry, and fishing	0.9	2.8	—	—	0.4	0.3	0.1
Precision production, craft and repair	1.7	2.8	2.4	2.0	1.8	2.5	3.8
Operators, fabricators, and laborers	6.6	15.2	10.8	10.6	19.2	11.5	9.1

SOURCE: U.S. Bureau of the Census, 1994a; 1995c.
*Occupations are organized as follows:
Managerial and professional specialty occupations:
   Executive, administrative, and managerial occupations
   Professional specialty occupations
Technical, sales, and administrative support occupations:
   Technicians, and related support occupations
   Sales occupations
   Administrative support occupations, including clerical

Service occupations:
   Private household occupations
   Protective service occupations
   Service occupations, except protective and household
Farming, forestry, and fishing occupations
Precision production, craft, and repair occupations
Operators, fabricators, and laborers:
   Machine operators, assemblers, and inspectors
   Transportation and material moving occupations
   Handlers, equipment cleaners, helpers, and laborers

Ethnic stratification in the United States was very easy to explain one hundred years or so ago. Any employer could discriminate against any racial or ethnic group without concern; it was perfectly legal. It remained legal up until the 1960s and, for all practical purposes, the 1970s. But those minorities discriminated against in the past were also unqualified for most well-paid, skilled work because of educational discrimination. These intertwined concerns produced the dual labor market and human capital perspectives as social scientists have attempted to unravel the forces that keep minorities down, in or out of the labor force. To what degree is discrimination still an active force in the American labor market? Would minorities attain occupational equality if they had more human capital to spend when arriving at personnel (see Smith, 1999)?

Before answering these questions in the next section, Figure 9.4 might provide some food for thought. Although it is impossible to hold all aspects of human capital constant, formal educational credentials have become increasingly important to opening occupational doors in the American economy. Figure 9.4 allows us to compare the yearly income of workers according to their race and ethnicity, their gender, and their years of education. As is obvious, education is the strongest predictor of income; the more of the first you have, the more of the second you will get. Second, it is obvious that gender is also extremely important. At the same levels of educational attainment, men of all races and ethnicity earn more than women (see Durden and Gaynor, 1998). White women with college degrees just barely earn more than white men with high school diplomas. Third, race and ethnicity is a factor. Keeping both gender and education constant, whites earn more than either Hispanics or blacks. This is much more true for men, however, than for women. Education brings rewards to white men over nonwhite men in increasing amounts—the more educated white men are, the better paid they are.

Before leaving Figure 9.4 (and this section) behind, we should note some alternative ways of viewing these same relationships. Figure 9.4 is limited to year-round, full-time workers. That characteristic is more likely to apply to men than women and to whites more than nonwhites (Iverson, 1995; Kilbourne et al., 1994). In short, more poorly rewarded women and minorities are not included in these data, giving white men the strongest competition we might give them. Another possibility would be to limit the age range of people included to just those between the ages of 25 to 34, for example, which would eliminate the advantages white men have gained through being in the labor market longer with fewer interruptions in their careers. The seniority they would gain should have an impact on their relative ranking. Surprisingly, limiting the ages of workers to only younger workers has virtually no impact on the overall outcome of Figure 9.4.

Overt and Institutional Discrimination in Employment

Economic advancement for minorities in the United States is probably best viewed as jumping a series of hurdles that are aligned in a very specific order. If the first or second is not cleared, the height of the seventh hurdle matters little. A child who is the offspring of a nonwhite, poor, central city, single mother on public assistance and who

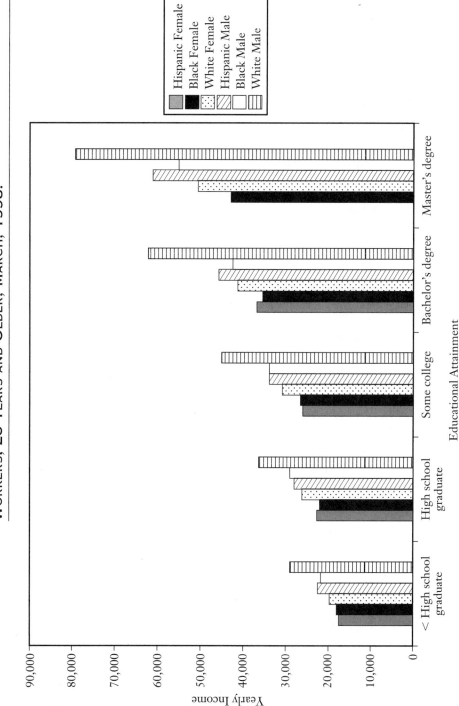

FIGURE 9.4 MEAN INCOME BY EDUCATIONAL ATTAINMENT FOR YEAR-ROUND, FULL-TIME WORKERS, 25 YEARS AND OLDER, MARCH, 1998.

Day and Curry, 1998.

faces daily violence and an inadequate school system will probably never become concerned with the test scores needed to enter an M.B.A. program. If he or she should surmount all those obstacles and enroll in that graduate program, then achieving employment and gaining promotions become issues. Although this is certainly an extreme example of social mobility, other hurdles lie in wait for much lower prestige employment. When we speak of discrimination in the workplace, one of the most difficult tasks is to keep our attention focused on only the latter hurdles. If discrimination occurs in the American system of education, it does not belong in this discussion. When education (and, more generally, human capital) issues arise, the concerns should be as follows: (1) Is educational attainment assessed differently according to the race, ethnicity or gender of the attainee? (2) Do we possess less tangible forms of human capital (e.g., family cultural background) that are independent of our educational attainment? (3) Are human capital requirements for employment and promotion clearly job related?

Educational Attainment, Occupational Traditions, and Discrimination   Figure 9.4 suggests that educational attainment somehow looks different on each person who attains it. Why else should white males get so much more return on their investment, even before they have worked long enough to achieve promotional advantages, if any? There are many facets to this explanation, but most of them wear the cloak of overt discrimination.

One of the strongest forces operating in the American occupational structure is the force of history. As a general rule of thumb, "If men do it, it's more important." In our earlier look at American history, men generally received greater occupational rewards than women *within* each racial or ethnic group in almost every period. As women moved increasingly into the labor force, they tended to be channeled into specific occupations that were quickly dubbed "women's work"; we also have seen minority males channeled into this employment sector throughout American history. In particular, any occupation with a nurturing aspect not only was set aside for women, but also it received an even greater penalty in wages. This is as true today as it was in 1900 or 1800 (Kilbourne et al., 1994). So when comparing college graduates, it also is necessary to realize that men are more likely to be receiving degrees in the more highly rewarded "male" occupations, while women are more likely to be majoring elsewhere. And why do they so choose, knowing the end results? This question leads us down the many paths of cultural socialization—a force that turns us in so many different directions. To understand where the engineers come from on graduation day, you must first ask which first-year students are most likely to have the mathematics and science background for such a program of study. Second, you must ask yourself why.

Institutional discrimination enters into the hiring equation through networks. Networks refer to having connections with people "on the inside" of an organization that may be hiring and/or connections with outside people who can exert influence on the insiders. The adage that "who you know" will determine how far you get still largely holds true at all levels of the occupational structure, even the lower end (Bowman, 1991a; Hondagneu-Sotelo, 1994; Stanton-Salazar and Dornbusch, 1995). As we will see in the following section, affirmative action programs have attempted to minimize the importance of networks in employment. Those higher in social class definitely have connections to more influential networks, which tends to perpetuate the status quo.

Beyond different fields of study in formal education, are racial, ethnic, and gender biases built into employer perceptions? What is the stereotypical image of a higher level private sector business manager? That image probably does not wear a dress or have nonwhite skin. Such perceptions among persons in charge of hiring or promotion can have a significant impact on minority careers. In what has been termed the "glass ceiling," minorities and women may hit an invisible barrier to promotion, which keeps the top positions in public and private organizations largely white and largely male (Reskin and Padavic, 1994). Smith (1997) found that promotions involving greater authority and responsibility were viewed more positively and were more highly rewarded when filled by white employees rather than black employees. Necessary qualities for promotion are usually even more diffuse than requirements for initial hiring. When overt discrimination exists in promotions, it is often hard to see anywhere but in its results. And even if top positions seem filled with the same kinds of people, the explanation typically offered is qualifications.

Intangible Human Capital   Intangible human capital refers to personal qualities, skills, and abilities that are usually highly related to educational attainment but often exist independently of them. It is best thought of as human capital without the formal credential. For example, growing up in a middle-class, European American, highly educated family exposes a child to experiences, ideas, modes of communication, and styles of personal interaction that will have an impact on anything he or she does (Mohr and DiMaggio, 1995). As noted, that background not only makes educational attainment easier, but also it could determine which of two college-educated individuals receives a promotion. The individual with those additional skills and abilities (ones we have all been taught to respect) might easily be perceived to be the more valuable worker.

How do we measure the intangible? Such a question may appear unanswerable, but people and social scientists do it all the time. The nonscientific measure is simply the impression one gets of another through interaction. If you walk away impressed, the other individual undoubtedly displayed many positively regarded qualities during the interchange. If you are a social scientist (or an employer searching for a systematic way to handle promotions), you might well devise a test in an attempt to measure these qualities. Not surprisingly, those who do well on such tests in formal educational environments also do well on them when applying for promotion.

A focus on intangible human capital automatically establishes a research agenda in which employee attributes are compared with the effects of discrimination, either overt or institutional. Much of the research on human capital depends on employee scores on standardized knowledge and skills tests such as the Armed Forces Qualification Test. Farkas and others (1997) note that differences in such scores explain much of the race and ethnic wage gap but do not explain the gender wage gap (see also Farkas and Vicknair, 1996). Kilbourne and coworkers (1994) find that African Americans tend to be concentrated in employment requiring less cognitive skill than their level of education would normally warrant. The latter finding, however, raises more questions than it answers. Are they in those jobs because their intangible human capital does not match their formal credentials or because they have faced discrimination?

Cancio and associates (1996) argue that "cognitive skills" as measured by standardized tests fall under the heading of institutional discrimination. Such tests, they maintain, are biased toward the knowledge and skill base of middle-class, European Americans and will always elicit lower scores from more culturally diverse populations. From their perspective, measures of cognitive skills are simply "proxies" or stand-ins for race in determining who gets ahead in the workplace (see also Maume et al., 1996).

Returning to Figure 9.4, which served as the springboard for this section's discussion, it appears that education does not buy success equally for different kinds of workers. The results depicted in Figure 9.4, particularly with regard to explaining race and ethnic income differentials, have appeared regularly in research that assesses the role of discrimination (Hsueh and Tienda, 1995; Thomas, 1993; Thomas et al., 1994). If discrimination exists, the next question concerns its type—is it overt or institutional? And if it is institutional (as in the case of a promotion denied because of a lacking cognitive skill), can that requirement be justified in terms of necessary job qualifications? These are just some of the questions that politicians and judges will have to wrestle with in the name of equal employment opportunity.

Equal Employment Opportunity: The Promise and the Complaints

Discussions about overt as opposed to institutional discrimination are a product of the modern era. When overt discrimination was perfectly legal and practiced regularly, there was really little point in discussing other hurdles. In the early twentieth century, Booker T. Washington was convinced that African Americans who were educated and culturally adapted to European Americans would be accepted. They were not. Ultimately, the Civil Rights Movement of the 1950s and 1960s would lead to changes in the law regarding overt discrimination, but attempting to remove the main hurdle only served to call attention to many other smaller hurdles that had been in place for many years.

Equal Employment Opportunity

The U.S. Congress passed the Civil Rights Act of 1964 in an effort to end overt discrimination on the basis of sex, race, or ethnicity. In particular Title VII of that law (Box 9.1) specifically prohibited discrimination in employment and promotion. Congress also created the Equal Employment Opportunity Commission (EEOC) to enforce Title VII. President Johnson added Executive Order 11246 to this legislation, prohibiting racial discrimination by federal contractors. (Since virtually every large organization, public or private, has economic connections with the federal

TITLE VII OF THE CIVIL RIGHTS ACTS OF 1964

§ 703 [2000e-2]. Unlawful employment practices.

(a) Employer practices. It shall be unlawful employment practice for an employer—

(1) to fail or refuse to hire or to discharge any individual, or otherwise to discriminate against any individual with respect to his compensation, terms, conditions of employment, because of such individual's race, color, religion, sex, or national origin; or

(2) to limit, segregate, or classify his employees or applicants for employment in any way which would deprive or tend to deprive any individual of employment opportunities or otherwise adversely affect his status as an employee, because of such individual's race, color, religion, sex, or national origin.

government, the scope of the executive order was indeed wide.) Throughout the remainder of the 1960s, the EEOC acted as arbitrator when complaints of employer discrimination were brought before it. The EEOC was granted teeth by the Equal Employment Opportunity Act of 1972, which allowed the commission to bring suit against employers. The initial plan for Title VII was to eliminate purposeful discrimination in the workplace. The effects of Title VII would expand, however, as the law changed via interpretation by the U.S. Supreme Court.

One of the most significant cases brought before the Supreme Court was *Griggs v. Duke Power Co.* The Duke Power Company had long discriminated against African American employees, segregating them in specific, low-level jobs in the organization. Following passage of the Civil Rights Act of 1964, Duke opened up new opportunities for African American workers but required high school diplomas and aptitude test scores for those workers to advance within the company. Since African American workers were less likely to have diplomas and tended to have lower test scores, the outcome of this policy was that most African Americans were denied advancement. Overt discrimination had been replaced by institutional discrimination. In its 1971 decision, the Supreme Court ruled unanimously that Title VII covered business practices having the outcome of discrimination *regardless of intent* unless the practices could be proven to be an essential "business necessity." In the case of *Griggs*, the Court felt that neither a diploma nor a high aptitude test score was a business necessity for those particular jobs (see Blumrosen, 1994; Rose, 1994).

Griggs opened two huge issues in widening the scope of Title VII. First, discrimination was broadened to include discriminatory results, regardless of purpose. Purposeful discrimination is extremely hard to document, whereas discriminatory results are obvious. This change would greatly increase the number of potential court challenges. Second, the idea of "business necessity" was introduced. Some requirements may seem ridiculous in relation to given job descriptions, but where does one draw the line and who is to draw it? The Supreme Court made it clear that organizations should draw those lines, but judges would ultimately determine whether they were drawn appropriately.

Title VII got a little more clout from the Supreme Court in *Albemarle Paper Co. v. Moody*. In 1975, the company was found in violation of Title VII, which included a provision that individuals shown to have faced purposeful discrimination should be granted back pay to make up for economic losses they suffered. In *Albemarle Paper Co. v. Moody*, the provision of back pay was expanded to include discrimination where intent was not proven. Once again, employers were reminded that all requirements must meet a clear business necessity criterion.

Affirmative Action

Affirmative action programs paralleled the actions of the EEOC. President Kennedy first used the term "affirmative action" in 1961, but it truly entered the national scene with President Johnson's Executive Order 11246. As the courts were moving toward defining employment discrimination in terms of outcomes, the federal government was mandating active measures from public and private organizations to alter those outcomes. Johnson made it clear that the point of affirmative action was to provide preferences to women and minorities as a means of countering damage done by past discrimination. In a June 4, 1965 commencement address at Howard University, he argued, "You do not take a person who, for years, has been hobbled by chains and liberate him, bring him up to the starting line of a race and then say, 'You are free to compete with all others,' and still justly believe you have been completely fair."

Affirmative action supporters speak out against California Proposition 209, which was passed in 1996.

Since affirmative action was to treat the outcome of discrimination—the absence of women and minorities in the workplace—it would have to go beyond the limits of the EEOC. An organization could argue in court that it would be happy to hire more female and minority electrical engineers if only there were more in the job market to employ. Provided the courts did not take the unlikely route of objecting to a college degree requirement ("business necessity"), the organization could not be held liable for discrimination. Affirmative action, on the other hand, would turn to the schools of engineering at centers of higher education and demand more women and minority graduates. The universities, in turn, would point fingers at secondary schools, complaining that women and minorities were not receiving the proper training to engage in engineering studies. In the midst of this finger pointing, federal government affirmative action guidelines began to change. Originally, the idea of affirmative action was for women and minorities to be educated/employed/promoted over white males, *all else being equal.* The problem was that things were not equal nor were they becoming more equal over time. As pressure to change outcomes grew, affirmative action guidelines began to look more like quotas with the pressure on organizations to find some way to reach those outcomes.

The first real court test of a quota-based affirmative action program came when Allan Bakke, a white male, was refused entrance to the medical school at the University of California at Davis in 1973 and 1974. The medical school admitted 100 students each year on the basis of undergraduate grades and admission test scores. Of the 100 slots, 16 were put aside for minority candidates. As it turned out, many white males did not have high enough scores for 84 regular slots, but they did have higher scores than some of the 16 minority candidates. Bakke sued the University of California.

In *Regents of the University of California v. Bakke*, a greatly split Supreme Court issued a somewhat confusing ruling. The 5-4 ruling in 1978 was that Bakke should have been admitted but that race could be taken into account in university admissions procedures. One response to *Bakke* was to make qualifications less clear. If, for example, an interview is added to any hiring, admission, or promotion criteria, discrimination becomes more difficult to show.

While affirmative action was being hotly debated in public forums, a changed Supreme Court reversed many of the 1970's decisions. In the 1989 *Wards Cove Packing, Co., Inc. v. Atonio*, for example, the Supreme Court reversed *Griggs*. The burden of proof regarding discrimination moved to the employee, and the employer would now face less strict rulings regarding business necessity. Congress responded to this and other similar court decisions with the Civil Rights Act of 1991, which essentially reversed five recent Supreme Court decisions. The Supreme Court followed the 1991 law with yet more decisions making employer discrimination harder to prove for plaintiffs.

As courts and legislators fought, some voters became directly involved. The 1996 passage of California Proposition 209 (the California Civil Rights Initiative) was clearly the largest and most far reaching example. The proposition requires the state of California to *not* use race, sex, color, ethnicity, or national origin in decisions of public employment, public education, or public contracting. After being passed by the voters, a coalition of civil rights groups fought the law, arguing that it left intact other kinds of preferences such as those favoring veterans and people with disabilities. In 1997, the 9th U.S. Circuit Court of Appeals denied the plea, making the proposition a state law. Affirmative action was essentially voted out in California. The battle continues.

Has it Worked? The Problems of Overt and Institutional Discrimination

Title VII of the 1964 Civil Rights Act coupled with the EEOC certainly achieved the goal of driving overt discrimination underground. There also clearly have been gains for women and minorities—especially for women—in education and employment. Beyond that, definitive assessments are difficult to make. Affirmative action programs aided certain groups while creating conflict and generating stereotypes. One particularly difficult stereotype is that many people assume that minorities were hired only because of their race or ethnicity and are by definition unqualified. Many minority students and employees find themselves in very unpleasant circumstances, facing the burden of having to prove themselves worthy. Probably more than anything, however, affirmative action programs have called attention to the problem of institutional discrimination. As we saw earlier in this chapter, the American economy has moved firmly toward increasingly highly skilled jobs. Access to the jobs requires previous access to and attainment of the necessary skills and credentials. That access still occurs along racial and ethnic lines.

Summary

African Americans, Native Americans, and Hispanics are more likely to be unemployed, dominate low-level occupations when employed, have low incomes, and live below the poverty level than other Americans. The gap between these groups and European Americans has altered remarkably little since the 1970s.

A significant factor in this economic gap is the industrial restructuring of the United States. In general, the United States has moved away from the primary and secondary sectors of the economy (raw material extraction and manufacturing, respectively) to the tertiary sector (service). The tertiary sector ranges from lawyers and investment bankers to domestic workers and fast-food workers. Unskilled minorities tend to cluster into the latter jobs, which pay little and have no opportunities for advancement. The key change here is the decline of the manufacturing sector—that part of the economy that previously provided a first foothold in the economy for unskilled immigrants and minorities. One result of this change has been growing income inequality in the United States.

While manufacturing employment has been decreasing overall in the United States, the change has been most dramatic in the cities of the northeast and north central United States. Jobs once available in the central cities of those areas are now either in the suburbs; in other regions of the United States; or, most likely, in other countries. Ethnic groups that are residentially concentrated in those central cities—African Americans and Puerto Ricans in particular—encounter primarily low-level tertiary sector employment in their environment. Poor public transportation makes even suburban employment impractical.

Such central city minorities have been described by some social scientists as the underclass—that group of people who live in neighborhoods of concentrated poverty in social isolation from other groups and who have a marginal attachment to the labor force. The lack of employment has a major impact on families because marriage can be an economic drain, particularly from a woman's standpoint. These social circumstances promote single-parent families. Additionally, the structure of public assistance in the United States (e.g., AFDC, Medicaid) tends to discourage those receiving it from venturing into the mainstream economy while encouraging them to move into the underground economy.

In the mainstream economy, African Americans, Native Americans, and Hispanics tend to face both overt and institutional discrimination. Much of the latter can be summed up under the heading of nonexistent human capital. Human capital consists of a wide variety of skills and abilities, ranging from those formally attained through educational institutions (tangible human capital) to personal qualities, skills, and abilities that are usually related to educational attainment but can exist independently of them (intangible human capital). Style of communication, for example, probably has more to do with one's family and community environment than with formal education, yet it can be a major factor in the outcome of a job interview.

The federal government created the Equal Employment Opportunity Commission (the EEOC) as the enforcement arm of the Civil Rights Act of 1964. The original goal was to monitor acts of overt discrimination. It became clear almost immediately, however, that forms of institutional discrimination could be just as effective in keeping specific minorities out of specific jobs; it was equally apparent that such requirements were springing up with the intention of doing just that. Ultimately, a three-way interaction began involving the EEOC, governmentally mandated affirmative action programs, and the U.S. Supreme Court. Initially, this combination responded forcefully to employers, requiring them to show clear business necessity for any occupational requirements. In addition, employers were held to have discriminated on the basis of discriminatory results even if no intent to discriminate could be shown.

By the 1980s, the Supreme Court changed (in both membership and decisions), taking a narrower view of employment discrimination. In particular, it became necessary for plaintiffs to prove employer discrimination. Simultaneously, local and state laws began to attack affirmative action programs. The best known and most far reaching was the passage in 1996 of Proposition 209 in California, which forbid the state to consider race, sex, color, ethnicity, or national origins in decisions regarding either education or public employment.

Chapter 9 Reading
Living Poor: Family Life Among Single Parent, African-American Women

Robin L. Jarrett

More single parent families are found among African Americans than any other racial or ethnic group in the United States. This chapter examined alternative social science explanations as to why this is true. Robin Jarrett decided it might be more useful to simply ask the women themselves. In the following excerpts from her article, Jarrett interviews young African American women who are single parents. She wants to know about their hopes and dreams and their problems and concerns.

Marriage, the Ideal: "Everybody Wants To Be Married"

Women consistently professed adherence to mainstream patterns. For virtually all of the women interviewed, legal marriage was the cornerstone of conventional family life. Marriage represented a complex of behaviors, including independent household formation, economic independence, compatibility, and fidelity and commitment that were generally associated with the nuclear family. Representative excerpts from group members illustrate:

Independent Household Formation

> We were talking about marriage and all of that...We was staying with his mother . . . I told him we'll get married and we'll get our own place.
>
> He lives with his grandmother. I don't want to move into his grandmother's home. I live at my mother's. I don't want him to move in there. When we get married, I want us to live in our own house, something we can call ours.

Economic Independence

> He asked me [to get married]. . . We never did, It's more like we waiting to get more financial.
>
> Charles, [my boyfriend] be half-stepping [financially]. That's why I'm not really ready for marriage.
>
> I plan on getting married. But I would rather wait. He said he wanted to wait until he made 22. He works two jobs, but he said he want to wait until he gets a better job, where he can support both of us.
>
> He's always nagging me to get married. I ask him: 'Are you going to be able to take me off aid and take care of all four of my children?' So when I say that he just laugh.

Compatibility

I think a person should never get married unless it's for love. [If] you want to spend the rest of your life with that person, you all [should] have a good understanding. If you marry somebody just because you pregnant, just because you have four or five kids by them, or because society or whoever pressured you into it, you goin' to become mean and resentful. And if that person turns out to not be what you thought or that marriage turns out to be something less than you hoped it would be, it's not goin' to be worth it.

I'm not married to him so I can do what I want to do. But when I get married, I can't do it at all. But it's not supposed to be like that. He says: 'I pay all the bills.' But you don't get to boss me.

I don't want to marry him 'cause me and him would never get along; but I like him. You know, I like him a whole lot. But then [my mother] say: 'Well then why you don't a marry him?' [It's] because . . . somehow our waves just won't click.

A lotta' time you can't get along with the children's father... Me and Carmen's father could not get along, point blank. [I]t wasn't the money. It's not 'cause I didn't have a father: he had a father. We came from good homes. We just could not get along. We don't even know how we made the baby. [laughter]

Fidelity and Commitment

If I get married, I believe in being all the way faithful. ·

I want you to take care of me, I'm not looking to jump into bed and call this a marriage. I want you to love me, care for me, be there when I need you because I'm going to be there for you when you need me.

As soon as [men] get married and things change and he's looking for somebody else. Man! Why didn't they find that person before they marry you and you start going through all those changes.

Nita, a mother of two children provided one of the most eloquent statements on the meaning of marriage. She said:

I would love to be married. . . I believe I would make a lovely wife. . . I would just love to have the experience of being there married with a man. I imagine me and my children, my son a basketball player . . . playing for the [Chicago] Bulls. My daughter . . . playing the piano, have a secretary job and going to college. . . Me, I'm at home playing the wifely duties. This man, not a boy, coming home with his manly odors. . . My husband comes home, takes off his work boots and have dinner. . . I would like to have this before I leave this earth, a husband, my home, my car.

Likewise, Charmaine, who despite her own unmarried status, firmly asserted:

I think everybody wants to get married. Everybody wants to have somebody to work with them . . . and go through life with. . . I would like to be married. . . I want to be married. I'm not gonna lie. I really do.

Women, despite their insistent statements concerning the importance of marriage as the cornerstone of mainstream family life, were well aware of the unconventionality of their actual behaviors. Women openly acknowledged that their single status, non-marital childbearing, and in some cases, female-headship, diverged from mainstream household and family formation patterns. Tisha said with a mixture of humor and puzzlement:

Is this what it's supposed to be like? So, I'm going backwards. Most people say: 'Well, you go to school, you get married, and you have kids.' Well, I had my kids. I'm trying to go to school and maybe, somewhere along the line, I'm going to catch up with everybody else.

Natty, the mother of an active preschooler who periodically appeared at the door of the meeting room, further observed:

I really would like to have two children but I'm not married… and I would like to be married before I do have another child… So maybe one day we might jump the broom or tie the knot or whatever.

Sherry's comments were similar:

I wanted to marry him because we had talked about it so long . . . we always talked about it . . . gettin' married, then have our kids and stuff and everything.

Tisha's, Natty's, and Sherry's observations indicate that the desired sequence of events entails economic independence, then marriage, and, finally, childbearing.

Women's observations in this study are consistent with past ethnographic research (Aschenbrenner 1975; Clark 1983; Holloman and Lewis 1978; Ladner 1971; Stack 1974; see also Anderson 1976). Even in Lee Rainwater's (1970) study of the purportedly notorious Pruitt-Igoe housing project in St. Louis, impoverished residents routinely professed adherence to mainstream values concerning marriage and family. He observed:

The conventionality and ordinariness of Pruitt-Igoeans' conception of good family life is striking. Neither in our questionnaires nor in open-ended interviews or observational contexts did we find any consistent elaboration of an unconventional ideal. In the working class, a good family life is seen to have at its core a stable marriage between two people who love and respect each other and who rear their children in an adequate home, preferably one that has its own yard. If only things went right, according to most Pruitt-Igoeans, their family life would not differ from that of most Americans (Rainwater 1970:48).

Marriage, the Reality: "That's a Little White Girl's Dream"

Women were pessimistic about actually contracting family roles as defined in the mainstream manner. Their aspirations for conventional family roles were tempered by doubt and, in some cases, outright pessimism.

Karen's comment reflected her sense of uncertainty:

> I would like to get married one day . . . to somebody that's as ready as I am. . . But it's so scary out here. You scared to have a commitment with somebody, knowing he's not on the level. . . They ready to get their life together; they looking for a future.

Denise and Chandra were more pessimistic about their chances for a conventional and stable family life:

> I used to have this in my head, all my kids got the same daddy, get married, have a house. That's a little white girl's dream. That stuff don't happen in real life. You don't get married and live happily ever after.
>
> It doesn't work that way. Just because you have a baby don't mean they gone stay with you. . . Even if you married, that don't mean he gone stay with you; he could up and leave.

Even Dee Dee's initially firm assertions were laced with doubt:

> I'm goin' to get married one day. I'm goin' to say I know I'm gettin' married one day, if it is just for a month. I'm gettin' married, I know that. [laughter] I know I am . . . well maybe.

Earlier in their lives most of these women assumed that their household and formation patterns would follow conventional paths. Remaining single, bearing children outside of marriage, and heading a household were not foregone conclusions. Rather, pessimism about the viability of mainstream patterns grew out of their first-hand experiences. Women related conflictual and depriving situations that caused them to reassess their expectations.

Andrea described her attempt to forge a long-term relationship and its disappointing outcome:

> I would rather live by myself, me and my two kids, because I used to stay with somebody. . . Me and him did not work out. We used to have to go scrape up some food to eat. I would rather stay by myself.

Both Pat and Lisa recount similar tribulations:

> It makes me angry to think about it. . . I go through changes [with him] and . . . sometimes I just throw up my hands in the air—excuse the expression—I just say, 'Fuck it! Had it! I'm tired! Sometime I say: 'Man disappear!'

[Men cause] a lot of headache and heartache. . . All the time you taking to set that man straight, you could be spending with your child. . . Instead of having time with your kids, you got to get him together.

Kara, like Pat and Lisa, expressed feelings of frustration;

You want to see [men] do something one way and they don't see it that way. They want to do it the way they want to do it. . . You get mad. You frustrated. It's just emotionally draining.

Women's experiences were augmented by the experiences of others. Through the processes of observation of and comparison to older women in the community, younger women gauged their chances of contracting ideal family forms. Comments from Regina and Tennye, respectively, illustrated this:

A good husband has a good job where I can stay home with the family, raise the kids like on TV. But then it's hard. You don't find too many, not like when our mothers was coming up,
   I don't think I'll ever find a husband because of the way I feel. I want it like my mother had it. [My father] took care of us. She been married to him since she was sixteen. He took care of her, took her out of her mother's house. She had four kids, he took care of all the kids.

These comments suggest that even as younger women compare themselves with older women, conventional patterns remain their reference point. Women's views also signal their awareness of declining opportunities for attaining mainstream family patterns within impoverished African-American communities.

Women's first-hand experiences indicate a more general point. Economic forces are not experienced in impersonal ways; nor are they experienced by solitary individuals, as implied by the structural perspective. Economic constraints are instead, mediated through social relationships and interaction processes. Individuals ponder their situations with others in similar circumstances. As a result of his own ethnographic work, Hannerz (1969) critiqued the mechanistic components of the structural argument:

[I]t is made to look as if every couple were left on its own to work out a new solution to problems which have confronted many of both their predecessors and their contemporaries in the black community (Hannerz 1969:76).

His comment also suggests that the generational persistence and reaffirmation of particular strategies occur because the socioeconomic conditions that support them are still operant (cf. Franklin 1988). This point is aptly illuminated by Myesha and Pam, whose circumstances mirrored their mothers':

My father wasn't around, But you know he tried. . . He calls [me] now. Well, with my boy friend, he [may] stay by my side. If he leaves, he just leave …
So, if my mama could do it, I know I can raise Daniel [my son].

My mother had eight of us. I sympathize with what she go through because she doesn't get any help. But she raised us all by herself and we doing okay. It's a lot of women that don't need no man to help raise her kids because I know I can take care of mine by myself.

Economic Impediments to Marriage: "I Could Do Bad By Myself"

The women's own interpretations concerning changes in household and family formation patterns are consistent with the structural explanation of poverty. Economic factors, according to women, played a prominent role in their decisions to forego marriage, bear children outside of marriage, and, in some cases, head households.

Iesha described how economic factors influenced her decisions. She said:

I had a chance to get married when I first had my two [children]. We had planned the date and everything, go down to city hall. . . When the day came along, I changed my mind. Right today I'm glad I did not marry him because he still ain't got no job. He still staying with his sister and look where I am. Ever since I done had a baby I been on my own. I haven't lived with no one but myself. I been paying bills now.

Renee, who was considering marriage to her current companion, also recounted how economic considerations influenced her decisions:

I could do bad by myself. . . If we get married and he's working, then he lose his job, I'm going to stand by him and everything. I don't want to marry nobody that don't have nothing going for themselves. . . I don't see no future. . . I could do bad by myself.

Cheryl echoed her views:

As far as I'm concerned about marriage and kids, I want to be married; but I also want to be married to somebody who is responsible, who can give me somethin' out of life. . . I would like that security.

Pat was even more direct in her preference for an economically stable mate:

If he's out of a job, he can't sit here too long. I can't do it alone. . . I got to see a place where he's helping me. But if you don't help, I got no time.

Tina, who was currently uninvolved ("on my own"), further described the link between male economic marginality and marriage:

I wanted to get married when I first found out I was pregnant, but he didn't want to get married. And I'm glad that he didn't. . . It would have been terrible; he wasn't working. Maybe that was one of the reasons why he did not want to get married.

Other qualitative and ethnographic studies also describe depressing effect[s] that economic pressures have on marriage among poor women and men (Aschenbrenner 1975; Liebow 1967: Hannerz 1969; Rainwater 1970; Stack 1974; Sullivan 1985). The absence of legal marriage or economically stable partnerships, however, did not preclude the formation of strong and stable male-female relationships. Many of the women were involved in a variety of unions. As previously described, some of these relationships were indeed conflictual. Others were remarkably stable, considering the economic constraints that both women and men faced. Several women described long-term relationships, some of which had endured for over a decade.

One said:

I'm not married. I got three kids. But their father is there with the kids. He been there since I was 16. . . I been with the same guy since I was 16 years old and I'm still with him now. I only had really one man in my life.

Another one echoed:

We been together for so many years; I really think we could work it out. . . I go over his house, me and the kids, and stay for weeks. Then we come back home.

Still another one underlined:

I been with my baby's father for 12 years. We still not married, so maybe one day we might jump the broom or tie the knot or whatever.

These comments are important because they identify the existence of strong alternative relationships that are not detected in demographic profiles that recognize only legal marriages. They also confirm the results of earlier ethnographic studies that identify a variety of male-female arrangements that exist outside of marriage (Aschenbrenner 1975; Jarrett 1992; Liebow 1967; Rainwater 1970; Schulz 1969; Stack 1974; Sullivan 1985). Such arrangements varied from casual friendships to fully committed partnerships. The information gathered from the focus groups and the detailed accounts resulting from ethnographic case studies suggest the need to explore the spousal and parental roles that men assume outside of marriage. These arrangements have significant implications for the support and well-being of women and children.

Women's decisions regarding household and family formation patterns were not surprising in light of the economic profiles of potential marital partners. Even when men worked, their employment options were limited. The prospective mates of the women interviewed were generally unemployed, underemployed, or relegated to the most insecure jobs in the secondary labor market. Within the context of the larger discussion on perceptions of social and economic opportunities, women described the types of jobs their male companions and friends assumed. They included car wash attendants; drug dealers; fast food clerks; grocery store stock and bag clerks; hustlers; informal car repairmen; lawn workers; street peddlers, and street salvage workers.

The focus group data thus confirm the structural explanation of poverty and its emphasis on economic factors, such as joblessness. But they also go beyond the primary concentration on the economic instability of men and its consequences for family maintenance. The focus group interviews indicate that women also considered their own resources in addition to those of the men. They assessed their own educational backgrounds, job experiences, welfare resources, and childcare arrangements. For example, women reviewed their educational qualifications and assessed their potential for economic independence.

Educational Attainment

As far as working, I have to be serious. I don't have any skills and I prefer to go to school . . . do something progressive, you know, to try to get off of [welfare].

Now I'm trying to go back to school 'cause when I dropped out . . . I was in the 11th grade and was pregnant. . . I was pregnant with her then, so I had to leave school. . . Now I'm trying to go back to school for nursing assistant, so I can get off all public aid: find somethin' else to do 'stead of being on welfare all my life.

I try to do what I can. And it's hard out there when you dropped out of high school or you may have a G.E.D. And you have a child . . . and then go and try to find a job.

Work Experiences

Contrary to common stereotypes, many of the women had worked. Women's past work experiences served to clarify the limitations of using the types of jobs available to them as a strategy of mobility. The women's comments focused on low wages, job access, and job inflexibility.

It don't make sense to go to McDonald's to make 3.35 an hour when you know you got to pay 4 dollars an hour to baby-sit and you got to have bus fare.

If you going to get something, you need something that's going to pay something, that's going to make a difference and not take away from it. And you know when they had that discussion like that on Oprah [Winfrey talk show], they don't really see that. They tell you get out there. One girl get on there talking about she'll scrub the floor for 3.50 [an hour], but what it's going to do for you? You still losing out. You not bringing in as much as you get if you were at home.

It was too far. . . I would have to get up at 4 o'clock in the morning in order to be at work at seven. [I] leave work at 3:30 and still wouldn't make it home until 8 o'clock. And it was too far when I wasn't making anything. . . I didn't have no time for my kids, no time for myself.

[I] miss[ed] a day on the weekend and they fire[d] me. I didn't understand. They call me, but I wouldn't go back, because ain't no telling when I get sick like I was sick then. I told them no I didn't want it. And I been looking, putting in applications hoping that somebody call.

Welfare Experiences

Welfare, like low-wage jobs, also represented an institutionalized impediment to mobility. The women's comments highlighted the need for benefits, the stigma of public aid, welfare regulations, and their need for childcare.

If [public aid is] going to do something, I prefer if they would take me off but leave my kids on. Because they would need it more and I figure I can take care of myself a little bit more than they can. You need that medical for them.

One reason, seriously . . . that I do not want [public aid] to take my check [is] because I need my medical card. They can take the money, but I need that medical card and I need those food stamps.

You got to go out there on your own not using [your] public aid background . . because a lot of companies not going to hire you because you coming from public aid.

They give you the runaround for nothing. . . This money not coming out... their pockets. . . [I]t's not like it's coming out they paycheck every week. . . It's coming from your parents paying they state taxes. . . You trying to take care of your children the best way you can and this is one of the ways that you can take care of your children.

How you goin' to get ahead? Somebody needs to explain it to me. . . I know a lot of people that graduated from college and stuff, they ain't got jobs. If you do get a job you got to know somebody. . . Soon as you get the job guess who be on your back? Mr. ADC.

They make you go through so many changes . . . so many changes for nothing... When I was goin' to school, they call [me for an appointment. I said:] 'Can I come after I get out of school?' [They said:] 'No, come now.' [I said:] 'I have finals.' [They said;] 'So, come or you will be cut off.'

Childcare Needs

Women, unlike men, had to factor childcare into their work schedules.

Well, I want to wait until my kids get about 5 [to work], so if something's going on [at the baby-sitter's] they can tell me. I don't want to be worried. I don't have nobody. I keep my own kids.

If I want to go out and get a job, I ain't going to pick any daycare in the city, because they ain't so safe either.

I just feel it was harder for a woman . . . with children . . . to find a job. When I was working it was always Keisha [my daughter], this, Keisha, that, Keisha this, that She did this today; she scribbled on my wall. . . So, my mother died. I quit working. . . I didn't have nobody to keep her. And so that was that.

As a result of their limited educational attainment, low-paying jobs, welfare disincentives, and childcare needs, most women came to perceive their economic options as severely limited. Consequently, when women sought other opportunities, they took both men's economic limitations and their own into account.

Alternatives to Conventional Marriage: "You Can Depend on Your Mama"

The focus group interviews expand on the structural explanation of poverty in yet another way. They serve to identify the strategic processes and sequences of events that follow women's decisions to forego marriage, bear children as single mothers, and in some cases. head households. Women responded to their poverty in three ways: they extended domestic and childcare responsibilities to multiple individuals; they relaxed paternal role expectations; and they assumed a flexible maternal role.

Domestic Kin Networks

The extension of domestic and childcare responsibilities beyond the nuclear family represented a primary response to economic marginality. Extended kin networks that centered around women provided assistance to single mothers and their children. For example, LaDawn, whose unintended pregnancy interrupted her plans to leave home, attend college, and get "real wild," described how living with her mother provides valuable support for her:

When your money is gone and you at home with your mama, you don't have to worry about where you getting your next meal from because mama is always going to figure out a way how you can get your next meal. . . And your mama would be there to depend on; you can depend on your mama.

Likewise, Rita, who currently lives alone with her son, also receives assistance from her mother and other female kin. She described the complex, but cooperative pattern, that characterizes the care of her child:

Well, on the days Damen has school, my mother picks him up at night and keeps him at her house. And then when she goes to work in the morning, she takes him to my grandmother's house.

And when my little sister gets out of school, she picks him up and takes him back to my mother's house. And then I go and pick him up.

Sheila, the mother of a preschooler and a newborn and who lives alone, described her situation:

I had a hard struggle. I had to ask my mama for a lot of help. . . I needed help for food. . . to go to school. . . help to watch my kids.

Ebony, who now lives alone, described the childcare benefits of living with her mother:

I'm on my own. . . I wish my mother would come stay with me . . . to help me out. Because when I was at home . . . it was things that she knew that I didn't know nothing about. Why the baby crying so much. Well, you had it outside [the blanket] with no covers on. Letting me know so when the next [child] came I knew not to do this.

Diane also described the childcare benefits of living with her mother. She further hinted how her mother's assistance facilitates Diane's role as the primary caregiver:

My mother gives me good advice . . . if something's wrong. [My twins] had the chicken pox. What am I gonna do?. . . They itching. What should I put on them? She helps me out that way. And I stays with my mother. Me and my mother sit down and talk. We don't have no kind of problems as far as her trying to raise [my kids].

The women's accounts in these focus group interviews are paralleled in similar ethnographic studies. Aschenbrenner (1975), Jarrett (1992), and Sullivan (1985), in their works, highlight the importance of grandmothers, as well as other women kin, in the lives of poor women and children. Grandmothers provide money on loan, childcare on a daily basis, and help with cooking and cleaning. These services allow some young mothers to finish school and get a job, staying off public assistance. Other qualitative studies provide comparable descriptions of supportive kin who provide care for poor children (Anderson 1990; Burton, 1991; Holloman and Lewis 1978; Liebow 1967; Stack 1974; Williams and Kornblum 1985; Zollar 1985). These examples are important in another way. They indicate that households labeled as female-headed are often embedded in larger kinship networks. Interhousehold family arrangements and the domestic activities shared between them are usually overlooked in quantitative studies. Consequently, female-headship as a living arrangement and family as a set of social relationships that may transcend household boundaries are often confounded (Jarrett 1992; Stack 1974; Yanagisako 1979).

10

Health: The Effects of Race, Ethnicity, and Class

An old saying popular with those who study race and ethnicity is that the quickest measure of discrimination is death rates. When members of one ethnic group tend to die at younger ages than the members of another, many of the causes are traceable to discrimination. Premature death may be traceable simply to poverty: the ethnic stratification produced by discrimination puts a higher proportion of a group in the lower classes, and poverty reduces life expectancy. Other causes of premature death may stem more directly from discrimination. Group members may be denied medical care or forced to live in unhealthy environments purely on the basis of their ethnicity (see Downey, 1998). Still other causes of premature death may stem from the interplay of poverty and discrimination, in that group members may adopt unhealthy lifestyles or turn to illegal (and dangerous) occupations as a result of their social position.

This chapter explores death and illness in the United States with an eye toward the effects of race, ethnicity, and social class on health. Obviously, the greatest cause of death is aging. Our concern is exclusively with premature, or what might be termed, "unnecessary" deaths. For example, the incidence of cancer increases greatly with age. In addition, some cancers can be successfully treated while others cannot. If we find certain groups with higher incidences of cancer at earlier ages or having lower survival rates with treatable cancers, we want to know why.

Our quest is not going to lead us down a simple road. As social scientists often lament, people do not live their lives in laboratories, waiting to be studied. In the real world, infinite numbers of factors can interact to produce better or poorer health in an individual. As we proceed to examine different groups, their environments, their lifestyles, and their health, our minimum task is to eliminate less plausible explanations while focusing our attention on the more plausible alternatives.

Death and Illness Rates in the United States: Race and Ethnic Variation

For all Americans in 1997, the single major cause of death was heart disease, accounting for 31 percent of all deaths. The second leading cause was cancer, accounting for an additional 23 percent of all deaths (National Center for Health Statistics, 1999). Although these two classifications actually consist of many different medical conditions, over half of all deaths resulted from cardiovascular problems or malignancies of some kind. Cerebrovascular disease is a distant third cause of death. These three leading causes were the same for both men and women. Beyond these, we do find gender differences in causes of death because men are more likely to acquire HIV infections or to be involved in accidents, suicides, and homicides; women are more likely to suffer from diseases such as pneumonia or Alzheimer's disease.

If we look at the causes of death from the perspective of differences among racial and ethnic groups, African Americans, Hispanics, and Native Americans are more likely to die from HIV infection, unintentional injuries, and homicide and legal intervention. By contrast, Asian or Pacific Islanders and European Americans have virtually identical patterns in causes of death, with injuries, HIV infection, and homicide and legal intervention either low on the list or off the list entirely (National Center for Health Statistics, 1999). However, the cause of death is not our only concern. We also need to examine differences in the ages at which death occurs.

Figure 10.1 offers a global perspective on death in the United States in 1997, comparing death rates by age and race and ethnicity. As with all the figures presented in this chapter, death *rates* are used so that racial and ethnic groups of different sizes can be directly compared. To summarize, death rates for African Americans are far higher at every age range than for any other group. The magnitude of this difference directs the focus of this chapter. Second, Native American death rates always exceed those of European Americans but by much less than African Americans. Third, Hispanics tend to have higher death rates than European Americans at earlier ages but tend to fare better at somewhat older ages. And fourth, Asians and Pacific Islanders have lower premature death rates than any other racial or ethnic group in the United States.

This section offers a brief overview of the data depicted in figures that appear later in this chapter where they are discussed in detail. They are referred to here to illustrate some generalities that can be made about death rates as they relate to race and ethnicity. Figure 10.2 illustrates death rates from the lone contagious disease among these causes—human immunodeficiency virus (HIV) infection. Death from acquired immunodeficiency syndrome (AIDS), a result of HIV infection, has an extreme impact on the African American community, affecting both men and women. Although the primary means of transmission in the United States are sexual intercourse and sharing needles among intravenous (IV) drug users, the disease originated among Americans through homosexual intercourse. Hence, we would expect lower rates for women than for men. The relatively high rates for African American and Hispanic women, especially at the younger age range, suggest a growing increase in the transmission of the disease through IV drug use, either as a mode of acquisition among

FIGURE 10.1 **DEATH RATES FROM ALL CAUSES, 1997**

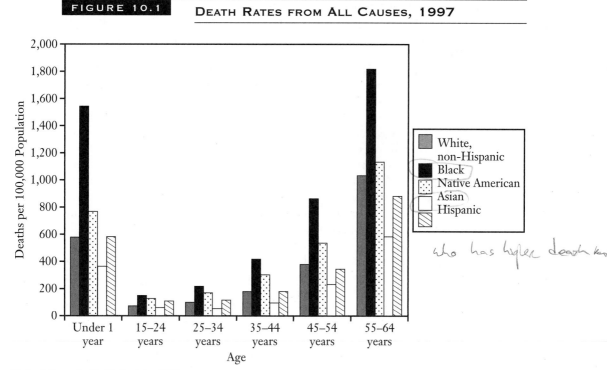

National Center for Health Statistics, 1999.

men and later passed on sexually, or directly acquired through those means by women. The relative absence of this disease among European American women suggests a decreasing attachment between the disease and homosexual activity.

As previously noted, heart disease and cerebrovascular disease are the number one and number three causes of death, respectively, in the United States; they also have a number of risk factors in common. Heart disease achieved its ranking because of its impact on European Americans—the largest single group within this comparison—whose death rates (until the age of 85) are exceeded only by African Americans, especially African American women. The elderly (over 85) are included in Figures 10.3 and 10.4 for two reasons: (1) The low death rates from heart disease among the Native American elderly show that this group is dying of other causes. Although heart disease is still the number one cause of death among Native Americans, they are the only racial or ethnic group in the United States to have death by unintentional accident account for over half as many deaths as heart disease. (2) Note the high rates of death from heart disease for European Americans and Asian and Pacific Islanders among the elderly. One interesting possible explanation for this is that groups with generally good health and good health care are more likely to reach old age; however, they may well be less healthy and strong than the elderly among groups who did not receive good health care and thus had to be strong to live so long (see Kunitz and Levy, 1989).

Cerebrovascular diseases (see Figures 10.5 and 10.6) present an interesting puzzle. Race and ethnicity do not appear to be factors except for African American men whose death rates are either double or triple the national average; African American women are not far behind.

Figures 10.7 and 10.8 provide an overview of cancer death rates from all forms of cancer. Cancer appears to be a considerable problem for European Americans and African Americans, affecting the latter noticeably more for both males and females. As with heart disease, cancer is not a cause of death that singles out Native Americans, Hispanics, or Asian and Pacific Islanders. Figures 10.9 through 10.11 show a breakdown of overall cancer rates into two of its forms. Respiratory cancer duplicates the overall rates of death for men but has both lower and more even rates for European American and African American women; breast cancer rates duplicate the overall death rates. Figure 10.11 is also significant in that African American women actually have a lower rate of breast cancer *incidence* than European American women—101.3 cases per 100,000 as opposed to 115 cases in 1995 (National Center for Health Statistics, 1999).

The figures on cancer death rates do not distinguish between incidence rates and survival rates. For virtually untreatable cancers, such as pancreatic cancer, the distinction is meaningless. But many forms of cancer have higher five-year survival rates through a combination of early detection and appropriate treatment. Table 10.1 shows differences in those rates in selected sites of cancer for European American and African American cancer victims. African Americans have a significantly lower five-year survival rate than European Americans, regardless of sex or cancer site (Chatters, 1991).

Homicide is largely a young male activity, wherever it occurs. The death rates in Figure 10.12 show the concentration of homicides among African American and, to a lesser extent, Hispanic and Native American young men; the same ratio holds true in Figure 10.13 for women in these two groups, but note that the overall number of

TABLE 10.1	FIVE YEAR SURVIVAL RATES (PERCENT OF PATIENTS SURVIVING) FOR SELECTED CANCER SITES, 1989–94.			
Cancer Site	**Whites**		**Blacks**	
	Male	**Female**	**Male**	**Female**
Breast		86.7		70.6
Colon	64.6	63.1	51.4	53.1
Lung and bronchus	13.0	16.5	9.7	13.9
Prostate	95.1		81.2	
Cervix uteri		71.5		59.0

National Center for Health Statistics, 1999.

deaths per 100,000 people is much lower. The gap between whites and nonwhites in regard to the homicide rate has been growing steadily for the last several decades.

Figures 10.14 and 10.15 display death rates from suicide. Again, note that the actual numbers of death are relatively low, especially for women. Nevertheless, some interesting trends beg explanation and are discussed later. Male suicide is highest among European Americans and Native Americans but in contrasting patterns by age. Native American men commit suicide at younger ages, but the rate decreases with age; conversely, the rate of suicide for European American men is lower at younger ages, but the rate increases with age. With women, suicide is virtually nonexistent among Native Americans; the European American and Asian and Pacific Islander rates parallel the European American male rate.

Finally, Figures 10.16 and 10.17 display motor vehicle–related death rates, which include drivers, passengers, and pedestrians. Beyond the connection with age for this form of death, the feature that claims our attention here is the very high death rate among Native Americans; these rates are higher among males than females, but both are in the same ratio to same sex comparisons with other racial and ethnic groups. A large percentage of all motor vehicle deaths are alcohol related; that is certainly true of Native Americans and figures prominently in our discussion later.

As we have seen through the data discussed in this overview, premature death is certainly more common among African Americans, Native Americans, and Hispanics than among European Americans or Asian and Pacific Islanders as Figure 10.1, which began the discussion, made quite clear. But in comparing differing causes of death, very different pictures appear for different groups. The following section provides a little perspective on different ways of viewing health and death and helps structure the explanations that follow.

Theories of Health

Health is frequently conceptualized in terms of four broad categories—human biology, environment, lifestyle, and health care—each of which plays a role in our quality of health (see Young, 1994; Polednak, 1989). First, human biology provides a fundamental starting point. The biology of our bodies predetermines that we will all die through the aging process. That same biology also may bring death or ill health sooner by containing genetic predispositions for various diseases or by providing a suitable home for viruses, bacteria, or protozoa that cause health problems.

Second, the environment interacts with biology in important ways. Contagious disease, for example, requires a carrier or medium for the virus or bacterium to spread. European migration to the New World served to introduce a wide variety of illnesses to the two continents. The carrier or medium also can be an insect, contaminated drinking water, or a wide variety of other sources. The environment plays a role with noncontagious disease as well. Smoking, air pollution, diet, and many other substances have all been connected with health problems such as cancer, heart disease, and cerebrovascular disease.

Third, (and highly related to the environment) is lifestyle. If your daily activities bring you into contact with substances that lessen the quality of your health, lifestyle is to blame. It is probably most useful to think of environment and lifestyle at opposite ends of a continuum, separated by the degree of individual choice involved. For example, Native Americans had no choice when small pox entered their environment: they could not avoid it. On the other hand, it is your choice to work in a factory that contains carcinogenic substances (although you have to work somewhere and your choices may be considerably limited). Cigarette smoking clearly falls at the lifestyle end of the continuum (although the addictive properties of nicotine may make not smoking a hard choice once you have started). Alcohol use is another choice that is addictive. On the other hand, alcohol companies have targeted many minority racial and ethnic groups with specific, culturally connected advertising (Alaniz and Wilkes, 1998). As we proceed to examine environmental and lifestyle causes of illness and death, keep this continuum firmly in mind. Some elements may affect everyone (as with Native Americans and small pox), while other risk factors may be real choices for the upper classes but not options for poorer people. Heating your home with a wood stove, for example, lowers your health through increased air pollution and the danger of fire. If you are poor, however, it may be the only choice available for heat.

Finally, we come to formal health care delivery systems—the world of physicians, nurses, hospitals, health insurance, drugs, and surgery that has health care as its goal. Being examined and treated by health care professionals improves health. Of the four areas discussed here, however, it is probably the least important. Any physician would much prefer proper sewage disposal and clean drinking water to the problems of treating cholera and typhoid. Similarly, an individual with a genetic predisposition toward acquiring cancer or heart disease may gain little ground from either a healthy lifestyle or subsequent medical treatment. Professional health care becomes extremely important, however, when the first three health factors are held more or less constant.

The four broad categories described here provide a basis for structuring our thinking about the causes of poor health. However, the question of how we acquire any particular health-related condition still remains. One of the most general approaches to this question is to organize health conditions into three general types—contagious disease, chronic (or noncontagious) disease, and injury. Contagious diseases include any health problem you can acquire from another person (directly or indirectly) carried by viruses or bacteria, for example. Chronic diseases relate more to internal human biology or to the interaction of a person with his or her environment. In either case, the ailment is not directly transferable to another. This category includes most of the major killers in the United States today such as heart disease, cancer, and so forth. Finally, injury is most related to environment and/or to lifestyle (or the lifestyles of others in the environment). If homicide rates and alcohol-impaired driving are frequent occurrences in an environment, all who live there will undoubtedly experience higher rates of death through injury.

These three categories of health problems also clearly relate to different types of societies. Nonindustrial societies have much higher death rates from contagion and injury. We would expect to find poor sanitation, poor nutrition (which lowers resistance), lack of vaccinations, and other conditions that leave the door open to contagion.

The number of injuries probably also would be higher. As just one example, earthquakes take a much higher toll in nonindustrial societies because of the poorer quality of building construction.

By contrast, industrial societies automatically create environmental changes such as sanitation improvements. Since contagion and injury do not respect age, inhabitants of industrial societies are much more likely to survive the first several decades of life and grow old enough to acquire the more age-related chronic diseases. Also, many of these diseases increase because of environmental changes brought about by industrial societies. For example, industrial societies increase levels of obesity, decrease levels of exercise, move people to diets much higher in animal fats (and away from vegetables), and introduce a host of environmental contaminants. As a result, the chronic diseases are often thought of as the diseases of modernization.

Minority Lifestyles and Experiences: Social Class and Ethnicity

This section examines the interplay among the health of racial and ethnic groups in the United States and explanations for variations that stem from biology, the environment, and lifestyle; the following section focuses on health care delivery systems. Of the three areas covered in this section, biological differences receive the least coverage because we know the least about them. Undoubtedly, part of the research vacuum stems from the many misuses of "race" over the years to explain complex human behavior. Health is a much more limited goal, however, in that (1) health is clearly biological and therefore intertwined with genetics, and (2) human populations that had been separated for centuries did develop biological differences appropriate to their respective environments. An example is the development of the sickle cell blood trait in areas with high levels of malaria, and we should expect to find additional biological differences that have an impact upon health.

Environmental and lifestyle differences have been studied much more fully. Keeping the previously described continuum in mind, we examine various illnesses and diseases in relation to ethnic stratification in the United States. Any ethnically stratified group under study therefore presents two questions at the outset: (1) How many health differences stem directly from the disproportionate poverty they face as a result of being ethnically stratified? The earlier example of the wood stove falls under this category. (2) How many health differences stem from distinctive cultural or lifestyle features of their particular group? Poorer people, for example, have to eat less healthy diets to some extent because healthier food is more expensive. Might a particular ethnic group have strong cultural dietary traditions that lead them to an unhealthy diet regardless of their income? Such interplay among variables is understandably difficult to sort out.

An additional issue in studying environment and lifestyle is discrimination. Environments and lifestyles can be chosen or they can be forced upon people through discrimination. To again use an earlier example, if the only available employment includes an unhealthy environment, does the worker really have a meaningful choice? An additional example is a poor mother who lives in an unhealthy environment and experiences poor nutrition before her child is born. Her child may subsequently be learning disabled all of his or her life. What part of the prenatal condition is chosen? Discrimination also may well produce conditions that are difficult to measure. Does discrimination, by itself, cause stress? And is stress related to hypertension, which is related to heart diseases and cerebrovascular diseases? As is evident, this section deals with some complex issues.

Contagious Diseases

As noted earlier, contagious diseases are not the major killers they once were in the United States. The only contagious diseases to appear among the ten leading causes of death for any racial or ethnic group in the United States are pneumonia/influenza and HIV infection, a contemporary viral infection. Historically, the major contagious killers were bacterial diseases; however, improved sanitation has done much to remove bacteria's mode of transmission, and medical science has been quite successful in finding vaccinations and treatments. The incidence of tuberculosis (TB), particularly newer, more virile strains that may be resistant to standard drug treatment, is increasing and appears to be having a greater impact on minorities (Seligsohn, 1994). The older variety, however, was at epidemic levels on Indian reservations and in urban minority communities a century ago. As recently as the mid-1950s, Native American deaths from TB were still much higher than for European Americans, but the gap has now essentially disappeared (Young, 1994). The incidence of TB among urban minorities showed similar patterns until the newer strains appeared. The incidence of pneumonia among Native Americans continues to be higher than average, especially among younger people. Possible causes are air pollution (wood stoves, smoking [both active and passive]), poor nutrition, and overcrowding (Young, 1994).

HIV infection is a very different matter. As a virus, it is much more difficult to control, and unlike TB, it has no cure. The most common means of transmission for European Americans is through homosexual activity (65 percent). That means of transmission also occurs for Native Americans (59 percent); they have a higher percentage of cases resulting from shared needles among IV drug users (Campbell, 1989a; Centers for Disease Control and Prevention, 1997; National Center for Health Statistics, 1999; Young, 1994). Their overall death rates from HIV infection, as noted earlier, are lower than for European Americans (Figure 10.2). In contrast, the much higher death rates for African Americans and Hispanics in 1997 are more the result of IV drug use, 31 percent and 28 percent, respectively (National Center for Health Statistics, 1999). The illegality of the behavior involved in HIV infection transmission makes intervention in its spread all the more difficult because federal and state governments are more willing to give away condoms than syringes (Stryker, 1989).

Women tend to be evenly split in all racial and ethnic groups between contracting the infection directly through IV drug use or through heterosexual intercourse with an infected partner. This indicates that disease spread through heterosexual transmission is becoming more common, although the numbers of cases for women in this category is still much smaller than total deaths for men (Centers for Disease Control and Prevention, 1997). The continuing link with IV drug use for both sexes

FIGURE 10.2   DEATH RATES FROM HIV INFECTION

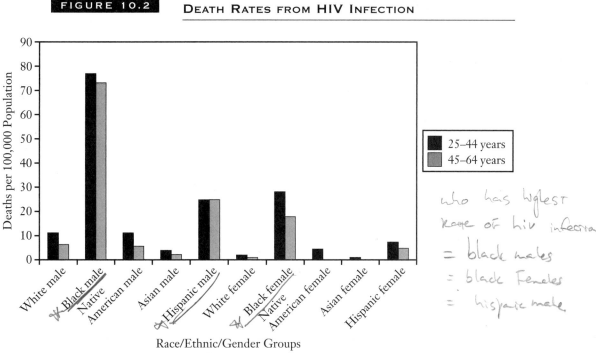

National Center for Health Statistics, 1999.

coupled with the higher use of IV drugs in minority communities has begun to turn AIDS into more of a "nonwhite" disease than a "homosexual" disease in the public mind. Because IV drug use is more an effect of lower class standing than race, ethnic stratification is once again in operation (Chatters, 1993; Quinn, 1993). Nevertheless, deaths from AIDS reached an all-time high in 1995, but dropped noticeably in 1996 (Centers for Disease Control and Prevention, 1997). Although this decrease is probably due to the increased effectiveness of drug treatment, that treatment remains expensive, which means the disease has a greater impact on minority communities.

Infant Mortality

Table 10.2 summarizes infant mortality in the United States for 1996. It is clear at a glance that infant mortality rates for all racial and ethnic groups are relatively similar with the exception of African Americans whose rates are more than twice as high. Infant mortality is commonly separated into neonatal mortality (from birth until 28 days of age) and postneonatal mortality (between 1 month and 1 year of age). As a general rule, neonatal mortality is attributed to something related to the mother and pregnancy (e.g., congenital anomalies, low birth weight, or respiratory distress syndrome), whereas postneonatal mortality is attributed to the newborn's environment

TABLE 10.2	INFANT MORTALITY IN THE UNITED STATES BY RACE AND ETHNICITY, 1996.	
	Neonatal Deaths per 1,000 Live Births	Postneonatal Deaths per 1,000 Live Births
White, non-Hispanic	3.9	2.1
Asian and Pacific Islander	3.3	1.9
Hispanic	4.0	2.1
Mexican American	3.8	2.1
Puerto Rican	5.6	3.0
Cuban	3.6	*
Central and South American	3.4	1.6
Black, non-Hispanic	9.4	4.8
Native American	4.7	5.3

(handwritten annotations: "in womb" above Neonatal; "already born" above Postneonatal)

National Center for Health Statistics, 1999.

*Postneonatal mortality rates for groups with fewer than 20,000 births are considered unreliable.

(e.g., infectious disease and sudden infant death syndrome [SIDS]) (Kramer, 1988). Low birth weights are related to poor prenatal care, alcoholism, smoking, and hypertension; SIDS is more common in low-income households (Kramer, 1988).

If we compare overall neonatal mortality with postneonatal mortality in the United States for 1996, we find that 4.8 neonatal deaths and 2.5 postneonatal deaths occurred in every 1,000 live births (National Center for Health Statistics, 1999). Table 10.2 provides a breakdown of these infant deaths by race and ethnicity. Both types of mortality occur more frequently among African American, Native American, and Puerto Rican infants than with other groups. In particular, African Americans have neonatal rates about twice most other racial and ethnic groups. With postneonatal deaths, African Americans and Native Americans are almost equal: both are about twice the national average (National Center for Health Statistics, 1999).

Table 10.2 suggests that neonatal mortality is largely responsible for the overall high rates of infant mortality among African Americans. If we consider low birth weight as a cause, we find that both African Americans and Native Americans have about twice the national average of low-birth-weight babies (less than 2,500 grams). With very low-birth-weight babies (less than 1,500 grams), however, Native Americans rates are actually a little lower than the national average while African American rates are more than twice the national average. Survival rates for infants weighing less than 1,500 grams are four times lower than for those weighing between 1,500 and 2,500 grams; this is a key factor in neonatal mortality (National Center for Health Statistics, 1999; VanLandingham and Hogue, 1995).

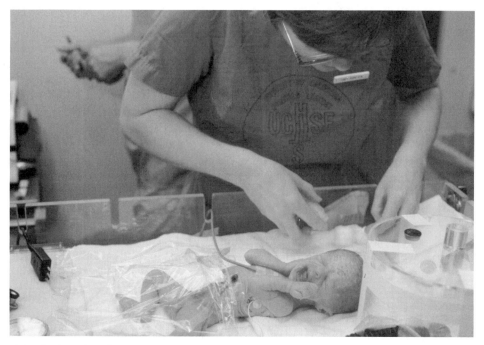

A low-birthweight infant receives hospital care.

If we turn our attention to the causes of low-birth-weight infants, some interesting differences between groups appear. Native American mothers are twice as likely to smoke during pregnancy as African American mothers. European American mothers are 1.6 times as likely to smoke during pregnancy (National Center for Health Statistics, 1997). Smoking among African American mothers seems to be an unlikely overall explanation. Prenatal nutrition is always a likely candidate but is unfortunately difficult to measure. It seems unlikely, however, that African American mothers suffer from substantially worse nutrition than Native American and Puerto Rican mothers. Alcohol consumption is another possibility, but again, this variable is difficult to measure precisely. Grouping all Native American peoples together for census purposes worsens the comparison because variation among peoples for alcohol consumption is extremely high, ranging from well below the national average among the Navajo to well above it for some Northern Plains peoples (Young, 1994).

Hypertension, on the other hand, appears to provide some possible explanations. Native American women have hypertension rates at the national average until they are well past child-bearing years (Young, 1994). African American women, however, are 35 percent more likely to suffer from hypertension than the national average (National Center for Health Statistics, 1999). Is social class a factor? Kramer (1988) points out that Native Americans receive health services directly from the federal government compared with a combination of insured, uninsured, and Medicaid-covered African

American women. On the other hand, while the U.S. census does not collect direct information on social class, some indirect measures such as educational attainment suggest that class may not be the culprit. African American women with more than 12 years of education have only slightly lower levels of infant mortality than those with less than a high school diploma (National Center for Health Statistics, 1999).

As the racial and ethnic variation in infant mortality dwindles, the strikingly high rates among African American infants call for special study. More than poverty is involved. Although we did not focus on Asian and Pacific Islanders in this section because of their low overall rates, the ethnic group among them with generally the lowest socioeconomic position—Indochinese refugees—has lower infant mortality rates than European Americans (Weeks and Rumbaut, 1991). In addition, even the elimination of low birth-weight infants among African Americans would not completely close the gap in infant mortality (Polednak, 1989). With the elimination of so many infectious diseases that used to kill so many infants, most indicators today suggest an investigation of African American women is in order with a focus on biology, lifestyle, and environment in the most general sense of the word (Edwards et al., 1994; Livingston, 1987).

Cardiovascular and Cerebrovascular Diseases

We now turn our attention to cardiovascular and cerebrovascular diseases, which represent two of the top three killers for all racial and ethnic groups, although they affect some groups more than others. Heart disease is the number one killer nationally, claiming 31 percent of all deaths in 1997. It is also the number one killer among Native Americans and Hispanics, resulting in 23 percent and 25 percent of all deaths, respectively (National Center for Health Statistics, 1999). This difference is reflected in Figures 10.3 and 10.4, which show that heart disease claims the highest percentage of its younger victims from African American and European Americans in that order, particularly males. Heart disease also is related to social class: individuals of any race or ethnicity in managerial and professional occupations are less than half as likely to die of heart disease as are operators, fabricators, and laborers (Navarro, 1991). As one of the "diseases of modernization," heart disease claims a higher percentage of European American lives than any other group, but it takes its toll primarily among the elderly; our concern with premature death directs our attention to African Americans.

The above pattern of premature death is repeated in an even stronger form with cerebrovascular disease. For both males and females, African Americans lead all three age groups depicted in Figures 10.5 and 10.6. Although the African American increased rate of death from heart disease is noticeable, their greater incidence of death from cerebrovascular disease is double and sometimes even triple that of other racial and ethnic groups. This extreme difference coupled with many shared risk factors leads to an examination of these diseases together.

How does modernization contribute to these diseases? What factors might contribute to the acquisition of heart and cerebrovascular diseases? Just a short list includes hypertension, diets high in animal fats (elevated cholesterol), lack of exercise, obesity, smoking, diabetes, alcohol consumption, and stress. Note also that many of these factors are interrelated in that stress can lead to hypertension and a poor diet can

FIGURE 10.3

DEATH RATES FROM HEART DISEASE, 1997—MALE

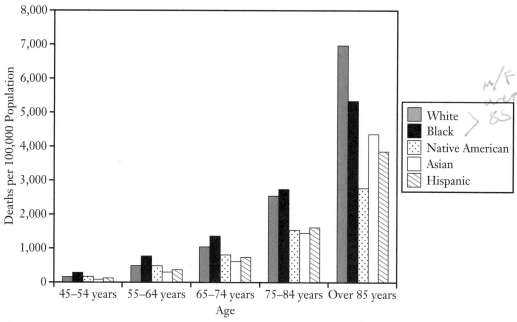

National Center for Health Statistics, 1999.

FIGURE 10.4

DEATH RATES FROM HEART DISEASE, 1997—FEMALE

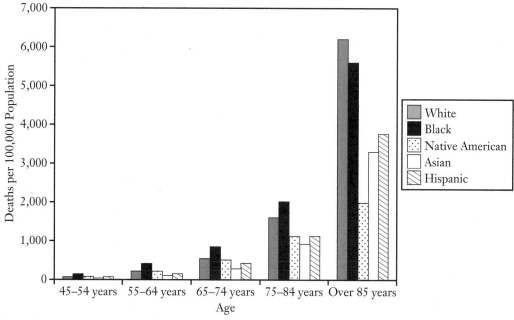

National Center for Health Statistics, 1999.

FIGURE 10.5

DEATH RATES FROM CEREBROVASCULAR DISEASE, 1997—MALE

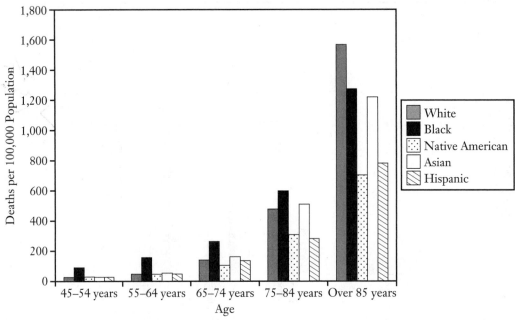

National Center for Health Statistics, 1999.

FIGURE 10.6

DEATH RATES FROM CEREBROVASCULAR DISEASE, 1997—FEMALE

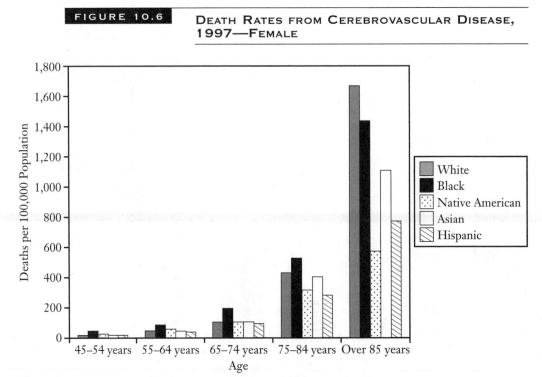

National Center for Health Statistics, 1999.

lead to obesity, which tends to promote diabetes and/or hypertension. The more risk factors present in an individual, the more the dangers multiply through such inter-relationships.

Hypertension is a natural place to start, particularly with regard to premature African American death. African Americans have higher rates of hypertension than any other racial or ethnic group in the United States. While non-Hispanic whites and Hispanics have hypertension rates between 19.3 percent and 25.2 percent (higher for men than women), African American men and women have rates at 35.0 percent and 34.2 percent, respectively. Although the incidence of hypertension has dropped for all Americans since 1960, the percentage gap between whites and blacks has actually increased (National Center for Health Statistics, 1999). Because hypertension alone is a major cause of cerebrovascular diseases, much of the difference in racial death rates can be traced to its door.

What increases hypertension? We have already explicitly noted stress, but most of the other causal factors previously listed are also causes of hypertension. (We could perhaps add sodium intake to that list; however, medical science is currently debating the strength of that connection.) Does that mean you can replace fatty diets, stress, and cigarettes with exercise and be free of hypertension? For some people, lifestyle changes can reduce hypertension, but not for all. African American men do smoke more than European American men, but the opposite is true for women in both groups. Lowering obesity tends to lower hypertension for European Americans, but weight loss has little impact on hypertension for African Americans (Dressler, 1993; Polednak, 1989). In addition, obesity levels in African American and European American men differ little; greater differences are found in obesity between European American and African American women. But Mexican American women have obesity rates almost as high as African American women, and Mexican American men have the highest obesity levels of all men. In spite of those figures, Mexican Americans do not suffer as much from these diseases (National Center for Health Statistics, 1999).

The drop in overall hypertension rates over the last several decades has often been attributed to a change in lifestyle as Americans have become better educated on health matters. Current educational attainment levels, however, have minimal impact on male European American hypertension and only a slightly higher impact on male African American hypertension. Even at the level of college graduation or higher, over 30 percent more African American men still have hypertension when compared with similar European American men (Dressler, 1993). Many researchers have argued that African Americans face higher stress levels in American society than other racial or ethnic groups. Such stress is unlikely to come solely from poverty produced by ethnic stratification, however, because the poor from both groups experience similar stress levels (Cockerham, 1990). Nevertheless, stress experienced by middle-class African Americans may well be an important factor in overall levels of hypertension (Dressler, 1993; Semmes, 1996). The persistence of these differences despite controlling for other known risk factors suggests that further research into biological predisposition for hypertension would prove useful (Polednak, 1989).

Although Hispanics and Native Americans have yet to face high levels of these diseases, the growth of Type II diabetes mellitus (adult-onset diabetes) in both groups suggests that the march to modernization may have begun (Campbell, 1989a; Markides et al., 1989). Native Americans were almost never troubled by diabetes before 1940; today, death rates among Native Americans from diabetes are almost twice the national average (National Center for Health Statistics, 1999).

Cancers

Cancer is another of the modernization diseases. In its various forms, lifestyle and environmental factors are increasingly singled out as primary causes. Because different types of cancer appear to be related to very different causes, care is needed when making generalizations. For example, Native Americans fall below the national average in cancer death rates yet, for some types of cancer, their death rates are far higher. You can learn much about a particular racial or ethnic group's lifestyle and environment by comparing rates for different sites of the disease.

In Figures 10.7 and 10.8, which provide summary data for all types of cancer in the United States, the same racial and ethnic distribution pattern appears as for heart and cerebrovascular diseases. African Americans lead all other groups in death rates at all of the nonelderly age groups, followed closely by European Americans; rates for all other groups are noticeably lower and become increasingly lower with age—the time in life when cancer incidence is most likely to grow.

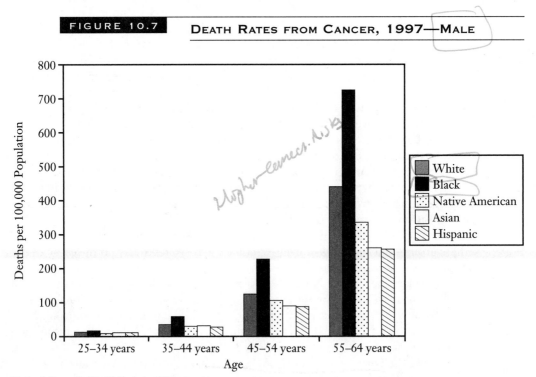

FIGURE 10.7   DEATH RATES FROM CANCER, 1997—MALE

National Center for Health Statistics, 1999.

FIGURE 10.8   DEATH RATES FROM CANCER, 1997—FEMALE

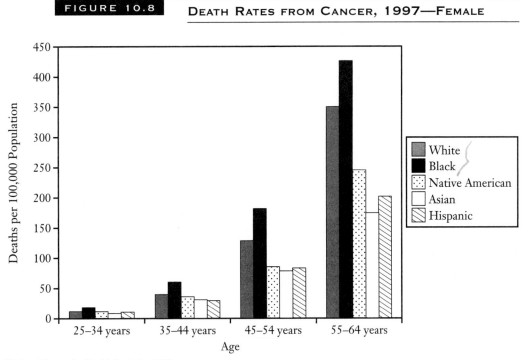

National Center for Health Statistics, 1999.

Figures 10.9 through 10.11 provide data on death rates from respiratory cancer and breast cancer. Respiratory cancer has been clearly linked to various unhealthy products we breathe into our lungs, ranging from tobacco smoke to numerous types of air pollution. The overall higher rates for all men can be explained by higher levels of smoking and to greater occupational exposure to pollutants. The fact that African American women—who smoke less than European American women—have equal rates of respiratory cancer at all ages should lead us to examine other environmental differences between the two groups. Of all causes, social class is definitely a major factor in developing respiratory cancer: the differences between African American and European American men disappear when they are compared at the same levels of education and income (Polednak, 1989).

Because respiratory cancers tend to have poor five-year survival rates, early detection is not a critical issue. Cancers are best thought of in terms of a continuum from the least treatable to the most treatable. We would expect to find few differences in survival rates for those cancers that are least treatable, and easily treatable cancers should produce minimal racial and ethnic differences because only the most poor and/or isolated from medical care should be affected. Cancer sites in the middle range of the continuum, however, are the most reflective of health differences because the time of detection and intensity of treatment both play major roles in survival rates. Breast cancer is a good example.

FIGURE 10.9 **DEATH RATES FROM RESPIRATORY CANCER, 1997—MALE**

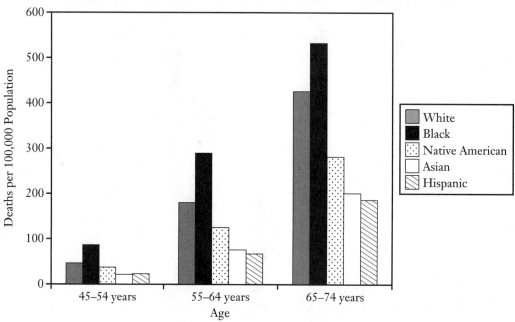

National Center for Health Statistics, 1999.

FIGURE 10.10 **DEATH RATES FROM RESPIRATORY CANCER, 1997—FEMALE**

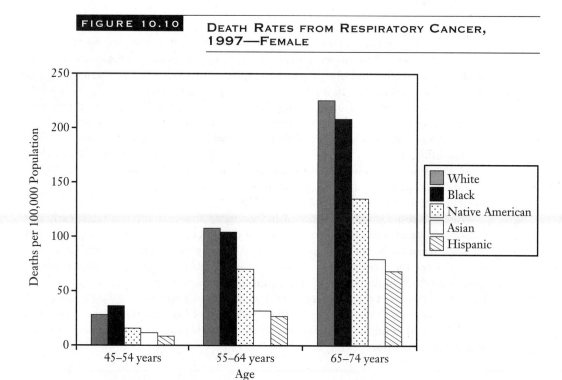

National Center for Health Statistics, 1999.

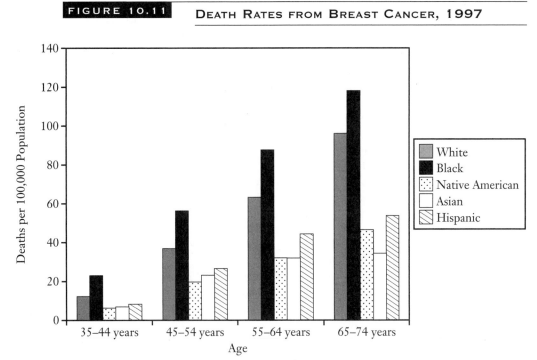

FIGURE 10.11 DEATH RATES FROM BREAST CANCER, 1997

National Center for Health Statistics, 1999.

Figure 10.11 shows the impact of breast cancer on African American and European American women. The gap between the two groups remains the same as individuals age, although the number of deaths increases greatly. Age is also a greater risk factor with these two groups when compared with Native Americans, Asian and Pacific Islanders, and Hispanics, all of whom face only minimally increased risks with age. As one of the cancer sites least understood in terms of causes, these data showing group differences should direct a line of future research.

In comparing European American and African American women with breast cancer, recall that African American women actually have a lower overall incidence of this disease than European American women—101.3 vs. 115 cases per 100,000 (National Center for Health Statistics, 1999). For women under 40, however, incidence rates between the two groups are somewhat similar (Polednak, 1986) Breast cancer metastasizes (spreads) quickly, so early detection is critical. Either African American women contract breast cancer that metastasizes more quickly—a subject of current research— or their cancer is detected later. Polednak (1986) conducted studies on breast cancer in both New York and Detroit, comparing European American and African American women. Cancer in African American women was more likely to be detected at the metastatic stage than in European American women in both cities. Additionally, those areas in New York City with the greatest income gap between African Americans and

European Americans also tended to have the greatest gap in the stage of detection. The 15 percent difference in five-year survival rates between the two groups is therefore not too surprising.

Cervical cancer rates also are related to social class but in ways different from breast cancer. As with breast cancer, not all the risk factors are known, but current research is focusing on the human papillomavirus (HPV), which is a sexually transmitted disease (STD). Rates of cervical cancer increase when women are exposed to HPV, and exposure is increased if they have had many sexual partners and/or their sexual partner has had many sexual partners. HPV is detected through Pap tests, and when detected early, it is highly treatable. Exposure to HPV *and* obtaining regular Pap tests are both related to social class. Cervical cancer is higher among African Americans, Native Americans, and Hispanics than among European Americans. Most of this difference disappears, however, when education is controlled for. More education—which is related to more income, greater knowledge of and access to Pap tests, and different levels of exposure to STDs—lessens a wide variety of risk factors in both incidence and death from this disease (Hampton, 1992; Horm and Burhansstipanov, 1992; Polednak, 1989; Young, 1994).

Other forms of cancer are related to race and ethnicity in various ways. Some are hidden by the way data are collected by the U.S. government. We noted earlier that Native Americans have relatively low rates of respiratory cancer, but Native Americans consist of many different ethnic groups. Urban Native Americans, who live predominantly in Los Angeles, Oklahoma City, and Tulsa, represent over half of all Native Americans. Their rate of smoking is 70 percent—twice as high as the national average (Hampton, 1992; National Center for Health Statistics, 1999). By contrast, most of the Southwest native peoples have very low rates of smoking, well below the national average. The Northern Plains tribes have death rates from lung cancer seven times that of the Southwestern tribes, but census figures only reflect the average (Hampton, 1992). When Southwestern tribes do have rising rates of lung diseases, it more often than not reflects the large number of past uranium miners rather than smokers among them (Polednak, 1989).

Stomach cancer is clearly a social class–related form of cancer. One of the major killers in the United States among all forms of cancer during the early part of the twentieth century, it is now very rare except among ethnically stratified minorities. Rates are higher among Native Americans and African Americans (Horm and Burhansstipanov, 1992; Polednak, 1989). The disappearance of stomach cancer coincided with the widespread refrigeration of food. Poorer people even today are more likely to come into contact with the carcinogenic bacteria that multiply with poor food preservation. An interesting picture of lifestyle change comes from early Japanese immigrants. The Issei from Japan suffered from high rates of stomach cancer but almost never from colon cancer. Their children, the Nisei, benefited from refrigeration but changed to an "American" diet with more fat and fewer vegetables. As a result, stomach cancer rates dropped while colon cancer rates rose (Polednak, 1989).

Although we have been focusing almost exclusively on lifestyle and environmental causes of cancer, biological investigation is still called for, as exemplified by the ongoing research into types of breast cancer noted earlier. Another area of similar research concerns the rate of gallbladder cancer among some Native Americans. In New Mexico, Native American rates of this cancer are ten times higher than for European Americans (Horm and Burhansstipanov, 1992). Lowenfels (1992) suggests that these peoples might have adapted centuries ago to food scarcity with a metabolism that today leads to obesity. That obesity in turn leads to gallstones, which are in turn related to gallbladder cancer.

As we have seen, cancer is in fact many different diseases. Most cancers are related to modernization and most have some connection to lifestyles and the environment, but this is not true for all. The great variation in cancer relative to cause, detection, and treatment makes this disease a good point of departure for understanding the complex interplay of factors that structure the racial and ethnic allocation of health in the United States. As that structure changes, cancer rates will reflect it.

Homicides, Suicides, and Accidents

This section is devoted to examining death rates by injury, both intentional and unintentional. Injuries are far from random occurrences in any society. The likelihood of murdering (or being murdered) varies immensely by race, ethnicity, and social class. Suicide is death by one's own hand, but only certain kinds of people choose to do it. Although accidents can happen to anyone, certain kinds of environments, both work and leisure, are conducive to accidental injury or death. If you live in such an environment, your odds of being injured grow accordingly. Our focus is the impact of race, ethnicity, and social class on death rates by injury.

Homicide and Legal Intervention Figures 10.12 and 10.13 portray homicide as largely limited to the world of African American and Hispanic young men. This subject receives far more attention in the following chapter on crime because homicide is typically viewed criminologically rather than medically. Nevertheless, and at the risk of repetition, conclusions reached in the next chapter need at least summary inclusion here.

Increased homicide rates are associated with low social class and attendant cultural values. Low social class standing in the United States places individuals in circumstances of limited economic opportunity, which produces criminal behavior, very negative relationships with law enforcement agencies, and a need for extralegal enforcement to protect illegal activities. With perhaps the exception of white-collar crime, making any behavior illegal automatically increases the level of violence associated with it. And, as we will see in the following chapter, the intervention of law enforcement can itself increase levels of community violence. Arresting drug traffickers, for example, creates turf wars among those still in business.

The cultural side of homicide (and violence in general) is less clear-cut but probably equally important. The highest levels of homicide have been always been in the American South. The media have focused American attention on urban violence, but the highest per capita homicide rates are in the South. Many reasons for this have been offered over the years, including the lower economic position of the region, but the cultural acceptance of violence in the South cannot be ignored. We see this clearly

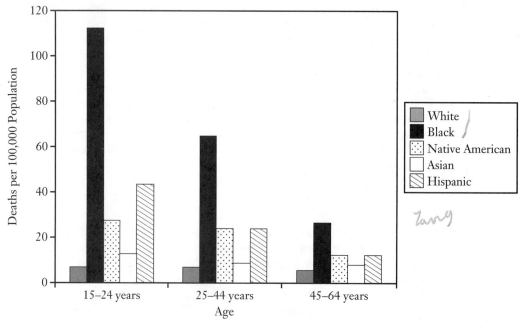

FIGURE 10.12

DEATH RATES FROM HOMICIDE AND LEGAL INTERVENTION, 1997—MALE

National Center for Health Statistics, 1999.

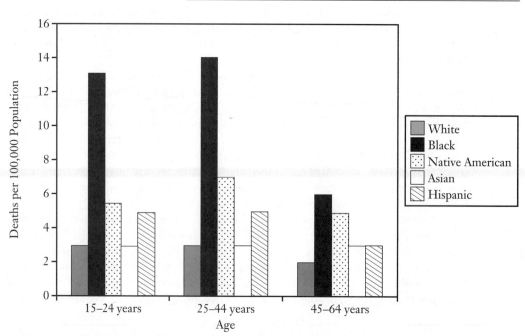

FIGURE 10.13

DEATH RATES FROM HOMICIDE AND LEGAL INTERVENTION, 1997—FEMALE

National Center for Health Statistics, 1999.

replicated with young, African American men in the urban North whose community position is intricately connected to their participation in violent activity.

Suicide   Suicide rates are even more complex than homicide rates. Some aspects of suicide have remained relatively stable, however, ever since the subject first entered the research literature with French sociologist Emile Durkheim's famous study of nineteenth century suicide (Durkheim, 1951 [1897]). Even in the last century, suicide was much more common among men than women (although more women attempt it), more common among older people than younger people, and more likely among the financially better off than the poor. As Figures 10.14 and 10.15 show, these patterns of suicide are generally true in the United States as European American men have higher rates than most groups, with rates increasing among them as their members age. By contrast, African American men and women and Hispanics have lower rates. The one change from this pattern occurs with Native Americans. Although Native American suicide occurs far more frequently with men than women, both gender groups dominate the death rates at the youngest ages and then the incidence of suicide progressively decreases with age. The number of suicides among native Americans older than the age of 65 is so low that they disappear from the chart. This is the same time of life when suicides increase for European Americans and Asian and Pacific Islanders.

Suicide rates among Native Americans have been much studied but remain largely a mystery. Various risk factors linked to Native American suicides include alcohol use, family disruption, feeling alienated from family and community, having friends who committed suicide, employment instability, and criminal convictions (Grossman et al., 1991; Hlady and Middaugh, 1988; Young, 1994). With the exception of higher levels of alcoholism among Native Americans, the other risk factors are present in about the same degrees with many African American and Hispanic younger people who have low suicide rates. As we have seen throughout this book, however, Native Americans have had a truly distinctive history as a minority in the United States. The roots of modern suicide among Native Americans may be quite deep indeed.

Accidents   Figures 10.16 and 10.17 illustrate death rates from motor vehicle–related injuries. The other major source of injury is occupation-related. Occupational injuries are social class related and a review of associated data would simply mirror the degree of ethnic stratification other death rate statistics in the United States have already revealed. Motor vehicles, however, can be driven in various ways, often related to cultural differences. Also of importance is the role of alcohol in such injuries.

As Figures 10.16 and 10.17 show, motor vehicle–related death rates affect men more than women, young people more than old people, and Native Americans more than any other racial or ethnic group in the United States. The first two observations are of no surprise to anyone who has paid for automobile insurance; the insurance companies have seen these data also. Native American death rates that are twice that of any other group might come as more of a surprise, and as high as they are, they have actually been dropping throughout the 1970s and 1980s (Hisnanick, 1994). The explanation for those high rates rests in the interaction among alcohol use, Native American cultural attitudes toward drinking, alcohol prohibition and the geographical isolation

FIGURE 10.14   **DEATH RATES FROM SUICIDE, 1997—MALE**

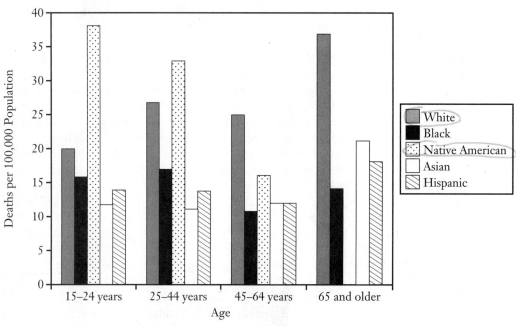

National Center for Health Statistics, 1999.

FIGURE 10.15   **DEATH RATES FROM SUICIDE, 1997—FEMALE**

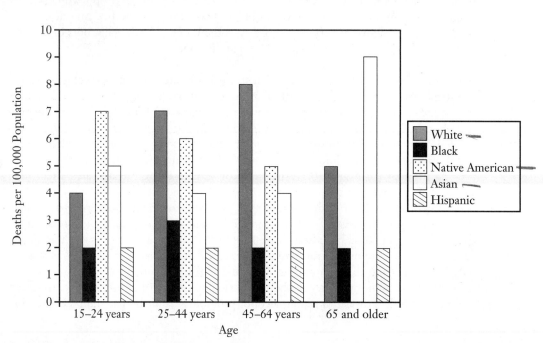

National Center for Health Statistics, 1999.

FIGURE 10.16 **DEATH RATES FROM MOTOR VEHICLE–RELATED INJURIES, 1997–MALE**

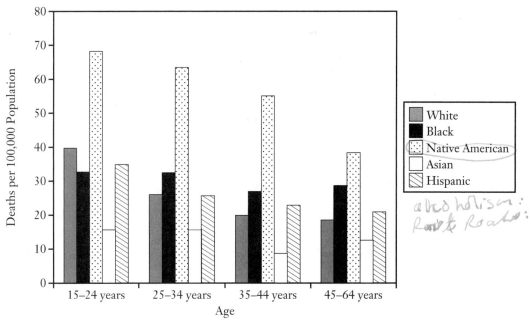

National Center for Health Statistics, 1999.

FIGURE 10.17 **DEATH RATES FROM MOTOR VEHICLE–RELATED INJURIES, 1997–FEMALE**

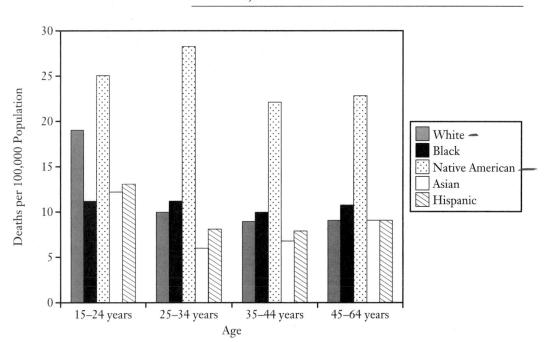

National Center for Health Statistics, 1999.

of many reservations, and the extreme diversity among native peoples with regard to all of these factors.

Unquestionably, alcohol abuse is a chronic problem among many native peoples. Native Americans are five times more likely to die of chronic liver disease and cirrhosis than the national average (National Center for Health Statistics, 1999). Considering that many tribes—notably the Navajo—have very low rates of alcohol consumption, other tribes obviously have serious alcohol problems. It is a common belief among most Americans, Native Americans included, that Native Americans react biologically differently to alcohol than do other racial and ethnic groups. The wide variation in alcohol use tends to contradict this notion, and most scientific studies have returned such a wide variety of findings that they too contradict it (May, 1994; Young, 1994). The stereotype of the Indian who cannot hold liquor is, indeed, a stereotype.

The variation among the native peoples has led many researchers to look at cultural factors. Alcohol is not the only drug abused by Native Americans; many reservations have high levels of varied substance abuse, particularly among younger people (Beauvais and Segal, 1992; Campbell, 1989a). May (1994) argues that sporadic alcohol abuse is more of a factor in accidental death than chronic alcoholism. He points to cultural patterns that exist with many tribes in which alcohol is consumed with expectations of becoming drunk followed by wild, risky behavior in automobiles (often with cultural prohibitions against wearing seatbelts) (see also O'Nell and Mitchell, 1996).

Another risk factor is the isolation of the western reservations, many of which prohibit alcohol on the reservation. This leads to the use of automobiles to obtain alcohol at some distance; the alcohol is often consumed off the reservation lands, and subsequently leads to driving while drunk to return (May, 1994; Weibel-Orlando, 1990). This also creates problems for Native American pedestrians who either may be hit by drunk drivers or are walking home themselves under the influence. If these pedestrians are not the victims of traffic accidents, they may become the victims of death by exposure. All of this occurs on reservation roads that are often in bad condition and poorly lit (Young, 1994). Reservation prohibition is a response to a real problem for Native Americans, but it may cause more problems than it solves in the long run (see also Olson et al., 1990).

Living in Poor Health: A Final Thought

The emphasis in this section on death rates has led us to ignore variations in health quality among the living. If racial, ethnic, and social class factors lead to premature death, it stands to reason that they might also lead to poor health before they result in death. We might also guess that the variation in health quality might be greater than the variation in death rates. We also give up some scientific precision in studying health quality, however. Death rates are easily measured while the quality of health is much more subjective on the part of the individual.

Krieger and Fee (1993) find that individuals in families earning less than $5,000 per year are almost four times more likely to have limited activities because of chronic conditions than those in families earning more than $25,000 per year. This also holds true in about the same proportion for individuals who label their health as "poor or fair." The figures are also virtually identical for European Americans and African Americans. One could argue that poor health might negatively affect income, turning the causal relationship around. Their case is strengthened by showing very similar variations in self-assessed health according to occupation (comparing professionals with private household workers) and educational attainment. With regard to the latter, only 4.2 percent of college graduates list their health as "poor or fair" as compared with 18.3 percent of those without a high school diploma. Navarro (1991) reports almost identical findings.

The risk factors for death rates examined throughout this section are identical to many of the causes of chronic ill health. The previously mentioned findings on social class and poor health should not really be much of a surprise. All of this material taken together suggests a complex interplay of the ways in which risk factors attach themselves to social classes and racial or ethnic groups. The only factor left to examine that might turn these risk factors around would be a positive experience with health care delivery in the United States.

Health Care Delivery in the United States

Health care delivery refers to that wide range of people, institutions, and technology that brings medical care to the population—drug companies, hospitals, insurance companies, private physicians, drugs, nurses, companies that produce medical technology, government research and social service agencies, and so on. All of these people, things, and agencies have a connection to the state of biological and psychological health. In almost all industrial countries, health care delivery is run or regulated by national governments and provided equally to all citizens. The unique connection between health care delivery and the economy as found in the United States means that individual access is related to one's economic position. It also may be related to one's race or ethnicity.

Medical Insurance and Employment: Who Is Covered?

The United States contains various options in regard to medical coverage. First, individuals may carry private medical insurance, which will pay some or all of various medical expenses, ranging from prescription drugs to major surgery. This type of coverage may be purchased individually, jointly by an employer and employee, or provided entirely by an employer. Individual purchase is extremely expensive. Insurance connected to employment is far superior for most individuals in that the overall cost of the coverage drops plus the employee receives some degree of help in paying for it. To obtain this coverage, however, you not only have to be employed but employed in a full-time capacity by a relatively large organization.

The federal government provides Medicare and Medicaid. Medicare is available to anyone who paid into the social security system during their working life and has reached the age of 65. Medicaid is a need-based program of coverage open to poor people and their dependents of all ages. Anyone eligible for AFDC, for example, is

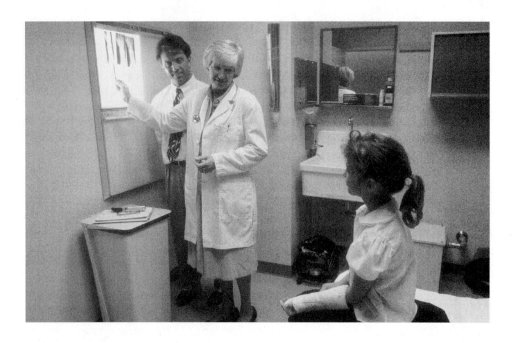

Differential access to health care can be seen when a modern private American medical setting is compared to a crowded, publicly run medical clinic.

almost certainly eligible for Medicaid. Unlike private insurance or Medicare, individuals covered by Medicaid will find many medical services and providers closed to them. They are much more likely to be serviced by clinics or hospital emergency rooms and they are also less likely to have the same physician on a regular basis (Fichtenbaum and Gyimah-Brempong, 1997; Quaye, 1994).

Finally, the obvious alternative to all of these options is to have no medical insurance whatsoever. This is a gamble that is seldom taken willingly. Ironically, it is forced on neither the middle class nor the very poor but on the lower middle class right between them—a class often referred to as the working poor. For a poor person who works toward social mobility, this economic minefield must be crossed first. Perhaps a better way to view this distinction is through employment in the lower classes. Among unemployed poor people, 40.4 percent were uninsured; of poor people with part-time employment, 44.8 percent were uninsured; of poor people with *full-time* employment, 47.5 percent were uninsured (U.S. Bureau of the Census, 1999d).

Overall in the United States in 1997, 70.7 percent of the population carried private insurance, most of whom had procured it through their employer. At the other extreme, another 17.5 percent carried no health insurance whatsoever (National Center for Health Statistics, 1999). Figure 10.18 shows health care coverage for selected

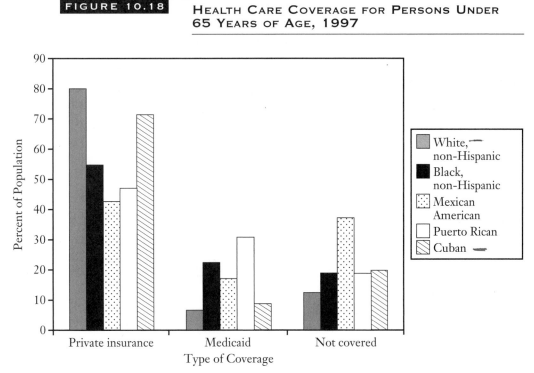

FIGURE 10.18 HEALTH CARE COVERAGE FOR PERSONS UNDER 65 YEARS OF AGE, 1997

National Center for Health Statistics, 1999.

racial and ethnic groups. Mexican Americans are the least likely to have private insurance and the most likely to have no coverage. Some of the noncoverage rates can be explained by the large numbers of new immigrants within that population who have no health insurance. In addition, the type of employment common with Mexican Americans tends to be outside the realm of employer coverage (Amey et al., 1995). Medicaid coverage occurs more often among the urban, black and Hispanic poor, with African Americans and Puerto Ricans dominating that form of coverage.

Variation in Patient Treatment: Race/Ethnicity, Social Class, or Insurance?

How does the state of an individual's medical coverage affect the early detection of potentially serious medical conditions? Table 10.3 shows the relationship between three standard medical tests—mammograms, clinical breast examination, and Pap tests—and the type of medical coverage women carried. In every case, private coverage produced the highest percentage of patients, followed by Medicaid and the uninsured. In just the few years since these data were collected, however, there has been a striking change in the percentages of African American women getting mammograms. In 1991, 56 percent of white, non-Hispanic women over 40 had mammograms compared with 48 percent of black, non-Hispanic women. By 1994, the percentages had risen to 61 and 64 percent respectively with black, non-Hispanic women leading for the first time (National Center for Health Statistics, 1997; Thomas et al., 1996). If this change produces positive results in mortality rates, they will not be apparent for several years.

Differences in treatment also exist between African Americans and European Americans with heart disease. When individuals from both groups complain of chest pain, European Americans are more likely to undergo more sophisticated tests such as arteriography; when a diagnosis is confirmed, European Americans are more than twice as likely to receive coronary artery bypass surgery (Dressler, 1993; Polednak, 1989). The association remains when controlling for the severity of heart disease, income, and insurance status (Ford et al., 1989; Wenneker and Epstein, 1989). Such findings are difficult to explain by social class.

| TABLE 10.3 | PERCENT OF WOMEN, 50-64 YEARS, WHO RECEIVED CANCER SCREENINGS BY TYPE OF MEDICAL COVERAGE, 1992. |

	Mammogram	Clinical Breast Exam	Pap Test
Private Medical Coverage	55.4	65.2	57.9
Medicaid	38.3	52.0	41.0
Uninsured	19.3	38.2	32.0

Makuc et al., 1994.

The amount of care received also varies by race and ethnicity. Andrews and others (1995) report that white patients hospitalized for mental illness and substance abuse receive longer treatment than do black or Hispanic patients. The authors note that type of coverage is the most likely predictor for most of the variation because whites with Medicare face quicker discharges than whites with private coverage. Both black and Hispanic patients in this study were far less likely to have private coverage when compared with whites.

With some of these variations, it is difficult to know exactly what medical bases may be justifying apparently discriminatory decision making by medical personnel. In the case of kidney transplants, however, a major part of the decision making is very clear. Donated kidneys are compared with potential recipients through a process called antigen matching. The closer the match, the greater the likelihood that the kidney will not be rejected by the host. There are enough biological differences between most African Americans and most European Americans to prevent close matches between the two groups. In effect, African Americans in need of a kidney need to wait for an African American kidney to become available. This would not a problem in itself except that African Americans have four times the rate of kidney failure as European Americans. Kidney failure can result from hypertension and diabetes, both of which are more common among African Americans than European Americans. Ayres and associates (1993) note that the higher demand coupled with the lower supply means that African Americans wait an average of 13.9 months for a kidney compared with 7.6 months for European Americans. They also argue that antigen matching is less critical today than it once was due to the development of drugs that lower rejection rates for donated organs.

As a final observation, African Americans regularly report that they receive a lower quality of care through interpersonal interactions with the medical community. Typical complaints include a lack of willingness among physicians to listen to them or to explain their conditions and/or test results fully (Bailey, 1991). Although many patients from all racial and ethnic groups voice similar complaints, such comments appear more frequently among minorities.

The difficulty of sorting out explanations for mortality rates should be obvious by now. Patients not only vary in race and ethnicity but they also bring differences in biology, lifestyle, social class, health insurance, education, home environment, and attitudes toward medical care. Certain medical problems may allow us to rule out half of that list as potential causes, but seldom are we able to absolutely pinpoint causes. In most cases, poor health stems from numerous causes and the difficulty in sorting out relative importance among them is immense. But one thing is clear—minority status does indeed increase the chances of suffering from poor health and ultimately takes years off your life.

Summary

People die prematurely or live in poor health for a wide variety of reasons. Genetics, lifestyle, the availability of medical treatment, and the environment all play roles.

Variations in each can produce higher or lower levels of particular diseases or health problems. Ethnic and racial minorities are, by definition, in the lower income brackets of society. Much of the disproportionate poor health minorities experience follows directly from poverty status. In addition, both overt and institutional discrimination are factors.

Most Americans die of heart diseases, cancers, and cerebrovascular diseases. These are sometimes described as the diseases of civilization. First, you have to live in a society modernized enough to avoid the many contagious (and usually earlier) killers that thrive in poor sanitation. Second (and related), you have to live long enough to develop these diseases because they are all more common among the aged.

Many minorities in the United States have higher death rates from contagious diseases and accidents/homicides. In addition, however, some groups also have higher and earlier death rates from the "civilized" diseases, particularly African Americans. As a general rule, however, European Americans and Asian Americans die from different causes than do African Americans, Native Americans, and Hispanics.

Of modern contagious diseases, HIV infection is clearly moving toward a minority disease. While incidence rates have been dropping within the gay community, they have been rising among IV drug users and the people with whom they have sexual relations.

Infant mortality is best thought of in two stages—the first month of life and the following eleven months of life. Death in the second stage generally stems from the infant's environment, particularly contagious diseases. Death in the first month is more often related to the health of the mother and/or low birth weights. African Americans are the one minority with significantly high levels of low birth weights and very low birth weights. This is apparently not related directly to poverty because other largely poor ethnic groups do not face this health risk. In addition, higher social class among African American women does not remove the risk. African American women do, however, have significantly higher rates of hypertension than do most other women.

Hypertension is also clearly a factor in cardiovascular and cerebrovascular diseases. Only African Americans have death rates from these causes that are similar to those of European Americans and Asian Americans. Hypertension has been related to genetics, lifestyle (what you eat and/or stress), and medical services (necessary to both diagnose and treat). We know little about the role of genetics, but African Americans clearly rank in a higher risk level than European Americans and Asian Americans in the other causes. Hispanics and Native Americans do not currently suffer high rates from these diseases.

Cancers are very similar to cardiovascular and cerebrovascular diseases in terms of the important roles of genetics, lifestyle, and medical services. They are also similar in their higher incidence rates among African Americans, European Americans, and Asian Americans. The latter two groups may well have old age to thank for these high rates. African Americans, on the other hand, face more threats from lifestyle and environmental factors; in addition, they are much less likely to experience early detection for those cancers that are treatable. These factors are also true, in general, for Native Americans and Hispanics, but their mortality rates from most cancers are still considerably lower.

African Americans, Native Americans, and Hispanics all have higher mortality rates from homicides, legal intervention, and accidents. Native Americans, in particular, have unusually high accident-related mortality rates, much of which is attributable to alcohol abuse. All of these groups have considerably lower rates of suicide than European Americans or Asian Americans.

A major issue with medical treatment concerns medical insurance. In the United States currently, most people who have such insurance either obtain it through their employment or receive it directly from the federal government as part of their public assistance. The uninsured are typically the working poor—people who are not on public assistance but whose jobs are too low level to provide fringe benefits. Because many minorities fit into this category, they are more likely to be uninsured than others in American society. Most research indicates that those with private insurance receive the best care followed by Medicaid patients and then the uninsured.

Research on noneconomic-based discrimination in health care delivery is generally inconclusive. European Americans are more likely to receive coronary bypass surgery than African Americans (controlling for severity of disease and insurance status). European Americans also receive more and longer treatment for mental illness and substance abuse than do Hispanics. The amount of anecdotal evidence concerning one-on-one relationships between medical personnel and minority patients is considerable, with many minority patients complaining that their doctors do not talk to them long enough or in enough detail.

Chapter 10 Reading
The Tuskegee Syphilis Experiment

James Jones

Syphilis is a strange disease that wreaks most of its havoc many years after contraction. The effects of syphilis had been known for centuries, but it was not until the 1940s when penicillin was used that an effective cure was found. Nevertheless, the Public Health Service continued the Tuskegee Syphilis Experiment until 1972. The treatment of study participants shows us how science can be as racist as any other segment of society.

In late July of 1972, Jean Heller of the Associated Press broke the story: for forty years the United States Public Health Service (PHS) had been conducting a study of the effects of untreated syphilis on black men in Macon County, Alabama, in and around the county seat of Tuskegee. The Tuskegee Study, as the experiment had come to be called, involved a substantial number of men: 399 who had syphilis and an additional 201 who were free of the disease chosen to serve as controls. All of the syphilitic men were in the late stage of the disease when the study began.

Under examination by the press the PHS was not able to locate a formal protocol for the experiment. Later it was learned that one never existed; procedures, it seemed, had simply evolved. A variety of tests and medical examinations were performed on the men during scores of visits by PHS physicians over the years, but the basic procedures called for periodic blood testing and routine autopsies to supplement the information that was obtained through clinical examinations. The fact that only men who had late, so-called tertiary, syphilis were selected for the study indicated that the investigators were eager to learn more about the serious complications that result during the final phase of the disease.

The PHS officers were not disappointed. Published reports on the experiment consistently showed higher rates of mortality and morbidity among the syphilitics than the controls. In fact, the press reported that as of 1969 at least 28 and perhaps as many as 100 men had died as a direct result of complications caused by syphilis. Others had developed serious syphilis-related heart conditions that may have contributed to their deaths.

The Tuskegee Study had nothing to do with treatment. No new drugs were tested; neither was any effort made to establish the efficacy of old forms of treatment. It was a non-therapeutic experiment, aimed at compiling data on the effects of the spontaneous evolution of syphilis on black males. The magnitude of the risks taken with the lives of the subjects becomes clearer once a few basic facts about the disease are known.

Syphilis is a highly contagious disease caused by the Treponema pallidum, a delicate organism that is microscopic in size and resembles a corkscrew in shape. The disease may be acquired or congenital. In acquired syphilis, the spirochete (as the Treponema pallidum is also called) enters the body through the skin or mucous membrane, usually during sexual intercourse, though infection may also occur from other forms of bodily contact such as kissing. Congenital syphilis is transmitted to the fetus in the infected mother when the spirochete penetrates the placental barrier.

From the onset of infection syphilis is a generalized disease involving tissues throughout the entire body. Once they wiggle their way through the skin or mucous membrane, the spirochetes begin to multiply at a frightening rate. First they enter the lymph capillaries where they are hurried along to the nearest lymph gland. There they multiply and work their way into the bloodstream. Within days the spirochetes invade every part of the body.

Three stages mark the development of the disease: primary, secondary, and tertiary. The primary stage lasts from ten to sixty days starting from the time of infection. During this "first incubation period," the primary lesion of syphilis, the chancre, appears at the point of contact, usually on the genitals. The chancre, typically a slightly elevated, round ulcer, rarely causes personal discomfort and may be so small as to go unnoticed. If it does not become secondarily infected, the chancre will heal without treatment within a month or two, leaving a scar that persists for several months.

While the chancre is healing, the second stage begins. Within six weeks to six months, a rash appears signaling the development of secondary syphilis. The rash may resemble measles, chicken pox, or any number of skin eruptions, though occasionally it is so mild as to go unnoticed. Bones and joints often become painful, and circulatory disturbances such as cardiac palpitations may develop. Fever, indigestion, headaches, or other nonspecific symptoms may accompany the rash. In some cases skin lesions develop into moist ulcers teeming with spirochetes, a condition that is especially severe when the rash appears in the mouth and causes open sores that are viciously infectious. Scalp hair may drop out in patches, creating a "moth-eaten" appearance. The greatest proliferation and most widespread distribution of spirochetes throughout the body occur in secondary syphilis.

Secondary syphilis gives way in most cases, even without treatment, to a period of latency that may last from a few weeks to thirty years. As if by magic, all symptoms of the disease seem to disappear, and the syphilitic patient does not associate with the disease's earlier symptoms the occasional skin infections, periodic chest pains, eye disorders, and vague discomforts that may follow. But the spirochetes do not vanish once the disease becomes latent. They bore into the bone marrow, lymph glands, vital organs, and central nervous systems of their victims. In some cases the disease seems to follow a policy of peaceful coexistence, and its hosts are able to enjoy full and long lives. Even so, autopsies in such cases often reveal syphilitic lesions in vital organs as contributing causes of death. For many syphilitic patients, however, the disease remains latent only two or three years. Then the delusion of a truce is shattered by the appearance of signs and symptoms that denote the tertiary stage.

It is during late syphilis, as the tertiary stage is also called, that the disease inflicts the greatest damage. Gummy or rubbery tumors (so-called gummas), the characteristic lesions of late syphilis, appear, resulting from the concentration of spirochetes in the body's tissues with destruction of vital structures. These tumors often coalesce on the skin forming large ulcers covered with a crust consisting of several layers of dried exuded matter. Their assaults on bone structure produce deterioration that resembles osteomyelitis or bone tuberculosis. The small tumors may be absorbed, leaving slightly scarred depressions, or they may cause wholesale destruction of the bone, such as the horrible mutilation that occurs when nasal and palate bones are eaten away. The liver may also be attacked; here the result is scarring and deformity of the organ that impede circulation from the intestines.

The cardiovascular and central nervous systems are frequent and often fatal targets of late syphilis. The tumors may attack the walls of the heart or the blood vessels. When the aorta is involved, the walls become weakened, scar tissue forms over the lesion, the artery dilates, and the valves of the heart no longer open and close properly and begin to leak. The stretching of the vessel walls may produce an aneurysm, a balloonlike bulge in the aorta. If the bulge bursts, and sooner or later most do, the result is sudden death.

The results of neurosyphilis are equally devastating. Syphilis is spread to the brain through the blood vessels, and while the disease can take several forms, the best known is paresis, a general softening of the brain that produces progressive paralysis and insanity. Tabes dorsalis, another form of neurosyphilis, produces a stumbling, foot-slapping gait in its victims due to the destruction of nerve cells in the spinal cord. Syphilis can also attack the optic nerve cells in the spinal cord. Syphilis can also attack the optic nerve, causing blindness, or the eighth cranial nerve, inflicting deafness. Since nerve cells lack regenerative power, all such damage is permanent.

The germ that causes syphilis, the stages of the disease's development, and the complications that can result from untreated syphilis were all known to medical science in 1932—the year the Tuskegee Study began.

Since the effects of the disease are so serious, reporters in 1972 wondered why the men agreed to cooperate. The press quickly established that the subjects were mostly poor and illiterate, and that the PHS had offered them incentives to participate. The men received free physical examinations, free rides to and from the clinics, hot meals on examination days, free treatment for minor ailments, and a guarantee that burial stipends would be paid to their survivors. Though the latter sum was very modest (fifty dollars in 1932 with periodic increases to allow for inflation), it represented the only form of burial insurance that many of the men had.

What the health officials had told the men in 1932 was far more difficult to determine. An officer of the venereal disease branch of the Center for Disease Control in Atlanta, the agency that was in charge of the Tuskegee Study in 1972, assured reporters that the participants were told what the disease could do to them, and that they were given the opportunity to withdraw from the program any time and receive treatment. But a physician with firsthand knowledge of the experiment's early years directly contradicted this statement. Dr. J. W. Williams, who was serving his internship at Andrews Hospital at the Tuskegee Institute in 1932 and assisted in the experiment's clinical work, stated that neither the interns nor the subjects knew what the study involved. "The people who came in were not told what was being done," Dr. Williams said. "We told them we wanted to test them. They were not told, so far as I know, what they were being treated for or what they were not being treated for." As far as he could tell, the subjects "thought they were being treated for rheumatism or bad stomachs." He did recall administering to the men what he thought were drugs to combat syphilis, and yet as he thought back on the matter, Dr. Williams conjectured that "some may have been a placebo." He was absolutely certain of one point; "We didn't tell them we were looking for syphilis. I don't think they would have known what that was."

A subject in the experiment said much the same thing. Charles Pollard recalled clearly the day in 1932 when some men came by and told him that he would receive a free physical examination if he appeared the next day at a nearby one-room school. "So I went on over and they told me I had bad blood," Pollard recalled. "And that's what they've been telling me ever since. They come around from time to time and check me over and they say, 'Charlie, you've got bad blood.'"

An official of the Center for Disease Control (CDC) stated that he understood the term "bad blood" was a synonym for syphilis in the black community. Pollard replied, "That could be true. But I never heard no such thing. All I knew was that they just kept saying I had the bad blood—they never mentioned syphilis to me, not even once." Moreover, he thought that he had been receiving treatment for "bad blood" from the first meeting on, for Pollard added: "They been doctoring me off and on ever since then, and they gave me a blood tonic."

The PHS's version of the Tuskegee Study came under attack from yet another quarter when Dr. Reginald G. James told his story to reporters. Between 1939 and 1941 he had been involved with public health work in Macon County—specifically the diagnosis and treatment of syphilis. Assigned to work with him was Eunice Rivers, a black nurse employed by the Public Health Service to keep track of the participants in the Tuskegee Study. "When we found one of the men from the Tuskegee Study," Dr. James recalled, "she would say, 'He's under study and not to be treated.' " These encounters left him, by his own description, "distraught and disturbed," but whenever he insisted on treating such a patient, the man never returned. "They were being advised they shouldn't take treatments or they would be dropped from the study," Dr. James stated. The penalty for being dropped, he explained, was the loss of the benefits that they had been promised for participating.

Once her identity became known, Nurse Rivers excited considerable interest, but she steadfastly refused to talk with reporters. Details of her role in the experiment came to light when newsmen discovered an article about the Tuskegee Study that appeared in Public Health Reports in 1953. Involved with the study from its beginning, Nurse Rivers served as the liaison between the researchers and the subjects. She lived in Tuskegee and provided the continuity in personnel that was vital. For while the names and faces of the "government doctors" changed many times over the years, Nurse Rivers remained a constant. She served as a facilitator, bridging the many barriers that stemmed from the educational and cultural gap between the physicians and the subjects. Most important, the men trusted her.

As the years passed the men came to understand that they were members of a social club and burial society called "Miss Rivers' Lodge." She kept track of them and made certain that they showed up to be examined whenever the "government doctors" came to town. She often called for them at their homes in a shiny station wagon with the government emblem on the front door and chauffeured them to and from the place of examination. According to the Public Health Reports article, these rides became "a mark of distinction for many of the men who enjoyed waving to their neighbors as they drove by." There was nothing to indicate that the members of "Miss Rivers' Lodge" knew they were participating in a deadly serious experiment.

Spokesmen for the Public Health Service were quick to point out that the experiment was never kept secret, as many newspapers had incorrectly reported when the story first broke. Far from being clandestine, the Tuskegee Study had been the subject of numerous reports in medical journals and had been openly discussed in conferences at professional meetings. An official told reporters that more than a dozen articles had appeared in some of the nation's best medical journals, describing the basic procedures of the study to a combined readership of well over a hundred thousand physicians. He denied that the Public Health Service had acted alone in the experiment, calling it a cooperative project that involved the Alabama State Department of Health, the Tuskegee Institute, the Tuskegee Medical Society, and the Macon County Health Department.

Apologists for the Tuskegee Study contended that it was at best problematic whether the syphilitic subjects could have been helped by the treatment that was available when the study began. In the early 1930s treatment consisted of mercury and two arsenic compounds called arsphenamine and neoarsphenamine, known also by their generic name, salvarsan. The drugs were highly toxic and often produced serious and occasionally fatal reactions in patients. The treatment was painful and usually required more than a year to complete. As one CDC officer put it, the drugs offered "more potential harm for the patient than potential benefits."

PHS officials argued that these facts suggested that the experiment had not been conceived in a moral vacuum. For if the state of the medical art in the early 1930s had nothing better than dangerous and less than totally effective treatment to offer, then in the balance, little harm was done by leaving the men untreated.

Discrediting the efficacy of mercury and salvarsan helped blunt the issue of withholding treatment during the early years, but public health officials had a great deal more difficulty explaining why penicillin was denied in the 1940s. One PHS spokesman ventured that it probably was not "a one-man decision" and added philosophically, "These things seldom are." He called the denial of penicillin treatment in the 1940s "the most critical moral issue about this experiment" and admitted that from the present perspective "one cannot see any reason that they could not have been treated at that time." Another spokesman declared; "I don't know why the decision was made in 1946 not to stop the program."

The thrust of these comments was to shift the responsibility for the Tuskegee Study to the physician who directed the experiment during the 1940s. Without naming anyone, an official told reporters: "Whoever was director of the VD section at that time, in 1946 or 1947, would be the most logical candidate if you had to pin it down." That statement pointed an accusing finger at Dr. John R. Heller, a retired PHS officer who had served as the director of the division of venereal disease between 1943 and 1948. When asked to comment, Dr. Heller declined to accept responsibility for the study and shocked reporters by declaring: "There was nothing in the experiment that was unethical or unscientific."

The current local health officer of Macon County shared this view, telling reporters that he probably would not have given the men penicillin in the 1940s either. He explained this curious devotion to what nineteenth-century physicians would have called "therapeutic nihilism" by emphasizing that penicillin was a new and largely untested drug in the 1940s. Thus, in his opinion, the denial of penicillin was a defensible medical decision.

A CDC spokesman said it was "very dubious" that the participants in the Tuskegee Study would have benefited from penicillin after 1955. In fact, treatment might have done more harm than good. The introduction of vigorous therapy after so many years might lead to allergic drug reactions, he warned. Without debating the ethics of the Tuskegee Study, the CDC spokesman pointed to a generation gap as a reason to refrain from criticizing it. "We are trying to apply 1972 medical treatment standards to those of 1932," cautioned one official. Another officer reminded the public that the study began when attitudes toward treatment and experimentation were much different. "At this point in time," the officer stated, "with our current knowledge of treatment and the disease and the revolutionary change in approach to human experimentation, I don't believe the program would be undertaken."

Journalists tended to accept the argument that the denial of penicillin during the 1940s was the crucial ethical issue. Most did not question the decision to withhold earlier forms of treatment because they apparently accepted the judgment that the cure was as bad as the disease. But a few journalists and editors argued that the Tuskegee Study presented a moral problem long before the men were denied treatment with penicillin. "To say, as did an official of the Center for Disease Control, that the experiment posed 'a serious moral problem' after penicillin became available is only to address part of the situation," declared the *St. Louis Post-Dispatch*. "The fact is in an effort to determine from autopsies what effects syphilis has on the body, the government from the moment the experiment began withheld the best available treatment for a particularly cruel disease. The immorality of the experiment was inherent in its premise."

Viewed in this light, it was predictable that penicillin would not be given to the men. *Time* magazine might decry the failure to administer the drug as "almost beyond belief or human compassion," but along with many other publications it failed to recognize a crucial point. Having made the decision to withhold treatment at the outset, investigators were not likely to experience a moral crisis when a new and improved form of treatment was developed. Their failure to administer penicillin resulted from the initial decision to withhold all treatment The only valid distinction that can be made between the two acts is that the denial of penicillin held more dire consequences for the men in the study. The *Chicago Sun Times* placed these separate actions in the proper perspective: "Whoever made the decision to withhold penicillin compounded the original immorality of the project."

The human dimension dominated the public discussions of the Tuskegee Study. The scientific merits of the experiment, real or imagined, were passed over almost without comment. Not being scientists, the journalists, public officials, and concerned citizens who protested the study did not really care how long it takes syphilis to kill people or what percentages of syphilis victims are fortunate enough to live to ripe old age with the disease. From their perspective the PHS was guilty of playing fast and loose with the lives of these men to indulge scientific curiosity.

Many physicians had a different view. Their letters defending the study appeared in editorial pages across the country, but their most heated counterattacks were delivered in professional journals. The most spirited example was an editorial in the *Southern Medical Journal* by Dr. R H. Kampmeir of Vanderbilt University's School of Medicine. No admirer of the press, he blasted reporters for their "complete disregard for their abysmal ignorance," and accused them of banging out "anything on their typewriters which will make headlines." As one of the few remaining physicians with experience treating syphilis in the 1930s, Dr. Kampmeir promised to "put this 'tempest in a teapot' into proper historical perspective."

Dr. Kampmeir correctly pointed out that there had been only one experiment dealing with the effects of untreated syphilis prior to the Tuskegee Study. A Norwegian investigator had reviewed the medical records of nearly two thousand untreated syphilitic patients who had been examined at an Oslo clinic between 1891 and 1910. A follow-up had been published in 1929, and it was the state of published medical experimentation on the subject before the Tuskegee Study began. Dr. Kampmeir did not explain why the Oslo Study needed to be repeated.

The Vanderbilt physician repeated the argument that penicillin would not have benefited the men, but he broke new ground by asserting that the men themselves were responsible for the illnesses and deaths they sustained from syphilis. The PHS was not to blame, Dr. Kampmeir explained, because "in our free society, antisyphilis treatment has never been forced." He further reported that many of the men in the study had received some treatment for syphilis down through the years and insisted that others could have secured treatment had they so desired. He admitted that the untreated syphilitics suffered a higher mortality rate than the controls, observing coolly: "This is not surprising. No one has ever implied that syphilis is a benign infection." His failure to discuss the social mandate of physicians to prevent harm and to heal the sick whenever possible seemed to reduce the Hippocratic oath to a solemn obligation not to deny treatment upon demand.

Journalists looked at the Tuskegee Study and reached different conclusions, raising a host of ethical issues. Not since the Nuremberg trials of Nazi scientists had the American people been confronted with a medical cause that captured so many headlines and sparked so much discussion. For many it was a shocking revelation of the potential for scientific abuse in their own country. "That it has happened in this country in our time makes the tragedy more poignant," wrote the editor of the *Philadelphia Inquirer*. Others thought the experiment totally "un-American" and agreed with Senator John Sparkman of Alabama, who denounced it as "absolutely appalling" and "a disgrace to the American concept of justice and humanity."

Memories of Nazi Germany haunted some people as the broader implications of the PHS's role in the experiment became apparent. A man in Tennessee reminded health officials in Atlanta that "Adolf Hitler allowed similar degradation of human dignity in inhumane medical experiments on humans living under the Third Reich," and confessed that he was "much distressed at the comparison." A New York editor had difficulty believing that "such stomach turning callousness could happen outside the wretched quackeries spawned by Nazi Germany."

The specter of Nazi Germany prompted some Americans to equate the Tuskegee Study with genocide. A civil rights leader in Atlanta, Georgia, charged that the study amounted to "nothing less than an official, premeditated policy of genocide." A student at the Tuskegee Institute agreed. To him, the experiment was "but another act of genocide by whites," an act that "again exposed the nature of whitey: a savage barbarian and a devil."

Most editors stopped short of calling the Tuskegee Study genocide or charging that PHS officials were little better than Nazis. But they were certain that racism played a part in what happened in Alabama. "How condescending and void of credibility are the claims that racial considerations had nothing to do with the fact that 600 [all] of the subjects were black," declared the *Afro-American* of Baltimore, Maryland. That PHS officials had kept straight faces while denying any racial overtones to the experiment prompted the editors of this influential black paper to charge "that there are still federal officials who feel they can do anything where black people are concerned."

The *Los Angeles Times* echoed this view. In deftly chosen words, the editors qualified their accusation that PHS officials had persuaded hundreds of black men to become "human guinea pigs" by adding: "Well, perhaps not quite that [human guinea pigs] because the doctors obviously did not regard their subjects as completely human." A Pennsylvania editor stated that such an experiment "could only happen to blacks." To support this view, the *New Courier* of Pittsburgh implied that American society was so racist that scientists could abuse blacks with impunity.

Other observers thought that social class was the real issue, that poor people, regardless of their race, were the ones in danger. Somehow people from the lower class always seemed to supply a disproportionate share of subjects for scientific research. Their plight, in the words of a North Carolina editor, offered "a reminder that the basic rights of Americans, particularly the poor, the illiterate, and the friendless, are still subject to violation in the name of scientific research." To a journalist in Colorado, the Tuskegee Study demonstrated that "the Public Health Service sees the poor, the black, the illiterate and the defenseless in American society as a vast experimental resource for the government." And the *Washington Post* made much the same point when it observed, "There is always a lofty goal in the research work of medicine but too often in the past it has been the bodies of the poor . . . on whom the unholy testing is done."

The problems of poor people in the rural South during the Great Depression troubled the editor of the *Los Angeles Times*, who charged that the men had been "trapped into the program by poverty and ignorance."

Yet poverty alone could not explain why the men would cooperate with a study that gave them so little in return for the frightening risks to which it exposed them. A more complete explanation was that the men did not understand what the experiment was about or the dangers to which it exposed them. Many Americans probably agreed with the *Washington Post's* argument that experiments "on human beings are ethically sound if the guinea pigs are fully informed of the facts and danger." But despite the assurances of PHS spokesmen that informed consent had been obtained, the Tuskegee Study precipitated accusations that somehow the men had either been tricked into cooperating or were incapable of giving informed consent.

An Alabama newspaper, the *Birmingham News*, was not impressed by the claim that the participants were all volunteers, stating that "the majority of them were no better than semiliterate and probably didn't know what was really going on." The real reason they had been chosen, a Colorado journalist argued, was that they were "poor, illiterate, and completely at the mercy of the 'benevolent' Public Health Service." And a North Carolina editor denounced "the practice of coercing or tricking human beings into taking part in such experiments."

The ultimate lesson that many Americans saw in the Tuskegee Study was the need to protect society from scientific pursuits that ignored human values. The most eloquent expression of this view appeared in the *Atlanta Constitution*. "Sometimes, with the best of intentions, scientists and public officials and others involved in working for the benefit of us all, forget that people are people," began the editor. "They concentrate so totally on plans and programs, experiments, statistics—on abstractions—that people become objects, symbols on paper, figures in a mathematical formula, or impersonal 'subjects' in a scientific study." This was the scientific blindspot to ethical issues that was responsible for the Tuskegee Study—what the *Constitution* called "a moral astigmatism that saw these black sufferers simply as 'subjects' in a study, not as human beings." Scientific investigators had to learn that "moral judgment should always be a part of any human endeavor," including "the dispassionate scientific search for knowledge."

11

Crime and Deviance: Who Goes to Prison in the United States?

Minority status can lower educational attainment; cause premature death; and increase unemployment or underemployment, the chances of poverty, and single-parenthood. In this chapter, we learn of some additional negatives. Minority status also can increase criminal (often violent) behavior, the chances of being imprisoned, and the danger of being a victim of crime. We examine the nature of crime and deviance in American society with a focus on how criminal behavior is defined and penalized. Compared with the more general topic of criminology, the distinguishing characteristic of this chapter is that the spotlight is always on racial and ethnic minorities. Major topics include what crimes they commit, why they commit the type and number of crimes they do, and what happens to them within the criminal justice system. Although most Americans are white, most prisoners are nonwhite. This chapter attempts to explain why.

Criminals and Victims: Racial and Ethnic Variation

Deviance is any social behavior that breaks social norms supported by established societal authorities at any given time. Many sets of social norms exist in any society, and they vary as you travel from group to group. Professional strippers do not live by the same social rules as ministers, yet members of both groups find their respective behaviors perfectly satisfactory. A member of either group would probably be uncomfortable spending much time in the company of the other. Which set of behaviors is normal and which is deviant? Most authorities in American society would side with the ministers. Without this stipulation, everyone is a deviant to someone else and the term degenerates into name-calling. Note, however, that our definition makes deviants of the Dutch family that hid Anne Frank and her parents from Nazis during World War II.

In addition, authorities can come and go, or they can change their minds about what is deviant. Either way, definitions of which norms are respectable (normal) and which are deviant will change. After World War II ended, the Dutch protectors of the Frank family changed from being viewed as lawbreakers to being viewed as heroes. Similarly, in the United States, cocaine was perfectly legal in the nineteenth and early twentieth centuries and was even used in various tonics and soft drinks. You may enjoy a Coca Cola from time to time, but it does not have the kick it had before 1903. For these reasons, we must always locate definitions of deviant behavior in particular times and places.

All of these limitations take on additional importance when we turn to the forms of deviant behavior that are additionally labeled "criminal" behavior. When societal authorities (in this case, legislators) become concerned enough about a particular norm, they declare it to be a law and specify punishment for its being broken. Unlike other deviance, which may result in strange looks or smirks from others, criminal deviance produces highly organized responses from specialized officials. Depending on one's society, those responses can vary in form and complexity. A medieval king, for example, might learn of wrongdoing and order a head removed within the hour. By contrast, modern American society maintains an elaborate criminal justice system, complete with legislators, police officers, lawyers, judges, courtrooms, and prisons. You may still wind up losing your head, but it takes longer.

This section of the chapter examines the interplay between racial and ethnic groups and criminal behavior, focusing on how members of specific groups become involved as perpetrators, victims, and sometimes both. We also take a close look at the legislative end of the criminal justice system along with its end product—jails and prisons filled largely with minority lawbreakers.

Crime and Victimization

The best way to begin looking at crime is through the experiences of its victims. First, we can learn who suffers the most from crime and what kinds of crime in particular. At the same time, we see the total picture of criminal behavior in the United States; victimization studies can tell us how many and which types of crimes occur during a given period. Finally, we also learn some things about the perpetrators of crime. What we learn varies from one type of crime to another—victims know more about robbers than burglars—but we do wind up with a fairly clear picture of violent crimes. With most violent crimes, the perpetrator has to be close to the victim. Often, the perpetrator is personally known to the victim, but failing that, violent crime victims generally know something about the sex, age, and race of the person who injures them.

Figures 11.1 and 11.2 show changes in violent and property crime, respectively, since 1973. Violent crimes have been relatively stable during this period, dropping somewhat in the mid-1980s and again in the mid-1990s. These types of crime are particularly related to the age of the perpetrator; when older people do commit crimes,

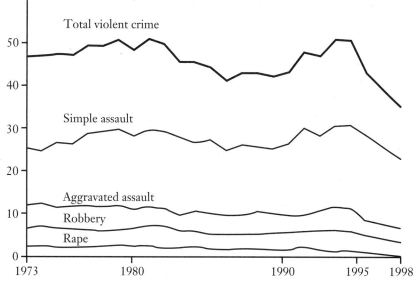

FIGURE 11.1 VIOLENT CRIME VICTIMIZATION RATES PER 1,000 PERSONS 1973–1998

U.S. Department of Justice, 1997e, 1999d.

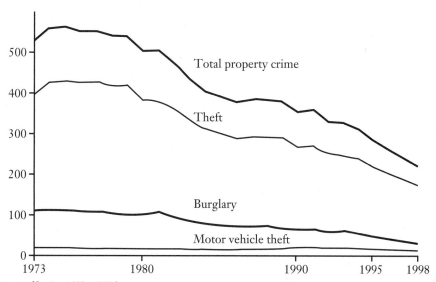

FIGURE 11.2 PROPERTY CRIME VICTIMIZATION RATES PER 1,000 PERSONS 1973–1998

U.S. Department of Justice, 1997e, 1999d.

they tend to be less violent. Violent crime rates are therefore very sensitive to the percentage of young people in a given population.

Figure 11.2 shows very different trends for property crimes. Since 1980, all property crimes have been dropping steadily. Only motor vehicle theft has fluctuated, rising in the late 1980s and early 1990s, but it stabilized in 1998 at its lowest level in two decades. Personal property is safer than it has been in quite some time. Like violent crimes, property crimes are also age related; however, almost all crime is age related because younger people are more willing to take chances. The overall aging of the American population therefore has something to do with these rates. The relative stability of violent crime, however, is more thoroughly examined later in this chapter.

Interestingly, American public opinion is not in keeping with these drops in crime rates. From 1982 to 1992, around 5 percent of Americans polled cited crime as the "most important problem facing this country today" when they were confronted with a list of approximately twenty social problems. In early 1994, when *all* crimes were dropping, that percentage shot up to 37 percent and peaked at 52 percent in the late summer of that year. By 1996, it had dropped to only 25 percent (U.S. Department of Justice, 1996a). Public opinion is apparently influenced by other forces than objective reality.

Who are the victims of these crimes? Table 11.1 provides a breakdown of violent crime victims separated by race/ethnicity and by gender for blacks and whites in 1997. (As always, remember that "Hispanics" may be either black or white so that

TABLE 11.1	ESTIMATED VICTIMIZATION RATES PER 1,000 POPULATION FOR PERSONS 12 AND OLDER, 1997				
	Male		**Female**		**Both Sexes**
Type of Crime	**White**	**Black**	**White**	**Black**	**Hispanic**
All personal crimes	45.8	61.3	33.9	44.7	45.6
Crimes of violence	44.3	59.0	32.6	40.6	43.1
Completed	12.4	20.4	10.6	15.0	15.2
Attempted/threatened	31.9	38.6	22.0	25.6	28.0
Rape/sexual assault	0.3	0.2	2.4	2.8	1.5
Robbery	5.3	11.9	2.4	3.6	7.3
Complete/property taken	3.3	7.8	1.5	2.7	4.5
Attempted	2.0	4.1	0.9	0.9	2.8
Assault	38.7	46.8	27.7	34.1	34.3
Aggravated	10.6	14.7	5.9	10.1	10.4
Simple	28.1	32.1	21.9	24.0	24.0

U.S. Department of Justice, 1999a.

column duplicates figures from the other columns.) These data show the victimization risk of both minority and gender status. Only 45.8 per every 1,000 white males were victimized compared with 61.3 per every 1,000 black males. Another way of thinking about these rates is in terms of percent of the whole population—approximately 4.6 percent of all white males in the United States were victimized in 1997 compared with slightly over 6 percent of all black males. A similar gap occurs between white and black females, although both had lower victimization rates than their male counterparts. Although the U.S. Department of Justice does not provide us with a gender breakdown for Hispanics, their 45.6 average rate for both sexes suggests a lower overall victimization rate than for blacks, but it is definitely higher than for whites.

Taking a closer look at Table 11.1, we see that black males are at particular risk as victims for the most violent crimes while white males have higher rates with less violent crimes. With black and white females, the gap in victimization is so large that black female rates are higher in every category. Because black females instigate relatively little violent crime, their high victimization rates show the danger inherent in the poverty status and crime rate of their community. And, as a final observation, note the high rates of robbery victimization for blacks and Hispanics as compared with whites: being robbed is often the result of being (or living) at the wrong place at the wrong time.

Table 11.2 provides us with a more general view of victimization between 1992 and 1996. These data show the extreme differences in victimization between Asian Americans and Native Americans with regard to violent crimes. While the Asian victimization rate is almost half that of white Americans, Native Americans face a victimization rate almost three times that of white Americans.

We noted earlier that victims of violent crime usually know something about their attackers. Since we later look at how minorities interact with the criminal justice system, this information is extremely useful in measuring the "justice" of that system. Michael Hindelang (1976) studied the match between the perpetrator's race supplied

TABLE 11.2   ANNUAL AVERAGE RATES OF VICTIMIZATION BY VIOLENT CRIME, 1992–1996, PER 1,000 PERSONS AGE 12 AND OLDER.

	All Races	Native American	White	Black	Asian
Violent Victimizations	50	124	49	61	29
Rape/sexual Assault	2	7	2	3	1
Robbery	6	12	5	13	7
Aggravated Assault	11	35	10	16	6
Simple Assault	31	70	32	30	15

U.S. Department of Justice, 1999b.

by the victim and the racial breakdown of arrests. For the violent crimes under study, the match was very close. This suggests that the disproportionate imprisonment of nonwhites for violent crimes occurs because they commit more violent crimes. We cannot use this approach, however, with most property crimes, because property crimes are rarely witnessed. Nor can we use it with drug possession or trafficking—offenses that become increasingly important in this chapter—because these crimes don't have "victims" who report a crime in the same sense that violent crimes do. The racial breakdowns of victimization rates do match self-report data provided by individuals on crimes they have committed (Sheley, 1993). All of this suggests that members of specific racial and ethnic groups are committing the acts for which they are arrested.

Blumstein (1982, 1993) approached the same question by comparing the racial breakdown of arrests with the proportions of European Americans and African Americans in prison. He concluded that differences in arrest rates by race explained 80 percent of the racial proportions in the prison population. The relationship was more true for some crimes than others, however. For two types of crime—larceny/auto theft and drug offenses—significantly more African Americans became prisoners than their arrest rates would predict. Still, 46 percent and 49 percent of prisoners, respectively, were explained by arrest rates for these two types of crime. Since neither of these two crimes is the kind of violent crime examined by Hindelang, there is no way of knowing if there is racial bias in the arrest process. We return to this question in the final section of the chapter.

Table 11.3 provides us with a more complete look at arrests in 1998, including racial and ethnic comparisons by types of offenses charged. Hispanics are mixed in with both white and black categories. Using the top percentage of arrests for each group as a starting point, it is easy to pick out offenses particularly common with certain groups. For example, Asian and Pacific Islanders make up 1.1 percent of all arrests but 4.8 percent of gambling arrests and 3.4 percent of juvenile runaways. (We should note that very few people are ever arrested for gambling.) Native Americans have a similar 1.2 percent overall arrest rate, but arrest rates for alcohol-related offenses are more than double (see Grobsmith, 1989). European Americans are highly represented in arson, vandalism, sex offenses, alcohol-related offenses, and some of the more minor juvenile offenses. African Americans, whose average arrest rate is two-and-one-half times their percentage of the population, appear in even higher numbers with murder, robbery, rape, gambling, vagrancy, and suspicion.

Ethnic stratification provides part of the explanation for arrest statistics. Poorer people are much more likely to be arrested for vagrancy (pretty much by definition) along with drunkenness; wealthier people become intoxicated indoors and, if caught, are usually in cars instead of on the sidewalk. We would also expect to find more poor robbers than poor embezzlers. However, more than just poverty is operating in these statistics: robbery is a more violent crime than embezzlement. However, the high percentages of African Americans and Puerto Ricans in other violent crime categories will lead us to more complex types of explanations when we look at the causes of crime in the middle section of this chapter (see Shai and Rosenwaike, 1988; Martinez, 1996).

Finally, where does all this crime occur? Personal crimes (including virtually all violent crimes) are more likely to occur in urban than suburban areas (46.3 compared

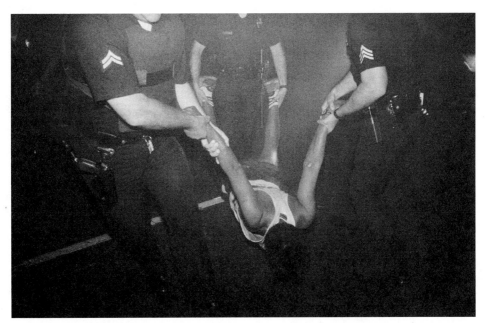

An urban arrest in progress.

with 35.5 per 1,000 population in 1998), and are more likely to occur in large urban areas and their suburbs than in and around smaller cities. Rural areas tend to have lower crime rates in almost all categories (U.S. Department of Justice, 1999d). Concentrated populations produce higher rates of crime; within that context, concentrations of poorer people in central cities produce higher crime rates. We return to this observation when we explore the causes of high crime rates.

Suburbs are safer, even in large metropolitan areas such as New York City, but they are not equally safer for all groups. Alba and others (1994) found that European Americans and Asian Americans in the New York City suburbs had lower crime victimization rates than Hispanics; suburban African Americans had the highest victimization rates. The race and ethnic differences lessened considerably, however, in suburbs with high household income, high rates of home ownership, and high English language proficiency. Being out of the central city lowers victimization rates, but being out of the city *and* in better economic circumstances is safer yet.

Perception and Fear of Crime

The Reverend Jesse Jackson commented in 1993, "There is nothing more painful to me at this stage in my life than to walk down the street and hear footsteps and start thinking about robbery—then look around and see somebody white and feel relieved" (quoted in Tonry, 1995:12). African Americans *do* commit a disproportionate amount of robbery; one could argue that the race of the person making those footsteps allows one to "play the odds" in how fearful one becomes. By the same token, however, the

TABLE 11.3   TOTAL ARRESTS IN THE UNITED STATES BY RACE, 1998.

Offense charged	Total Arrests					Percent Distribution				
	Total	White	Black	American Indian or Alaskan Native	Asian or Pacific Islander	Total	White	Black	American Indian or Alaskan Native	Asian or Pacific Islander
TOTAL	10,225,920	6,957,337	3,033,710	122,879	111,994	100.0	68.0	29.7	1.2	1.1
Murder and nonnegligent manslaughter	12,318	5,478	6,580	139	121	100.0	44.5	53.4	1.1	1.0
Forcible rape	21,646	13,022	8,112	241	271	100.0	60.2	37.5	1.1	1.3
Robbery	86,926	37,370	48,086	550	920	100.0	43.0	55.3	.6	1.1
Aggravated assault	358,506	220,777	130,018	3,645	4,066	100.0	61.6	36.3	1.0	1.1
Burglary	232,545	158,925	68,508	2,355	2,757	100.0	68.3	29.5	1.0	1.2
Larceny-theft	933,285	609,326	296,886	11,317	15,756	100.0	65.3	31.8	1.2	1.7
Motor vehicle theft	106,607	61,827	41,808	1,162	1,810	100.0	58.0	39.2	1.1	1.7
Arson	12,045	8,881	2,916	109	139	100.0	73.7	24.2	.9	1.2
Violent crime	479,396	276,647	192,796	4,575	5,378	100.0	57.7	40.2	1.0	1.1
Property crime	1,284,482	838,959	410,118	14,943	20,462	100.0	65.3	31.9	1.2	1.6
Crime Index total	1,763,878	1,115,606	602,914	19,518	25,840	100.0	63.2	34.2	1.1	1.5
Other assaults	943,293	604,058	317,905	11,697	9,633	100.0	64.0	33.7	1.2	1.0
Forgery and counterfeiting	80,979	53,801	25,717	484	977	100.0	66.4	31.8	.6	1.2
Fraud	267,447	179,870	84,266	1,306	2,005	100.0	67.3	31.5	.5	.7
Embezzlement	12,199	7,698	4,275	77	149	100.0	63.1	35.0	.6	1.2
Stolen property; buying, receiving, possessing	98,144	56,216	40,112	812	1,004	100.0	57.3	40.9	.8	1.0

Vandalism	212,136	156,150	51,298	2,638	2,050	100.0	73.6	24.2	1.2	1.0
Weapons; carrying, possessing, etc.	135,485	80,311	52,911	891	1,372	100.0	59.3	39.1	.7	1.0
Prostitution and commercialized vice	68,425	41,344	25,698	414	969	100.0	60.4	37.6	.6	1.4
Sex offenses (except forcible rape and prostitution)	65,841	49,224	15,050	754	813	100.0	74.8	22.9	1.1	1.2
Drug abuse violations	1,104,934	680,069	411,854	5,989	7,022	100.0	61.5	37.3	.5	.6
Gambling	9,213	2,753	5,985	36	439	100.0	29.9	65.0	.4	4.8
Offenses against the family and children	99,097	67,547	28,903	922	1,725	100.0	68.2	29.2	.9	1.7
Driving under the influence	950,938	826,117	99,794	13,611	11,416	100.0	86.9	10.5	1.4	1.2
Liquor laws	443,185	378,472	49,427	11,456	3,830	100.0	85.4	11.2	2.6	.9
Drunkenness	509,257	418,308	77,052	11,917	1,980	100.0	82.1	15.1	2.3	.4
Disorderly conduct	499,434	322,046	167,682	6,120	3,586	100.0	64.5	33.6	1.2	.7
Vagrancy	21,940	11,620	9,648	552	120	100.0	53.0	44.0	2.5	.5
All other offenses (except traffic)	2,683,508	1,715,532	905,171	31,170	31,635	100.0	63.9	33.7	1.2	1.2
Suspicion	3,792	2,752	965	50	25	100.0	72.6	25.4	1.3	.7
Curfew and loitering law violations	135,983	97,006	36,217	1,305	1,455	100.0	71.3	26.6	1.0	1.1
Runaways	116,812	90,837	20,866	1,160	3,949	100.0	77.8	17.9	1.0	3.4

U.S. Department of Justice, 1999c.

vast majority of African Americans do not commit acts of robbery, and Jesse Jackson knows that. In 1998, only 4 of every 1,000 Americans were robbed by anyone of any race at any time (U.S. Department of Justice, 1999d). Jackson is pained that he too has accepted a dominant stereotype of race and violent crime. But fear and perceptions do not always follow the reality of a threat. More Americans die every year from bee stings than shark attacks, but it is hard to remember that while paddling about in salt water.

Public perceptions of crime are influenced more by politicians and the media than by objective reality. Probably the best known example from politics was the use of Willie Horton in the 1988 Presidential election between George Bush and Michael Dukakis. Horton was a state prisoner in Massachusetts (Dukakis's home state) who committed a rape and a knife attack while on furlough from prison. Lee Atwater, Bush's political strategist, used Horton's case as symbolic of Dukakis being "soft on crime" and too liberal to be elected. The fact that Horton was African American made this symbolic point easier to accept for most European American voters. The nineteenth century fixation on African American males raping European American women was still alive and well. In point of fact, furloughs are commonplace in prisons and Horton had had nine previous uneventful furloughs. Yet the image of the crime as presented succeeded in increasing European American fears of African American criminals.

The media spread the Willie Horton story through political advertising and speech coverage. The media also influence how Americans view all crime through their standard treatment of the news. Johnstone and coworkers (1994) studied newspaper coverage of homicides in Chicago during 1987. Of the 684 homicides committed, 33 percent appeared in the papers. Homicides most likely to receive coverage were multiple victim homicides and when males murdered females; newspapers were less likely to cover homicides if the victim was either African American or Hispanic. In addition, lower socioeconomic status victims of any race or ethnicity were less likely to receive coverage (Hawkins et al., 1995). Overall newspaper behavior would suggest that European American women of higher social status were the most common murder victims. As we have seen, however, European American women, particularly those of higher socioeconomic status, have the lowest victimization rates from all forms of violent crime when compared with all males or nonwhite women (U.S. Department of Justice, 1999d). Reading the Chicago newspapers (and presumably most American newspapers) leaves the opposite impression from the reality of crime; the only accurate impression provided is that men commit more homicides than women.

An examination of individual perceptions and fears regarding crime shows that not all Americans believe everything they read. One parallel between newspaper accounts and actual fear, however, is that women tend to report higher levels of fear of crime than men (Parker and Ray, 1990; Parker et al., 1993; Pryor and McGarrell, 1993). The elderly also report greater fear of crime even though their actual victimization rates are lower than for younger people (Parker and Ray, 1990; Pryor and McGarrell, 1993; U.S. Department of Justice, 1996a). Beyond these findings, most people seem to have a pretty clear idea of when they are in danger. Urban residents report varying levels of fear of crime but often believe their own neighborhoods are safer than some other (presumably more dangerous) neighborhoods elsewhere (Pryor and McGarrell, 1993). In viewing European American fear of victimization from African Americans, that fear increases with levels of European American prejudice (St. John and Heald-Moore, 1996); however, both prejudice *and* fear of victimization decrease for European Americans who live in close proximity to African Americans (Skogan, 1995). Another quite common finding that matches the reality of victimization is that African Americans and Hispanics report higher fear of crime than do European Americans (Parker, 1988; Parker and Ray, 1990; Parker et al., 1993; Pryor and McGarrell, 1993; Skogan, 1995).

Fear and perceptions regarding crime are important in that beliefs about crime often find their way into law. There is some question about whether politicians shape public opinion or vice-versa—we will return to that question shortly—but the two are clearly related. In 1900, most Americans believed alcohol to be a more dangerous substance than cocaine. Less then two decades later, alcohol would become illegal in the United States. Today, most people believe the opposite and the laws are also reversed. Since the current state of laws and sentencing determines what kinds of people meet the criminal justice system, we cannot separate an examination of the two. The following section examines incarcerated Americans; the subsequent section examines that result in terms of the U.S. legal system.

Race and Ethnicity in the American Prison Population

We face the same problems of categorizing race and ethnicity with prisoners that we have faced throughout this book. The census often compares Hispanics with blacks and whites even though nearly all Hispanics are distributed between those two latter categories. At other times, Hispanics are excluded, leaving comparisons of non-Hispanic whites with non-Hispanic blacks. The second type of count is more useful for our purposes, but the federal government does not always provide it. In addition, we have to note that individuals sometimes classify themselves differently than authorities; some even change their self-labels during their lives if they have multiethnic or multiracial ancestry. We encounter all of these problems in trying to learn who occupies prison cells in the United States.

Incarcerated individuals in the United States may find themselves in various locations, ranging from local jails to military imprisonment. For the most part, we limit numerical comparisons to state and federal prisons. Between them, they hold most incarcerated individuals and their statistics are more reliable than for jails. The U.S. Department of Justice maintains prison records along both kinds of racial and ethnic categorization described above, only rarely including Native Americans and Asian and Pacific Islanders as specific categories.

Native Americans and Asian and Pacific Islanders are typically found lumped together as "other races." In 1996, "other races" comprised 2.6 percent of the state and federal prison population (U.S. Department of Justice, 1999e). U.S. population estimates for 1997 place combined Native Americans and Asian and Pacific Islanders (both not Hispanic) at 4.3 percent of the population with Asian and Pacific Islanders the larger at 3.6 percent (U.S. Bureau of the Census, 1999a). If we turn to arrest statistics for 1998, each of these two groups comprised a little over 1 percent of total arrests (U.S. Department of Justice, 1999e). Beyond arrest, Native Americans made up 1.8 percent of the 1997 state and federal prison population—more than double their percentage in the population—while Asian and Pacific Islanders comprised only 0.8 percent of that prison population. These data strongly suggest that Asian and Pacific Islanders are considerably underrepresented in the U.S. prison population while Native Americans are overrepresented: Asian and Pacific Islander arrests totaled less than a third of their percentage in the population while Native American arrests were double their percentage.

Non-Hispanic whites were 72.3 percent of the total U.S. population in 1998 but comprised only 36 percent of all state and federal prisoners. As with Asian and Pacific Islanders, they are underrepresented in the prison population. By contrast, non-Hispanic blacks comprise 12.1 percent of the population but 48 percent of the prison population. Hispanics are 11.2 percent of the population but 15.8 percent of the prison population (U.S. Bureau of the Census, 1999a; U.S. Department of Justice, 1997f). As with Native Americans, both of these groups are overrepresented in terms of their percentage of the overall population.

Non-Hispanic blacks are by far the most overrepresented group in the prison population, and their presence is increasing. Since 1985, non-Hispanic black males in the prison population increased 143 percent compared with 103 percent for non-Hispanic white males (U.S. Department of Justice, 1997f). (Women prisoners in both groups increased at even greater rates, but women are only about 6 percent of the total prison population; still, the rate of increase is worth noting.) Hispanics, on the other hand, increased their percentage of the prison population by 219 percent since 1985—the greatest increase of any group (U.S. Department of Justice, 1999e). While their overall percentage is 15.8 percent, it was less than 11 percent in 1985.

Figure 11.3 provides a little historical perspective to this picture. The African American percentage of the population has not changed drastically during the twentieth century. Their birth rate has been high, but other immigrants have helped keep their overall percentage about the same. In state prisons in 1926, African Americans were overrepresented but only by about two times their population percentage. Since then, the progression toward equal numbers with white prisoners in state prisons has been steady. Although this historical information ends in 1986, African Americans in the 1990s (including those of Hispanic origin) comprise more than half of the state prison population—the location of most of America's prisoners.

Figure 11.4 illustrates some of these comparisons concerning over- and underrepresentation of various groups in the prison population. As we saw in 1998, African Americans were almost four times as likely to be incarcerated compared with their overall population percentage, while European Americans were about half as likely to reach the same end. As Figure 11.4 clearly shows, African American males were about seven times as likely to be imprisoned in 1996 as European American men. If we expand "imprisoned" to include all men under criminal justice system control (those on probation and parole, for example), we encounter even higher numbers. In 1990, 23 percent of African American men between the ages of 20 and 29 were under criminal justice system supervision in some form. In Washington, D.C., that percentage rose to 42 percent and in Baltimore, to 56 percent (Tonry, 1995).

Figure 11.4 also shows the rate of change in these statistics since 1985. By combining these two statistics, we can more clearly the see the overall impact on the

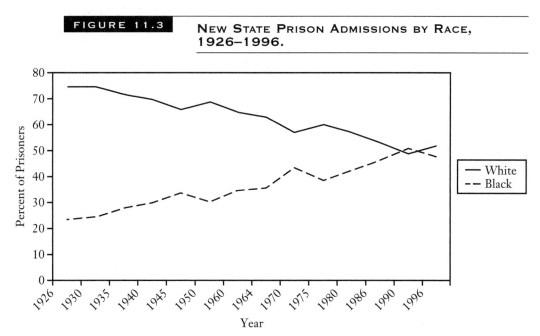

FIGURE 11.3 NEW STATE PRISON ADMISSIONS BY RACE, 1926–1996.

U.S. Department of Justice, 1999e.

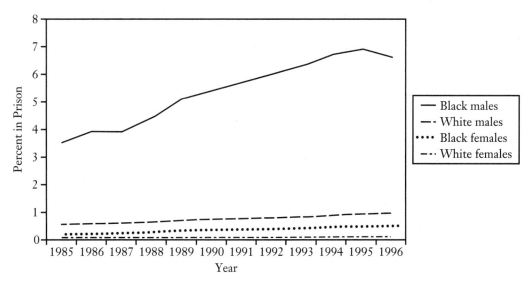

FIGURE 11.4 PERCENT OF ADULT POPULATION IN STATE OR FEDERAL PRISONS OR IN LOCAL JAILS, 1985–1996.

U.S. Department of Justice, 1999e.

African American community. While non-Hispanic blacks in the prison population increased 143 percent compared with 103 percent for non-Hispanic whites, there is a big numbers difference in the two communities. For African Americans, moving from 3.5 percent of the black male community to almost 7 percent is a very noticeable change that will have an impact on almost everyone in the community. For European Americans, a change from 0.5 percent to slightly less than 1 percent means that many European Americans will not even know an individual who is incarcerated.

Before leaving this overview of the U.S. prison population, a final but very important point is in order. Prisoners of any race or ethnicity are extremely likely to come from the bottom end of the social stratification system. In a 1993 study of the 1991 state prison population, only 34 percent had received high school diplomas (another 25 percent achieved a general equivalency degree [GED]), 33 percent were unemployed when arrested, and over half earned less than $10,000 during the year previous to arrest (U.S. Department of Justice, 1993). All of these figures are well below the national averages for that year. Ethnic stratification is obviously an important cause of the high percentages of nonwhites in prison. Any group more likely to be poor is more likely to have more members in prison. Is race or ethnicity a factor? Any reasonable explanation of incarceration needs to look at both race/ethnicity and social class.

The Politics of Crime: Defining the Act

Both crimes and acts of deviance are matters of social definition (see Curra, 1994). No behavior is ever either inherently deviant or inherently criminal (Winther, 1996). Based on our earlier definitions of these terms, we must turn to the influence of appropriate societal authorities to find the clout that forms and maintains these social definitions. Although this is true for both kinds of acts, it is especially true of criminal acts, since the process of social definition is so formal. Elected representatives of the people separate criminal acts from ongoing social behavior by declaring and voting them so. To the extent that we trust our representatives (and perhaps believe that they know more than we do), we may tend to follow their judgment. Returning to the alcohol and cocaine example, few of us are experts on the effects of such substances on the human body. Before we can study crime, therefore, we must study those who make the laws, the information they have at hand, and the relation between those officials and the general public.

Index Crime vs. White Collar and Corporate Crime   Criminal behavior is divided into endless subcategories by government officials and researchers. The Federal Bureau of Investigation creates a subset of crime known as index crime. **Index crime** includes murder, forcible rape, robbery, aggravated assault, burglary, larceny-theft, motor vehicle theft, and arson (U.S. Department of Justice, 1999c). The first four are considered violent crimes, while the last four are considered property crimes. When journalists report that crime rates are rising or falling, they are generally reporting FBI statistics on these eight offenses, usually separated into the violent and property categories.

In 1998, 17.2 percent of all arrests in the United States were for index crime offenses (U.S. Department of Justice, 1999c). Of the other offenses, over a million arrests

White-collar crime is viewed and treated very differently from other kinds of crime, but its impact is well documented.

each occurred for drug offenses, driving under the influence, and assaults other than aggravated. There were also numerous acts of vandalism, drunkenness, disorderly conduct, fraud, and liquor law offenses—the list is almost endless. These non-index offenses—all nonviolent offenses—comprised over 80 percent of all arrests. Combined with property index crime, 95 percent of all arrests are for nonviolent offenses.

The people who commit these different crimes are not randomly distributed throughout the population. **White collar crime**, for example, is a term made popular by Sutherland (1949) to describe a variety of on-the-job offenses by middle-class people who are in a position of trust. Employee theft is a classic example of such crimes. Unlike regular thieves, an employee has a key to the warehouse and is trusted to enter. For our purposes, white collar crime is distinguished by (1) the social class of the criminal, and (2) the nonviolence of the criminal act. The two go together: an individual with a key to the warehouse is very unlikely to commit a robbery; a lower class person who lacks that key has fewer criminal options available.

On average, the four types of violent offenses produce a much greater likelihood of serving time in prison than nonviolent offenses and about twice the average sentence length (U.S. Department of Justice, 1997c; 1997d). In 1998, 93 percent of the violent crime arrests were for two of the four offenses—robbery and aggravated assault (U.S. Department of Justice, 1999c). Black Americans, both Hispanic and non-Hispanic, made up 55.3 percent and 36.3 percent of the arrests, respectively, for those two offenses that year. If members of a given group are active in the crimes most punished, we would certainly expect to find many of them in prison.

The next logical question is to consider why some crimes are more greatly punished than others and who decides that? Doing physical damage to other people would seem to rate high on the list of punishable crimes. Few in the United States would disagree. But which is more important: the amount of physical damage caused or the circumstances (and people) through which it occurs? In short, how important are our images of criminals in determining levels of punishment? Is a poor nonwhite criminal acting alone more worthy of punishment than executives of a large corporation? A look at corporate crime in the United States will shed some light on how criminal behavior is defined.

Corporate crime refers to illegal actions taken by corporations to increase their profits. Such crimes as price fixing, bid rigging, environmental damage, and ignoring safety regulations fall under this heading. Although decisions to commit these acts are made by individuals (who will presumably profit personally if their corporation does so), the acts are clearly made in the name of the corporation with the sole purpose of increasing its profits. Price fixing, for example, involves a decision among executives from several corporations to sell similar products at a fixed price with no competition. Every corporation involved profits at the expense of those who consume those products. Victims of price fixing understandably prefer it to burglary, for example, in that the property stolen is usually spread out over a large percentage of the population; no one person suffers a major property loss. But what happens when illegal corporate actions result in death?

Paul Brodeur (1985) examined the history of the asbestos industry in the United States. Asbestos is a mineral and one of the best insulators ever found. The problem for people is that asbestos fibers can be inhaled into the lungs. Once there, they bounce around forever until they finally cause asbestosis or some form of lung cancer. Washing the clothes of an asbestos worker is a major health risk. Asbestos is far more deadly than any cigarette ever rolled. All of these risks became common knowledge in the United States by the 1980s, but how long had asbestos manufacturers known it?

In the course of many personal injury lawsuits brought against asbestos manufacturers (Johns-Manville was the largest), evidence arose that the dangers of the mineral were well known to executives within the industry for decades; some had known as much as a half century earlier. Did they tell their workers, introduce safety devices into factories, or just stop manufacturing this product? They bought liability insurance instead. As lawsuits mounted and corporations moved toward declaring bankruptcy, the courts stepped in. Funds were established to provide some relief for asbestos workers and the corporations were allowed to continue in operation, making other products. No corporate executive was fined or incarcerated. Overall, the asbestos manufacturers profited by their decisions. Many thousands of American workers died.

The story of the asbestos industry informs us that corporate crime can not only be violent but also can cause death beyond the wildest imaginations of any serial killer. Yet it is not punished. One could even argue that it is rewarded in some circumstances. It is not criminal behavior because it is not defined as criminal behavior by those authorities who make laws. This raises questions concerning the relation between behavior and the people doing it. Is death resulting from actions of a poor minority automatically viewed differently from death caused by the behavior of a white, upper class business executive? Are the latter's actions automatically viewed more positively because of higher social status? The asbestos story raises these questions. Perhaps more important for our purposes, however, the handling of drug offenses in the United States also raises these questions.

Drug Laws Earlier in this chapter, we examined the increasing proportions of African Americans and Hispanics in state and federal prisons. Much of this increase occurred just since the 1980s. Except for homicide rates, which have increased for African Americans and Puerto Ricans, members of these groups have not engaged in additional criminal activity through either violent or property crime (Tonry, 1995). What *has* changed is the American criminal justice system. Drug offenses began receiving very different treatment in arrest, prosecution, and sentencing beginning in the early 1980s (Byrne and Taxman, 1994; Cohn et al., 1991; Dressel, 1994; Scheingold, 1995; Tonry, 1994).

It is not disputed that drug trafficking among minority youth did increase in the 1980s (Fagan, 1992; Saner et al., 1995; Skogan and Lurigio, 1992). Much of the growth occurred in central cities, which brings to mind Wilson's notion of the underclass discussed earlier (Wilson, 1987, 1993). That growth was dwarfed, however, by increased antidrug activities from the criminal justice system. Drug arrests rose throughout the late 1960s and peaked in 1975. At that time, arrest rates were running about 0.6 percent in the African American community and 0.3 percent in the European American community. Starting in the early 1980s, arrest rates for African Americans began to rise sharply, reaching five times the European American rate by 1988 (Tonry, 1995).

The increase in arrests led to an increase in sentenced drug offenders. Between 1985 and 1995, the number of European American drug offenders sentenced to state prisons rose 306 percent; during that same period, the number of African American drug offenders increased 707 percent (U.S. Department of Justice, 1997f). Figure 11.5 shows these changes in comparison with other offenses in percentages of individuals incarcerated each year from 1980 to 1996. The dramatic rise in incarcerated drug offenders between 1984 and 1990 stands out against the relative stability of or decreases in other offenses. The Anti-Drug Abuse Act of 1986 mandated stiffer federal sentencing for drug possession and trafficking, in addition to the sharp rises in arrest rates previously noted. The arrest rates brought more drug offenders to court and the sentencing rules sent more of them to prison. In 1980, drug offenders comprised 6.8 percent of the state prison population in the United States; by 1996, they had risen to 30.2 percent (U.S. Department of Justice, 1999e). In federal prisons by 1991, 56 percent of all prisoners were drug offenders, up from 25 percent in 1980 (Tonry, 1995).

The Anti-Drug Abuse Act of 1986 specified minimum federal sentences for given amounts of various drugs found in the possession of an apprehended individual. This tended to increase incarceration for many drug possessors and traffickers, but the law made significant distinctions for different drugs. One of the most notable and controversial concerned powdered cocaine and crack cocaine. An individual convicted of selling 500 grams of powdered cocaine would receive a minimum five-year sentence; another individual convicted of selling 5 grams of crack cocaine would receive the same sentence. The same 100 to 1 ratio continued for increased amounts. Selling

FIGURE 11.5 PERCENT OF SENTENCED PRISONERS ADMITTED TO STATE PRISONS BY OFFENSE TYPE, 1980–1996

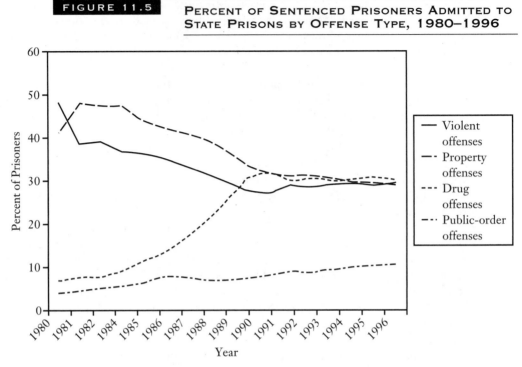

U.S. Department of Justice, 1999e.

5,000 grams of powdered cocaine and 50 grams of crack cocaine would both bring the same minimum ten-year sentence. This law was followed by the Anti-Drug Abuse Act of 1988, which provided mandatory incarceration for first offense possession of crack cocaine—the only controlled substance ever punished in that way in the United States (U.S. Sentencing Commission, 1997). Congress justified the difference on the grounds that crack was more addictive. If there is irony in these sentencing distinctions, it is that all crack cocaine is manufactured from powdered cocaine. Without traffickers in the latter, there would never be the former.

How do such sentencing guidelines differentially affect racial and ethnic groups? By 1996, black Americans were receiving 56 percent of all felony convictions in state courts for drug trafficking. This percentage was exceeded only by their robbery convictions of 66 percent (U.S. Department of Justice, 1999f). Drug trafficking sentences for state prisons are very similar in length to those given to burglars; drug trafficking sentences in federal prisons are much closer in length to those given for violent index crime (U.S. Department of Justice, 1999c). Given the sentencing guidelines previously discussed, our next question concerns the drugs that are being trafficked.

In 1991, 90 percent of crack arrests were of African Americans while 75 percent of powdered cocaine arrests were of European Americans (Miller, 1996). We see similar figures at the end of the criminal justice line. In 1993, 88.3 percent of all federal convictions for crack cocaine involved African Americans (U.S. Sentencing Commission,

FIGURE 11.6   REPORTED DRUG USE IN PREVIOUS 12 MONTHS BY COLLEGE STUDENTS AND ATTITUDES ABOUT DRUG ABUSE

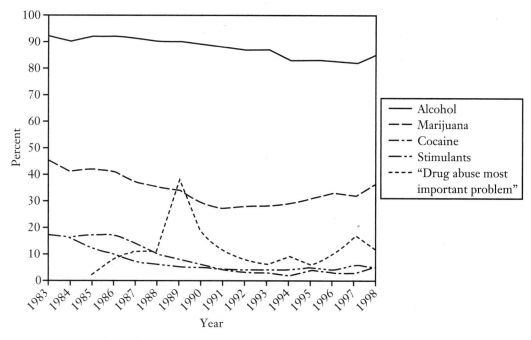

U.S. Department of Justice, 1999a.

1997). At the same time, most users of crack cocaine are European American (U.S. Sentencing Commission, 1997). Surprisingly, racial and ethnic groups abuse most substances in roughly the same percentage as their makeup of the overall population (Miller, 1996).

Political changes in the definition of criminal severity regarding drug offenses were justified as a response to public opinion. The assumption was that American voters wanted a "get tough" policy on drugs. Figure 11.6 raises some interesting questions about that assumption because it shows that the use of virtually all drugs, including legal drugs such as alcohol, has been steadily declining since 1979 for American college students, leveling off in recent years. Almost identical statistics appear for high school students. Drug abuse, in short, is decreasing and has been for some time (Tonry, 1995). Overlaid on those figures is the variation over the years in responses to the Gallup Poll question, "What do you think is the most important problem facing the country?" From a list of almost twenty possible social problems, drugs had never figured prominently until the mid-1980s, reaching a peak in 1989. By the time Americans began expressing great concern about the drug problem in the United States, legal intervention in drug offenses was already long under way. Such data suggest that politicians may fuel public opinion as often as the reverse.

This section's focus on drug offenses and the criminal justice system claims neither that drug traffickers are wonderful people nor that legislators had racist intent when writing laws. Rather, the intent is to show how policy can differentially affect racial and ethnic groups. Lawmakers also notice these things. The Violent Crime Control and Law Enforcement Act of 1994 directed the United States Sentencing Commission to investigate current drug sentencing and make recommendations. The Commission noted the racial differences in outcomes and found not enough persuasive information regarding differences in the drugs to warrant current practices. Their recommendation was that cocaine was cocaine, regardless of its form, and that sentencing should reflect that equality (U.S. Sentencing Commission, 1997).

Causes of Minority Crime

The previous section showed that race and ethnicity do make a difference in the criminal justice system. Minorities are much more likely to enter that system than other Americans. Most research indicates, however, that the overrepresentation of minorities in the prison system occurs because they commit more criminal acts; the race/ethnic breakdown of victimizers, arrestees, and convicts stays fairly constant. That is not to say that the criminal justice system is totally color blind—we examine that more closely in the next and final section of this chapter—but we have to look beyond the fate of Rodney King to understand minority crime.

Minorities do commit more criminal acts. With any given offense, we want to know if that commission occurred because of some factor in the minority community, some aspect of poverty that accompanies minority status, or a combination of the two. In addition, we need to examine the criminal act itself. Are the criminal acts of poor, nonwhite individuals punished more severely than the criminal acts of middle-class, white individuals? Do the criminal acts of "respectable" people just look more respectable because of the individual involved? Can a convicted respectable person "suffer enough" by losing so much respectability that prison is deemed unnecessary? Employee thieves are punished far less than shoplifters (if, indeed, they are punished at all beyond being fired). Corporate crime, as we have seen, is handled more like civil cases in court even when it causes death. We need to be conscious of all these questions in this section.

It would seem to be common sense that poverty alone would be the greatest predictor of crime. After all, those with the least property and the fewest options should have the greatest need. Poverty certainly explains crime selection—poor people cannot commit white collar crime—but it does not provide a full explanation for crime motivation. Yzaguirre (1987) notes the impact of rising poverty and unemployment rates on Hispanic crime. Krivo and Peterson (1996) compare European American and African American neighborhoods in Columbus, Ohio. They report equally high rates of violent crime in all disadvantaged neighborhoods, regardless of the race and ethnic composition of the neighborhood (see also Bankston and Zhou, 1997). Myers (1990) provides a longer view on poverty, showing how increased poverty can change crime definitions. She notes that lower cotton prices over the years in the South were associated with higher percentages of African Americans in prison. Unlike the other studies here, however, Myers does not attribute this to rising African American crime but to European American economic fears and their need to remove African Americans from the labor force when work is hard to find.

Other research on poverty and crime rates includes related (intervening) factors that affect whether poverty causes crime. Most research in this general area emphasizes the development of central city subcultures that condone or even promote crime and violence. This approach is very much in keeping with Wilson's idea of the underclass introduced in Chapter 9 (Wilson, 1987, 1993). Bursik and Grasmick (1993), for example, examine crime rates in Chicago. They report higher crime rates in the most disadvantaged African American neighborhoods but only in those most socially isolated. Neighborhoods with political "brokers" in them—people who can mediate between the community and city hall and/or the suburbs—have lower crime rates (see also Shihadeh and Graham, 1996; Vowell and Howell, 1998).

Additional research in the Wilson tradition focuses more closely on central city subcultures. Wilson (1987, 1993) notes that such subcultures develop through social isolation, which promotes a permanent lifestyle outside of the traditional labor market and a high dependence on public assistance. Such subcultures also can promote criminal behavior. Juvenile delinquency is related to poverty, particularly when the juvenile associates with aggressive peers and learns cultural norms supportive of violence (Heimer, 1997). Very similar social processes occur with African American adult criminals (Harer and Steffensmeier, 1992; Sheley, 1993). Shihadeh and Steffensmeier (1994) attribute much of this subculture to the family disruption characteristic of central city poverty. Lemelle (1995) argues that violent crime from young, male African Americans is actually a political statement on racism in America.

The Wilson tradition also emphasizes the importance of growing income inequality and the flight to the suburbs for those who can afford it. The suburban flight certainly adds to the social isolation of the central city, but the emphasis on income inequality more than poverty provides a slightly new approach. The implication is that poverty alone does not cause as much crime as increasing gaps between those who have and those who have not (Shihadeh and Flynn, 1996). By contrast, Liska and Bellair (1995) argue that the flight to the suburbs, particularly for European Americans, is not the cause of central city crime but rather the result of it.

Martinez (1996) finds that a higher income gap *within* the Latino community predicts higher homicide rates more accurately than overall poverty levels among Latinos. Harer and Steffensmeier (1992) compare inequality and race in all possible combinations, which allows them to compare violent crime rates according to inequality within the European American population, the African American population, and inequality between racial groups. They conclude that inequality is a major factor in rising crime rates, but more so for European Americans than for African Americans. This is true regardless of whether the inequality occurs only with the European American population or between the two races. Braun (1995) also finds income inequality to be a cause of crime but more so within the African American community than the European American community. Obviously, more work needs to be done here, but the role of income inequality seems pretty clear.

The Criminal Justice System in the United States

The criminal justice system in the United States is perhaps best thought of as a long conveyor belt. People are placed on the belt at one end upon arrest. The belt then moves them along, placing them in one legal circumstance after another. At any point in this travel, the individual may exit from the belt. If, for example, it becomes apparent to the court that prosecution will be difficult or impossible, a case may be dismissed early. The same individual may find himself or herself in court, farther down the belt, and have a judge dismiss a case for various reasons. Later, a jury may acquit. For individuals who do not exit the belt (that is, those who plead guilty or are found guilty), the criminal justice system may fine them, place them on probation, or incarcerate them.

Leaving the belt early in the process may not have much to do with the crime charged. If a case is thrown out because of illegally obtained evidence, for example, it does not matter whether the crime was murder or shoplifting. Although minorities do commit more serious crimes than their proportion of the population would warrant, our focus here is the impact, if any, of minority status on the ease of exiting the conveyor belt. And because minorities are disproportionately poor, we also are interested in the impact of poverty, if any, on the ease of exiting; again, anything that happens to poor people will happen to a greater degree with those minorities ethnically stratified.

As a final introductory note to this section, recall that we have already covered the most important questions regarding racial or ethnic status and the criminal justice system: all evidence suggests that the same percentages among racial and ethnic groups persist for those who commit crime, those arrested for crime, and those convicted for crime. Although the evidence is more convincing for violent as opposed to property crime, we should not expect to find too many surprises in this section. People who wind up in prison are there primarily because they engage more in behavior defined as criminal. When racial or ethnic bias enters the criminal justice system, it is most likely to occur with less serious crimes and with juvenile offenders. It also can occur in small doses during the many stages of prosecution, adding up to noticeable bias. A 1989 New Jersey survey of 169 superior court judges and 113 court managers found that 98 percent of respondents saw racial or ethnic incremental bias in their court system (Miller, 1996). Social class bias, however, is built into the system from one end of the conveyor belt to the other. In that regard, the criminal justice system is much like other institutions in American society.

Arrest: Entry into the System

Contemporary images of American police relating to minorities include the video recording of Rodney King receiving a beating on the streets of Los Angeles. They may also include the audio recording of Los Angeles police officer Mark Fuhrman making racist comments, used during the O.J. Simpson trial to impugn his testimony (Mathis, 1995; Gibbs, 1996). And not too many years have passed since the beating death of African American Arthur McDuffie by police in Miami in 1980 (Porter and Dunn, 1984). As with Rodney King, the beating followed a high-speed chase. Also, in both cases, the acquittal of the police officers involved led to riots. Both are extreme examples of what criminologists often refer to as "street justice" where police make sure punishment occurs in case the courts decide otherwise (Mann, 1993). Does this occur more often with African Americans than European Americans? Of course it must. They are involved in so many arrests. Does it occur more often to African Americans than European Americans when both are in the process of arrest? Most research indicates that police are more likely to show bias based on the social class or demeanor of arrestees than on their race or ethnicity; research results conflict even in this area, however (Klinger, 1994; Lundman, 1994, 1996; Sheley, 1993).

How does the African American community view the police in general? Table 11.4 provides African American and European American responses to the Gallup Poll question, "How much confidence [do] you, yourself, have in [the police]?" While there are clear differences between the two groups, 74 percent of African Americans have either "some" or a "great deal" of confidence in the police. As we noted earlier, African Americans are the primary victims of most crimes. Considering the group differences in confidence, are African American negative responses best explained by differential police bias or simply the greater degree of contact? We have already noted the larger numbers of arrests; African Americans also are more likely to be stopped and questioned by police than are European Americans (Mann, 1993). The amount of interaction seems to be important when we turn to the second question in Table 11.4 regarding the criminal justice system in general. European Americans may have less contact with police, but they have considerable contact with other legal matters. They appear to be less than satisfied customers (see Thomas and Rappaport, 1996).

TABLE 11.4   ATTITUDES OF AFRICAN AMERICANS AND EUROPEAN AMERICANS TOWARD THE POLICE AND THE CRIMINAL JUSTICE SYSTEM, 1999.

Question: How much confidence [do] you, yourself, have in [the police]?

	Great deal/quite a lot	Some	Very little	None
White	59%	33%	8%	*
Black	40%	34%	26%	0%

* Less than 0.5%

Question: How much confidence [do] you, yourself, have in [the criminal justice system]?

	Great deal/quite a lot	Some	Very little	None
White	22%	42%	33%	3%
Black	28%	31%	39%	2%

U.S. Department of Justice, 1999a.

Decisions concerning where police officers patrol directly affect who gets arrested. The more police in a neighborhood, the greater the number of arrests in that neighborhood. In general, police tend to patrol poorer and nonwhite neighborhoods. Part of the reason for this is that violent crime rates are higher in those neighborhoods. There are also large numbers of crack cocaine dealers—a profession associated with increased neighborhood violence (Tonry, 1995; Fagan, 1992). Ironically, arresting crack cocaine dealers tends to add to the violence because the removal of dealers produces turf wars among those remaining (Miller, 1996).

Central city crack cocaine dealers are particularly vulnerable to arrest because much of their trade is conducted on the streets. The "public" aspect of their crime not only adds visibility and increases public complaints but also makes them more suitable for arrest than suburban drug dealers. The latter are much more likely to do business indoors, which makes them harder to find and procuring admissible evidence more difficult (Tonry, 1995). Although nonwhites no doubt dominate the crack cocaine direct sale trade, higher percentages of them are likely to be arrested because of the way they do business and the frequency of police patrols.

Police discretion is most evident in marginal criminal behavior where penalties are less. This is particularly noticeable with juvenile justice. Since juveniles are handled through a totally separate court system from adults (with a clear emphasis on rehabilitation rather than punishment), the stakes are low. Almost all research on juvenile justice finds a greater likelihood of African American juveniles receiving harsher dispositions. The "conveyor belt" here includes the decision to arrest; the decision to place in immediate detention; and, ultimately, court decisions regarding probation or out-of-home placement. Juveniles from poorer and/or single-parent households may face bias concerning official evaluations of their home environment; this can produce increased numbers of out-of-home placements. However, most research finds much more bias in the juvenile justice system at the initial stages where police are most involved. Decisions about whether to arrest and detain are typically biased against nonwhite juveniles (Miller, 1996; Pope and Feyerherm, 1995; Sampson and Laub, 1993; Sheley, 1993; Tonry, 1995; U.S. Department of Justice, 1996a).

The Court System

The court system enters the picture immediately upon arrest. Not surprisingly, what follows is extremely complicated and contains a great many steps. At almost every step, an opportunity arises for the arrested individual to leave the system. Before taking a closer examination of the more important steps, an overview of the process might be helpful.

While the arrested individual is in the process of either hiring a lawyer or being assigned a public defender, the prosecution must decide whether it will charge the individual and, if so, with what. If the individual is to be charged, he or she may be released until a court appearance is scheduled, either on their own recognizance or on bail. The former allows the individual to remain free until the court appearance at no cost. This is usually granted for lesser crimes and/or to individuals of higher social status with known assets. Bail involves leaving something of value with the court to insure appearance, at which time the bail is returned. The court's concern here is that the individual not flee the jurisdiction before the court appearance.

Differences in the prosecution and sentencing of minority vs. white defendants hampers the American court system.

The next stage involves lawyers on both sides. The prosecution considers the range of possibilities from no prosecution through the most serious reasonable charge for the crime. Causing the death of someone, for example, can range from first-degree murder (planned and intentional) to manslaughter (unplanned and unintentional). Regardless of what the prosecution believes occurred, the charge must be based on the quality of evidence available. Meanwhile, lawyers for the defense must decide whether to plead guilty or not guilty. This decision also will be based on what evidence the defense believes the prosecution has. It also may be influenced by a plea bargain.

A **plea bargain** is literally a bargain in which the individual arrested agrees to plead guilty to a lesser charge than the most serious reasonable charge available to the prosecution. The arrested individual bargains away potential freedom for a guarantee of a lesser punishment. A very high percentage of criminal cases is handled through the plea bargain. As with any bargain, success depends on the existence of benefits to both sides.

Guilty pleas go straight to a judge for sentencing. Typically, the judge is somewhat limited by sentencing guidelines for given offenses. Not guilty pleas are of course typically mediated through a jury. Whatever feelings, beliefs, and experiences individual jurors bring to their assignments will understandably be important in their final decisions. It is in the interest of lawyers on both sides that the jurors who appear in the final panel are those with experiences beneficial to their respective goals. Either side may challenge the seating of an individual juror based on various causes.

After all of this, cases may be appealed. The criminal justice system never sleeps. This overview provides only the briefest of introductions to the criminal court system and necessarily omits many important steps. Our concern is with the most important steps and how they affect poor individuals and/or minority individuals.

Prosecutors and Defenders   Arrest for a criminal offense creates numerous and interrelated problems for the individual affected. Two of the first (and most important) problems are securing a defense lawyer and gaining freedom from incarceration between arrest and court appearance. Since that latter interval could be a long time, freedom not only brings obvious enjoyment but also, as we will see, provides greater opportunities for an improved defense.

Public defenders are court-appointed lawyers whose salaries are paid by the jurisdiction of the court. Ethnic stratification invariably sends more minority defendants to public defenders than to private lawyers. Unlike private lawyers, who are paid more when they do more work, public defenders receive the same salary regardless of the time they invest in their clients' cases. Different incentives do not guarantee different outcomes, but in general, those who can afford private lawyers are well advised to do so: defendants with public defenders have less positive outcomes (Sheley, 1993). It might seem logical that a salaried defender would have more interest in a plea bargain than in taking a case to jury—a plea bargain is much less work and the incentives are equal. In a study of why individuals plead guilty, however, Albonetti (1990) found that having a public defender increased the odds of an African American pleading guilty, but had no impact on European American defendants. In addition, the plea bargains were less beneficial to African American defendants. Hagan and Albonetti (1982) note that African American defendants tend to shy away from plea bargaining, having less trust in court officials and defense attorneys than do European Americans.

For their part, prosecutors tend to pursue cases with European American victims more forcefully than cases with minority victims. This is especially true if the defendant is African American (Sheley, 1993). Miller (1996) notes the impact of federal sentencing guidelines on prosecution practices. Guidelines were designed to remove bias, if any, from sentencing, but the impact occurs earlier in the process, with prosecutors less likely to offer plea bargains. Miller emphasizes the political nature of the prosecutor's job. The potential for mandatory lengthy sentences for more serious charges made the plea bargain less attractive for prosecutors. The result in federal courts during the early 1990s was that average sentences for African Americans rose 55 percent while average sentences for European Americans rose 7 percent. Note also that a prosecutor's decision to charge a defendant with a more serious crime also has an impact on the setting of bail.

The amount of bail and possibility of pretrial release are important in defendant outcomes. Again, ethnic stratification is a major factor in that poorer people are less able to afford bail and less likely to be released on their recognizance. Individuals who receive a pretrail release are more able to take an active part in their own defense before trial. These individuals obtain more favorable outcomes (Holmes and Daudistel, 1984; Sheley, 1993).

While bail is clearly related to the defendant's monetary resources, is it related to race or ethnicity? Sheley (1993) notes that African Americans are less likely to obtain pretrial release than European Americans, but this could be due to either different financial circumstances or the severity of the crime charged. Courts typically set bail amounts based on the judge's opinion of the defendant's pretrial flight likelihood: the greater the likelihood of that flight, the higher the bail to decrease that likelihood.

The obvious major factor here is the severity of crime charged. Individuals who face stiffer penalties would understandably be expected to forfeit higher amounts of bail to avoid trial. Ayres and Waldfogel (1994) studied the bail bond market in New Haven, Connecticut, comparing African American, Hispanic, and European American defendants. Given the same objective degree of pretrial flight probabilities, bail was consistently set higher for African Americans and male Hispanics. Albonetti and others (1989) found similar results but noted that the perceived dangerousness of the defendant also played a role.

Judges and Juries We have already encountered judges in the criminal justice process with the subject of bail. Although there is some evidence of judicial bias, judges in general are probably the most stable and least biased part of the criminal justice system (Welch et al., 1984). When sentencing bias does occur, it is most common for lesser offenses. For example, with less serious crimes, European Americans are more likely to receive probation than minorities (Miller, 1996). Holmes and associates (1993) compare Hispanic and European American judges in El Paso County, Texas, where Mexican Americans outnumber European Americans. They found that both groups of judges sentenced Hispanic defendants equally but that European American judges showed more lenience toward European American defendants. Gorton and Boies (1999) note that race is an important factor in the length of felony sentencing, mediated only when strict sentencing guidelines are in place.

Often, perceived bias in judicial sentencing has other roots than a personal bias by the judge. Holmes and Daudistel (1984) compare felony sentencing in El Paso, Texas, and Tucson, Arizona, for Hispanic, African American, and European American defendants. Greater racial and ethnic sentencing disparities in cases tried before juries appeared in El Paso than in Tucson, with both Hispanic and African American defendants receiving longer sentences for similar crimes. Significant racial and ethnic differences appeared in both cities, however. In Texas, those juries played a role in sentencing as well as determining guilt. In addition, plea bargains were highly restricted in El Paso. Although these factors were not true in Tucson, judges did impose stiffer sentences on minority defendants found guilty by juries. Of judges and juries, however, the former was clearly the less biased.

The minority percentage of a state's population often appears to play a role in sentencing decisions. Sheley (1993) notes that African Americans are most likely to receive stiffer sentences in states that are largely urban and have relatively small African American populations. Holmes and Daudistel (1984) found just the opposite in their comparison of El Paso with Tucson. Given that every state is different in terms of sentencing guidelines and court procedures (Byrne and Taxman, 1994), comparisons across state lines should always be viewed with caution.

Juries probably should not be held to the same standards as other participants in the criminal justice system. Their members, by definition, are not professionals in this area. Jurors bring their stereotypes with them into the jury box. Gordon and others (1988) and Gordon (1990) examine the interplay between stereotypes and decisions by placing subjects in simulated jury situations, facing cases varying in race, ethnicity, and type of crime. Such "juries" tend to punish European American embezzlers more than African American embezzlers but recommend the reverse of that punishment when the crime charged is burglary. Individuals stereotypically linked with a specific type of crime tend to receive less leniency.

Juror stereotyping is probably most evident with the most serious cases—capital punishment cases. Granted, attorneys attempt to promote such stereotyping when possible by being selective in seating jurors. In a capital case with an African American defendant and a European American victim, for example, prosecutors will definitely strive for a heavily European American jury while defense attorneys will seek the reverse (Fukurai et al, 1993; Lowery, 1991; see also Moore and Moore, 1997; see Stewart, 1995). All research on jury decisions in capital cases turn up the same result: the death penalty is much more likely to be employed when the victim is European American (Sheley, 1993).

Capital punishment was banned in the United States between 1968 and 1976. Both the banning and the reinstatement of capital punishment came from the U.S. Supreme Court. Since 1976, 60 percent of the murder victims in Alabama have been African American, but 80 percent of the death penalty outcomes occurred when the victim was European American. In Georgia during the same period, African Americans were the victims in 63 percent of the homicides, yet 13 of 14 executions in that state involved homicides with European American victims (Lowery, 1991). Overall, 50 percent of all homicide victims are European American, but 85 percent of death penalty cases involve European American victims (Jackson, 1996).

History presents a very similar picture. Radelet (1989) tracks the 15,978 executions held in the United States since 1608. Of those executed, only 30 were European Americans convicted for crimes against African Americans. By contrast, African Americans have received over 52 percent of all executions in the United States since 1930, including 89 percent of executions for the crime of rape (U.S. Department of Justice, 1997a). If there is an irony in the American criminal justice system, it is that minority defendants seem to face the most racial and ethnic bias when charged with the least serious crimes *and* the most serious crimes.

Summary

Racial and ethnic minorities are disproportionately involved in all aspects of crime and criminal justice. They vary from each other and from European Americans in the commission of crime, the types of crime committed, crime victimization, arrest rates, conviction rates, and their presence in the prison population.

Since 1980, property crime rates and arrest rates in the United States have dropped considerably. In the same period, violent crime rates dropped also but to a lesser degree. The one exception to this trend is drug-related crimes. The change with these crimes is more the result of changes in laws and sentencing regarding drugs.

Of all racial and ethnic minorities, African Americans are far more likely to engage in criminal activities and to be victimized by criminal activity. Rates are also high for Puerto Ricans. Hispanics in general fall in between African Americans and

European Americans with regard to victimization rates. For all groups, men are more likely to be victimized than women. African American involvement with crime and victimization is particularly true with regard to violent crimes. Most studies concerning violent crime show considerable consistency in the racial and ethnic breakdown of self-reported criminal activity, reports by victims, and arrest rates.

Public perceptions of crime typically do not match the statistics. Politicians often promote fear of crime to create political issues on which to run. The most famous such case was the use of Willie Horton by the 1988 Bush campaign for the Presidency. Newspapers tend to overreport violent crimes against higher class European American women—a group with the lowest victimization rates in the country. Additionally, general fears about drug-related crime have coincided with a several decades long drop in drug use throughout American society.

Both European Americans and Asian Americans are underrepresented in the U.S. prison population. Native Americans and Hispanics are imprisoned at approximately twice their percentages of the overall population; of all racial and ethnic groups, Hispanics have had the greatest increase in their percentage of the prison population since 1985. Nevertheless, African Americans clearly dominate the prison population, occupying cells at four times their percentage of the population—a rise from twice their percentage in 1926.

Much of the minority overrepresentation in arrest rates and the prison population is related to their social class and the types of crimes they commit. Poorer people are much more likely to commit the index crimes (including most one-on-one violent crime), which are the most severely punished crimes in American society. By contrast, middle-class people and above are more highly involved in white collar crime and corporate crime—crimes where the rewards are greater but the punishments are lower.

One of the most extreme variations in law that affects minorities is found in the drug laws regarding cocaine. An individual must possess one hundred times the amount of powdered cocaine to receive the same sentence as an individual possessing crack cocaine. Because those who sell powdered cocaine are largely European Americans (due to the high capital investment) and those selling crack are largely African American, the sentencing differential results in filling the prisons with African Americans.

Much of the research on causes of high minority crime rates focuses on high unemployment, income inequality, housing concentrations in central cities, and subcultural values that support criminal behavior.

The criminal justice system affects the outcome of minority crime primarily through legislation and sentencing guidelines such as those concerning specific drugs. The remainder of the system, ranging from police officers to public defenders to judges to prison wardens, largely reflects that legislation and those guidelines. Most research on the criminal justice system indicates high levels of discrimination against poor offenders and minimal (but present) levels of discrimination against minority offenders. Key factors in this research are levels of police patrolling in minority neighborhoods, the expense of hiring private lawyers, the system by which bail and pretrial release operate, the results of plea bargaining, and bias in jury decisions.

Chapter 11 Reading
Race and the Death Penalty

Anthony G. Amsterdam

Picking up where Chapter 11 left off, Anthony Amsterdam makes an impassioned plea against the death penalty in the United States. His argument is not directed against the penalty per se but rather against the racial bias in its implementation. Much of his argument focuses on the case of an African American from Georgia—Warren McCleskey—who received the death penalty for the homicide of a European American police officer. The Supreme Court decided in McCleskey v. Kemp that racism was not a factor in the jury's decision. Read on and reach your own conclusions.

There are times when even truths we hold self-evident require affirmation. For those who have invested our careers and our hopes in the criminal justice system, this is one of those times. Insofar as the basic principles that give value to our lives are in the keeping of the law and can be vindicated or betrayed by the decisions of any court, they have been sold down the river by a decision of the Supreme Court of the United States less than a year old.

I do not choose by accident a metaphor of slavery. For the decision I am referring to is the criminal justice system's *Dred Scott* case. It is the case of Warren McCleskey, a black man sentenced to die for the murder of a white man in Georgia. The Supreme Court held that McCleskey can be constitutionally put to death despite overwhelming unrebutted and unexplained statistical evidence that the death penalty is being imposed by Georgia juries in a pattern which reflects the race of convicted murderers and their victims and cannot be accounted for by any factor other than race.

This is not just a case about capital punishment. The Supreme Court's decision, which amounts to an open license to discriminate against people of color in capital sentencing, was placed upon grounds that implicate the entire criminal justice system. Worse still, the Court's reasoning makes us all accomplices in its toleration of a racially discriminatory administration of criminal justice.

Let us look at the *McCleskey* case. His crime was an ugly one. He robbed a furniture store at gunpoint, and he or one of his accomplices killed a police officer who responded to the scene. McCleskey may have been the triggerman. Whether or not he was, he was guilty of murder under Georgia law.

But his case in the Supreme Court was not concerned with guilt. It was concerned with why McCleskey had been sentenced to death instead of life imprisonment for his crime. It was concerned with why, out of seventeen defendants charged with the killings of police officers in Fulton County, Georgia, between 1973 and 1980, only Warren McCleskey—a black defendant charged with killing a white officer—had been chosen for a death sentence. In the only other one of these seventeen cases in which the predominantly white prosecutor's office in Atlanta had pushed for the death penalty, a black defendant convicted of killing a black police officer had been sentenced to life instead.

It was facts of that sort that led the NAACP Legal Defense Fund to become involved in McCleskey's case. They were not unfamiliar facts to any of the lawyers who, like myself, had worked for the Legal Defense Fund for many years, defending blacks charged with serious crimes throughout the South. We knew that in the United States black defendants convicted of murder or rape in cases involving white victims have always been sentenced to death and executed far out of proportion to their numbers, and under factual circumstances that would have produced a sentence of imprisonment—often a relatively light sentence of imprisonment—in identical cases with black victims or white defendants or both.

Back in the mid-sixties the Legal Defense Fund had presented to courts evidence of extensive statistical studies conducted by Dr. Marvin Wolfgang, one of the deans of American criminology, showing that the grossly disproportionate number of death sentences which were then being handed out to black defendants convicted of the rape of white victims could not be explained by any factor other than race. Prosecutors took the position then that these studies were insufficiently detailed to rule out the influence of every possible nonracial factor, and it was largely for that reason that the courts rejected our claims that our black death-sentenced clients had been denied the Equal Protection section of the Laws. Fortunately, in 1972 we had won a Supreme Court decision that saved the lives of all those clients and outlawed virtually every death penalty statute in the United States on procedural grounds; and when the States enacted new death-penalty laws between 1973 and 1976, only three of them reinstated capital punishment for rape. Now that it no longer mattered much, the prosecutors could afford to take another tack. When we argued against the new capital murder statutes on the ground that the Wolfgang studies had shown the susceptibility of capital sentencing laws to racially discriminatory application, the Government of the United States came into the Supreme Court against us saying, Oh, yes, Wolfgang was "a careful and comprehensive study, and we do not question its conclusion that during the twenty years between [1945 and 1965] . . ., in southern states, there was discrimination in rape cases." However, said the Government, this "research does not provide support for a conclusion that racial discrimination continues, . . . or that it applies to murder cases."

So we were well prepared for this sort of selective agnosticism when we went to court in the *McCleskey* case. The evidence that we presented in support of McCleskey's claim of racial discrimination left nothing out. Our centerpiece was a pair of studies conducted by Professor David Baldus, of the University of Iowa and his colleagues, which examined 2,484 cases of murder and non-negligent manslaughter that occurred in Georgia between 1973, the date when its present capital murder statute was enacted, and 1979, the year after McCleskey's own death sentence was imposed. The Baldus team got its data on these cases principally from official state records, supplied by the Georgia Supreme Court and the Georgia Board of Pardons and Paroles.

Through a highly refined protocol, the team collected information regarding more than five hundred factors in each case—information relating to the demographic and individual characteristics of the defendant and the victim, the circumstances of the crime and the strength of the evidence of guilt, and the aggravating and mitigating features of each case: both the features specified by Georgia law to be considered in capital sentencing and every factor recognized in the legal and criminological literature as theoretically or actually likely to affect the choice of life or death. Using the most reliable and advanced techniques of social-science research, Baldus processed the data through a wide array of sophisticated statistical procedures, including multiple-regression analyses based upon alternative models that considered and controlled for as few as 10 in as many as 230 sentencing factors in each analysis. When our evidentiary case was presented in court, Baldus reanalyzed the data several more times to take account of every additional factor, combination of factors, or model for analysis of factors suggested by the State of Georgia's expert witnesses, its lawyers, and the federal trial judge. The Baldus study has since been uniformly praised by social scientists as the best study of any aspect of criminal sentencing ever conducted.

What did it show? That death sentences were being imposed in Georgia murder cases in a clear, consistent pattern that reflected the race of the victim and the race of the defendant and could not be explained by any non-racial factor. For example:

(1) Although less than 40 percent of Georgia homicide cases involve white victims, in 87 percent of the cases in which a death sentence is imposed, the victim is white. White-victim cases are almost eleven times more likely to produce a death sentence than are black-victim cases.

(2) When the race of the defendant is considered too, the following figures emerge: 22 percent of black defendants who kill white victims are sentenced to death; 8 percent of white defendants who kill white victims are sentenced to death; 1 percent of black defendants who kill black victims are sentenced to death; 3 percent of white defendants who kill black victims are sentenced to death. It should be noted that out of the roughly 2,500 Georgia homicide cases found, only 64 involved killings of black victims by white defendants, so the 3 percent death-sentencing rate in this category represents a total of two death sentences over a six-year period. Plainly, the reason why racial discrimination against black defendants does not appear even more glaringly evident is that most black murderers kill black victims; almost no identified white murderers killed black victims; and virtually nobody is sentenced to death for killing a mere black victim.

(3) No non-racial factor explains these racial patterns. Under multiple regression analysis, the model with the maximum explanatory power shows that after controlling for legitimate non-racial factors, murderers of white victims are still being sentenced to death 4.3 times more often than murderers of black victims. Multiple regression analysis also shows that the race of the victim is as good a basis for predicting whether or not a murderer will be sentenced to death as are the aggravating circumstances which the Georgia statute explicitly says should be considered in favor of a death sentence, such

as whether the defendant has a prior murder conviction, or whether he is the primary actor in the present murder.

(4) Across the whole universe of cases, approximately 5 percent of Georgia killings result in a death sentence. Yet when more than 230 non-racial variables are controlled for, the death sentencing rate is 6 percentage points higher in white-victim cases than in black-victim cases. What this means is that in predicting whether any particular person will get the death penalty in Georgia, it is less important to know whether or not he committed a homicide in the first place than to know whether, if he did, he killed a white victim or a black one.

(5) However, the effects of race are not uniform across the entire range of homicide cases. As might be expected, in the least aggravated sorts of cases, almost no one gets a death sentence; in the really gruesome cases, a high percentage of both black and white murderers get death sentences; so it is in the mid-range of cases—cases like McCleskey's—that race has its greatest impact. The Baldus study found that in these mid-range cases the death sentencing rate for killers of white victims is 34 percent as compared to 14 percent for killers of black victims. In other words, out of every thirty-four murderers sentenced to death for killing a white victim, twenty of them would not have gotten death sentences if their victims had been black.

The bottom line is this; Georgia has executed eleven murderers since it passed its present statute in 1973. Nine of the eleven were black. Ten of the eleven had white victims. Can there be the slightest doubt that this revolting record is the product of some sort of racial bias rather than a pure fluke?

A narrow majority of the Supreme Court pretended to have such doubts and rejected McCleskey's Equal-Protection challenge to his death sentence. It did not question the quality or the validity of the Baldus study, or any of the findings that have been described here. It admitted that the manifest racial discrepancies in death sentencing were unexplained by any non-racial variable, and that Baldus's data pointed to a "likelihood" or a "risk" that race was at work in the capital sentencing process. It essentially conceded that if a similar statistical showing of racial bias had been made in an employment-discrimination case or in a jury-selection case, the courts would have been required to find a violation of the Equal Protection Clause of the Fourteenth Amendment. But, the Court said, racial discrimination in capital sentencing cannot be proved by a pattern of sentencing results: a death-sentenced defendant like McCleskey must present proof that the particular jury or the individual prosecutor, or some other decision-maker in his own case, was personally motivated by racial considerations to bring about his death. Since such proof is never possible to obtain, racial discrimination in capital sentencing is never possible to prove.

The Court gave four basic reasons for this result. First, since capital sentencing decisions are made by a host of different juries and prosecutors, and are supposed to be based upon "innumerable factors that vary according to the characteristics of the individual defendant and the facts of the particular capital offense," even sentencing patterns that are explicable by race and inexplicable except by race do not necessarily show that any single decision-maker in the system is acting out of a subjective purpose to discriminate. Second, capital punishment laws are important for the protection of society; the "[i]mplementation of these laws necessarily requires discretionary judgments"; and, "[b]ecause discretion is essential to the criminal justice process, we [sh]ould demand exceptionally clear proof before we... infer that the discretion has been abused" Third, this same respect for discretionary judgments makes it imprudent to require juries and prosecutors to explain their decisions, so it is better to ignore the inference of racial discrimination that flows logically from their behavior than to call upon them to justify such behavior upon non-racial grounds.

Fourth, more is involved than capital punishment. "McCleskey's claim... throws into serious question the principles that underlie our entire criminal justice system." This is so because "the Baldus study indicates a discrepancy that appears to correlate with race," and "[a]pparent disparities in sentencing are an inevitable part of our criminal justice system." "Thus," says the Court, "if we accepted McCleskey's claim that racial bias has impermissibly tainted the capital sentencing decision, we could soon be faced with similar claims as to other types of penalty. Moreover, the claim that... sentence rests on the irrelevant factor of race easily could be extended to apply to claims based on unexplained discrepancies that correlate to membership in other minority groups, and even to gender"—and even to claims based upon "the defendant's facial characteristics, or the physical attractiveness of the. . . victim." In other words, if we forbid racial discrimination in meting out sentences of life or death, we may have to face claims of discrimination against Blacks, or against women, or perhaps against ugly people, wherever the facts warrant such claims, in the length of prison sentences, in the length of jail sentences, in the giving of suspended sentences, in the making of pre-trial release decisions, in the invocation of recidivist sentencing enhancements, in the prosecutor's decisions whether to file charges, and how heavily to load up the charges, against black defendants as compared with white defendants or against ugly defendants as compared with ravishingly beautiful defendants; and of course the whole criminal justice system will then fall down flat and leave us in a state of anarchy. In thirty years of reading purportedly serious judicial opinions, I have never seen one that came as close to Thomas De Quincy's famous justification for punishing the crime of murder: "If once a man indulges himself in murder, very soon he comes to think little of robbing; and from robbing he next comes to drinking and Sabbath-breaking, and from that to incivility and procrastination."

Notice that the Court's version of this slippery-slope argument merely makes explicit what is implied throughout its opinion in the *McCleskey* case. Its decision is not limited to capital sentencing but purports to rest on principles which apply to the whole criminal justice system. Every part of that system from arrest to sentencing and parole, in relation to every crime from murder to Sabbath-breaking, involves a multitude of separate decision-makers making individualized decisions based upon "innumerable [case-specific] factors." All of these decisions are important for the protection of society from crime. All are conceived as "necessarily require[ing] discretionary judgments." In making these discretionary judgments, prosecutors and judges as well as jurors have traditionally been immunized from inquiry into their motives. If this kind of discretion implies the power to treat black people differently from white people and to escape the responsibility for explaining why one is making life-and-death decisions in an apparently discriminatory manner, it implies a tolerance for racial discrimination throughout the length and breadth of the administration of criminal justice. What the Supreme Court has held, plainly, is that the very nature of the criminal justice system requires that its workings be excluded from the ordinary rules of law and even logic that guarantee equal protection to racial minorities in our society.

And it is here, I suggest, that any self-respecting criminal justice professional is obliged to speak out against this Supreme Court's conception of the criminal justice system. We must reaffirm that there can be no justice in a system which treats people of color differently from white people, or treats crimes against people of color differently from crimes against white people.

We must reaffirm that racism is itself a crime, and that the toleration of racism cannot be justified by the supposed interest of society in fighting crime. We must pledge that when anyone even a majority of the Supreme Court—tells us that a power to discriminate on grounds of race is necessary to protect society from crime, we will recognize that we are probably being sold another shipment of propaganda to justify repression. Let us therefore never fail to ask the question whether righteous rhetoric about protecting society from crime really refers to protecting only white people. And when the answer, as in the McCleskey case, is that protecting only white people is being described as "protecting society from crime," let us say that we are not so stupid as to buy this version of the Big Lie, nor so uncaring as to let it go unchallenged.

Let us reaffirm neither the toleration of racism by the Supreme Court nor the pervasiveness of racism in the criminal justice system can make it right, and that these things only make it worse. Let us reaffirm that racism exists, and is against the fundamental law of this Nation, whenever people of different races are treated differently by any public agency or institution as a consequence of their race and with no legitimate non-racial reason for the different treatment. Let us dedicate ourselves to eradicating racism, and declaring it unlawful, not simply in the superficial, short-lived situation where we can point to one or another specific decision-maker and show that his decisions were the product of conscious bigotry, but also in the far more basic, more intractable, and more destructive situation where hundreds upon hundreds of different public decision-makers, acting like Georgia's prosecutors and judges and juries—without collusion and in many cases without consciousness of their own racial biases—combine to produce a pattern that bespeaks the profound prejudice of an entire population.

Also, let us vow that we will never claim—or stand by unprotestingly while others claim for us—that, because our work is righteous and important, it should be above the law. Of course, controlling crime is vital work; that is why we give the agencies of criminal justice drastic and unique coercive powers, including the powers of imprisonment and death. And of course discretion in the execution of such powers is essential. But it is precisely because the powers that the system regulates are so awesome, and because the discretion of its actors is so broad, that it cannot be relieved of accountability for the exercise of that discretion. Nor can it be exempted from the scrutiny that courts of law are bound to give to documented charges of discrimination on the ground of race by any agency of government. Let us declare flatly that we neither seek nor will accept any such exemption, and that we find it demeaning to be told by the Supreme Court that the system of justice to which we have devoted our professional lives cannot do its job without a special dispensation from the safeguards that assure to people of every race the equal protection of the law.

This is a stigma criminal justice practitioners do not deserve. Service in the criminal justice system should be a cause not for shame but for pride. Nowhere is it possible to dedicate one's labors to the welfare of one's fellow human beings with a greater sense that one is needed and that the quality of what one does can make a difference. But to feel this pride, and to deserve it, we must consecrate ourselves to the protection of all people, not a privileged few. We must be servants of humanity, not of caste. Whether or not the Supreme Court demands this of us, we must demand it of ourselves and of our coworkers in the system. For this is the faith to which we are sworn by our common calling: that doing justice is never simply someone else's job; correcting injustice is never simply someone else's responsibility.

Acculturation   The acquisition of one ethnic group's culture by another ethnic group (see *behavioral assimilation*).

Affirmative Action Programs   Programs initiated by the executive branch of the federal government originally designed to provide preferences for equally qualified women and minorities as compared to other applicants in applying for education and/or employment.

Assimilation   The process by which minority group members come to acquire the dominant group's culture and receive acceptance to higher social status within that society (see *acculturation, behavioral assimilation,* and *structural assimilation*).

Behavioral Assimilation   The efforts by minority group members to acquire the culture of a dominant group (see *acculturation*).

Caste Society   A society with a system of social stratification that is relatively closed to social mobility among its members.

Class Society   A society with a system of social stratification that is relatively open to social mobility among its members.

Concept   An intellectual construct that specifies (or directs attention to) certain similarities among different observations.

Coolie Labor   A slang term for Chinese indentured laborers who came to the New World during the nineteenth century. They were primarily in demand in South America and the West Indies.

Corporate Crime   Illegal actions taken by corporations to increase their profits. Such crimes as price fixing, bid rigging, environmental damage, and ignoring safety regulations fall under this heading.

Culture   Everything that people create, share with one another, and pass along to the next generation (see *nonmaterial culture* and *material culture*).

Discrimination   Any actions designed to hinder the competitive abilities of members of another group (see *overt discrimination* and *institutional discrimination*).

Dominant Group   An ethnic or racial group that occupies the dominant positions in a society, monopolizing positions of power and wealth. Among other things, it is responsible for the creation of minority group status for other racial or ethnic groups (see *minority group*).

***Encomienda* System**   Spanish labor system in the New World in which Native Americans were in virtual slavery and placed into forced labor.

Ethnic Group   A group of people who (1) share a common culture and/or ancestry, or (2) are defined by themselves or others as sharing a common culture and/or ancestry.

Ethnic Stratification A particular form of social stratification in which particular ethnic or racial groups are clustered at particular levels within the system.

Ethnocentrism The belief that one's own culture is superior to all others.

Expulsion Actions taken by the dominant group that forcibly remove the members of a given minority group beyond the borders of the dominant group's country.

Genocide Actions taken by the dominant group that serve to cause the deaths of many or all members of a given minority group.

Gini Index A statistic that measures the degree of inequality in a society. Ranging from a value of zero to a value of 1, zero would mean that every household in a society had the same income; a value of 1 would mean that one household received 100 percent of the income.

Human Capital An attribute of a worker that can create other kinds of capital. Education can therefore be conceptualized as human capital, which can be acquired by an individual and subsequently invested in the economy.

Indentured Servitude A labor system in which the laborer enters into a voluntary contract with an employer, selling future labor for a specified period of time for compensation at the contract signing.

Index Crime The Federal Bureau of Investigation creates a subset of crime known as index crime. This subset includes murder, forcible rape, robbery, aggravated assault, burglary, larceny-theft, motor vehicle theft, and arson.

Institutional Discrimination Discrimination in which the criteria for selection consist of skills, abilities, or attributes that individuals in particular groups may or may not possess (see *discrimination*).

Labor Exploitation A labor system in which the laborers produce appreciably more value through their labor than they receive in compensation; this applies particularly to laborers whose options in the labor force have been severely limited through discrimination.

Landsmannshaften Jewish organizations in the United States formed in the nineteenth century with membership of immigrants from particular regions in Europe. The primary function of these organizations was to help Jewish immigrants become settled and find employment in the United States.

Manifest Destiny A popular belief during the midnineteenth century that it was the will of God for European Americans to populate and dominate North America.

Material Culture Any cultural creations that are physical. Examples include clothing, buildings, books, tools, or art.

Melting Pot The joining (or amalgamating) of two or more ethnic or racial groups into one through intermarriage and the subsequent blending of all the cultures involved into one, wholly new culture.

Minority Group An ethnic or racial group that occupies a subordinate position in a society, unable to gain access to wealth or influence. Their subordinate status is maintained through limitations placed on them by the structure of their society.

Mulatto An individual with both European and African ancestry.

Nationalism Political efforts undertaken by a minority group to achieve separatism (see *separatism*).

Nonmaterial Culture   Any cultural creations that are not physical. Examples would include language, religious beliefs, food preferences, or expectations of how others will behave.

Overt Discrimination   Discrimination in which the criterion for selecting those affected is simply their group membership (see *discrimination*).

Plea Bargain   A legal system bargain in which an accused individual agrees to plead guilty to a lesser charge than the most serious reasonable charge available to the prosecution.

Pluralism   Actions taken by a minority group to maintain their cultural distinctiveness and achieve general economic and political acceptance of their group from the dominant group.

Political Machine   An urban political organization that seeks to obtain and maintain political dominance over city offices and employment. They were most common in the United States between the midnineteenth and midtwentieth centuries.

Prejudice   Negative attitudes held concerning the members of another group.

Primary (Economic) Sector   Economic activities associated with the production of raw materials.

Pull   Positive social, economic, and political factors in a potential host society that serve as a "magnet" for immigration (see *push*).

Push   Negative social, economic, and political factors in a society that encourage societal members to emigrate, seeking more favorable circumstances elsewhere (see *pull*).

Racial Group   Ethnic groups that have been labeled as biologically different from other ethnic groups.

Redemptioners   Indentured laborers who sign a promissory note with a ship's captain in return for passage in lieu of contracting directly with a future employer. The redemptioner is then responsible for obtaining a labor contract to pay the promissory note (see *indentured servitude*).

Secondary (Economic) Sector   Economic activities related to manufacturing.

Separatism   The political goal of a minority group to achieve independence from a dominant group through the creation of a separate state for their group. This state may be created through the partitioning of the original multicultural society or through the mass exodus of minority members to land elsewhere.

Social Category   A socially significant shared characteristic that creates the grouping of individuals by whether or not they share that characteristic. Examples are gender, ethnicity, race, or educational level.

Social Class   Economic position within a system of social stratification (as in economic rewards attached to particular positions).

Social Mobility   The movement of individuals or groups within the ranking systems of social stratification. It can be **intragenerational** (a change in social class that occurs within the lifetime of one individual) or **intergenerational** (change that occurs across generations).

Social Movement   The organized effort by a group of people to bring about social change, either in the members themselves or in members and society alike.

Social Power   The ability to influence the behavior of others that derives from positions within a system of social stratification.

Social Status   The prestige or relative importance of social position attached to a level of the social stratification system.

Social Stratification   The arrangement of different activities of a society into a hierarchy whereby activities ranked high are highly rewarded and activities ranked low are poorly rewarded.

Socioeconomic Status (SES)   A measure of social position that combines an individual's occupational prestige, educational attainment, and income.

Stereotype   An overgeneralization about a group of people that may range from being completely inaccurate to often inaccurate.

Structural Assimilation   The dominant group's acceptance of a newly behaviorally assimilated minority group. This acceptance may be limited to economic, political, and the more impersonal social arenas of a society (structural assimilation at the level of secondary relationships) or it may include full acceptance including intermarriage (structural assimilation at the level of primary relationships).

Tertiary (Economic) Sector   The production and distribution of services in an economy.

Theory (Scientific)   An explanation of observed behavior that specifies relations among relevant concepts, predicting the degree to which they affect each other and why.

Underclass   Those individuals in the United States who reside in central cities (especially those in the North) in neighborhoods of concentrated poverty in social isolation from other groups, maintain marginal (or nonexistent) ties to the labor force, and share a culture (or "social milieu") that reinforces their labor force marginality.

White Collar Crime   A term made popular by Sutherland (1949) to describe a variety of on-the-job offenses by middle-class people who are in a position of trust. Employee theft is a classic example of such crimes.

Adorno, T.W., Frenkel-Brunswik, E., Levinson, Daniel, & Sanford, R. Nevitt. (1950). *The Authoritarian Personality*. New York, NY: Harper & Row.

Agbayani-Siewert, Pauline, & Revilla, Linda. (1995). Filipino Americans. In Pyong Gap Min (Ed.), *Asian Americans: Contemporary Trends and Issues* (pp. 134–168). Thousand Oaks, CA: Sage.

Alaniz, Maria Luisa, & Wilkes, Chris. (1998). Pro-Drinking Messages and Message Environments for Young Adults: The Case of Alcohol Industry Advertising in African American, Latino & Native American Communities. *Journal of Public Health Policy, 19* (4), 447–471.

Alba, Richard D. (1999). Immigration and the American Realities of Assimilation and Multiculturalism. *Sociological Forum, 14* (1), 3–25.

Alba, Richard D., Logan, John R., & Bellair, Paul E. (1994). Living with Crime: The Implications of Racial/Ethnic Differences in Suburban Location. *Social Forces, 73* (2), 395–434.

Alba, Richard D., Logan, John R., Stults, Brian J., Marzan, Gilbert, & Zhang, Wenquan. (1999). Immigrant Groups in the Suburbs: A Reexamination of Suburbanization and Spacial Assimilation. *American Sociological Review, 64* (3), 446–460.

Albonetti, Celesta A. (1990). Race and the Probability of Pleading Guilty. *Journal of Quantitative Criminology, 6* (3), 315–334.

Albonetti, Celesta A., Hauser, Robert M., Hagan, John & Nagel, Ilene H. (1989). Criminal Justice Decision Making as a Stratification Process: The Role of Race and Stratification Resources in Pretrial Release. *Journal of Quantitative Criminology, 5* (1), 57–82.

Alderman, Clifford Lindsey. (1975). *Colonists for Sale: The Story of Indentured Servants in America*. New York, NY: Macmillan.

Alexander, Ann Field. (1992). Like an Evil Wind: The Roanoke Riot of 1893 and the Lynching of Thomas Smith. *The Virginia Magazine of History and Biography, 100* (2), 173–206.

Allport, Gordon. (1954). *The Nature of Prejudice*. New York, NY: Addison-Wesley.

Amey, Cheryl., Seccombe, Karen, & Duncan, R. Paul. (1995). Health Insurance Coverage of Mexican American Families in the U.S.: The Effect of Employment Context and Family Structure in Rural and Urban Settings. *Journal of Family Issues, 16* (4), 488–510.

Amsterdam, Anthony G. Race and the Death Penalty. *Criminal Justice Ethics* 7 (1) (Winter–Spring 1998), pp. 2, 84–86.

Anderson, Elijah. (1993). Sex Codes and Family Life. In William Julius Wilson (Ed.), *The Ghetto Underclass: Social Science Perspectives* (pp. 76–95). Newbury Park, CA: Sage.

Anderson, Jervis. (1981). *This Was Harlem: A Cultural Portrait, 1900–1950*. New York, NY: Farrar Straus Giroux.

Andrews, R.M., Harris, D.R., & Elixhauser, A. (1995). *Gender, Race/Ethnicity, and Treatment of Adults in Hospitals by Diagnosis*. Agency for Health Care Policy and Research. Washington, D.C.: United States Government Printing Office.

Aptheker, Herbert. (1963). *American Negro Slave Revolts*. New York, NY: International Publishers.

Austen, Ralph A. (1988). The 19th Century Islamic Slave Trade from East Africa (Swahili and Red Sea Coasts): A Tentative Census. *Slavery & Abolition, 9* (3), 21–44.

Austen, Ralph A. (1992). The Mediterranean Islamic Slave Trade out of Africa: A Tentative Census. *Slavery & Abolition, 13* (1), 214–248.

Austen, Ralph A., & Smith, Woodruff D. (1990). Private Tooth Decay as Public Economic Virtue: The Slave-Sugar Triangle, Consumerism, and European Industrialization. *Social Science History, 14* (1), 95–115.

Ayres, Ian, Dooley, Laura G., & Gaston, Robert S. (1993). Unequal Racial Access to Kidney Transplantation. *Vanderbilt Law Review, 46* (4), 805–863.

Azuma, Eiichiro. (Winter, 1994). A History of Oregon's Issei, 1880–1952. *Oregon Historical Quarterly, 84* (4), 315–367.

Bacon, Jean (1999). Constructing Collective Ethnic Identities: The Case of Second Generation

Asian Indians. *Qualitative Sociology, 22* (2), 141–160.

Baiamonte, John V., Jr. (1989). Community Life in the Italian Colonies of Tangipahoa Parish, Louisiana, 1890–1950. *Louisiana History, 30* (4), 365–397.

Bailey, E.J. (1991). Hypertension: An Analysis of Detroit African American Health Care Treatment Patterns. *Human Organization, 50,* 287–297.

Bailey, Ronald. (1990). The Slave Trade and the Development of Capitalism in the United States: The Textile Industry in New England. *Social Science History, 14* (3), 373–414.

Baker, Houston A., Jr. (1987). *Modernism and the Harlem Renaissance.* Chicago, IL: The University of Chicago Press.

Balboni, Alan. (1991a). From America's Little Italys to the Boomtown in the Desert: Italian-Americans in Las Vegas, 1947–1970. *Nevada Historical Society Quarterly, 34* (3), 379–399.

Bankston, Carl L. III, & Caldas, Stephen J. (1996). Majority African American Schools and Social Injustice: The Influence of De Facto Segregation on Academic Achievement. *Social Forces, 75* (2), 535–555.

Bankston, Carl L. III, & Zhou, Min. (1997). Valedictorians and Delinquents: The Bifurcation of Vietnamese American Youth. *Deviant Behavior, 18* (4), 343–364.

Barkai, Avraham. (1994). *Branching Out: German-Jewish Immigration to the United States, 1820–1914.* New York, NY: Holmes & Meier.

Barrera, Mario. (1979). *Race and Class in the Southwest: A Theory of Racial Inequality.* Notre Dame, IN: University of Notre Dame Press.

Barsh, Russell Lawrence. (1987). Plains Indian Agrarianism and Class Conflict. *Great Plains Quarterly, 7* (2), 83–90.

Barth, Fredrik. (1969). Introduction. In Fredrik Barth (Ed.), *Ethnic Groups and Boundaries: The Social Organization of Culture Difference* (pp. 9–38). Boston, MA: Little, Brown and Company.

Bates, Timothy. (1995). Rising Skill Levels and Declining Labor Force Status among African American Males. *The Journal of Negro Education, 64* (3), 373–383.

Baugher, Eleanor, & Lamison-White, Leatha. (1996). *Poverty in the United States: 1995.* U.S. Bureau of the Census, Current Population Reports, Series P60–194. Washington, D.C.: United States Government Printing Office.

Bean, Frank D., Berg, Ruth R., & Van Hook, Jennifer V. W. (1996). Socioeconomic and Cultural Incorporation and Marital Disruption among Mexican Americans. *Social Forces, 75* (2), 593–617.

Beauvais, Fred, & Segal, Bernard. (1992). Drug Use Patterns among American Indian & Alaskan Native Youth: Special Rural Populations. *Drugs & Society, 7* (1–2), 77–94.

Beck, E.M., & Tolnay, Stewart E. (1990). The Killing Fields of the Deep South: The Market for Cotton and the Lynching of Blacks, 1882–1930. *American Sociological Review, 55* (4), 526–539.

Beckles, Hilary McD. (1985). Plantation Production and White Proto Slavery: White Indentured Servants and the Colonization of the English West Indies, 1624–1645. *Americas, 41* (3), 21–45.

Beckles, Hilary McD., & Downes, Andrew. (1987). The Economics of Transition to the Black Labor System in Barbados, 1630–1680. *Journal of Interdisciplinary History, 18* (2), 225–247.

Bennett, Claudette E. (1995). *The Black Population in the United States: March 1994 and 1993.* U.S. Bureau of the Census, Current Population Reports P20–480. Washington, D.C.: United States Government Printing Office.

Benson, Nettie Lee. (1987). Texas Viewed from Mexico, 1820–1834. *Southwestern Historical Quarterly, 90* (3), 219–291.

Berger, David. (Ed.). (1983). *The Legacy of Jewish Migration: 1881 and Its Impact.* New York, NY: Brooklyn College Press.

Berleant-Schiller, Riva. (1989). Free Labor and the Economy in Seventeenth-Century Montserrat. *William and Mary Quarterly, 46* (3), 539–564.

Berlin, Ira. (1974). *Slaves without Masters: The Free Negro in the Antebellum South.* New York, NY: Pantheon Books.

Billings, Warren M. (1991). The Law of Servants and Slaves in Seventeenth-Century Virginia. *Virginia Magazine of History and Biography, 99* (1), 45–62.

Blackley, Paul R. (1990). Spatial Mismatch in Urban Labor Markets: Evidence from Large US Metropolitan Areas. *Social Science Quarterly, 71* (1), 39–52.

Blakely, Robert L., & Detweiler-Blakely, Bettina. (1989). The Impact of European Diseases in the Sixteenth-Century Southeast: A Case Study. *Midcontinental Journal of Archaeology, 14* (1), 62–89.

Blau, Peter M., & Duncan, Otis Dudley. (1967). *The American Occupational Structure*. New York, NY: Wiley.

Blauner, Robert. (1972). *Racial Oppression in America*. New York, NY: Harper and Row.

Blouet, Olwyn Mary. (1990). Slavery and Freedom in the British West Indies, 1823–33: The Role of Education. *History of Education Quarterly*, *30* (4), 625–643.

Blumrosen, Alfred W. (1994). Strangers in Paradise: Griggs v. Duke Power Co. and the Concept of Employment Discrimination. In Paul Burstein (Ed.), *Equal Employment Opportunity: Labor Market Discrimination and Public Policy* (pp. 105–120). New York, NY: Aldine de Gruyter.

Blumstein, Alfred. (1982). On the Racial Disproportionality of United States' Prison Populations. *Journal of Criminal Law and Criminology*, *73*, 1259–81.

Blumstein, Alfred. (1993). Racial Disproportionality of U.S. Prison Populations Revisited. *University of Colorado Law Review*, *64*, 743–60.

Boehm, Randolph H. (1990). Historical Records of the U.S. Department of Justice at the National Archives as Illustrated in the Peonage Files and the Strike Files. *Microform Review*, *19* (4), 214–219.

Boles, John B. (1983). *Black Southerners 1619–1869*. Lexington, KY: The University of Kentucky Press.

Bonacich, Edna. (1972). A Theory of Ethnic Antagonism: The Split Labor Market. *American Sociological Review*, *37*, 547–559.

Bonacich, Edna. (1973). A Theory of Middleman Minorities. *American Sociological Review*, *37*, 547–559.

Bonacich, Edna, & Modell, John. (1980). *The Economic Basis of Ethnic Solidarity: Small Business in the Japanese American Community*. Berkeley, CA: University of California Press.

Boocock, Sarene Spence. (1978). The Social Organization of the Classroom. In Ralph Turner, James Coleman, & Renee C. Fox (Eds.), *Annual Review of Sociology* (pp. 1–28). Palo Alto, CA: Annual Reviews.

Boogaart, Ernst van den. (1992). The Trade between Western Africa and the Atlantic World, 1600–90: Estimates of Trends in Composition and Value. *Journal of African History*, *33* (3), 369–385.

Borjas, George J. (1990). *Friends or Strangers: The Impact of Immigrants on the U.S. Economy*. New York, NY: Basic Books.

Boseker, Barbara J. (1994). The Disappearance of American Indian Languages. *Journal of Multilingual & Multicultural Development*, *15* (2–3), 147–160.

Bosher, J.F. (1995). Huguenot Merchants and the Protestant International in the Seventeenth Century. *William and Mary Quarterly*, *52* (1), 77–102.

Bowman, Phillip J. (1991a). Work Life. In James S. Jackson (Ed.), *Life in Black America* (pp. 124–155). Newbury Park, CA: Sage.

Bowman, Phillip J. (1991b). Joblessness. In James S. Jackson (Ed.), *Life in Black America* (pp. 156–178). Newbury Park, CA: Sage.

Braddock, Jomills Henry, II, Dawkins, Marvin P., & Trent, William. (1994). Why Desegregate? The Effect of School Desegregation on Adult Occupational Desegregation of African Americans, Whites and Hispanics. *International Journal of Contemporary Sociology*, *31* (2), 273–283.

Braddock, Jomills Henry, II, & McPartland, James M. (1989). Social-Psychological Processes That Perpetuate Racial Segregation: The Relationship between School and Employment Desegregation. *Journal of Black Studies*, *19* (3), 267–289.

Brandes, Joseph. (1976). From Sweatshop to Stability: Jewish Labor between Two World Wars. *Yivo Annual of Jewish Social Science*, *16*, 1–149.

Brass, Paul R. (1985). *Ethnic Groups and the State*. London, G.B.: Croom Helm.

Braun, Denny. (1995). Negative Consequences to the Rise of Income Inequality. *Research in Politics and Society*, *5*, 3–31.

British Parliamentary Papers, Shipping Casualties (Loss of the Steamship *Titanic*), 1912, cmd. 6352. In *Report of a Formal Investigation into the Circumstances Attending the Foundering on the 15th April, 1912, of the British Steam-ship "Titanic," of Liverpool, after Striking Ice in or near Latitude 41 Degrees 46' N., Longitude 50 Degrees 14' W., North Atlantic Ocean, Whereby Loss of Life Ensued*, 42. London: His Majesty's Stationery Office, 1912.

Brodeur, Paul. (1985). *Outrageous Misconduct: The Asbestos Industry on Trial*. New York, NY: Pantheon Books.

Broussard, C. Anne, & Joseph, Alfred L. (1998). Tracking: A Form of Educational Neglect? *Social Work in Education*, *20* (2), 110–120.

Brown, Bernard. (1985). Head Start: How Research Changed Public Policy. *Young Children*, *40*, 9–13.

Brown, Dee. (1970). *Bury My Heart at Wounded Knee: An Indian History of the American West*. New York, NY: Holt, Rinehart & Winston, Inc.

Brown, Mary Elizabeth. (1989). Competing to Care: Aiding Italian Immigrants in New York Harbor, 1890's–1930's. *Mid-America, 71* (3), 137–151.

Brundage, David. (1992). Respectable Radicals: Denver's Irish Land League in the Early 1880s. *Journal of the West, 31* (2), 52–58.

Bryson, Ken. (1996). *Household and Family Characteristics: March 1995*. U.S. Bureau of the Census, Current Population Reports P20–488. Washington, D.C.: United States Government Printing Office.

Buckser, Andrew S. (1992). Lynching as Ritual in the American South. *Berkeley Journal of Sociology, 37*, 11–28.

Bursik, Robert J., & Grasmick, Harold G. (1993). Economic Deprivation and Neighborhood Crime Rates, 1960–1980. *Law & Society Review, 27* (2), 263–283.

Byrne, James M., & Taxman, Faye S. (1994). Crime Control Policy and Community Corrections Practice: Assessing the Impact of Gender, Race, and Class. *Evaluation and Program Planning, 17* (2), 227–233.

Cancer Statistics Branch, Division of Cancer Prevention and Control. National Cancer Institute. (1996). *Cancer: Rates and Risks*. Washington, D.C.: United States Government Printing Office.

Cancian, Maria, Danziger, Sheldon & Gottschalk, Peter. (1993). Working Wives and Family Income Inequality among Married Couples. In Sheldon Danziger & Peter Gottschalk (Eds.), *Uneven Tides: Rising Inequality in America* (pp. 195–222). New York, NY: Russell Sage Foundation.

Cancio, A. Silvia, Evans, T. David, & Maume, David J. Jr. (1996). Reconsidering the Declining Signficance of Race: Racial Differences in Early Career Wages. *American Sociological Review, 61* (4), 541–556.

Capeci, Dominic, Jr., & Knight, Jack C. (1990). Reactions to Colonialism: The North American Ghost Dance and East African Maji-Maji Rebellions. *Historian, 52* (4), 584–601.

Caputo, Richard K. (1995b). Income Inequality and Family Poverty. *Families in Society, 76* (10), 604–615.

Cardoso, Lawrence A. (1980). *Mexican Emigration to the United States: 1897–1931: Socio-Economic Patterns*. Tucson, AZ: The University of Arizona Press.

Carew, Jan. (1992). United We Stand! Joint Struggles of Native Americans and African Americans in the Columbian Era. *Monthly Review, 44* (3), 103–127.

Carlson, Alvar W. (1994). America's New Immigration: Characteristics, Destinations, and Impact, 1970–1989. *Social Science Journal, 31* (3), 213–236.

Carter, Bob, Green, Marci, & Halpern, Rick. (1996). Immigration Policy and the Racialization of Migrant Labour: The Construction of National Identities in the USA and Britain. *Ethnic and Racial Studies, 19* (1), 135–157.

Cartmill, Matt. (1998). The Status of the Race Concept in Physical Anthropology. *American Anthropologist, 100* (3), 651–660.

Case, Charles E., Greeley, Andrew M., & Fuchs, Stephen. (1989). Social Determinants of Racial Prejudice. *Sociological Perspectives, 32* (1), 469–83.

Cazden, Robert E. (1987). The American Liberal 1854–55: Radical Forty-Eighters Attempt to Breach the Language Barrier. *Yearbook of German-American Studies, 22*, 91–99.

Celano, Marianne P., & Tyler, Forrest B. (1991). Behavioral Acculturation among Vietnamese Refugees in the United States. *Journal of Social Psychology, 131* (3), 373–385.

Center for Disease Control and Prevention, (1997). *HIV/AIDS Surveillance Report: U.S. HIV and AIDS Cases Reported through June 1997*. Atlanta, GA: Center for Disease Control and Prevention.

Chan, Sucheng. (1991). The Exclusion of Chinese Women. In Sucheng Chan (Ed.), *Entry Denied: Exclusion and the Chinese Community in America, 1882–1943* (pp. 94–146). Philadelphia, PA: Temple University Press.

Chang, Edward T. (1993). Jewish and Korean Merchants in African American Neighborhoods: A Comparative Perspective. *Amerasia Journal, 19* (2), 5–21.

Chatters, Linda M. (1991). Physical Health. In James S. Jackson (Ed.), *Life in Black America* (pp. 199–220). Newbury Park, CA: Sage.

Chatters, Linda M. (1993). HIV/AIDS within African American Communities: Diversity and Interdependence. *Health Education Quarterly, 20* (3), 321–327.

Chiswick, Barry R. (1992). Jewish Immigrant Wages in America in 1909: An Analysis of the Dillingham Commission Data. *Explorations in Economic History, 29* (3), 274–289.

Clarence-Smith, William Gervase. (1984). The Portuguese Contribution to the Cuban Slave and Coolie Trades in the Nineteenth Century. *Slavery & Abolition, 5* (1), 24–33.

Clarence-Smith, William Gervase. (1988). The Economics of the Indian Ocean and Red Sea Slave Trades in the 19th Century: An Overview. *Slavery & Abolition, 9* (3), 1–20.

Clarence-Smith, William Gervase. (1989). The Economics of the Indian Ocean and Red Sea Slave Trades in the 19th Century: An Overview. In William Gervase Clarence-Smith (Ed.), *The Economics of the Indian Ocean Slave Trade in the Nineteenth Century* (pp. 1–20). London: Frank Cass & Co. Ltd.

Clark, Dennis. (1991). *Erin's Heirs: Irish Bonds of Community*. Lexington, KY: The University Press of Kentucky.

Cockerham, William C. (1990). A Test of the Relationship between Race, Socioeconomic Status, and Psychological Distress. *Social Science and Medicine, 31* (12), 1321–1326.

Cohen, Miriam. (1992). *Workshop to Office: Two Generations of Italian Women in New York City, 1900–1950*. Ithaca, NY: Cornell University Press.

Cohen, Naomi Wiener. (1984). *Encounter with Emancipation: The German Jews in the United States, 1830–1914*. Philadelphia, PA: Jewish Publication Society.

Cohen, Naomi Wiener. (1992). *Jews in Christian America: The Pursuit of Religious Equality*. New York, NY: Oxford University Press.

Cohn, Raymond L. (1989). Maritime Mortality in the Eighteenth and Nineteenth Centuries: A Survey. *International Journal of Maritime History, 1* (1), 159–191.

Cohn, Raymond L. (1995). A Comparative Analysis of European Immigrant Streams to the United States during the Early Mass Migration. *Social Science History, 19* (1), 63–89.

Cohn, Samuel, & Fossett, Mark. (1995). Why Racial Employment Inequality Is Greater in Northern Labor Markets: Regional Differences in White–Black Employment Differentials. *Social Forces, 74* (2), 511–542.

Cohn, Steven F., Barkan, Steven E., & Halteman, William A. (1991). Punitive Attitudes toward Criminals: Racial Consensus or Racial Conflict? *Social Problems, 38* (2), 287–296.

Colchester, Marcus. (1993). Slave and Enclave: Towards a Political Ecology of Equatorial Africa. *The Ecologist, 23* (5), 166–173.

Coleman, James S., Campbell, Ernest Q., Hobson, Carol J., McPartland, James, Mood, Alexander M., Weinfeld, Fredric D., & York, Robert L. (1966). *Equality of Educational Opportunity*. Washington, D.C.: U.S. Government Printing Office.

Coleman, James S., & Hoffer, Thomas. (1987). *Public and Private High Schools: The Impact of Communities*. New York, NY: Basic Books.

Coleman, James S., Hoffer, Thomas, & Kilgor, Sally. (1982). *High School Achievement: Public, Catholic, and Other Private Schools Compared*. New York, NY: Basic Books.

Collins, Randall. (1979). *The Credential Society: An Historical Sociology of Education and Stratification*. New York, NY: Academic Press.

Cookson, Peter W., Jr., & Persell, Caroline Hodges. (1991). Race and Class in America's Elite Preparatory Boarding Schools: African Americans as the Outsiders Within. *The Journal of Negro Education, 60* (2), 219–228.

Cotts Watkins, Susan, & London, Andrew S. (1994). Personal Names and Cultural Change: A Study of the Naming Patterns of Italians and Jews in the United States in 1910. *Social Science History, 18* (2), 169.

Crofts, Daniel W. (1995). From Slavery to Sharecropping. *Reviews in American history, 23* (3), 458–463.

Crowther, Edward R. (1994). Mississippi Baptists, Slavery, and Secession, 1806–1861. *The Journal of Mississippi History, 56* (2), 129–148.

Crozier, Alan. (1984). The Scotch-Irish Influence on American English. *American Speech, 59* (4), 310–331.

Curra, John. (1994). *Understanding Social Deviance: From the Near Side to the Outer Limits*. New York, NY: HarperCollins.

Curtin, Philip. (1969). *The Atlantic Slave Trade: A Census*. Madison, WI: University of Wisconsin Press.

Curtin, Philip. (1975). *Economic Change in Precolonial Africa: Senegambia in the Era of the Slave Trade*. Madison, WI: University of Wisconsin Press.

Daniel, Pete. (1972). *The Shadow of Slavery: Peonage in the South, 1901–1969*. New York, NY: Oxford University Press.

Daniel, Pete. (1979). The Metamorphosis of Slavery, 1865–1900. *Journal of American History, 66* (1), 88–99.

Daniels, Christine. (1995). Without any Limitation of Time: Debt and Servitude in

Colonial America. *Labor History*, *36* (2), 232–250.

Daniels, Roger. (1975). *The Decision to Relocate the Japanese Americans*. Philadelphia, PA: J.B. Lippincott Company.

Daniels, Roger. (1988). *Asian America: Chinese and Japanese in the United States since 1850*. Seattle, WA: University of Washington Press.

Darity, William, Jr. (1990). British Industry and the West Indies Plantations. *Social Science History*, *14* (1), 117–149.

Darity, William, Jr. (1992). A Model of "Original Sin": Rise of the West and Lag of the Rest. *American Economic Review*, *82* (2), 162–167.

Davis, James, & Smith, Tom. (1984). *General Social Survey Cumulative File, 1972–1982*. Ann Arbor, MI: Inter-University Consortium for Political and Social Research.

Davis, James Earl. (1994). College in Black and White: Campus Environment and Academic Achievement of African American Males. *The Journal of Negro Education*, *63* (4), 620–633.

Davis, James Earl, & Jordan, Will J. (1994). The Effects of School Context, Structure, and Experiences on African American Males in Middle and High School. *The Journal of Negro Education*, *63* (4), 570–587.

Dawkins, Marvin P., & Braddock, Jomills Henry, II. (1994). The Continuing Significance of Desegregation: School Racial Composition and African American Inclusion in American Society. *The Journal of Negro Education*, *63* (3), 394–405.

Day, Jennifer, & Curry, Andrea. (1998). *Educational Attainment in the United States: March 1998*. U.S. Bureau of the Census, Current Population Reports P20–513. Washington, D.C.: United States Government Printing Office.

De Marco, William M. (1981). *Ethnics and Enclaves: Boston's Italian North End*. Ann Arbor, MI: UMI Research Press.

Decker, Peter R. (1979). Jewish Merchants in San Francisco: Social Mobility on the Urban Frontier. *American Jewish History*, *68* (4), 396–407.

Devens, Carol. (1992). If We Get the Girls, We Get the Race: Missionary Education of Native American Girls. *Journal of World History*, *3* (2), 219–237.

Diamond, Jared. (1994). Race without Color. *Discover*, *15* (11), 82–91.

Diner, Hasia. (1993). A Time for Gathering: The Second Migration. *American Jewish History*, *81* (1), 22–33.

Dixon, Blase. (1973). The Catholic University of America and the Racial Question, 1914–1918. *Records of the American Catholic Historical Society of Philadelphia*, *84* (4), 221–224.

Dobyns, Henry F. (1966). Estimating Aboriginal American Population: An Appraisal of Techniques with a New Hemispheric Estimate. *Current Anthropology*, *7*, 395–416.

Dobyns, Henry F. (1983). *Their Number Become Thinned: Native American Population Dynamics in Eastern North America*. Knoxville, TN: University of Tennessee Press.

Dolan, Jay P. (1972). Immigrants in the City: New York's Irish and German Catholics. *Church History*, *41* (3), 354–368.

Dollard, John, Miller, Neil, Doob, Leonard, Mowrer, O.H., & Sears, Robert. (1939). *Frustration and Aggression*. New Haven, CT: Yale University Press.

Domino, George. (1992). Cooperation and Competition in Chinese and American Children. *Journal of Cross-Cultural Psychology*, *23* (4), 456–467.

Donato, Katharine M., & Massey, Douglas S. (1993). Effect of the Immigration Reform and Control Act on the Wages of Mexican Migrants. *Social Science Quarterly*, *74* (3), 523–541.

Douglass, William. (1972). *Summary, Historical and Political, Of the First Planting, Progressive Improvements, and Present State of the British Settlements in North America*. New York: Arno Press.

Downey, Geraldine, & Moen, Phyllis. (1987). Personal Efficacy, Income, and Family Transitions: A Longitudinal Study of Women Heading Households. *Journal of Health and Social Behavior*, *28* (3), 320–333.

Downey, Liam. (1998). Environmental Injustice: Is Race or Income a Better Predictor? *Social Science Quarterly*, *79* (4), 766–778.

Dressel, Paula L. (1994). And We Keep on Building Prisons: Racism, Poverty, and Challenges to the Welfare State. *Journal of Sociology & Social Welfare*, *21*, 7–30.

Dressler, William W. (1993). Health in the African American Community: Accounting for Health Inequalities. *Medical Anthropology Quarterly*, *7* (4), 325–345.

Du Bois, W.E.B. (1969). *The Black North in 1901: A Social Study*. New York, NY: Arno Press.

Durden, Garey C., & Gaynor, Patricia E. (1998). More on the Cost of Being Other than White and Male: Measurement of Race, Ethnic, and Gender Effects on Yearly Earnings. *The American Journal of Economics and Sociology, 57* (1), 95–103.

Durkheim, Emile. (1951 [1897]). *Suicide*. Translation by John A. Spaulding and George Simpson. New York: NY: The Free Press.

Dusek, Jerome B. (Ed.). (1985). *Teacher Expectancies*. Hillsdale, NJ: Erlbaum.

Dyer, James, Vedlitz, Arnold, & Worchel, Stephen. (1989). Social Distance among Racial and Ethnic Groups in Texas: Some Demographic Correlates. *Social Science Quarterly, 70,* 607–16.

Edwards, Cecile, Knight, Enid, et al. (1994). Multiple Factors as Mediators of the Reduced Incidence of Low Birth Weight in an Urban Clinic. *Journal of of Nutrition, 124,* 927–926.

Ehrenreich, Barbara. (1990). *Fear of Falling: The Inner Life of the Middle Class*. New York, NY: Harper Perennial.

Ehrlich, Walter. (1988). The First Jews of St. Louis. *Missouri Historical Review, 83* (1), 57–76.

Eitches, Edward. (1971). Maryland's Jew Bill. *American Jewish Historical Quarterly, 60* (3), 258–279.

Ekirch, A. Roger. (1984). Great Britain's Secret Convict Trade to America, 1783–1784. *American Historical Review, 89* (5), 1285–1291.

Eltis, David. (1986). Slave Departures from Africa, 1811–1867: An Annual Time Series. *African Economic History, 15,* 143–171.

Eltis, David. (1989b). Trade between Western Africa and The Atlantic World Before 1870: Estimates of Trends in Value, Composition and Direction. *Research in Economic History, 12,* 197–239.

Eltis, David. (1991). Precolonial Western Africa and the Atlantic Economy. In Barbara L Solow (Ed.), *Slavery and the Rise of the Atlantic System* (pp. 97–119). New York, NY: Cambridge University Press.

Eltis, David, & Jennings, Lawrence C. (1988). Trade between Western Africa and the Atlantic World in the Pre-Colonial Era. *American Historical Review, 93* (4), 936–959.

Emmer, Pieter. (1991). The Dutch and the Making of the Second Atlantic System. In Barbara L Solow (Ed.), *Slavery and the Rise of the Atlantic System* (pp. 75–96). New York, NY: Cambridge University Press.

Emmons, David M. (1989). *The Butte Irish: Class and Ethnicity in an American Mining Town, 1875–1925*. Urbana, IL: University of Illinois Press.

Enchautegui, Maria E. (1992). Geographical Differentials in the Socioeconomic Status of Puerto Ricans: Human Capital Variations and Labor Market Characteristics. *International Migration Review, 26* (4[100]), 1267–1290.

Engerman, Stanley L. (1986). Servants to Slaves to Servants: Contract Labour and European Expansion. In P. C. Emmer (Ed.), *Colonialism and Migration: Indentured Labour Before and After Slavery* (pp. 263–294). Dordrecht, The Netherlands: Martinus Nijhoff Publishers.

Enloe, Cynthia H. (1973). *Ethnic Conflict and Political Development*. Boston, MA: Little, Brown and Company.

Erie, Steven P. (1988). *Rainbow's End: Irish Americans and the Dilemmas of Urban Machine Politics, 1840–1985*. Berkeley, CA: University of California Press.

Espiritu, Yen Le. (1996). Colonial Oppression, Labour Importation, and Group Formation: Filipinos in the United States. *Ethnic and Racial Studies, 19* (1), 29–48.

Ewald, Janet J. (1992). Slavery in Africa and the Slave Trades from Africa. *American Historical Review, 97* (2), 465–485.

Faber, Eli. (1993). The Formative Era of American Jewish History. *American Jewish History, 81* (1), 9–21.

Fagan, Jeffrey. (1992). Drug Selling and Licit Income in Distressed Neighborhoods: The Economic Lives of Street-Level Drug Users and Dealers. In Adele V. Harrell & George E. Peterson (Eds.), *Drugs, Crime, and Social Isolation: Barriers to Urban Opportunity* (pp. 99–180). Washington, D.C.: The Urban Institute Press.

Fage, J. D. (1989). African Societies and the Atlantic Slave Trade. *Past & Present, 125,* 97–115.

Falola, Toyin. (1994). Slavery and Pawnship in the Yoruba Economy of the Nineteenth Century. *Slavery & Abolition, 15* (2), 221–245.

Falsey, Barbara, & Heyns, Barbara. (1984). The College Channel: Private and Public School Reconsidered. *Sociology of Education, 57,* 111–122.

Farkas, George, England, Paula, Vicknair, Keven, and Kilbourne, Barbara Stanek. (1997). Cognitive Skill, Skill Demands of Jobs, and Earnings among Young European American, African American, and Mexican American Workers. *Social Forces, 75* (3), 913–940.

Farkas, George, & Vicknair, Keven. (1996). Appropriate Tests of Racial Wage Discrimination Require Controls for Cognitive Skill: Comment on Cancio, Evans, and Maume. *American Sociological Review, 61* (4), 557–560.

Fasenfest, David, & Perrucci, Robert. (1994). Changes in Occupation and Income, 1979–1989: An Analysis of the Impact of Place and Race. *International Journal of Contemporary Sociology, 31* (2), 203–233.

Fein, Steven, & Spencer, Steven J. (1997). Prejudice as Self-image Maintenance Affirming the Self through Derogating Others. *Journal of Personality and Social Psychology, 73*, 31–44.

Fejgin, Naomi. (1995). Factors Contributing to the Academic Excellence of American and Asian Students. *Sociology of Education, 68* (1), 18–30.

Feldman, Glenn. (1995). Lynching in Alabama, 1889–1921. *The Alabama Review, 48* (2), 114–141.

Fichtenbaum, Rudy, & Gyimah-Brempong, Kwabena. (1997). The Effects of Race on the Use of Physicians' Services. *International Journal of Health Services, 27* (1), 139–156.

Finkelman, Paul. (1993). The Crime of Color. *Tulane Law Review, 67* (6), 2063–2112.

Fogel, Robert William, & Engerman, Stanley L. (1974). *Time on the Cross: The Economics of American Negro Slavery*. Boston, MA: Little, Brown and Company.

Fogleman, Aaron S. (1981–82). An Inquiry into the Loyalties of the Germans in Central North Carolina during the American Revolution. *E.C. Barksdale Student Lectures, 7*, 72–88.

Foner, Eric. (1988). *Reconstruction: America's Unfinished Revolution, 1863–1877*. New York, NY: Harper & Row.

Forbes, Ella. (1990). African American Resistance to Colonization. *Journal of Black Studies, 21* (2), 210–228.

Forbes, Jack D. (1984). Mulattoes and People of Color in Anglo-North America: Implications for Black-Indian Relations. *Journal of Ethnic Studies, 12* (2), 17–61.

Forbes, Jack D. (1992). The Hispanic Spin: Party Politics and Governmental Manipulation of Ethnic Identity. *Latin American Perspectives, 19* (4), 59–78.

Ford, Earl, Cooper, Richard, Castaner, Angel, Simmons, Brian, & Mar, Maxine. (1989). Coronary Arteriography and Bypass Surgery among Whites and Other Racial Groups Relative to Hospital-Based Incidence Rates for Coronary Artery Disease: Findings from NHDS. *American Journal of Public Health, 79*, 437–440.

Fordham, Signithia, & Ogbu, John U. (1986). Black Students' School Success: Coping with the Burden of 'Acting White'. *Urban Review, 24* (1), 176–206.

Franco, Jeré. (1990). Bringing Them in Alive: Selective Service and Native Americans. *Journal of Ethnic Studies, 18* (3), 1–27.

Friedman-Kasaba, Kathie. (1996). *Memories of Migration: Gender, Ethnicity, and Work in the Lives of Jewish and Italian Women in New York, 1870–1924*. Albany, NY: State University of New York Press.

Fukurai, Hiroshi, Butler, Edgar W., & Krooth, Richard. (1993). *Race and the Jury: Racial Disenfranchisement and the Search for Justice*. New York, NY: Plenum.

Gaines, Stanley O., Jr., & Reed, Edward S. (1995). Prejudice from Allport to DuBois. *American Psychologist, 50*, 96–103.

Galenson, David W. (1977). Immigration and the Colonial Labor System: An Analysis of the Length of Indenture. *Explorations in Economic History, 14* (4), 360–377.

Galenson, David W. (1981b). The Market Evaluation of Human Capital: The Case of Indentured Servitude. *Journal of Political Economy, 89* (3), 446–467.

Galenson, David W. (1981d). White Servitude and the Growth of Black Slavery in Colonial America. *Journal of Economic History, 41* (1), 39–49.

Galenson, David W. (1982). The Atlantic Slave Trade and the Barbados Market, 1673–1723. *Journal of Economic History, 42* (3), 491–511.

Galenson, David W. (1991). Economic Aspects of the Growth of Slavery in the Seventeenth-Century Chesapeake. In Barbara L Solow (Ed.), *Slavery and the Rise of the Atlantic System* (pp. 265–292). New York, NY: Cambridge University Press.

Gambino, Richard. (1991). Italian American Religious Experience: An Evaluation. *Italian Americana, 10* (1), 38–60.

Garibaldi, Antoine M. (1991). The Role of Historically Black Colleges in Fascilitating Resilience among African American Students. *Education and Urban Society, 18* (4), 103–112.

Garroni, Maria Susanna. (1991). Italian Parishes in a Burgeoning City: Buffalo, 1880–1920. *Studi Emigrazione, 28* (103), 351–368.

Gemery, Henry A. (1986). Markets for Migrants: English Indentured Servitude and Emigration in the Seventeenth and Eighteenth Centuries. In P. C. Gemery (Ed.), *Colonialism and Migration: Indentured Labour before and after Slavery* (pp. 33–54). Dordrecht, The Netherlands: Martinus Nijhoff Publishers.

Genovese, Eugene D. (1974). *Roll, Jordan, Roll: The World the Slaves Made*. New York, NY: Random House.

Genovese, Eugene D., & Fox-Genovese, Elizabeth. (1979). The Slave Economies in Political Perspective. *Journal of American History, 66* (1), 7–23.

Gibbs, Jewelle Taylor. (1996). *Race and Justice: Rodney King and O.J. Simpson in a House Divided*. San Francisco, CA: Jossey-Bass.

Gimlin, Debra. (1996). Pamela's Place: Power and Negotiation in the Hair Salon. *Gender & Society, 10* (5), 505–526.

Glanz, Rudolph. (1976). Some Remarks on Jewish Labor and American Public Opinion in the Pre-World War I Era. *Yivo Annual of Jewish Social Science, 16*, 178–201.

Glenn, Susan A. (1990). *Daughters of the Shtetl: Life and Labor in the Immigrant Generation*. Ithica, NY: Cornell University Press.

Goldberg, Arthur. (1975). The Jew in Norwich, Connecticut: A Century of Jewish Life. *Rhode Island Jewish Historical Notes, 7* (1), 79–103.

Goldstein, Michael. (1992). Alone, a Jew is Nothing—Jewish Community in Providence in the Middle to Late Nineteenth Century. *Rhode Island Jewish Historical Notes, 11* (2), 218–232.

Goldstein, Sidney, & Kosmin, Barry. (1992). Religious and Ethnic Self-Identification in the United States, 1989–90: A Case Study of the Jewish Population. *Ethnic Groups, 9* (4), 219–245.

Gordon, Milton M. (1964). *Assimilation in American Life: The Role of Race, Religion, and National Origins*. New York, NY: Oxford University Press.

Gordon, Milton M. (1978). *Human Nature, Class, and Ethnicity*. New York, NY: Oxford University Press.

Gordon, Randall A. (1990). Attributions for Blue-Collar and White-Collar Crime: The Effects of Subject and Defendant Race on Simulated Juror Decisions. *Journal of Applied Social Psychology, 20* (12), 971–983.

Gordon, Randall A., Bindrim, Thomas A., McNicholas, Michael L., & Walden, Teresa L. (1988). Perceptions of Blue-Collar and White-Collar Crime: The Effect of Defendant Race on Simulated Juror Decisions. *Journal of Social Psychology, 128* (2), 191–197.

Gorton, Joe, & Boies, John L. (1999). Sentencing Guidelines and Racial Disparity across Time: Pennsylvania Prison Sentences in 1977, 1983, 1992, and 1993. *Social Science Quarterly, 80* (1), 37–54.

Gottschalk, Peter, & Danziger, Sheldon. (1993). Family Structure, Family Size, and Family Income: Accounting for Changes in the Economic Well-Being of Children, 1968–1986. In Sheldon Danziger & Peter Gottschalk (Eds.), *Uneven Tides: Rising Inequality in America* (pp. 167–194). New York, NY: Russell Sage Foundation.

Gould, Stephen Jay (1994). The Geometer of Race. *Discover, 15* (11), 64–69.

Grant, Madison. (1921). *The Passing of the Great Race (or The Racial Basis of European History)*. London, G.B.: G. Bell and Sons, Ltd.

Greeley, Andrew M. (1982). *Catholic High Schools and Minority Students*. New Brunswick, NJ: Transaction Books.

Greenberg, Mark I. (1993). Ambivalent Relations: Acceptance and Anti-Semitism in Confederate Thomasville. *American Jewish Archives, 45* (1), 13–29.

Grinde, Donald, Jr. (1977). Native American Slavery in the Southern Colonies. *Indian History, 10* (2), 38–42.

Griswold del Castillo, Richard. (1990). *The Treaty of Guadalupe Hidalgo: A Legacy of Conflict*. Norman, OK: University of Oklahoma Press.

Grobsmith, Elizabeth S. (1989). The Relationship between Substance Abuse and Crime among Native American Inmates in the Nebraska Department of Corrections. *Human Organization, 48* (4), 285–298.

Grossman, D.C., Milligan, C., & Deyo, R.A. (1991). Risk Factors for Suicide Attempts among Navajo Adolescents. *American Journal of Public Health, 81*, 870–74.

Grubb, Farley. (1985a). The Incidence of Servitude in Trans-Atlantic Migration, 1771–1804. *Explorations in Economic History, 22* (3), 316–339.

Grubb, Farley. (1992). The Long-Run Trend in the Value of European Immigrant Servants, 1654–1831: New Measurements and Interpretations. *Research in Economic History, 14*, 167–240.

Grubb, Farley. (1994b). The End of European Immigrant Servitude in the United States: An Economic Analysis of Market Collapse, 1772–1835. *Journal of Economic History, 54* (4), 794–824.

Guerin-Gonzales, Camille, & Story, Victor O. (1995). Mexican Workers and American Dreams: Immigration, Repatriation and California Farm Labor, 1900–1939. *The International Migration Review, 29* (4), 1066.

Guerra, Francisco. (1993). The European-American Exchange. *History and Philosophy of the Life Sciences, 15* (3), 313–327.

Hagan, John, & Albonetti, Celesta. (1982). Race, Class and the Perception of Criminal Injustice in America. *American Journal of Sociology, 88,* 329–356.

Hall, Tony. (1993). A Canadian Perspective in Native Studies. *Canadian Review of Studies in Nationalism, 20* (1–2), 79–86.

Halpern, Rick. (1992). Race, Ethnicity, and Union in the Chicago Stockyards, 1917–1922. *International Review of Social History, 37* (1), 25–58.

Hampton, James W. (1992). Cancer Prevention and Control in American Indians Alaska Natives. *American Indian Culture and Research Journal, 16* (3), 41–49.

Hannan, Michael T. (1979). The Dynamics of Ethnic Boundaries in Modern States. In John W. Meyer & Michael T. Hannan (Eds.), *National Development and the World System: Educational, Economic, and Political Change, 1950–1970* (pp. 253–275). Chicago, IL: University of Chicago Press.

Hanson, Sandra L. (1994). Lost Talent: Unrealized Educational Aspirations and Expectations among U.S. Youths. *Sociology of Education, 67* (3), 159–183.

Harer, Miles D., & Steffensmeier, Darrell. (1992). The Differing Effects of Economic Inequality on Black and White Rates of Violence. *Social Forces, 70* (4), 1035–1054.

Harries, Patrick. (1981). Slavery, Social Incorporation and Surplus Extraction: The Nature of Free and Unfree Labour in South-East Africa. *Journal of African History, 22* (3), 309–330.

Harry, Beth, & Anderson, Mary G. (1994). The Disproportionate Placement of African American Males in Special Education Programs: A Critique of the Process. *The Journal of Negro Education, 63* (4), 602–619.

Hartley, E.L. (1946). *Problems in Prejudice.* New York, NY: Kings Crown.

Hauser, Robert M., & Featherman, David L. (1976). *Occupations and Social Mobility in the U.S.* Washington, D.C: U.S. Government Printing Office.

Hawkins, Darnell F., Johnstone, John W. C., & Michener, Arthur. (1995). Race, Social Class,

and Newspaper Coverage of Homicide. *National Journal of Sociology, 9* (1), 113–140.

Hearn, James C. (1991). Academic and Nonacademic Influences on the College Destinations of 1980 High School Graduates. *Sociology of Education, 64* (3), 158–171.

Heavner, Robert O. (1978). Indentured Servitude: The Philadelphia Market, 1771–1773. *Journal of Economic History, 38* (3), 701–713.

Hechter, Michael (1975). *Internal Colonialism: The Celtic Fringe in British National Development, 1536–1966.* Berkeley, CA: University of California Press.

Hechter, Michael. (1978). Group Formation and the Cultural Division of Labor. *American Journal of Sociology, 84* (2), 293–318.

Hechter, Michael, & Levi, Margaret. (1979). The Comparative Analysis of Ethnoregional Movements. *Ethnic and Racial Studies, 2,* 260–274.

Heidler, David S. (1993). The Politics of National Aggression: Congress and the First Seminole War. *Journal of the Early Republic, 13* (4), 501–530.

Heimer, Karen. (1997). Socioeconomic Status, Subcultural Definitions, and Violent Delinquency. *Social Forces, 75* (3), 799–833.

Henderson, Thomas M. (1976). *Tammany Hall and the New Immigrants: The Progressive Years.* New York, NY: Arno Press.

Hendrick, Irving G. (1976). Federal Policy Affecting the Education of Indians in California, 1849–1934. *History of Education Quarterly, 16* (2), 163–186.

Henry, Sheila E. (1999). Ethnic Identity, Nationalism, and International Stratification: The Case of the African American. *Journal of Black Studies, 29* (3), 438–454.

Higham, John. (1981). *Strangers in the Land: Patterns of American Nativism 1860–1925.* New York, NY: Atheneum.

Hindelang, Michael. (1976). *Criminal Victimization in Eight American Cities: A Descriptive Analysis of Common Theft and Assault.* Washington, D.C.: Law Enforcement Assistance Administration.

Hisnanick, John J. (1994). Comparative Analysis of Violent Deaths in American Indians and Alaska Natives. *Social Biology, 41* (1–2), 96–109.

Hittman, Michael. (1992). The 1890 Ghost Dance in Nevada. *American Indian Culture and Research Journal, 16* (4), 123–166.

Hlady, W.G., & Middaugh, J.P. (1988). Suicides in Alaska: Firearms and Alcohol. *American Journal of Public Health, 78,* 179–80.

Hoerder, Dirk. (1994). Changing Paradigms in

Migration History: From "To America" to World-Wide Systems. *Canadian Review of American Studies, 24* (2), 105–126.

Hoerig, Karl A. (1994). The Relationship between German Immigrants and the Native Peoples in Western Texas. *Southwestern Historical Quarterly, 97* (3), 423–451.

Hofstra, Warren R. (1990). Land, Ethnicity, and Community at the Opequon Settlement, Virginia, 1730–1800. *Virginia Magazine of History and Biography, 98* (3), 423–448.

Holm, Tom. (1979). Indian Lobbyists: Cherokee Opposition to the Allotment of Tribal Lands. *American Indian Quarterly, 5* (2), 115–134.

Holmes, Malcolm D., & Daudistel, Howard C. (1984). Ethnicity and Justice in the Southwest: The Sentencing of Anglo, Black, and Mexican Origin Defendants. *Social Science Quarterly, 65*, 265–77.

Holmes, Malcolm D., Hosch, Harmon M., Daudistel, Howard C., Perez, Dolores A., & Graves, Joseph B. (1993). Judges' Ethnicity and Minority Sentencing: Evidence Concerning Hispanics. *Social Science Quarterly, 74* (3), 496–506.

Holmes, William F. (1969). Whitecapping: Agarian Violence in Mississippi, 1902–1906. *Journal of Southern History, 35*, 165–185.

Hondagneu-Sotelo, Pierrette. (1994). Regulating the Unregulated? Domestic Workers' Social Networks. *Social Problems, 41* (1), 50–64.

Hood, Denice W. (1992). Academic and Noncognitive Factors Affecting the Retention of Black Men at a Predominantly White University. *Journal of Negro Education, 61* (1), 12–23.

Horm, John W., & Burhansstipanov, Linda. (1992). Cancer Incidence, Survival, and Mortality among American Indians and Alaska Natives. *American Indian Culture and Research Journal, 16* (3), 21–40.

Howe, Irving. (1976). *World of Our Fathers*. New York, NY: Simon and Schuster.

Hsia, Jayjia, & Hirano-Nakanishi, Marsha. (1989). The Demographics of Diversity: Asian Americans and Higher Education. *Change, 21* (6), 20–27.

Hsueh, Sheri, & Tienda, Marta. (1995). Earnings Consequences of Employment Instability among Minority Men. *Research in Social Stratification and Mobility, 14*, 39–69.

Huggins, Nathan Irvin (1971). *Harlem Renaissance*. New York, NY: Oxford University Press.

Hymowitz, Kay S. (1992). Self-Esteem and Multiculturalism in the Public Schools. *Dissent, 39*, 23–29.

Ichioka, Yuji. (1979). Asian Immigrant Coal Miners and United Mine Workers of America: Race and Class at Rock Springs, Wyoming, 1907. *Amerasia Journal, 6* (2), 1–23.

Ichioka, Yuji. (1980). Japanese Immigrant Labor Contractors and the Northern Pacific and the Great Northern Railroad Companies, 1898–1907. *Labor History, 21* (3), 325–350.

Ichioka, Yuji. (1983). Recent Japanese Scholarship on the Origins and Causes of Japanese Immigration. *Immigration History Newsletter, 15* (2), 2–6.

Inikori, Joseph E. (1981). Market Structure and the Profits of the British African Trade in the Late Eighteenth Century. *Journal of Economic History, 41* (4), 745–776.

Inikori, Joseph E. (1988–89). Slavery and Capitalism in Africa. *Indian Historical Review, 15* (1–2), 137–151.

Inozemtsev, Vladislav. (1999). Work, Creativity and the Economy. *Society, 36* (2), 45–54.

Iverson, Roberta Rehner. (1995). Poor African American Women and Work: The Occupational Attainment Process. *Social Problems, 42* (4), 554–573.

Jackson, Jesse. (1996). *Legal Lynching: Racism, Injustice and the Death Penality*. New York, NY: Marlowe & Co.

Jackson, Juanita, Slaughter, Sabra, & Blake, J. Herman. (1974). The Sea Islands as a Cultural Resource. *The Black Scholar, 5* (6), 32–39.

Jaret, Charles. (1991). Recent Structural Change and US Urban Ethnic Minorities. *Journal of Urban Affairs, 13* (3), 307–336.

Jargowsky, Paul A. (1996). Take the Money and Run: Economic Segregation in U.S. Metropolitan Areas. *American Sociological Review, 61* (6), 984–998.

Jarrett, Robin L. (1994). Living Poor: Family Life among Single Parent, African American Women. *Social Problems, 41* (1), 30–49.

Jencks, Christopher, Bartlett, Susan, Corcoran, Mary, Crouse, James, Eaglesfield, David, Jackson, Gregory, McCelland, Kent, Mueser, Peter, Olneck, Michael, Swartz, Joseph, Ward, Sherry, & Williams, Jill. (1979). *Who Gets Ahead? The Determinants of Economic Success in America*. New York, NY: Basic Books.

Jennings, Francis. (1992). Intertribal Relationships in Pre-Columbian North America. *Towson State Journal of International Affairs, 27* (1), 41–50.

Jiobu, Robert M. (1988). Ethnic Hegemony and the Japanese of California. *American Sociological Review, 53* (3), 353–367.

Jo, Moon H. (1992). Korean Merchants in the Black Community: Prejudice among the Victims of Prejudice. *Ethnic and Racial Studies, 15* (3), 395–411.

Johnstone, John W. C., Hawkins, Darnell F., & Michener, Arthur. (1994). Homicide Reporting in Chicago Dailies. *Journalism Quarterly, 71* (4), 860–872.

Jones, Richard C. (1995). Immigration Reform and Migrant Flows: Compositional and Spatial Changes in Mexican Migration after the Immigration Reform Act of 1986. *Annals of the Association of American Geographers, 85* (4), 715–730.

Jordan, Will J., Lara, Julia, & McPartland, James M. (1996). Exploring the Causes of Early Dropout among Race-Ethnic and Gender Groups. *Youth and Society, 28* (1), 62–94.

Jwaideh, Albertine, & Cox, J.W. (1989). The Black Slaves of Turkish Arabia during the 19th Century. In William Gervase Clarence-Smith (Ed.), *The Economics of the Indian Ocean Slave Trade in the Nineteenth Century* (pp. 45–59). London: Frank Cass & Co. Ltd.

Kaganoff, Nathan M. (1976). The Business Career of Haym Salomon as Reflected in His Newspaper Advertisements. *American Jewish Historical Quarterly, 66* (1), 35–49.

Kallen, Horace M. (1915). Democracy versus the Melting Pot. *The Nation, 100,* 190–194, 217–222.

Kallen, Horace M. (1924). *Culture and Democracy in the United States* . New York, NY: Liveright.

Kalmijn, Matthijs. (1996). The Socioeconomic Assimilation of Caribbean American Blacks. *Social Forces, 74* (3), 911–930.

Kanjanapan, Wilawan. (1995). The Immigration of Asian Professionals to the United States: 1988–1990. *International Migration Review, 29* (109), 7–32.

Kao, Grace, Tienda, Marta, & Schneider, Barbara. (1996). Racial and Ethnic Variation in Academic Performance. *Research in Sociology of Education and Socialization, 11,* 263–297.

Kasarda, John D. (1990). Structural Factors Affecting the Location and Timing of Underclass Growth. *Urban Geography, 11,* 234–264.

Kasarda, John D. (1992). The Severely Distressed in Economically Transforming Cities. In Adele V. Harrell & George E. Peterson (Eds.), *Drugs, Crime, and Social Isolation: Barriers to Urban Opportunity* (pp. 45–98). Washington, D.C.: The Urban Institute Press.

Kasarda, John D. (1993). Urban Industrial Transition and the Underclass. In William Julius Wilson (Ed.), *The Ghetto Underclass: Social Science Perspectives* (pp. 43–64). Newbury Park, CA: Sage.

Kasarda, John D. (1995). Industrial Restructuring and the Changing Location of Jobs. In Reynolds Farley (Ed.), *State of the Union: America in the 1990s.* (pp. 215–268). New York, NY: Russell Sage Foundation.

Kershaw, Terry. (1992). The Effects of Educational Tracking on the Social Mobility of African Americans. *Journal of Black Studies, 23* (1), 152–169.

Kilbourne, Barbara, England, Paula, & Beron, Kurt. (1994). Effects of Individual, Occupational, and Industrial Characteristics on Earnings: Intersections of Race and Gender. *Social Forces, 72* (4), 1149–1176.

Kim, Kwang Chung, & Hurh, Won Moo. (1993). Beyond Assimilation and Pluralism: Syncretic Sociocultural Adaptation of Korean Immigrants in the US. *Ethnic and Racial Studies, 16* (4), 696–713.

Kitano, Harry H.L., & Daniels, Roger. (1988). *Asian Americans: Emerging Minorities.* Englewood Cliffs, NJ: Prentice Hall.

Klein, Martin A. (1990). The Impact of the Atlantic Slave Trade on the Societies of the Western Sudan. *Social Science History, 14* (2), 231–253.

Klinger, David A. (1994). Demeanor or Crime? Why "Hostile" Citizens Are More Likely to Be Arrested. *Criminology, 32,* 475–93.

Knapp, Jeffrey. (1988). Elizabethan Tobacco. *Representations, 21,* 26–66.

Knobel, Dale T. (1986). *Paddy and the Republic: Ethnicity and Nationality in Antebellum America.* Middletown, CT: Wesleyan University Press.

Knowlton, Clark. (1967). Land-Grant Problems among the State's Spanish-Americans. *New Mexico Business, 20,* 1–13.

Korus, Paula. (1992). Mui Tsai: Chinese Slave Girls in the Inland Northwest. *Pacific Northwest Forum, 6* (1), 38–43.

Kramer, Joyce M. (1988). Infant Mortality and Risk Factors among American Indians Compared to Black and White Rates: Implications for Policy Change. In Winston A. Van Horne (Ed.), *Ethnicity and Health* (pp. 89–115). Milwaukee, WI: The University of Wisconsin Press.

Krieger, Nancy, & Fee, Elizabeth. (1993). What's Class Got to Do With It? The State of Health

Data in the United States Today. *Socialist Review*, 23, 59–82.

Krivo, Lauren J., & Peterson, Ruth D. (1996). Extremely Disadvantaged Neighborhoods and Urban Crime. *Social Forces, 75* (2), 619–648.

Kubitschek, Warren N., & Hallinan, Maureen T. (1996). Race, Gender, and Inequity in Track Assignments. *Research in Sociology of Education and Socialization, 11*, 121–146.

Kunitz, Stephen J., & Levy, Jerrold E. (1989). Aging and Health among Navajo Indians. In Kyriakos S. Markides (Ed.), *Aging and Health: Perspectives on Gender, Race, Ethnicity, and Class* (pp. 211–246). Newbury Park, CA: Sage.

Kwong, Peter. (1987). *The New Chinatown*. New York, NY: Hill and Wang.

Laermans, Rudi. (1993). Learning to Consume: Early Department Stores and the Shaping of the Modern Consumer Culture. *Theory, Culture & Society, 10* (4), 79–102.

Landers, Jane. (1993). Black-Indian Interaction in Spanish Florida. *Colonial Latin American Historical Review, 2* (2), 141–162.

Law, Robin. (1991c). *The Slave Coast of West Africa, 1550–1750: The Impact of the Atlantic Slave Trade on an African Society*. New York, NY: Oxford University Press.

Lay, Kenneth James. (1993). Sexual Racism: A Legacy of Slavery. *National Black Law Journal, 13* (1–2), 165–183.

Leahy, Peter J., Buss, Terry F., & Quane, James M. (1995). Time on Welfare: Why Do People Enter and Leave the System? *The American Journal of Economics and Sociology, 54* (1), 33–46.

Leashore, Bogart R. (1984). Black Female Workers: Live-in Domestics in Detroit, Michigan, 1860–1880. *Phylon, 45* (2), 111–120.

Lee, Carol D. (1992). Literacy, Cultural Diversity, and Instruction. *Education and Urban Society, 24*, 279–291.

Lee, Stacey J. (1994). Behind the Model-Minority Stereotype: Voices of High- and Low-Achieving Asian American Students. *Anthropology and Education Quarterly, 25* (4), 413–429.

Lemelle, Jr., Anthony J. (1995). *Black Male Deviance*. Westport, CT.: Praeger.

Leung, Edwin Pak-Wah. (1986). From Prohibition to Protection: Ch'ing Government's Changing Policy toward Chinese Emigration. *Asian Profile, 14* (6), 485–491.

Levin, Jack. (1975). *The Functions of Prejudice*. New York, NY: Harper & Row.

Levine, Barry B. (1985). Miami: The Capital of Latin America. *Wilson Quarterly, 9* (5), 46–69.

Levy, Frank. (1987). *Dollars and Dreams: The Changing American Income Distribution*. New York, NY: Russell Sage Foundation.

Lewis, Sol. (Ed.). (1973). *The Sand Creek Massacre: A Documentary History*. New York, NY: Sol Lewis.

Lewis, Bernard. (1990). *Race and Slavery in the Middle East: An Historical Enquiry*. New York, NY: Oxford University Press.

Lewis, David Rich. (1993). Still Native: The Significance of Native Americans in the History of the Twentieth-Century American West. *Western Historical Quarterly, 24* (2), 203–227.

Lewis, Oscar. (1965). *La Vida: A Puerto Rican Family in the Culture of Poverty*. New York, NY: Random House.

Lieberson, Stanley. (1961). A Societal Theory of Race and Ethnic Relations. *American Journal of Sociology, 26*, 902–10.

Light, Ivan, Bernard, Richard B., & Kim, Rebecca. (1999). Immigrant Incorporation in the Garment Industry of Los Angeles. *International Migration Review, 33* (1), 5–25.

Light, Ivan, & Rosenstein, Carolyn. (1995). *Race, Ethnicity, and Entrepreneurship in Urban America*. New York, NY: Aldine de Gruyter.

Light, Ivan, Sabagh, Georges, Bozorgmehr, Mehdi, & Der-Martirosian, Claudia. (1994). Beyond the Ethnic Enclave Economy. *Social Problems, 41* (1), 65–80.

Lindenthal, Jacob Jay. (1981). Health and the Eastern European Jewish Immigrant. *American Jewish History, 70* (4), 420–441.

Lintelman, Joy K. (1991). Our Serving Sisters: Swedish-American Domestic Servants and Their Ethnic Community. *Social Science History, 15* (3), 381–395.

Liska, Allen E. (1994). Ghost Dancers Rise. *Humanity and Society, 18* (4), 55–65.

Liska, Allen E., & Bellair, Paul E. (1995). Violent-Crime Rates and Racial Composition: Convergence over Time. *American Journal of Sociology, 101* (3), 578–610.

Livingston, Ivor L. (1987). Blacks, Life-Style and Hypertension: The Importance of Health Education. *Humboldt Journal of Social Relations, 14* (1–2), 195–213.

Lord, Walter. (1955). *A Night to Remember*. New York, NY: Holt, Rinehart and Winston.

Lovejoy, Paul E., & Richardson, David. (1995). British Abolition and Its Impact on Slave Prices

Along the Atlantic Coast of Africa, 1783–1850. *Journal of Economic History, 55* (1), 98–119.

Lowe, Lydia. (1992). Paving the Way: Chinese Immigrant Workers and Community-based Labor Organizing in Boston. *Amerasia Journal, 18* (1), 39–48.

Lowenfels, Albert B. (1992). Gallstones and Gallbladder Cancer in Southwestern Native Americans. *American Indian Culture and Research Journal, 16* (3), 77–85.

Lowery, Joseph E. (1991). *Hearings before the Subcommittee on Civil and Constitutional Rights of the Committe on the Judiciary, House of Representatives.* Washington, D.C.: U.S. Government Printing Office.

Luhman, Reid. (1990). Appalachian English Stereotypes: Language Attitudes in Kentucky. *Language in Society, 19,* 331–348.

Lundberg, Shelly J. (1994). Equality and Efficiency: Antidiscrimination Policies in the Labor Market. In Paul Burstein (Ed.), *Equal Employment Opportunity: Labor Market Discrimination and Public Policy* (pp. 85–100). New York, NY: Aldine de Gruyter.

Lundman, Richard J. (1994). Demeanor or Crime? The Midwest City Police-Citizen Encounters Study. *Criminology, 32,* 631–56.

Lundman, Richard J. (1996). Demeanor and Arrest: Additional Evidence from Previously Unpublished Data. *Journal of Research in Crime and Delinquency, 33,* 306–23.

Mageean, Deirdre. (1984). Perspectives on Irish Migration Studies. *Ethnic Forum, 4* (1–2), 36–48.

Mahadi, Abdullahi. (1992). The Aftermath of the Jihd in the Central Sudan as a Major Factor in the Volume of the Trans-Saharan Slave Trade in the Nineteenth Century. *Slavery & Abolition, 13* (1), 111–128.

Mahard, Rita E., & Crain, Robert L. (1983). Research on Minority Achievement in Desegregated Schools. In Christine H. Rossell & Willis D. Hawley (Eds.), *The Consequences of School Desegregation* (pp. 103–125). Philadelphia, PA: Temple University Press.

Maloney, Thomas N. (1995). Degrees of Inequality: The Advance of Black Male Workers in the Northern Meat Packing and Steel Industries before World War II. *Social Science History, 19* (1), 31–64.

Mann, Coramae Richey. (1993). *Unequal Justice: A Question of Color.* Bloomingham, IN: Indiana University Press.

Manning, Patrick. (1983). Contours of Slavery and Social Change in Africa. *American Historical Review, 88* (4), 835–857.

Manning, Patrick. (1990a). Slavery and the Slave Trade in Colonial Africa. *Journal of African History, 31* (1), 135–140.

Manning, Patrick. (1990b). The Slave Trade: The Formal Demography of a Global System. *Social Science History, 14* (2), 255–279.

Mantell, Martin E. (1973). *Johnson, Grant, and the Politics of Reconstruction.* New York, NY: Columbia University Press.

Markides, Kyriakos S., Coreil, Jeannine, & Rogers, Linda Perkowski. (1989). Aging and Health among Southwestern Hispanics. In Kyriakos S. Markides (Ed.), *Aging and Health: Perspectives on Gender, Race, Ethnicity, and Class* (pp. 177–210). Newbury Park, CA: Sage.

Martinez, Ramiro, Jr. (1996). Latinos and Lethal Violence: The Impact of Poverty and Inequality. *Social Problems, 43* (2), 131–146.

Marx, Jonathan, & Solomon, Jennifer Crew. (1993). Health and School Adjustment of Children Raised by Grandparents. *Sociological Focus, 26* (1), 81–86.

Massey, Douglas S., & Denton, Nancy A. (1988). Suburbanization and Segregation in U.S. Metropolitan Areas. *American Journal of Sociology, 94,* 592–626.

Massey, Douglas S., & Eggers, Mitchell L. (1990). The Ecology of Inequality: Minorities and the Concentration of Poverty, 1970–1980. *American Journal of Sociology, 95* (5), 1153–1188.

Massey, Douglas S., & Fong, Eric. (1990). Segregation and Neighborhood Quality: Blacks, Hispanics, and Asians in the San Francisco Metropolitan Area. *Social Forces, 69* (1), 15–32.

Massey, Douglas S., Gross, Andrew B., & Shibuya, Kumiko. (1994). Migration, Segregation, and the Geographic Concentration of Poverty. *American Sociological Review, 59* (3), 425–445.

Mathis, Deborah. (September 18, 1995). To Blacks, Fuhrman Tapes Are No Surprise. *Indianapolis Business Journal, 16* (26), 7B.

Maume, David J., Cancio, A. Silvia, & Evans, T. David. (1996). Cognitive Skills and Racial Wage Inequality: Reply to Farkas and Vicknair. *American Sociological Review, 61* (4), 561–564.

May, Philip A. (1994). The Epidemiology of Alcohol Abuse among American Indians: The

Mythical and Real Properties. *American Indian Culture and Research Journal, 18* (2), 121–143.

McCaffrey, James M. (1992). *Army of Manifest Destiny: The American Soldier in the Mexican War, 1846–1848*. New York, NY: New York University Press.

McCaffrey, Lawrence John. (1976). *The Irish Dispora in America*. Bloomington, IN: Indiana University Press.

McCaffrey, Lawrence John. (1992a). Irish Textures in American Catholicism. *Catholic Historical Review, 78* (1), 1–18.

McCaffrey, Lawrence John. (1992b). *Textures of Irish America*. Syracuse, NY: Syracuse University Press.

McCarthy, Cameron. (1990). Multicultural Approaches to Racial Inequality in the United States. *Curriculum and Teaching, 5*, 25–35.

McCarthy, John, & Zald, Mayer. (1977). Resource Mobilization and Social Movements: A Political Theory. *American Journal of Sociology, 82*, 1212–1241.

McClain, Charles J., Jr. (1990). Chinese Immigration: A Comment on Cloud and Galenson. *Explorations in Economic History, 27* (3), 363–378.

McClatchy, Valentine Stuart. (Ed.). (1978). *Four Anti-Japanese Pamphlets*. New York, NY: Arno Press.

McDaniel, Antonio. (1994). Historical Racial Differences in Living Arrangements of Children. *Journal of Family History, 19* (1), 57–77.

McDougall, E. Ann. (1992). Salt, Saharans, and the Trans-Saharan Slave Trade: Nineteenth-Century Developments. *Slavery & Abolition, 13* (1), 61–88.

McGowan, Winston. (1990). African Resistance to the Atlantic Slave Trade in West Africa. *Slavery & Abolition, 11* (1), 5–29.

McKee, Samual, Jr. (1965). *Labor in Colonial New York, 1664–1776*. Port Washington, NY: Ira J. Friedman, Inc.

McLanahan, Sara, & Garfinkel, Irwin. (1993). Single Mothers, the Underclass, and Social Policy. In William Julius Wilson (Ed.), *The Ghetto Underclass: Social Science Perspectives* (pp. 109–121). Newbury Park, CA: Sage.

McLanahan, Sara, & Sandefur, Gary. (1994). *Growing Up with a Single Parent*. Cambridge, MS: Harvard University Press.

McNall, Miles, Dunnigan, Timothy, & Mortimer, Jeylan T. (1994). The Educational Achievement of the St. Paul Hmong. *Anthropology and Education Quarterly, 25* (1), 44–65.

McNeal, Ralph B., Jr. (1995). Extracurricular Activities and High School Dropouts. *Sociology of Education, 68* (1), 62–80.

McNeill, William H. (1976). *Plagues and Peoples*. Garden City, NY: Anchor Press/Doubleday.

McPike, Dan. (1991). Native American Women of the West. *Gilcrease Magazine of American History and Art, 13* (3), 28–30.

Mehan, Hugh. (1997). Tracking Untracking: The Consequences of Placing Low-Track Students in High-Track Classes. In Peter M. Hall (Ed.), *Race, Ethnicity, and Multiculturalism: Policy and Practice* (pp. 115–150). New York, NY: Garland Publishing, Inc.

Meier, August, & Rudwick, Elliott. (1976). *From Plantation to Ghetto*. New York, NY: Hill and Wang.

Meier, Matt S., & Ribera, Feliciano. (1993). *Mexican Americans–American Mexicans*. New York, NY: Hill and Wang.

Meléndez, Edwin, & Figueroa, Janis Barry. (1992). The Effects of Local Labor Market Conditions on Labor Force Participation of Puerto Rican, White, and Black Women. *Hispanic Journal of Behavioral Sciences, 14* (1), 76–90.

Menton, Linda K. (1994). Research Report: Nisei Soldiers at Dachau, Spring 1945. *Holocaust and Genocide Studies, 8* (2), 258–274.

Mercer, Jane. (1973). *Labeling the Mentally Retarded*. Berkeley, CA: University of California Press.

Merton, Robert. (1968). *Social Theory and Social Structure*. New York, NY: Free Press.

Miller, Jerome G. (1996). *Search and Destroy: African American Males in the Criminal Justice System*. New York, NY: Cambridge University Press.

Miller, Kathleen Atkinson. (1991). The Ladies and the Lynchers: A Look at the Association of Southern Women for the Prevention of Lynching. *Southern Studies, 2* (3–4), 261–280.

Miller, Kerby A. (1985). *Emigrants and Exiles: Ireland and the Irish Exodus to North America*. New York, NY: Oxford University Press.

Miller, Kerby A., Boling, Bruce, & Doyle, David N. (1980). Emigrants and Exiles: Irish Cultures and Irish Emigration to North America 1790–1922. *Irish Historical Studies, 22* (86), 97–125.

Mills, Jon K., McGrath, Donna, & Sobkoviak, Patti. (1995). Differences in Expressed Racial Prejudice and Acceptance of Others. *The Journal of Psychology, 129*, 357–379.

Milne, Ann M., Myers, David E., Rosenthal, Alvin S., & Ginsburg, Alan. (1986). Single Parents, Working Mothers, and the Educational Achievement of School Children. *Sociology of Education, 59* (3), 125–139.

Min, Pyong Gap. (1990). Problems of Korean Immigrant Entrepreneurs. *International Migration Review, 24* (3[91]), 436–455.

Min, Pyong Gap. (1995a). An Overview of Asian Americans. In Pyong Gap Min (Ed.), *Asian Americans: Contemporary Trends and Issues* (pp. 10–37). Thousand Oaks, CA: Sage.

Min, Pyong Gap. (1995b). Major Issues Relating to Asian American Experiences. In Pyong Gap Min (Ed.), *Asian Americans: Contemporary Trends and Issues* (pp. 38–57). Thousand Oaks, CA: Sage.

Min, Pyong Gap. (1995c). Korean Americans. In Pyong Gap Min (Ed.), *Asian Americans: Contemporary Trends and Issues* (pp. 199–231). Thousand Oaks, CA: Sage.

Miron, Louis F., & Lauria, Mickey. (1995). Identity Politics and Student Resistance to Inner-City Public Schooling. *Youth and Society, 27* (1), 29–54.

Mitchell, Brian C. (1988). *The Paddy Camps: The Irish of Lowell, 1821–61.* Urbana, IL: University of Illinois Press.

Mohr, John, & DiMaggio, Paul. (1995). The Intergenerational Transmission of Cultural Capital. *Research in Social Stratification and Mobility, 14,* 167–199.

Moore, Joan. (1989). Is There a Hispanic Underclass? *Social Science Quarterly, 70* (2), 265–284.

Moore, Michael C., & Moore, Lynda J. (1997). Fall from Grace: Implications of the O. J. Simpson Trial for Postmodern Criminal Justice. *Sociological Spectrum, 17* (3), 305–322.

Morales, Rebecca, & Bonilla, Frank. (1995). Restructuring and the New Inequality. In Rebecca Morales & Frank Bonilla (Eds.), *Latinos in a Changing U.S. Economy: Comparative Perspectives on Growing Inequality* (pp. 1–27). Newbury Park, CA: Sage.

Morales, Rebecca, & Ong, Paul M. (1995). The Illusion of Progress: Latinos in Los Angeles. In Rebecca Morales & Frank Bonilla (Eds.), *Latinos in a Changing U.S. Economy: Comparative Perspectives on Growing Inequality* (pp. 55–84). Newbury Park, CA: Sage.

Morgan, Kenneth. (1985). The Organization of the Convict Trade to Maryland: Stevenson, Randolph & Cheston, 1768–1775. *William and Mary Quarterly, 42* (2), 201–227.

Morris, Aldon D. (1984). *The Origins of the Civil Rights Movement: Black Communities Organizing for Change.* New York, NY: The Free Press.

Morris, Michael. (1989). From the Culture of Poverty to the Underclass: An Analysis of a Shift in Public Language. *The American Sociologist, 20* (2), 123–133.

Morris, Michael. (1996). Culture, Structure, and the Underclass. In M. Brinton Lykes, Ali Banuazizi, Ramsay Liem, & Michael Morris (Eds.), *Myths about the Powerless* (pp. 34–49). Philadelphia, PA: Temple University Press.

Moynihan, Daniel P. (1965). *The Negro Family: The Case for National Action.* Washington, D.C.: U.S. Department of Labor.

Mulkey, Lynn M., Crain, Robert L, & Harrington, Alexander J. C. (1992). One-Parent Households and Achievement: Economic and Behavioral Explanations of a Small Effect. *Sociology of Education, 65* (1), 48–65.

Mullis, Ina V.S., Dossey, John A., Foertsch, Mary A., Jones, Lee R., & Gentile, Claudia A. (1991). *Trends in Academic Progress.* Educational Testing Service, U.S. Department of Education, Washington, D.C.: U.S. Government Printing Office.

Murdock, Steve H., Zhai, Nanbin, & Saenz, Rogelio. (1999). The Effect of Immigration on Poverty in the Southwestern United States, 1980–1990. *Social Science Quarterly, 80* (2), 310–340.

Murphy, Kevin M., & Welch, Finis. (1993). Industrial Change and the Rising Importance of Skill. In Sheldon Danziger & Peter Gottschalk (Eds.), *Uneven Tides: Rising Inequality in America* (pp. 101–132). New York, NY: Russell Sage Foundation.

Myers, Martha A. (1990). Black Threat and Incarceration in Postbellum Georgia. *Social Forces, 69* (2), 373–393.

Myrdal, Gunnar. (1944). *An American Dilemma: The Negro Problem and Modern Democracy.* New York, NY: Harper & Brothers Publishers.

Nagel, Joane. (1986). The Political Construction of Ethnicity. In Susan Olzak & Joane Nagel (Eds.), *Competitive Ethnic Relations* (pp. 93–113). Orlando, FL: Academic Press.

Nagel, Joane, & Olzak, Susan. (1982). Ethnic Mobilization in New and Old States: An Extension of the Competition Model. *Social Problems, 30* (2), 127–143.

Nairn, Tom. (1977). *The Break-up of Britain: Crisis and Neo-Nationalism.* London, G.B.: New Left Books.

Nakanishi, Don T. (1989). A Quota on Excellence? The Asian American Admissions Debate. *Change, 21* (6), 39–47.

Nakanishi, Don T. (1993). Surviving Democracy's "Mistake": Japanese Americans & the Enduring Legacy of Executive Order 9066. *Amerasia Journal, 19* (1), 7–35.

National Center for Health Statistics, (1997). *Health, United States, 1996–97 and Injury Chartbook.* Hyattsville, MD: National Center for Health Statistics.

National Center for Health Statistics, (1999). *Health, United States, 1999 with Health and Aging Chartbook.* Hyattsville, MD: National Center for Health Statistics.

National Commission on Excellence in Education, (1983). *A Nation at Risk: The Imperative for Educational Reform.* Washington, D.C.: U.S. Government Printing Office.

Navarro, Vicente. (1991). Race or Class or Race and Class: Growing Mortality Differentials in the United States. *International Journal of Health Services, 21* (2), 229–235.

Newman, Dorothy, Amidei, Nancy, Carter, Barbara, Day, Dawn, Kruvant, William, & Russell, Jack. (1978). *Protest, Politics, and Prosperity: Black Americans and White Institutions, 1940–75.* New York, NY: Pantheon.

Newsinger, John. (1996). The Great Irish Famine: A Crime of Free Market Economics. *Monthly Review, 47* (11), 11–19.

Nishi, Setsuko Matsunaga. (1995). Japanese Americans. In Pyong Gap Min (Ed.), *Asian Americans: Contemporary Trends and Issues* (pp. 95–133). Thousand Oaks, CA: Sage.

Noel, Donald L. (1968). A Theory of the Origin of Ethnic Stratification. *Social Problems, 16,* 157–172.

Noguera, Pedro A. (1995). Preventing and Producing Violence: A Critical Analysis of Responses to School Violence. *Harvard Educational Review, 65* (2), 189–212.

Norgren, Jill. (1994). The Cherokee Nation Cases of the 1830s. *Journal of Supreme Court History,* 65–82.

Nuñez, Michael J. (1992). Violence at Our Border: Rights and Status of Immigrant Victims of Hate Crimes and Violence Along the Border between the United States and Mexico. *Hastings Law Journal, 43* (6), 1573.

O'Connor, Thomas H. (1979). The Irish in Boston. *The Urban and Social Change Review, 12* (2), 19–23.

O'Connor, Thomas H. (1995). *The Boston Irish: A Political History.* Boston, MA: Northeastern University Press.

O'Nell, Theresa D., & Mitchell, Christina M. (1996). Alcohol Use among American Indian Adolescents: The Role of Culture in Pathological Drinking. *Social Science and Medicine, 42* (4), 565–578.

Ogbu, John U. (1988). Class Stratification, Racial Stratification, and Schooling. In Lois Weis (Ed.), *Class, Race, and Gender in American Education* (pp. 163–182). Albany, NY: State University of New York Press.

Olson, Lenora M., Becker, Thomas M., Wiggins, Charles L., Key, Charles R., & Samet, Jonathan M. (1990). Injury Mortality in American Indian, Hispanic, and Non-Hispanic White Children in New Mexico 1958–1982. *Social Science and Medicine, 30* (4), 479–486.

Olzak, Susan. (1992). *The Dynamics of Ethnic Competition and Conflict.* Stanford, CA: Stanford University Press.

Orlans, Harold. (1989). The Revolution at Gallaudet. *Change, 21* (1), 9–18.

Pan, Lynn (1990). *Sons of the Yellow Emperor: A History of the Chinese Diaspora.* Boston, MA: Little, Brown and Company.

Pardo, Mary. (1991). Creating Community: Mexican American Women in Eastside Los Angeles. *Aztlan, 20* (1–2), 39–69.

Park, Edward J.W. (1999). Friends or Enemies?: Generational Politics in the Korean American Community in Los Angeles. *Qualitative Sociology, 22* (2), 161–175.

Parker, Keith D. (1988). Black–White Differences in Perceptions of Fear of Crime. *Journal of Social Psychology, 128* (4), 487–494.

Parker, Keith D., McMorris, Barbara J., Smith, Earl, & Murty, Komanduri S. (1993). Fear of Crime and the Likelihood of Victimization: A Bi-Ethnic Comparison. *Journal of Social Psychology, 133* (5), 723–732.

Parker, Keith D., & Ray, Melvin C. (1990). Fear of Crime: An Assessment of Related Factors. *Sociological Spectrum, 10* (1), 29–40.

Patthey-Chavez, G. Genevieve. (1993). High School as an Arena for Cultural Conflict and Acculturation for Latino Angelinos. *Anthropology and Education Quarterly, 24* (1), 33–60.

Pearson, Willie, Jr. (1987). The Flow of Black Scientific Talent: Leaks in the Pipeline. *Humboldt Journal of Social Relations, 14* (1–2), 44–61.

Pérez-Stable, Marifeli, & Uriarte, Miren. (1995). Cubans and the Changing Economy of Miami. In Rebecca Morales & Frank Bonilla (Eds.),

Latinos in a Changing U.S. Economy: Comparative Perspectives on Growing Inequality (pp. 133–159). Newbury Park, CA: Sage.

Persons, Georgia A. (1996). Is Racial Separation Inevitable and Legal? *Society, 33* (3[221]), 19–24.

Pfeffer, Max J. (1994). Low-Wage Employment and Ghetto Poverty: A Comparison of African American and Cambodian Day-Haul Farm Workers in Philadelphia. *Social Problems, 41* (1), 9–29.

Phillips, William D., Jr. (1991). The Old World Background of Slavery in the Americas. In Barbara L Solow (Ed.), *Slavery and the Rise of the Atlantic System* (pp. 43–61). New York, NY: Cambridge University Press.

Piersen, William D. (1996). *From Africa to America: African American History from the Colonial Era to the Early Republic, 1526–1790.* New York: Twayne Publishers.

Pleck, Elizabeth Hafkin (1979). *Black Migration and Poverty: Boston 1865–1900.* New York, NY: Academic Press.

Polednak, Anthony P. (1986). Breast Cancer in Black and White Women in New York State: Case Distribution and Incidence Rates by Clinical Stage at Diagnosis. *Cancer, 58,* 807–815.

Polednak, Anthony P. (1989). *Racial & Ethnic Differences in Disease.* New York, NY: Oxford University Press.

Pope, Carl E., & Feyerherm, William, U.S. Department of Justice. (1995). *Minorities and the Juvenile Justice System: Research Summary.* Washington, D.C.: Office of Juvenile Justice and Delinquency Prevention.

Porter, Bruce, & Dunn, Marvin. (1984). *The Miami Riot of 1980: Crossing the Bounds.* Lexington, MS: D.C. Heath.

Portes, A., & Jensen, L. (1989). The Enclave and the Entrants: Patterns of Ethnic Enterprise in Miami before and after Mariel. *American Sociological Review, 54,* 929–949.

Portes, A., & Manning, R.D. (1986). The Immigrant Enclave: Theory and Empirical Examples. In Susan Olzak & Joane Nagel (Eds.), *Competitive Ethnic Relations* (pp. 47–68). Orlando, FL: Harcourt-Brace-Jovanovich.

Pryor, Douglas W., & McGarrell, Edmund F. (1993). Public Perceptions of Youth Gang Crime: An Exploratory Analysis. *Youth and Society, 24* (4), 399–418.

Quaye, Randolph. (1994). The Health Care Status of African Americans. *Black Scholar, 24* (2), 12–18.

Quinn, Sandra. (1993). AIDS and the African American Woman: The Triple Burden of Race, Class, and Gender. *Health Education Quarterly, 20* (3), 304–320.

Quintana, Frances Leon. (1990). Land, Water, and Pueblo-Hispanic Relations in Northern New Mexico. *Journal of the Southwest, 32* (3), 288–299.

Radelet, Michael L. (1989). Executions of Whites for Crimes against Blacks: Exceptions to the Rule? *The Sociological Quarterly, 30* (4), 529–544.

Rader, Maryann T. (1991). Indians and Education in the Pacific Northwest. *Pacific Northwest Forum, 4* (1), 102–107.

Ragin, Charles. (1980). Celtic Nationalism in Britain: Political and Structural Bases. In T. Hopkins & I. Wallerstein (Eds.), *Processes of the World System* (pp. 249–265). Beverly Hills, CA: Sage.

Ragin, Charles. (1986). The Impact of Celtic Nationalism on Class Politics in Scotland and Wales. In Susan Olzak & Joane Nagel (Eds.), *Competitive Ethnic Relations* (pp. 199–221). Orlando, FL: Academic Press.

Randolph, Suzanne M. (1995). African American Children in Single-Mother Families. In Bette J. Dickerson (Ed.), *African American Single Mothers: Understanding Their Lives and Families* (pp. 117–145). Thousand Oaks, CA: Sage.

Rao, V. Nandini, Rao, V. V. Prakasa, & Fernandez, Marilyn. (1990). An Exploratory Study of Social Support among Asian Indians in the U.S.A. *International Journal of Contemporary Sociology, 27* (3–4), 229–245.

Ravenstein, E.G. (1889). The Laws of Migration. *Journal of the Royal Statistical Society,* 241–301.

Reich, Jerome R. (1994). *Colonial America (3rd ed.).* Englewood Cliffs, NJ: Prentice Hall.

Reid, Russell M. (1988). Church Membership, Consanguineous Marriage, and Migration in a Scotch-Irish Frontier Population. *Journal of Family History, 13* (4), 397–414.

Reischauer, Robert D. (1993). Immigration and the Underclass. In William Julius Wilson (Ed.), *The Ghetto Underclass: Social Science Perspectives* (pp. 137–148). Newbury Park, CA: Sage.

Renault, François. (1988). The Structures of the Slave Trade in Central Africa in the 19th Century. *Slavery & Abolition, 9* (3), 146–165.

Repak, Terry A. (1994). Labor Recruitment and

the Lure of the Capital: Central American Migrants in Washington, D.C. *Gender & Society, 8* (4), 507–524.

Reskin, Barbara, & Padavic, Irene. (1994). *Women and Men at Work*. Thousand Oaks, CA: Pine Forge Press.

Richardson, David. (1987). The Costs of Survival: The Transport of Slaves in the Middle Passage and the Profitability of the 18th-Century British Slave Trade. *Explorations in Economic History, 24* (2), 178–196.

Richardson, David. (1989a). Slave Exports from West and West-Central Africa, 1700–1810: New Estimates of Volume and Distribution. *Journal of African History, 30* (1), 1–22.

Richardson, David. (1991). Slavery, Trade, and Economic Growth in Eighteenth-Century New England. In Barbara L Solow (Ed.), *Slavery and the Rise of the Atlantic System* (pp. 237–264). New York, NY: Cambridge University Press.

Ricks, Thomas M. (1989). Slaves and Slave Traders in the Persian Gulf, 18th and 19th Centuries: An Assessment. In William Gervase Clarence-Smith (Ed.), *The Economics of the Indian Ocean Slave Trade in the Nineteenth Century* (pp. 60–70). London: Frank Cass & Co. Ltd.

Rist, Ray C. (1996). Color, Class, and the Realities of Inequality. *Society, 33* (3[221]), 32–36.

Rivkin, Steven G. (1994). Residential Segregation and School Integration. *Sociology of Education, 67* (4), 279–292.

Roark, James L. (1977). *Masters without Slaves: Southern Planters in the Civil War and Reconstruction*. New York, NY: W.W. Norton & Co.

Rodriguez, Havidan. (1992). Household Composition, Employment Patterns, and Income Inequality: Puerto Ricans in New York and Other Areas of the U.S. Mainland. *Hispanic Journal of Behavioral Sciences, 14* (1), 52–75.

Rogers, George C., Jr. (1988). Who Is a South Carolinian? *South Carolina Historical Magazine, 89* (1), 3–12.

Rokkan, S., & Urwin, D. (1983). *Economy, Territory, Identity: Politics of West European Peripheries*. London, G.B.: Sage.

Roper, John Herbert, & Brockington, Lolita G. (1984). Slave Revolt, Slave Debate: A Comparison. *Phylon, 45* (2), 98–110.

Rose, David L. (1994). Twenty-Five Years Later: Where Do We Stand on Equal Employment Opportunity Law Enforcement? In Paul Burstein (Ed.), *Equal Employment Opportunity: Labor Market Discrimination and Public Policy* (pp. 39–52). New York, NY: Aldine de Gruyter.

Rosenbaum, J.E. (1991). Black Pioneers: Do Their Moves to the Suburbs Increase Economic Opportunity for Mothers and Children? *Housing Policy Debate, 2,* 1179–1213.

Rosenthal, Robert, & Jacobson, Lenore. (1968). *Pygmalion in the Classroom*. New York, NY: Holt, Rinehart and Winston.

Rosenwaike, Ira. (1994). Characteristics of Baltimore's Jewish Population in a Nineteenth-Century Census. *American Jewish History, 82* (1–4), 123–139.

Royce, Edward. (1993). *The Origins of Southern Sharecropping*. Philadelphia, PA: Temple University Press.

Rumbaut, Rubén G. (1995). Vietnamese, Laotian, and Cambodian Americans. In Pyong Gap Min (Ed.), *Asian Americans: Contemporary Trends and Issues* (pp. 232–270). Thousand Oaks, CA: Sage.

Salinger, Sharon V. (1981). Colonial Labor in Transition: The Decline of Indentured Servitude in Late Eighteenth-Century Philadelphia. *Labor History, 22* (2), 165–191.

Salinger, Sharon V. (1983). Send No More Women: Female Servants in Eighteenth-Century Philadelphia. *Pennsylvania Magazine of History and Biography, 107* (1), 29–48.

Salinger, Sharon V. (1987). *To Serve Well and Faithfully: Labor and Indentured Servants in the Pennsylvania, 1682–1800*. New York, NY: Cambridge University Press.

Samford, Patricia. (1996). The Archaeology of African-American Slavery and Material Culture. *The William and Mary Quarterly, 53* (1), 87–114.

Sampson, Robert J., & Laub, John H. (1993). Structural Variations in Juvenile Court Processing: Inequality, the Underclass, and Social Control. *Law and Society Review, 27* (2), 285–311.

Sandefur, Gary D. (1986). American Indian Migration and Economic Opportunities. *International Migration Review, 20* (1), 55–68.

Saner, Hilary, MacCoun, Robert, & Reuter, Peter. (1995). On the Ubiquity of Drug Selling among Youthful Offenders in Washington, D.C., 1985–1991: Age, Period, or Cohort Effect? *Journal of Quantitative Criminology, 11* (4), 337–362.

Sankowski, Edward. (1996). Racism, Human Rights, and Universities. *Social Theory and Practice, 22* (2), 225–249.

Santiago, Anna M., & Galster, George. (1995). Puerto Rican Segregation in the United States: Cause or Consequence of Economic Status? *Social Problems, 42* (3), 361–389.

Scheingold, Stuart A. (1995). Politics, Public Policy, and Street Crime. *The Annals of the American Academy of Political and Social Science, 539*, 155–168.

Schermerhorn, Richard A. (1970). *Comparative Ethnic Relations: A Framework for Theory and Research*. New York, NY: Random House.

Schiele, Jerome H. (1994). Afrocentricity: Implications for Higher Education. *Journal of Black Studies, 25* (2), 150–169.

Schuler, Monica. (1986). The Recruitment of African Indentured Labourers for European Colonies in the Nineteenth Century. In P. C. Schuler (Ed.), *Colonialism and Migration: Indentured Labour before and after Slavery* (pp. 125–162). Dordrecht, The Netherlands: Martinus Nijhoff Publishers.

Seligsohn, Diane. (1994). The New Underclass and Re-emerging Diseases. *World Health, 47*, 25–27.

Seltzer, Richard, Frazier, Michael, & Ricks, Irelene. (1995). Multiculturalism, Race, and Education. *The Journal of Negro Education, 64* (2), 124–140.

Semmes, Clovis E. (1996). *Racism, Health, and Post-Industrialism: A Theory of African American Health*. Westport, CT: Praeger.

Serwatka, Thomas S., Deering, Sharian, & Grant, Patrick. (1995). Disproportionate Representation of African Americans in Emotionally Handicapped Classes. *Journal of Black Studies, 25* (4), 492–506.

Shai, Donna, & Rosenwaike, Ira. (1988). Violent Deaths among Mexican-, Puerto Rican and Cuban-Born Migrants in the United States. *Social Science and Medicine, 26* (2), 269–276.

Sheak, Robert J., & Dabelko, David D. (1993). The Declining Middle Class: More Evidence. *Free Inquiry in Creative Sociology, 21* (1), 29–35.

Sheley, Joseph F. (1993). Structural Influences on the Problem of Race, Crime, and Criminal Justice Discrimination. *Tulane Law Review, 67* (6), 2273–2292.

Sheriff, Abdul. (1989). Localisation and Social Composition of the East African Slave Trade, 1858–1873. In William Gervase Clarence-Smith (Ed.), *The Economics of the Indian Ocean Slave Trade in the Nineteenth Century* (pp. 131–145). London: Frank Cass & Co. Ltd.

Sheth, Manju. (1995). Asian Indian Americans. In Pyong Gap Min (Ed.), *Asian Americans: Contemporary Trends and Issues* (pp. 169–198). Thousand Oaks, CA: Sage.

Shibutani, Tamotsu, & Kwan, Kian M. (1965). *Ethnic Stratification: A Comparative Approach*. New York, NY: The Macmillan Company.

Shihadeh, Edward, & Steffensmeier, Darrell. (1994). Economic Inequality, Family Disruption, and Urban Black Violence. *Social Forces, 73* (2), 729–752.

Shihadeh, Edward S., & Flynn, Nicole. (1996). Segregation and Crime: The Effect of Black Social Isolation on the Rates of Black Urban Violence. *Social Forces, 74* (4), 1325–1352.

Shihadeh, Edward S, & Graham, Ousey. (1996). Metropolitan Expansion and Black Social Dislocation: The Link between Suburbanization and Center-City Crime. *Social Forces, 75* (2), 649–666.

Sickels, Robert J. (1972). *Race, Marriage, and the Law*. Albuquerque, NM: University of New Mexico Press.

Siegel, Jill. (1975). Occupational and Geographic Mobility in San Francisco, 1870–1890: An Empirical Inquiry. *Sociological Focus, 8* (3), 223–241.

Skogan, Wesley G. (1995). Crime and the Racial Fears of White Americans. *The Annals of the American Academy of Political and Social Science, 539*, 59–71.

Skogan, Wesley G., & Lurigio, Arthur J. (1992). The Correlates of Community Antidrug Activism. *Crime and Delinquency, 38* (4), 510–521.

Smith, Abbot Emerson. (1947). *Colonists in Bondage: White Servitude and Convict Labor in America, 1607–1776*. Chapel Hill, NC: The University of North Carolina Press.

Smith, Norman W. (1978). The Ku Klux Klan in Rhode Island. *Rhode Island History, 37* (2), 35–45.

Smith, Ryan A. (1997). Race, Income, and Authority at Work: A Cross-Temporal Analysis of Black and White Men (1972–1994). *Social Problems, 44* (1), 19–37.

Smith, Ryan A. (1999). Racial Differences in Access to Hierarchical Authority: An Analysis of Change over Time, 1972–1994. *The Sociological Quarterly, 40* (3), 367–395.

Solorzano, Daniel G. (1992). An Exploratory Analysis of the Effects of Race, Class, and Gender on Student and Parent Mobility Aspirations. *The Journal of Negro Education, 61* (1), 30–44.

Sorenson, Elaine, & Bean, Frank D. (1994). The

Immigration Reform and Control Act and the Wages of Mexican Origin Workers: Evidence from Current Population Surveys. *Social Science Quarterly, 75* (1), 1–17.

Souden, David. (1978). `Rogues, Whores and Vagabonds'? Indentured Servant Emigrants to North America, and the Case of Mid-Seventeenth-Century Bristol. *Social History, 3* (1), 23–41.

South, Scott J., & Crowder, Kyle D. (1997). Escaping Distressed Neighborhoods: Individual, Community, and Metropolitan Influences. *American Journal of Sociology, 102* (4), 1040–1084.

Spener, David. (1988). Transitional Bilingual Education and the Socialization of Immigrants. *Harvard Educational Review, 58* (2), 133–153.

Spener, David, & Bean, Frank D. (1999). Self-Employment Concentration and Earnings among Mexican Immigrants in the U.S. *Social Forces, 77* (3), 1021–1047.

Spicer, Edward H. (1980). American Indians. In Stephen Thernstrom, Ann Orlov, & Oscar Handlin (Eds.), *Harvard Encyclopedia of American Ethnic Groups* (pp. 58–122). Cambridge, MA: Harvard University Press.

Spickard, Paul R. (1996). *Japanese Americans: The Formation and Transformations of an Ethnic Group.* New York, NY: Twayne Publishers.

Spreitzer, Elmer, & Snyder, Eldon E. (1991). Sports within the Black Subculture: A Matter of Social Class or a Distinctive Subculture? *Journal of Sport and Social Issues, 14* (1), 48–58.

St. John, Craig, & Heald-Moore, Tamara. (1996). Racial Prejudice and Fear of Criminal Victimization by Strangers in Public Settings. *Sociological Inquiry, 66* (3), 267–284.

Stack, Carol. (1974). *All Our Kin: Strategies for Survival in a Black Community.* New York, NY: Harper and Row.

Stanton-Salazar, Ricardo D., & Dornbusch, Sanford M. (1995). Social Capital and the Reproduction of Inequality: Information Networks among Mexican-Origin High School Students. *Sociology of Education, 68* (2), 116–135.

Stehr, Nico. (1999). The Future of Social Inequality. *Society, 36* (5), 54–59.

Sterba, James P. (1996). Understanding Evil: American Slavery, the Holocaust, and the Conquest of the American Indians. *Ethics: An International Journal of Social, Political and Legal Philosophy, 106* (2), 424–448.

Stidham, Ronald, & Carp, Robert A. (1995). Indian Rights and Law before the Federal District Courts. *Social Science Journal, 32* (1), 87–100.

Stratton, David H. (1983). The Snake River Massacre of Chinese Miners, 1887. In Duane A. Smith (Ed.), *A Taste of the West: Essays in Honor of Robert G. Athearn* (pp. 109–129). Boulder, CO: Pruett.

Strobel, Frederick R., & Peterson, Wallace C. (1997). Class Conflict, American Style: Distract and Conquer. *Journal of Economic Issues, 31* (2), 433–443.

Strong, John A. (1990). The Pigskin Book: Records of Native-American Whalemen, 1696–1721. *Long Island Historical Journal, 3* (1), 17–28.

Stryker, Jeff. (1989). IV Drug Use and AIDS: Public Policy and Dirty Needles. *Journal of Health Politics, 14* (4), 719–740.

Stuart, Paul. (1977). United States Indian Policy: From the Dawes Act to the American Indian Policy Review Commission. *Social Service Review, 51* (3), 451–463.

Susser, I. (1996). The Construction of Poverty and Homelessness in US Cities. *Annual Review of Anthropology, 25,* 411–435.

Sutherland, Edwin. (1949). *White Collar Crime.* New York, NY: Dryden.

Suzuki, Masao. (1995). Success Story? Japanese Immigrant Economic Achievement and Return Migration, 1920–1930. *The Journal of Economic History, 55* (4), 889–901.

Swadesh, Frances Leon. (1974). *Los Primeros Pobladores: Hispanic Americans of the Ute Frontier.* Notre Dame, IN: University of Notre Dame Press.

Szymanski, Albert. (1976). Racial Discrimination and White Gain. *American Sociological Review, 41,* 403–414.

Tamura, Eileen H. (1995). Gender, Schooling and Teaching, and the Nisei in Hawaii: An Episode in American Immigration History, 1900–1940. *Journal of American Ethnic History, 14* (4), 3–26.

Tang, Joyce. (1993). Whites, Asians, and Blacks in Science and Engineering: A Reconsideration of Their Economic Prospects. *Research in Social Stratification and Mobility, 12,* 249–291.

Taylor, Marylee C. (1998). How White Attitudes Vary with the Racial Composition of Local Populations Numbers Count. *American Sociological Review, 63* (4), 512–535.

Thernstrom, Stephen. (1964). *Poverty and Progress: Social Mobility in a Nineteenth-Century City.* Cambridge, MA: Harvard University Press.

Thomas, Laurine R., Fox, Sarah A., Leake, Barbara G., & Roetzheim, Richard G. (1996). The Effects of Health Beliefs on Screening Mammography Utilization among a Diverse Sample of Older Women. *Women and Health*, *24* (3), 77–94.

Thomas, Melvin E. (1993). Race, Class, and Personal Income: An Empirical Test of the Declining Significance of Race Thesis, 1968–1988. *Social Problems, 40* (3), 328–342.

Thomas, Melvin E., Herring, Cedric, & Horton, Hayward Derrick. (1994). Discrimination over the Life Course: A Synthetic Cohort Analysis of Earnings Differences between Black and White Males, 1940–1990. *Social Problems, 41* (4), 608–628.

Thomas, R. Elizabeth, & Rappaport, Julian. (1996). Art as Community Narrative: A Resource for Social Change. In M. Brinton Lykes, Ali Banuazizi, Ramsay Liem, & Michael Morris (Eds.), *Myths about the Powerless* (pp. 317–336). Philadelphia, PA: Temple University Press.

Thompson, Aaron, & Luhman, Reid. (1997). Familial Predictors of Educational Attainment: Regional and Racial Variations. In Peter M. Hall (Ed.), *Race, Ethnicity, and Multiculturalism: Policy and Practice* (pp. 63–88). New York, NY: Garland Publishing, Inc.

Thornton, Russell. (1987). *American Indian Holocaust and Survival: A Population History since 1492*. Norman, OK: University of Oklahoma Press.

Tienda, Marta. (1993). Puerto Ricans and the Underclass Debate. In William Julius Wilson (Ed.), *The Ghetto Underclass: Social Science Perspectives* (pp. 122–136). Newbury Park, CA: Sage.

Tienda, Marta, & Jensen, L. (1988). Poverty and Minorities: A Quarter-Century Profile of Color and Socioeconomic Disadvantage. In Gary Sandefur & Marta Tienda (Eds.), *Divided Opportunities* (pp. 23–61). New York, NY: Plenum Press.

Tolnay, Stewart E., & Beck, E.M. (1990). Black Flight: Lethal Violence and the Great Migration, 1900–1930. *Social Science History, 14* (3), 347–370.

Tomaskovic-Devey, Donald, & Roscigno, Vincent J. (1996). Racial Economic Subordination and White Gain in the U.S. South. *American Sociological Review, 61* (4), 565–589.

Tonry, Michael. (1994). Racial Politics, Racial Disparities, and the War on Crime. *Crime and Delinquency, 40* (4), 475–494.

Tonry, Michael. (1995). *Malign Neglect: Race, Crime, and Punishment in America*. New York, NY: Oxford University Press.

Toro-Morn, Maura I. (1995). Gender, Class, Family, and Migration: Puerto Rican Women in Chicago. *Gender and Society, 9* (6), 712–726.

Torres, Andres, & Bonilla, Frank. (1995). Decline within Decline: The New York Perspective. In Rebecca Morales & Frank Bonilla (Eds.), *Latinos in a Changing U.S. Economy: Comparative Perspectives on Growing Inequality* (pp. 85–108). Newbury Park, CA: Sage.

Tracy, Sharon. (1993). The Evolution of the Hong Triads/Tongs into the Current Drug Market. *Journal of Third World Studies, 10* (1), 22–36.

Tsai, Shih-Shan Henry. (1986). *The Chinese Experience in America*. Bloomington, IN: Indiana University Press.

Tsukamoto, Mary, & Pinkerton, Elizabeth. (1988). *We the People: A Story of Internment in America*. Elk Grove, CA: Laguna Publishers.

Tuan, Mia. (1999). Neither Real Americans nor Real Asians? Multigeneration Asian Ethnics Navigating the Terrain of Authenticity. *Qualitative Sociology, 22* (2), 105–125.

Turner, Margery Austin, Edwards, John G., & Mikelsons, Maris. (1991). *Housing Discrimination Study: Analyzing Racial and Ethnic Steering*. U.S. Department of Housing and Urban Development. Washington, D.C: United States Government Printing Office.

Tushnet, Mark V. (1987). *The NAACP's Legal Strategy against Segregated Education, 1925–1950*. Chapel Hill, NC: The University Press of North Carolina.

U.S. Bureau of the Census, 1990 Census of the Population. CP-3-7. (1990). *Characteristics of American Indians by Tribe and Language*. Washington, D.C.: United States Government Printing Office.

U.S. Bureau of the Census, (1993a). *1990 Census of the Population: Asians and Pacific Islanders in the United States*. CP-3-5. Washington, D.C.: United States Government Printing Office.

U.S. Bureau of the Census, (1993b). *Population Profile of the United States, 1993*. Washington, D.C.: United States Government Printing Office.

U.S. Bureau of the Census, (1993c). *Statistical Abstract of the United States: 1993*. Washington, D.C.: U.S. Government Printing Office.

U.S. Bureau of the Census, Current Population Reports, Series P20-475. (1994a). *The Hispanic Population in the United States: March, 1993.* Washington, D.C.: United States Government Printing Office.

U.S. Bureau of the Census, Current Population Reports, Series P20-478. (1994b). *Marital Status and Living Arrangements: March 1993.* Washington, D.C.: United States Government Printing Office.

U.S. Bureau of the Census, Current Population Reports, Series P23-189. (1995a). *Population Profile of the United States: 1995.* Washington, D.C.: United States Government Printing Office.

U.S. Bureau of the Census, Current Population Reports, Series P20-486. (1995b). *The Foreign Born Population: 1994.* Washington, D.C.: United States Government Printing Office.

U.S. Bureau of the Census, (1995c). *The Nation's Asian and Pacific Islander Population—1994.* Washington, D.C.: United States Government Printing Office.

U.S. Bureau of the Census, Current Population Reports, P60-193. (1996a). *Money Income in the United States: 1995.* Washington, D.C.: United States Government Printing Office.

U.S. Bureau of the Census, Current Population Reports, Series P23-191. (1996b). *How We're Changing: Demographic State of the Nation: 1996.* Washington, D.C.: United States Government Printing Office.

U.S. Bureau of the Census, Current Population Reports, P70-57. (1996c). *A Perspective on Low-Wage Workers.* Washington, D.C.: United States Government Printing Office.

U.S. Bureau of the Census, Current Population Reports, P70-58. (1996d). *Who Gets Assistance?* Washington, D.C.: United States Government Printing Office.

U.S. Bureau of the Census, Current Population Reports, P70-56. (1996e). *Moving Up and Down the Income Ladder.* Washington, D.C.: United States Government Printing Office.

U.S. Bureau of the Census, Current Population Reports, P70-55. (1996f). *Who Stays Poor? Who Doesn't?* Washington, D.C.: United States Government Printing Office.

U.S. Bureau of the Census, Current Population Reports, P60-191. (1996g). *A Brief Look at Postwar U.S. Income Inequality.* Washington, D.C.: United States Government Printing Office.

U.S. Bureau of the Census, Current Population Reports, P60-195. (1996h). *Health Insurance Coverage: 1995.* Washington, D.C.: United States Government Printing Office.

U.S. Bureau of the Census, (1996i). *March 1996 Current Population Survey.* Washington, D.C.: United States Government Printing Office.

U.S. Bureau of the Census, Current Population Reports, Series P23-193. (1997a). *How We're Changing: Demographic State of the Nation: 1997.* Washington, D.C.: United States Government Printing Office.

U.S. Bureau of the Census, Current Population Reports, Series P20-494. (1997b). *The Foreign Born Population: 1996.* Washington, D.C.: United States Government Printing Office.

U.S. Bureau of the Census, Current Population Reports, Series P25-1131. (1997c). *Population Projections: States, 1995-2025.* Washington, D.C.: United States Government Printing Office.

U.S. Bureau of the Census, (1997d). *Statistical Abstract of the United States, 1996.* Washington, D.C.: U.S. Government Printing Office.

U.S. Bureau of the Census, (1999a). *Statistical Abstract of the United States: 1999.* Washington, D.C.: U.S. Government Printing Office.

U.S. Bureau of the Census, Current Population Reports, Series P60-207. (1999b). *Poverty in the United States: 1998.* Washington, D.C.: United States Government Printing Office.

U.S. Bureau of the Census, Current Population Reports, Series P23-195. (1999c). *Profile of the Foreign Born Population in the United States: 1997.* Washington, D.C.: United States Government Printing Office.

U.S. Bureau of the Census, Current Population Reports, P60-208. (1999d). *Health Insurance Coverage: 1998.* Washington, D.C.: United States Government Printing Office.

U.S. Department of Education, National Center for Education Statistics. (1989). *Digest of Education Statistics-1989.* Washington, D.C.: U.S. Government Printing Office.

U.S. Department of Education, National Center for Education Statistics. (1990). *The Condition of Education-1990: Volume One: Elementary and Secondary Education.* Washington, D.C.: U.S. Government Printing Office.

U.S. Department of Education, National Center for Education Statistics. (1996a). *The Condition of Education 1996.* NCES 96-304, by Thomas M. Smith. Washington, D.C.: U.S. Government Printing Office.

U.S. Department of Education, National Center for Education Statistics. (1996b). *Youth Indicators, 1996*. NCES 96-027, by Thomas Snyder and Linda Shafer. Washington, D.C.: United States Government Printing Office.

U.S. Department of Education, National Center for Education Statistics. (1997). *The Condition of Education 1997*. Washington, D.C.: U.S. Government Printing Office.

U.S. Department of Education, National Center for Education Statistics. (1998). *Digest of Education Statistics-1998*. Washington, D.C.: U.S. Government Printing Office.

U.S. Department of Education, National Center for Education Statistics. (1999). *The Condition of Education 1999*. Washington, D.C.: U.S. Government Printing Office.

U.S. Department of Justice, Bureau of Justice Statistics. (1993). *Survey of State Prison Inmates, 1991*. Washington, D.C.: U.S. Government Printing Office.

U.S. Department of Justice, Bureau of Justice Statistics. (1994a). *Capital Punishment, 1993*. Washington, D.C.: U.S. Government Printing Office.

U.S. Department of Justice, Bureau of Justice Statistics. (1994b). *Prisoners in 1993*. Washington, D.C.: U.S. Government Printing Office.

U.S. Department of Justice, Bureau of Justice Statistics. (1995). *Drugs and Crime Facts, 1994*. Washington, D.C.: U.S. Government Printing Office.

U.S. Department of Justice, Bureau of Justice Statistics. (1996a). *Sourcebook of Criminal Justice Statistics 1995*. Washington, D.C.: U.S. Government Printing Office.

U.S. Department of Justice, Bureau of Justice Statistics. (1996b). *Uniform Crime Reports, 1995*. Washington, D.C.: U.S. Government Printing Office.

U.S. Department of Justice, Bureau of Justice Statistics. (1997a). *Correctional Populations in the United States, 1995*. Washington, D.C.: United States Government Printing Office.

U.S. Department of Justice, Bureau of Justice Statistics. (1997b). *Fiscal Year 1997: At a Glance*. Washington, D.C.: U.S. Government Printing Office.

U.S. Department of Justice, Bureau of Justice Statistics. (1997c). *Felony Sentences in State Courts, 1994*. Washington, D.C.: U.S. Government Printing Office.

U.S. Department of Justice, Bureau of Justice Statistics. (1997d). *Felony Sentences in the United States, 1994*. Washington, D.C.: U.S. Government Printing Office.

U.S. Department of Justice, Bureau of Justice Statistics. (1997e). *Criminal Victimization, 1973-95*. Washington, D.C.: U.S. Government Printing Office.

U.S. Department of Justice, Bureau of Justice Statistics. (1997f). *Prisoners in 1996*. Washington, D.C.: U.S. Government Printing Office.

U.S. Department of Justice, Bureau of Justice Statistics. (1999a). *Sourcebook of Criminal Justice Statistics 1999*. Washington, D.C.: U.S. Government Printing Office.

U.S. Department of Justice, Bureau of Justice Statistics. (1999b). *American Indians and Crime*. Washington, D.C.: U.S. Government Printing Office.

U.S. Department of Justice, Bureau of Justice Statistics. (1999c). *Uniform Crime Reports, 1998*. Washington, D.C.: U.S. Government Printing Office.

U.S. Department of Justice, Bureau of Justice Statistics. (1999d). *Criminal Victimization 1998*. Washington, D.C.: U.S. Government Printing Office.

U.S. Department of Justice, Bureau of Justice Statistics. (1999e). *Correctional Populations in the United States, 1996*. Washington, D.C.: United States Government Printing Office.

U.S. Department of Justice, Bureau of Justice Statistics. (1999f). *Felony Sentences in State Courts, 1996*. Washington, D.C.: U.S. Government Printing Office.

U.S. Department of Justice, Bureau of Justice Statistics. (1999g). *Prisoners in 1998*. Washington, D.C.: U.S. Government Printing Office.

U.S. Equal Employment Opportunity Commission, (1995). *Indicators of Equal Employment Opportunity-Status and Trends*. Washington, D.C.: U.S. Government Printing Office.

U.S. Immigration and Naturalization Service., (1993). *Statistical Yearbook of the Immigration and Naturalization Service*. Washington, D.C.: U.S. Government Printing Office.

U.S. Immigration and Naturalization Service., (1997a). *Illegal Alien Resident Population*. Washington, D.C.: U.S. Government Printing Office.

U.S. Immigration and Naturalization Service., (1997b). *Fact Sheet: Operation Gatekeeper: Three Years of Results At-a-Glance*. Washington, D.C.: U.S. Government Printing Office.

U.S. Immigration and Naturalization Service., (1999). *1997 Statistical Yearbook of the*

Immigration and Naturalization Service. Washington, D.C.: U.S. Government Printing Office.

Van den Berghe, Pierre L. (1967). *Race and Racism: A Comparative Perspective.* New York, NY: John Wiley.

Van Ness, John R., &, Van Ness, Christine M. (Eds.). (1980). *Spanish and Mexican Land Grants in New Mexico and Colorado.* Santa Fe, NM: Center for Land Grant Studies.

VanLandingham, Mark J., & Hogue, Carol J.R. (1995). Birthweight-Specific Infant Mortality Risks for Native Americans and Whites, United States, 1960 and 1984. *Social Biology, 42* (1–2), 83–94.

Voeks, Robert. (1993). African Medicine and Magic in the Americas. *Geographical Review, 83* (1), 66–78.

Von Hassell, Malve. (1993). Issei Women: Silences and Fields of Power. *Feminist Studies, 19* (3), 549–569.

Vowell, Paul R., & Howell, Frank M. (1998). Modeling Delinquent Behavior: Social Disorganization, Perceived Blocked Opportunity, and Social Control. *Deviant Behavior, 19* (4), 361–395.

Wabeke, Bertis Harry. (1983). Dutch Emmigration to British America, 1664–1776. *De Halve Maen, 57* (3), 1–4.

Wacquant, Löic J.D., & Wilson, William Julius. (1993). The Cost of Racial and Class Exclusion in the Inner City. In William Julius Wilson (Ed.), *The Ghetto Underclass: Social Science Perspectives* (pp. 25–42). Newbury Park, CA: Sage.

Wagley, C., & Harris, H. (1958). *Minorities in the New World.* New York, NY: Columbia University Press.

Wallace, R. Stuart. (1985). The Development of the Scotch-Irish Myth in New Hampshire. *Historical New Hampshire, 40* (3–4), 109–134.

Wallerstein, Immanual (1974). *The Modern World System: Capitalist Agriculture and the Origins of European World-Economy in the Sixteenth Century.* New York, NY: Academic Press.

Warren, James Francis. (1989). Karayuki-San of Singapore: 1877–1941. *Journal of the Malaysian Branch of the Royal Asiatic Society, 69* (2), 45–80.

Warren, Robert, & Passel, Jeffrey S. (1987). A Count of the Uncountable: Estimates of Undocumented Aliens Counted in the 1980 United States Census. *Demography, 24,* 375–93.

Watkins, Mel. (1994). *On the Real Side: Laughing, Lying, and Signifying: The Underground Tradition of African American Humor That Trans-* *formed American Culture from Slavery to Richard Pryor.* New York, NY: Simon & Shuster.

Watts, David S., & Watts, Karen M. (1991). The Impact of Female-Headed Single Parent Families on Academic Achievement. *Journal of Divorce and Remarriage, 17* (1–2), 97–114.

Wayne, Stephen J. (1996). *The Road to the Whitehouse, 1996.* New York, NY: St. Martin's Press.

Webb, James L. A., Jr. (1993). The Horse and Slave Trade between the Western Sahara and Senegambia. *Journal of African History, 34* (2), 221–246.

Weeks, John R., & Rumbaut, Ruben G. (1991). Infant Mortality among Ethnic Immigrant Groups. *Social Science and Medicine, 33* (3), 327–334.

Weibel-Orlando, Joan. (1990). American Indians and Prohibition: Effect or Affect? Views from the Reservation and the City. *Contemporary Drug Problems, 17* (2), 293–322.

Weinberg, Sydney Stahl. (1988). *The World of Our Mothers: The Lives of Jewish Immigrant Women.* Chapel Hill, NC: University of North Carolina Press.

Welch, Susan, Gruhl, John, & Spohn, Cassia. (1984). Dismissal, Conviction, and Incarceration of Hispanic Defendants: A Comparison with Anglos and Blacks. *Social Science Quarterly, 65,* 257–276.

Wenneker, Mark B., & Epstein, Arnold M. (1989). Racial Inequalities in the Use of Procedures for Patients with Ischemic Heart Disease in Massachusetts. *Journal of the American Medical Association, 261,* 253–257.

Westphall, Victor. (1965). *The Public Domain in New Mexico 1854–1891.* Albuquerque, NM: University of New Mexico Press.

Whatley, Warren C. (1993). African American Strikebreaking from the Civil War to the New Deal. *Social Science History, 17* (4), 525–558.

White, Karl R. (1982). The Relation between Socioeconomic Status and Academic Achievement. *Psychological Bulletin, 91,* 461–481.

Whitfield, Stephen J. (1982). Commercial Passions: The Southern Jew as Businessman. *American Jewish History, 71* (3), 342–357.

Wilkinson, Doris Y. (1996). Integration Dilemmas in a Racist Culture. *Society, 33* (3[221]), 27–31.

Williams, Teresa Kay, & Thornton, Michael C. (1998). Social Construction of Ethnicity versus Personal Experience: The Case of Afro-Amerasians. *Journal of Comparative Family Studies, 29* (2), 255–267.

Wilson, Brian, & Sparks, Robert. (1996). "It's Gotta Be the Shoes": Youth, Race, and Sneaker Commericals. *Sociology of Sport Journal, 13* (4), 398–427.

Wilson, Patricia M., & Wilson, Jeffrey R. (1992). Environmental Influences on Adolescent Educational Aspirations: A Logistic Transform Model. *Youth and Society, 24* (1), 52–70.

Wilson, William J. (1981). The Black Community in the 1980s: Questions of Race, Class, and Public Policy. *Annals of the American Academy of Political & Social Science, 454,* 26–41.

Wilson, William J. (1987). *The Truly Disadvantaged: The Inner City, the Underclass, and Public Policy.* Chicago, IL: University of Chicago Press.

Wilson, William J. (1993). The Underclass: Issues, Perspectives, and Public Policy. In William Julius Wilson (Ed.), *The Ghetto Underclass: Social Science Perspectives* (pp. 1–24). Newbury Park, CA: Sage.

Wilson-Sadberry, Karen R., Winfield, Linda F., & Royster, Deirdre A. (1991). Resilience and Persistence of African-American Males in Postsecondary Enrollment. *Education and Urban Society, 24* (1), 87–102.

Winther, Paul. (1996). The Killing of Neni Bai. In Reid Luhman (Ed.), *The Sociological Outlook: A Text with Readings* (pp. 106–111). San Diego, CA: Collegiate Press.

Wojtkiewicz, Roger A., & Donato, Katharine M. (1995). Hispanic Educational Attainment: The Effects of Family Background and Nativity. *Social Forces, 74* (2), 559–574.

Wolff, Edward N. (1995). *Top Heavy: A Study of the Increasing Inequality of Wealth in America.* New York, NY: The Twentieth Century Fund Press.

Wong, Morrison G. (1995). Chinese Americans. In Pyong Gap Min (Ed.), *Asian Americans: Contemporary Trends and Issues* (pp. 58–94). Thousand Oaks, CA: Sage.

Wong, Paul, Applewhite, Stephen, & Daley, J. Michael. (1990). From Despotism to Pluralism: The Evolution of Voluntary Organizations in Chinese American Communities. *Ethnic Groups, 8* (4), 215–233.

Wood, Forrest G. (1991). *The Arrogance of Faith: Christianity and Race in America from the Colonial Era to the Twentieth Century.* Boston, MA: Northeastern University Press.

Wood, Peter H. (1989). The Changing Population of the Colonial South: An Overview by Race and Region, 1685–1790. In Peter H. Wood, Gregory A. Waselkov, & M. Thomas Hatley (Eds.), *Powhatan's Mantle: Indians in the Colonial Southeast* (pp. 35–103). Lincoln, NE: University of Nebraska Press.

Worcester, Donald E., & Schilz, Thomas F. (1984). The Spread of Firearms among the Indians on the Anglo-French Frontiers. *American Indian Quarterly, 8* (2), 103–115.

Wright, Mary C. (1981). Economic Development and Native American Women in the Early Nineteenth Century. *American Quarterly, 33* (5), 525–536.

Wyman, Mark. (1984). *Immigrants in the Valley: Irish, Germans, and Americans in the Upper Mississippi Country, 1830–1860.* Chicago, IL: Nelson-Hall.

Yamato, Alexander. (1994). Racial Antagonism and the Formation of Segmented Labor Markets: Japanese Americans and Their Exclusion from the Work Force. *Humboldt Journal of Social Relations, 20* (1), 31–63.

Yinger, John. (1991). *Housing Discrimination Study: Incidence of Discrimination and Variation in Discriminatory Behavior.* U.S. Department of Housing and Urban Development. Washington, D.C.: United States Government Printing Office.

Yoon, In-Jin. (1995). The Growth of Korean Immigrant Entrepreneurship in Chicago. *Ethnic and Racial Studies, 18* (2), 315–335.

Young, T. Kue. (1994). *The Health of Native Americans: Toward a Biocultural Epidemiology.* New York, NY: Oxford University Press.

Yzaguirre, Raul. (1987). Public Policy, Crime, and the Hispanic Community. *The Annals of the American Academy of Political and Social Science, 494,* 101–104.

Zhou, Min, & Logan, John R. (1991). In and Out of Chinatown: Residential Mobility and Segregation of New York City's Chinese. *Social Forces, 70* (2), 387–408.

Zhou, Min, & Kamo, Yoshinori. (1994). An Analysis of Earnings Patterns for Chinese, Japanese, and Non-Hispanic White Males in the United States. *The Sociological Quarterly, 35* (4), 581–602.

Chapter 1 Reading Excerpted and reprinted from the Introduction and the chapter titled "'I Don't Feel at Home Anymore': Social and Cultural Change," included in *The First Suburban Chinatown: The Remaking of Monterey Park, California* by Timothy P. Fong, by permission of Temple University Press. © 1994 by Temple University. All Rights Reserved.

Chapter 2 Reading "Imagine a Country" by Holly Sklar, *Z Magazine*, July/Aug., 1997, pp. 65–71. Copyright by Holly Sklar. Reprinted by permission of the author.

Chapter 4 Reading "Celia, a Slave" by Melton A. McLaurin. Copyright © 1991 by University of Georgia Press. Reprinted by permission.

Chapter 5 Reading "Boy, You Better Learn to Count Your Money" by Aaron Thompson. Reprinted by permission of the author.

Chapter 6 Reading 1 From 'The Ethics of Living Jim Crow" in *Uncle Tom's Children* by Richard Wright. Copyright 1937 by Richard Wright. Copyright renewed 1965 by Ellen Wright. Reprinted by permission of HarperCollins Publishers, Inc.

Chapter 6 Reading 2 "We the People: A Story of Internment in America" by Mary Tsukamoto and Elizabeth Pinkerton. Copyright © 1988 Laguna Publishers. Reprinted by permission of the author.

Chapter 7 Reading "The Real Immigrant Story: Making it Big in America" by Denise M. Topolnicki and Kim Jeanhee, *Money*, 24 (1), pp. 128–138, 1995. Reprinted from the 1995 issue of *Money* by special permission; copyright © 1995 by Time, Inc.

Chapter 8 Reading "No Can Geeve Up" by Harold Murai. Copyright by Harold M. Murai. Reprinted by permission of the author.

Chapter 9 Reading "Living Poor: Family Life Among Single Parent, African-American Women" by Robin L. Jarrett, *Social Problems*, 41 (1), pp. 30–49, 1994. Copyright © 1994 by The Society for the Study of Social Problems. Reprinted by permission.

Chapter 10 Reading Reprinted with the permission of The Free Press, a Division of Simon & Schuster, Inc., from *Bad Blood: The Tuskegee Syphilis Experiment* by James H. Jones. Copyright © 1981, 1993 by The Free Press.

Chapter 11 Reading "Race and the Death Penalty" by Anthony G. Amsterdam, *Criminal Justice Ethics*, 7 (1) (Winter/Spring 1988), pp. 2, 84–86. Copyright © 1988 by The Institute for Criminal Justice Ethics. Reprinted by permission of The Institute for Criminal Justice Ethics, 555 West 57th St., Suite 601, New York, NY 10019-1029.

Page iii: Corbis/Bettmann Archive

Chapter 1   Page 11: AP/Wide World Photos; **Page 13:** AP/Wide World Photos.

Chapter 2   Page 37: AP/Wide World Photos; **Page 42, top:** © Steve Chenn/ Corbis; **Page 42, bottom:** AP/Wide World Photos; **Page 46:** © Stephanie Maze/Corbis.

Chapter 3 · Page 70: © Owen Franken/Corbis; **Page 77:** © Richard Bickel/Corbis.

Chapter 4   Page 97: North Wind Picture Archives; **Page 100:** Reproduced from the Collections of the Library of Congress; **Page 111:** Corbis/Baldwin H. Ward; **Page 125:** © Bettmann/Corbis.

Chapter 5   Page 144: Reproduced from the Collections of the Library of Congress; **Page 145:** "Yard in Jersey Street" Circa 1898. Museum of the City of New York. The Jacob A. Riis Collection © Museum of the City of New York.; **Page 157:** The Granger Collection, New York; **Page 158:** California History Section, California State Library.

Chapter 6   Page 180: Museum of the City of New York, The Jacob A. Riie Collection, #155; **Page 184:** Brown Brothers; **Page 190:** © Corbis; **Page 193:** Reprinted from the Collections of the Library of Congress; **Page 195:** © Bettmann/Corbis; **Page 198:** Photographs and Prints Division, Schomburg Center for Research in Black Culture, The New York Public Library, Astor, Lenox, and Tilden Foundations; **Page 201:** © 1963 Bob Adelman/ Magnum Photos; **Page 203:** AP/Wide World Photos; **Page 208:** Brown Brothers; **Page 210:** National Archives (210-G-2C-160).

Chapter 7   Page 241: © Stephanie Maze/Corbis; **Page 252:** © Owen Franken/Corbis; **Page 256:** AP/Wide World Photos.

Chapter 8   Page 280: AP/Wide World Photos; **Page 295:** AP/Wide World Photos.

Chapter 9   Page 314, top: AP/Wide World Photos; **Page 314, bottom:** AP/Wide World Photos; **Page 318:** © Owen Franken/Corbis; **Page 330:** AP/Wide World Photos.

Chapter 10   Page 355: © Ted Streshinsky/Corbis; **Page 372, top:** © Warren Morgan/Corbis; **Page 372, bottom:** AP/Wide World Photos.

Chapter 11   Page 393: © Adamsmith Productions/Corbis; **Page 401:** © Peter Turnley/Corbis; **Page 411:** © Steve Chenn/Corbis.